Surfactants in Tribology

Volume 5

Surfactants in Tribology

Volume 5

Edited by
Girma Biresaw and K.L. Mittal

CRC Press
Taylor & Francis Group
Boca Raton London New York

CRC Press is an imprint of the
Taylor & Francis Group, an **informa** business

CRC Press
Taylor & Francis Group
6000 Broken Sound Parkway NW, Suite 300
Boca Raton, FL 33487-2742

First issued in paperback 2020

© 2017 by Taylor & Francis Group, LLC
CRC Press is an imprint of Taylor & Francis Group, an Informa business

No claim to original U.S. Government works

ISBN-13: 978-0-367-57296-9 (pbk)
ISBN-13: 978-1-4987-3479-0 (hbk)

Visit the Taylor & Francis Web site at
http://www.taylorandfrancis.com

and the CRC Press Web site at
http://www.crcpress.com

Contents

SECTION I Ionic Liquids and Polymeric Amphiphiles in Tribology

SECTION II Nanomaterials in Tribology

SECTION III   Tribological Applications in Automotive, Petroleum Drilling, and Food

SECTION IV   Biobased Amphiphilic Materials in Tribology and Related Fields

Preface

Surfactants perform a wide range of functions in tribology. These include basic lubrication functions, such as control of friction and wear, as well as controlling a wide range of lubricant properties. Examples of lubricant properties that can be modified with the application of surfactants include emulsification/demulsification, bioresistance, oxidation resistance, and rust/corrosion prevention. Surfactants also spontaneously form a wide range of organized assemblies in polar and nonpolar solvents. Examples of organized assemblies include monolayers, normal/reverse micelles, o/w and w/o microemulsions, hexagonal and lamellar lyotropic liquid crystals, and uni- and multilamellar vesicles. However, the tribological properties of these organized surfactant structures are not yet fully investigated or understood. Recently, there has been a great deal of interest in certain organized assemblies such as self-assembled monolayers (SAMs). These structures are expected to play critical roles in the lubrication of a wide range of products, including microelectromechanical systems (MEMS) and nanoelectromechanical systems (NEMS).

Whereas there is a great deal of literature on the topics of surfactants and tribology separately, there is very little information on the subject of surfactants and tribology together. This is despite the fact that surfactants play many critical roles in tribology and lubrication. In order to fill this gap in the literature linking surfactants and tribology, we decided to organize the first symposium on "Surfactants in Tribology" as a part of the Sixteenth International Symposium on Surfactants in Solution (SIS) in Seoul, South Korea, June 4–9, 2006 (SIS-2006).

The SIS series of biennial symposia began in 1976 and have since been held in many corners of the globe and are attended by "who is who" in the surfactant community. These meetings are recognized by the international community as the premier forum for discussing the latest research findings on surfactants in solution. In keeping with the SIS tradition, leading researchers from around the world engaged in unraveling the importance and relevance of surfactants in tribological phenomena were invited to present their latest findings at the premier "Surfactants in Tribology" symposium. We decided to invite also leading scientists working in this area, who may or may not have participated in the symposium, to submit written accounts (chapters) on their recent research findings in this field, which culminated in the publication of the first volume in the series Surfactants in Tribology in 2008.

Since the first symposium, interest in the relevance of surfactants in tribology has continued to grow among scientists and engineers working in the areas of both surfactants and tribology. Therefore, we decided to organize follow-up symposia on this topic at subsequent SIS meetings. Concomitantly, "Surfactants in Tribology" symposia were held during SIS-2008 (Berlin, Germany), SIS-2010 (Melbourne, Australia), SIS-2012 (Edmonton, Canada), and SIS-2014 (Coimbra, Portugal). Each of these symposia has been followed by the publication of subsequent volumes in the series Surfactants in Tribology, Volumes 2, 3, 4, and 5, respectively.

Volume 5 (the current volume), based on the symposium held in Coimbra, Portugal, in 2014, comprises a total of 16 chapters dealing with various aspects of surfactants and tribology, some of which had not been covered at all in previous volumes in this series. These 16 chapters have been logically grouped into four theme areas as follows. Section I consists of four chapters dealing with ionic liquids and polymeric amphiphiles in tribology. Topics covered in Section I include tribochemical reaction of ionic liquids on steel sliding surfaces, friction of polyethylene-b-poly(ethylene glycol) diblock copolymers on model substrates, aqueous lubrication with polyelectrolytes, and use of polymers in viscosity index modification and pour point depression Section II consists of three chapters dealing with nanomaterials in tribology. Topics discussed in Section II include polyethylene nanotubes formed by mechanochemical fragments, nanotechnology and performance development of cutting fluids, and silver nanoparticles colloidal dispersions. Section III comprises three chapters dealing with tribological applications in automotive, petroleum drilling, and food. Topics discussed in Section III include correlating engine dynamometer fuel economy to time-dependent tribology, formation of tribofilms on steel surfaces from surfactants with different degrees of ethoxylation, and load- and velocity-dependent friction behavior of milk fat. Section IV deals with biobased amphiphilic materials in tribology and related fields. Topics discussed in Section IV include test methods for biodegradability of lubricants; biosynthesis and derivatization of microbial glycolipids and potential applications in tribology; biofuels from vegetable oils as alternative fuels: advantages and disadvantages; field pennycress: a new oilseed crop for the production of biofuels, lubricants, and high-quality proteins; biobased lubricant additives; and assessment of agricultural wastes as biosorbents for heavy metal ions removal from wastewater.

Surface science and tribology play many critical roles in various industries. Manufacture and use of consumer and industrial products rely on the application of advanced surface and tribological knowledge. Examples of major industrial sectors that rely on these two disciplines include mining, agriculture, manufacturing (metals, plastics, wood, automotive, computers, MEMS, NEMS, appliances, planes, rails, etc.), construction (homes, roads, bridges, etc.), transportation (cars, boats, rails, airplanes), and medicine (instruments and diagnostic devices; transplants for knee, hips, and other body parts). The chapters in *Surfactants in Tribology, Volume 5* discuss some of the underlying tribological and surface science issues relevant to many situations in diverse industries. We believe that the information compiled in this book will be a valuable resource to scientists and technologists working in or entering the fields of tribology and surface science.

This volume and its predecessors (volumes 1–4) contain bountiful information and reflect the latest developments highlighting the relevance of surfactants in various tribological phenomena pertaining to many different situations. As we learn more about the connection between surfactants and tribology, new and improved ways to control lubrication, friction, and wear utilizing surfactants will emerge.

Now it is our pleasant task to thank all those who helped in materializing this book. First and foremost, we are very thankful to the contributors for their interest, enthusiasm, and cooperation as well as for sharing their findings, without which this book could not be born. Also we extend our appreciation to Barbara (Glunn) Knott at Taylor & Francis Group for her steadfast interest in and unwavering support for this book project and to the staff at Taylor & Francis Group for giving this book a body form.

Girma Biresaw
K.L. Mittal

Editors

Girma Biresaw received a PhD in physical organic chemistry from the University of California, Davis. As a postdoctoral research fellow at the University of California, Santa Barbara, he investigated reaction kinetics and products in surfactant-based organized assemblies for 4 years. As a scientist at the Aluminum Company of America (Alcoa), he conducted research in tribology, surface/colloid science, and adhesion for 12 years. Dr. Biresaw is currently a research chemist/lead scientist at the Agricultural Research Service (ARS) of the U.S. Department of Agriculture in Peoria, IL, where he is conducting research in tribology, adhesion, and surface/colloid science in support of programs aimed at developing biobased products from farm-based raw materials. He has authored/coauthored more than 300 scientific publications, including more than 85 peer-reviewed manuscripts, 6 patents, and 7 edited books. Dr. Biresaw is a fellow of the STLE and an editorial board member for the *Journal of Biobased Materials and Bioenergy.*

Kashmiri Lal (Kash) Mittal received his PhD from the University of Southern California in 1970 and was associated with IBM Corp. from 1972 to 1994. He is currently teaching and consulting worldwide in the areas of adhesion and surface cleaning. He is the editor of 125 published books, as well as others that are in the process of publication, within the realms of surface and colloid science and of adhesion. He has received many awards and honors and is listed in many biographical reference works. Dr. Mittal was a founding editor of the *Journal of Adhesion Science and Technology* and was its editor-in-chief until April 2012. He has served on the editorial boards of a number of scientific and technical journals. He was recognized for his contributions and accomplishments by the international adhesion community that organized the *First International Congress on Adhesion Science and Technology* in Amsterdam in 1995 on the occasion of his 50th birthday (235 papers from 38 countries were presented). In 2002, he was honored by the global surfactant community, which instituted the Kash Mittal Award in the surfactant field in his honor. In 2003, he was honored by the Maria Curie-Sklodowska University, Lublin, Poland, which awarded him the title of doctor *honoris causa.* In 2010, he was honored by both adhesion and surfactant communities on the occasion of the publication of his 100th edited book. In 2012, he initiated a new journal titled *Reviews of Adhesion and Adhesives.* In 2014, two books entitled *Recent Advances in Adhesion Science and Technology and Surfactants Science and Technology: Retrospects and Prospects* were published in his honor.

Contributors

Ali A. Abd-Elaal
Petrochemicals Department
Egyptian Petroleum Research Institute
Nasr City, Egypt

Maram T.H. Abou Kana
National Institute of LASER Enhanced
    Science
Cairo University
Giza, Egypt

J. de Vicente Álvarez
Faculty Sciences
Department of Applied Physics
University of Granada
Granada, Spain

Mert Arca
Department of Chemical Engineering
Pennsylvania State University
University Park, Pennsylvania

J.E. Arellano
Production Department of Strategic
    Research
PDVSA, Intevep
Los Teques, Venezuela

Richard D. Ashby
Eastern Regional Research Center
Agricultural Research Service
U.S. Department of Agriculture
Wyndmoor, Pennsylvania

Lia Beraldo da Silveira Balestrin
Institute of Chemistry
University of Campinas
Campinas, Sao Paulo, Brazil

Grigor B. Bantchev
Bio-Oils Research Unit
National Center for Agricultural
    Utilization Research
Agricultural Research Service
U.S. Department of Agriculture
Peoria, Illinois

Cristina Bignardi
Department of Mechanical and
    Aerospace Engineering (DIMEAS)
Polytechnic University of Turin
Torino, Italy

Sophie Bistac
Laboratory of Photochemistry and
    Macromolecular Engineering
University of Mulhouse
Mulhouse, France

Maurice Brogly
Laboratory of Photochemistry and
    Macromolecular Engineering
University of Mulhouse
Mulhouse, France

Thiago Augusto de Lima Burgo
Department of Physics
Federal University of Santa Maria
Santa Maria, Brazil

Steven C. Cermak
Bio-Oils Research Unit
National Center for Agricultural
    Utilization Research
Agricultural Research Service
U.S. Department of Agriculture
Peoria, Illinois

David L. Compton
Renewable Products Research Unit
National Center for Agricultural
  Utilization Research
Agricultural Research Service
U.S. Department of Agriculture
Peoria, Illinois

Frank J. DeBlase
Chemtura Corporation Part of the
  LANXESS Group
Naugatuck, Connecticut

Sevim Z. Erhan
Eastern Regional Research Center
Agricultural Research Service
U.S. Department of Agriculture
Wyndmoor, Pennsylvania

Roque L. Evangelista
Bio-Oils Research Unit
National Center for Agricultural
  Utilization Research
Agricultural Research Service
U.S. Department of Agriculture
Peoria, Illinois

Kervin O. Evans
Renewable Products Research Unit
National Center for Agricultural
  Utilization Research
Agricultural Research Service
United States Department of Agriculture
Peoria, Illinois

Diane Fischer
Laboratory of Photochemistry and
  Macromolecular Engineering
University of Mulhouse
Mulhouse, France

Fernando Galembeck
Institute of Chemistry
University of Campinas
Campinas, Brazil

Gerhard Gaule
Hermann Bantleon GmbH
Ulm, Germany

J.M. González
Production Department of Strategic
  Research
PDVSA, Intevep
Los Teques, Venezuela

Dogan Grunberg
Food and Industrial Oil Research Unit
National Center for Agricultural
  Utilization Research
Agricultural Research Service
U.S. Department of Agriculture
Peoria, Illinois

David R.K. Harding
Department of Chemistry
Institute of Fundamental Sciences
Massey University
Palmerston North, New Zealand

Hassan H.H. Hefni
Petrochemicals Department
Egyptian Petroleum Research Institute
Nasr City, Egypt

Milagros P. Hojilla-Evangelista
Plant Polymer Research Unit
National Center for Agricultural
  Utilization Research
Agricultural Research Service
U.S. Department of Agriculture
Peoria, Illinois

Yusuke Ichise
Department of Mechanical
  Engineering
Tokyo University of Science
Tokyo, Japan

Terry A. Isbell
Bio-Oils Research Unit
National Center for Agricultural
    Utilization Research
Agricultural Research Service
U.S. Department of Agriculture
Peoria, Illinois

Nadia G. Kandile
Department of Chemistry
Faculty for Women
Ain Shams University
Heliopolis, Cairo, Egypt

Shouhei Kawada
Department of Mechanical
    Engineering
Tokyo University of Science
Tokyo, Japan

and

Japan Society for the Promotion
    of Science
Tokyo, Japan

Seunghwan Lee
Department of Mechanical
    Engineering
Technical University of Denmark
Kongens Lyngby, Denmark

Zengshe Liu
Bio-Oils Research Unit
National Center for Agricultural
    Utilization Research
Agricultural Research Service
U.S. Department of Agriculture
Peoria, Illinois

Mona Y. Mohamed
Faculty of Agriculture
Cairo University
Giza, Egypt

Bryan R. Moser
Bio-Oils Research Unit
National Center for Agricultural
    Utilization Research
Agricultural Research Service
U.S. Department of Agriculture
Peoria, Illinois

Ben Müller-Zermini
Hermann Bantleon GmbH
Ulm, Germany

Rex E. Murray
Bio-Oils Research Unit
National Center for Agricultural
    Utilization Research
Agricultural Research Service
U.S. Department of Agriculture
Peoria, Illinois

Nabel A. Negm
Petrochemicals Department
Egyptian Petroleum Research Institute
Nasr City, Egypt

F. Quintero
Production Department of Strategic
    Research
PDVSA, Intevep
Los Teques, Venezuela

S. Rosales
Production Department of Strategic
    Research
PDVSA, Intevep
Los Teques, Venezuela

Shinya Sasaki
Department of Mechanical Engineering
Tokyo University of Science
Tokyo, Japan

Brajendra K. Sharma
Illinois Sustainable Technology Center
Prairie Research Institute
University of Illinois
Urbana Champaign, Illinois

Douglas Soares da Silva
Institute of Chemistry
University of Campinas
Campinas, Brazil

Daniel K.Y. Solaiman
Eastern Regional Research Center
Agricultural Research Service
U.S. Department of Agriculture
Wyndmoor, Pennsylvania

Chiharu Tadokoro
Department of Mechanical
    Engineering
Saitama University
Saitama, Japan

Teresa Tomasi
Center for Innovation
Ducom Instruments Europe B.V.
Groningen, The Netherlands

and

Polytechnic University of Turin
Torino, Italy

Angela M. Tortora
Center for Innovation
Ducom Instruments Europe B.V.
Groningen, The Netherlands

Dan Vargo
Functional Products Inc.
Macedonia, Ohio

Deepak H. Veeregowda
Center for Innovation
Ducom Instruments Europe B.V.
Groningen, The Netherlands

Seiya Watanabe
Department of Mechanical Engineering
Tokyo University of Science
Tokyo, Japan

and

Japan Society for the Promotion of
    Science
Tokyo, Japan

Mona A. Youssif
Egyptian Petroleum Research Institute
Nasr City, Egypt

Section I

Ionic Liquids and Polymeric Amphiphiles in Tribology

1 Tribochemical Reaction of Ionic Liquids as Lubricants on Steel Sliding Surfaces

Shouhei Kawada, Yusuke Ichise, Seiya Watanabe, Chiharu Tadokoro, and Shinya Sasaki

CONTENTS

1.1 INTRODUCTION

Ionic liquids are organic salts comprising a cation and an anion. They exist in the liquid state at room temperature and are expected to function as novel lubricant ingredients because of their attractive characteristics, such as high thermal stability, low vapor pressure, low melting point, high ion conductivity, and wide electrochemical window [1–12]. In the field of tribology, ionic liquids can be used as novel base oils and/or additives for extreme environments such as high vacuum (e.g., spacecraft) and high temperature [13–17].

It has been reported that fluorinated ionic liquids show a low friction coefficient and wear by forming a reactive film on the sliding surface [3,18–20]. The commercial application of ionic liquids in lubricants requires understanding their detailed lubricating mechanism on sliding surfaces. In most investigations of such mechanism, surface analysis techniques such as x-ray photoelectron spectroscopy (XPS) and time-of-flight secondary ion mass spectroscopy (ToF-SIMS) are used. However, in situ observations during sliding tests are difficult to perform using these two techniques.

To investigate the mechanism of reactive film formation in situ, in this study, we used a quadrupole mass spectrometer (Q-mass). The method can measure the partial pressure of a gas in a vacuum chamber and allows to quantitatively follow the progress of the tribochemical decomposition process of ionic liquids on the steel sliding surfaces in real time [21]. In addition, the friction coefficient and wear scar of the ionic liquids were evaluated. The worn surfaces from sliding experiments were analyzed using optical microscopy and XPS. The results of tribochemical decomposition processes, the lubricating property, and the surface analysis of the ionic liquids are analyzed to propose a detailed lubricating mechanism.

1.2   EXPERIMENTAL SETUP

1.2.1   MATERIALS

1.2.1.1   Ionic Liquids

Four kinds of ionic liquids were investigated. Table 1.1 lists the chemical names and molecular structures of the ionic liquids: 1-butyl-3-methylimidazolium tetrafluoroborate ([BMIM][BF_4]), 1-butyl-3-methylimidazolium bis(trifluoromethylsulfonyl) imide ([BMIM][TFSI]), 1-butyl-3-methylimidazolium hexafluorophosphate ([BMIM][PF_6]), and 1-ethyl-3-methylimidazolium tetracyanoborate ([EMIM] [TCB]). [BF_4], [TFSI], and [PF_6] anions are fluorine containing and their structures are well known. On the other hand, the [TCB] anion is fluorine-free, and its molecular structure resembles that of [BF_4]. These ionic liquids were purchased from Merck Chemicals, Germany, in high purity >98%. Table 1.2 lists the physical properties of these ionic liquids. Viscosity was measured using a tuning-fork vibration viscometer (SV-1A, A&D Company, Japan). The oscillator was made of titanium, which has a high corrosion resistance.

1.2.1.2   Test Specimen

The specifications of the test specimens were as follows: disk diameter × thickness = 24 × 7.9 mm; ball diameter = 4 mm; the disk and the ball were made of bearing steel AISI 52100, Rockwell hardness (HRC) of 60. Before the sliding test, the specimens were cleaned twice with a mixed solvent of petroleum benzene and acetone for 20 min.

1.2.2   FRICTION AND DECOMPOSITION

1.2.2.1   Instrument

The tribochemical decomposition of the ionic liquids was investigated using an atmosphere-controlled ball-on-disk friction tester, as shown in Figure 1.1. The frictional force and the outgassing were measured with a load cell and quadrupole mass spectrometer (Q-mass, MKS Instruments, Inc., Tokyo, Japan), respectively.

TABLE 1.1

Structure of Ionic Liquids Investigated in This Work

1-Butyl-3-methylimidazolium tetrafluoroborate [BMIM][BF$_4$]

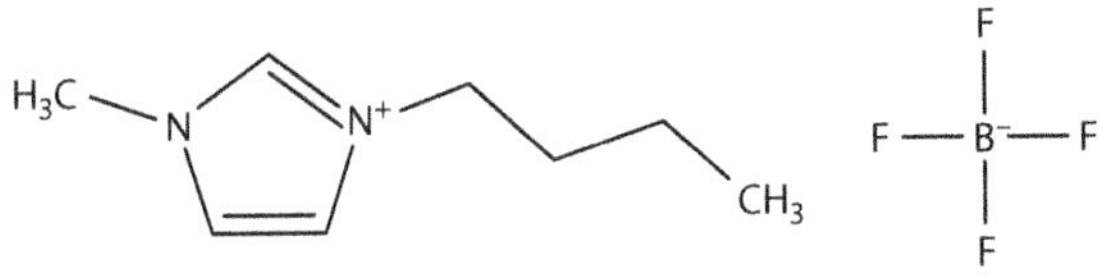

1-Butyl-3-methylimidazolium bis(trifluoromethylsulfonyl)imide [BMIM][TFSI]

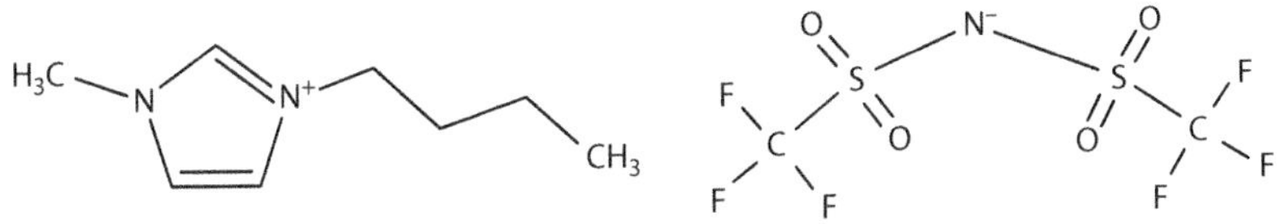

1-Butyl-3-methylimidazolium hexafluorophosphate [BMIM][PF$_6$]

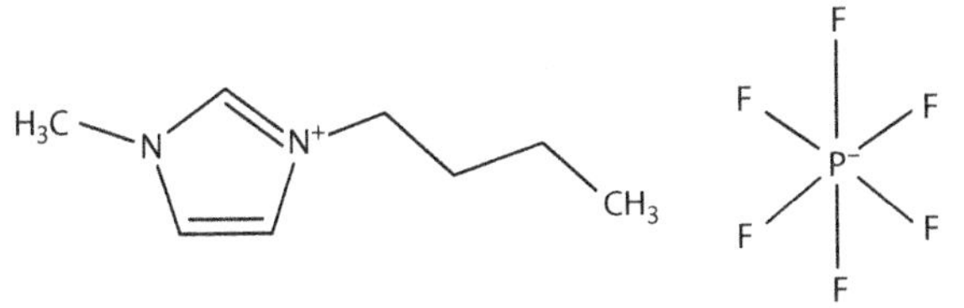

1-Ethyl-3-methylimidazolium tetracyanoborate [EMIM][TCB]

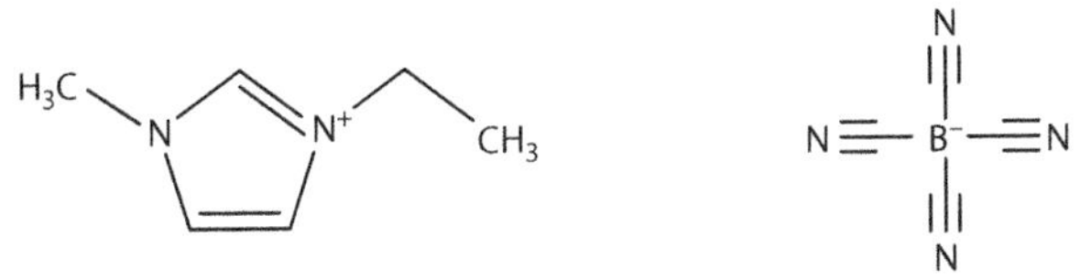

TABLE 1.2

Physical Properties of Ionic Liquids Investigated in This Work

Ionic Liquid	Melting Point (°C)	Viscosity (mPa s)			Viscosity Index
		40°C	70°C	100°C	
[BMIM][BF$_4$]	<−50	46.15	17.94	8.62	152
[BMIM][TFSI]	<0	40.99	16.92	8.99	179
[BMIM][PF$_6$]	13	116.91	45.32	18.06	155
[EMIM][TCB]	13	8.88	4.42	2.69	144

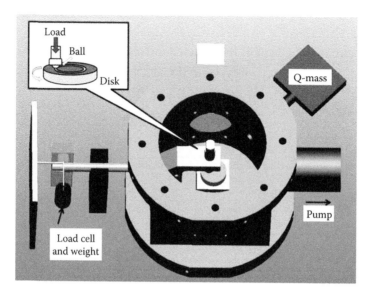

FIGURE 1.1   Schematic of sliding tester used.

The detectable range of the Q-mass is 1–200 (m/e) using a sensitivity of the partial pressure meter of 3.8×10^{-7} A/Pa with the secondary electron multiplier. The partial pressures of the ions were converted from the ion currents using the conversion rate for N_2 since it is difficult to calibrate the conversion rate of each ion fragment.

1.2.2.2   Procedure

The test conditions are listed in Table 1.3. When the vacuum reached the set value, the Q-mass software was activated, and vacuuming continued for two more hours to remove the residual gas from the Q-mass before the test was performed.

1.2.3   SURFACE ANALYSIS

The specimens, after each sliding test, were ultrasonically cleaned with a mixed solvent of petroleum benzene and acetone for 10 min. The wear state of the specimens and wear scar diameter of the ball were recorded using an optical microscopy

TABLE 1.3
Test Conditions

Applied load	4.5 N
Chamber pressure	2.0×10^{-5} Pa
Speed	100 rpm
Lubricants volume	30 μL
Chamber temperature	25°C
Test duration time	2 h

(OM, VHX-100, Keyence, Japan). The worn surfaces were analyzed using XPS (QUANTERAII, ULVAC-PHI, Inc., Japan) with a monochromatic AlKα x-ray source (1486.6 eV). The diameter of the spot size of the x-ray source was set at 300 μm for the analysis. All spectra presented here are referenced to the C1s peak (285.0 eV).

1.3   RESULTS AND DISCUSSION

1.3.1   LUBRICATING PROPERTY

Figure 1.2 shows the average friction coefficient recorded during the last 5 min of each sliding experiment and the wear scar diameter of the sliding ball for each ionic liquid. Figure 1.3 shows images of the worn surface after the sliding tests. The friction coefficients and the wear behaviors differ depending on the anion of the ionic

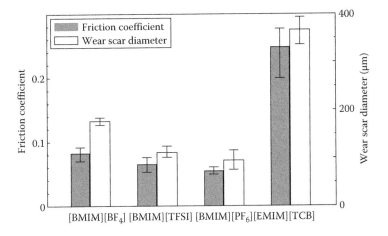

FIGURE 1.2   Average wear scar diameter and friction coefficient for the last 5 min lubricated with each ionic liquid.

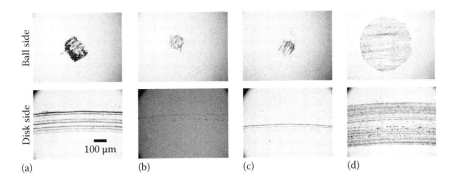

FIGURE 1.3   Worn surface optical images for (a) [BMIM][BF$_4$], (b) [BMIM][TFSI], (c) [BMIM][PF$_6$], and (d) [EMIM][TCB].

liquid used. For [BMIM][BF₄], the friction coefficient had a low value of ~0.08 and a wear diameter of ~200 μm. For [BMIM][TFSI], the friction coefficient was lower than that of [BMIM][BF₄], and the wear diameter was ~100 μm. For [BMIM][PF₆], the friction coefficient was ~0.05, and the wear diameter was <100 μm, the smallest among the investigated ionic liquids. Finally, in the case of the fluorine-free ionic liquid [EMIM][TCB], the friction coefficient was the highest among the investigated ionic liquids. In addition, the wear scar diameter of [EMIM][TCB] was the largest among the investigated ionic liquids, four times that of [BMIM][PF₆]. These results show that fluorinated ionic liquids provide the lowest friction and wear scar diameter because they form reactive films on the sliding surface.

1.3.2 FRICTION AND DECOMPOSITION BEHAVIOR

We anticipated the nature of gas generated from the structures of the ionic liquids. The outgassed hydrocarbons originated from the cations, while, HF, H_2S, CF_3, PH_3, and CN originated from the anions. The outgassing of [EMIM] (mass-to-charge ratio, $m/e = 111$), [BMIM] ($m/e = 139$), [BF₄] ($m/e = 87$), [PF₆] ($m/e = 145$), and [TCB] ($m/e = 111$) could not be detected. [TFSI] ($m/e = 220$) could not be detected as well because its m/e value exceeded the detectable range. Figures 1.4 through 1.7

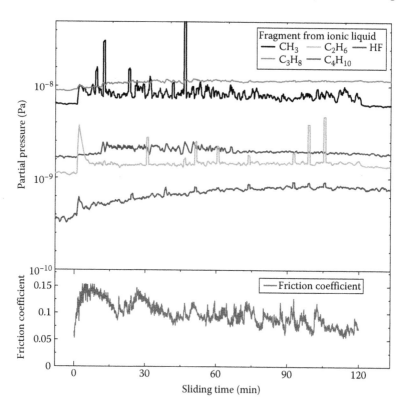

FIGURE 1.4 Partial pressure and friction coefficient behavior of [BMIM][BF₄].

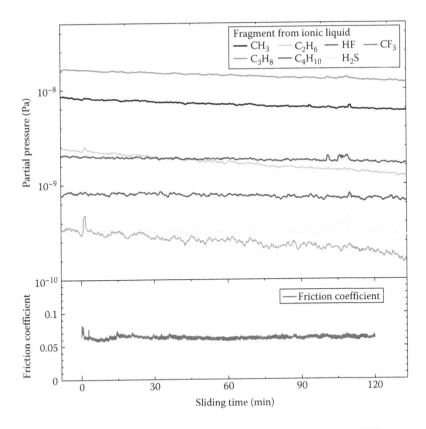

FIGURE 1.5 Partial pressure and friction coefficient behavior of [BMIM][TFSI].

show the evolution of the friction and the partial pressure behaviors during the experiment for each ionic liquid.

The friction behavior of [BMIM][BF$_4$] was very unstable. The friction coefficient was in the range 0.05–0.15. The increase in partial pressure from this ionic liquid was confirmed because it decomposed during sliding probably due to catalytic decomposition on the nascent AISI 52100 surface [13,22]. The fragments from the cations were detected immediately after the onset of the experiment. However, those derived from the anions were detected after 15 min. We believe that this outgassing behavior indicates that the anion reacted with the nascent AISI 52100 surface in the first 15 min. The reactive film was removed during sliding after 15 min, resulting in the release of anion fragments.

The friction behavior of [BMIM][TFSI] was very stable (Figure 1.5). The increase in partial pressure from this ionic liquid was detected as soon as the sliding started, and the friction coefficient during this period was high (0.07–0.09). However, outgassing was not observed after 3 min, and the friction coefficient was a stable value (~0.06) during this period. During the first 3 min, the fragments derived from both cations and anions were detected. It is assumed that the partial pressure behavior

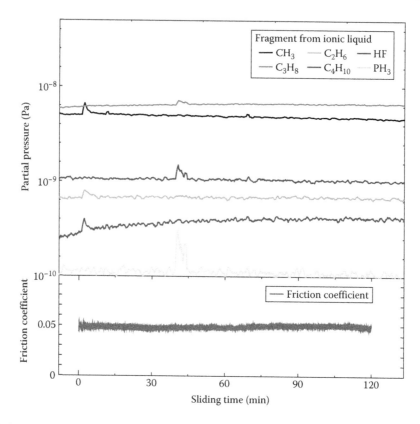

FIGURE 1.6 Partial pressure and friction coefficient behavior of [BMIM][PF$_6$].

indicates the reactive films derived from [BMIM][TFSI] formed at 3 min, and served as protective films for the AISI 52100 surface.

The friction behavior of [BMIM][PF$_6$] was very stable. The increase in partial pressure from this ionic liquid was detected as soon as the sliding started but stopped after 3 min. During this period, only the fragments derived from the cations were detected. However, after about 40 min, fragments from the anions abruptly appeared. It is assumed that the reactive films of [BMIM][PF$_6$] were formed at 3 min and were removed by sliding after 40 min. However, it is assumed that the restoration of the reactive film was fast because the friction behavior did not change during this period. Thus, we believe that a reactive film derived from [BMIM][PF$_6$] worked as the protective film for the AISI 52100 surface.

Finally, for [EMIM][TCB], the friction coefficient was initially high, and increases in partial pressure from both the cation and the anion were detected throughout the experiment. This result shows that [EMIM][TCB] underwent catalytic decomposition on the nascent AISI 52100 surface throughout the experiment. The result indicates that [EMIM][TCB] did not form a reactive film with the nascent AISI 52100 surface, which explains the large friction coefficient.

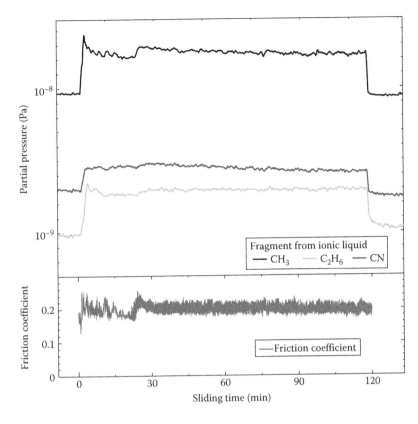

FIGURE 1.7 Partial pressure and friction coefficient behavior of [EMIM][TCB].

1.3.3 XPS ANALYSIS

The Q-mass results indicated that a surface-protective film derived from the fluorinated ionic liquids is formed on the sliding surface. To obtain information on the chemical structure of the sliding surface, we conducted a surface analysis of the disk specimen after each test using XPS.

For [BMIM][BF_4], boron (B) and fluorine (F) XPS spectra are shown in Figures 1.8 and 1.9, respectively. For the B1s spectrum shown in Figure 1.8, the signal-to-noise (S/N) ratio is too weak. The peak of boron nitride (189.9 eV) was not observed. In the F1s spectrum shown in Figure 1.9, the peak of FeF_2 and/or FeF_3 (684.7 eV) was observed [22–24]. These results indicate that the reactive film produced was derived from [BF_4]. Since the reactive film improved lubricating property, we believe that the lubricating property improved because of the formation of FeF_2 and/or FeF_3 films on the surface [19,20,25].

For [BMIM][TFSI], sulfur (S) and fluorine (F) XPS spectra are shown in Figures 1.10 and 1.11, respectively. In the S2p spectrum shown in Figure 1.10, the peaks of $FeSO_4$ (168.9 eV) and FeS and/or FeS_2 (161.9 eV) were observed [23,26–28], while in the F1s spectrum shown in Figure 1.11, the peaks of FeF_2 and/or FeF_3

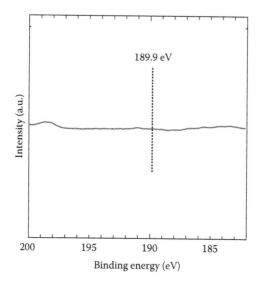

FIGURE 1.8   B1s XPS spectrum for [BMIM][BF$_4$].

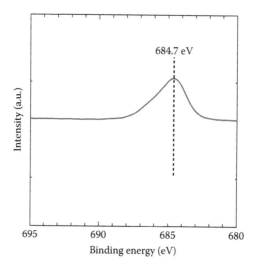

FIGURE 1.9   F1s XPS spectrum for [BMIM][BF$_4$].

(684.7 eV) and [TFSI] (688.2 eV) were observed [22–24]. These results indicate that the reactive film produced was derived from [TFSI]. In addition, [TFSI] adsorbed on the AISI 52100 surface from the peak of CF$_3$. Compared to [BMIM][BF$_4$], [BMIM] [TFSI] produced not only fluoride but also sulfide. It is believed that these reactive films considerably reduced the friction coefficient and the wear scar diameter than films from [BMIM][BF$_4$]. From the Q-mass results, this reactive film was formed after 3 min and covered the nascent AISI 52100 surface. Hence, the AISI 52100 surface became inert and the increase in partial pressure was not detected.

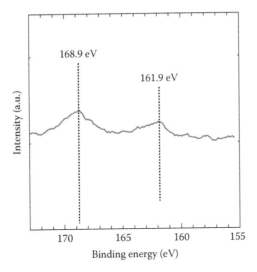

FIGURE 1.10 S2p XPS spectrum for [BMIM][TFSI].

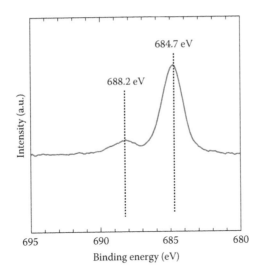

FIGURE 1.11 F1s XPS spectrum for [BMIM][TFSI].

For [BMIM][PF$_6$], phosphorus (P) and fluorine (F) XPS spectra are shown in Figures 1.12 and 1.13, respectively. In the P2p spectrum shown in Figure 1.12, the peak of FePO$_4$ (133.6 eV) was observed [23,29,30], and in the F1s spectrum shown in Figure 1.13, the peaks of FeF$_2$ and/or FeF$_3$ (684.7 eV) and [PF$_6$] (687.1 eV) were observed. These results indicate that the reactive film produced was derived from [PF$_6$] and that [PF$_6$] was adsorbed on the AISI 52100 surface. Thus, [BMIM][PF$_6$] produced not only fluoride but also phosphide fragments. The excellent tribological

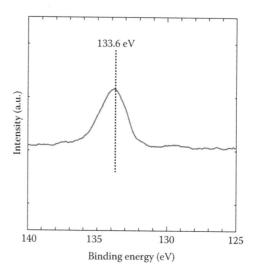

FIGURE 1.12   P2p XPS spectrum for [BMIM][PF$_6$].

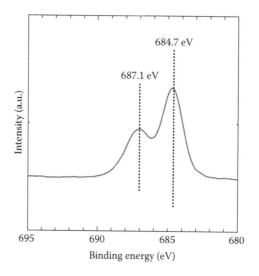

FIGURE 1.13   F1s XPS spectrum for [BMIM][PF$_6$].

behavior and formation of FePO$_4$ were found for the steel/steel contact [31]. Oxygen was almost nonexistent because these experiments were conducted in vacuum. However, Suzuki et al. reported that FePO$_4$ was produced in the vacuum [19]. It is assumed that phosphorus reacted with the oxidation film of iron. We believe that these reactive films reduced the friction coefficient and the wear scar diameter as compared to [BMIM][BF$_4$]. From the Q-mass results, this reactive film was formed after 3 min and covered the nascent AISI 52100 surface.

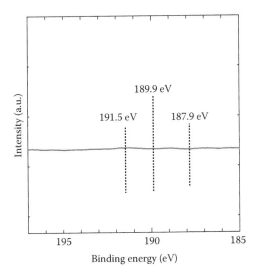

FIGURE 1.14 B1s XPS spectrum for [EMIM][TCB].

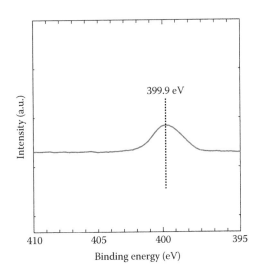

FIGURE 1.15 N1s XPS spectrum for [EMIM][TCB].

For [EMIM][TCB], boron (B) and nitrogen (N) XPS spectra are shown in Figures 1.14 and 1.15, respectively. In the B1s spectrum shown in Figure 1.14, the peaks of FeB and/or Fe_2B (187.9 eV), BN (189.9 eV), and [TCB] (191.5 eV) were not observed. In the N1s spectrum shown in Figure 1.15, the peaks of CN (399.9 eV) were observed. These results indicate that no reactive film derived from [TCB] was produced and that CN was adsorbed on the AISI 52100 surface. Hence, this ionic liquid showed poor lubricating property, that is, high friction coefficient and high wear scar diameter.

1.4 CONCLUSIONS

In this chapter, the lubricating property and the tribochemical decomposition of ionic liquids on the AISI 52100 steel surfaces were investigated. The worn surfaces after the sliding test were analyzed by using XPS. The main conclusions from this investigation are as follows:

1. Fluorinated ionic liquids, [BMIM][BF$_4$], [BMIM][TFSI], and [BMIM][PF$_6$], provided low friction and wear, which were attributed to the formation of films on the nascent AISI 52100 surface. On the other hand, the fluorine-free ionic liquid, [EMIM][TCB], produced high friction and wear.
2. Ionic liquids decomposed on the nascent AISI 52100 steel surface because of catalytic decomposition. Hydrocarbon fragments derived from the cations of fluorinated ionic liquids were detected during the sliding. Increased partial pressures from the ionic liquids [BMIM][TFSI] and [BMIM][PF$_6$] were not detected during the first 3 min, because the reactive film was being formed during these 3 min and covered the nascent AISI 52100 surface. For [EMIM][TCB], the fragments derived from both the cation and anion of the ionic liquids were detected during the entire duration of the experiment. It can be concluded that [EMIM][TCB] did not form a reactive film on the nascent AISI 52100 steel surface.
3. Depending on the type of reactive film, the fluorinated ionic liquids showed different friction and wear properties. For the three fluorinated ionic liquids studied in this work, the reactive film was composed of FeF$_2$ and/or FeF$_3$. In addition, [BMIM][TFSI] and [BMIM][PF$_6$] produced sulfide- and phosphide-containing films, respectively. They also deactivated the nascent AISI 52100 steel surface. [EMIM][TCB] formed no such reactive film, and CN was adsorbed on the AISI 52100 surface. The adsorption film was easily removed during sliding.

ACKNOWLEDGMENT

This work was supported by a Grant-in-Aid from the Japan Society for the Promotion of Science (JSPS) Fellows (No. 15J05958).

REFERENCES

1. M.D. Bermudez, A.E. Jiménez, J. Sanes, and F.J. Carrion, Ionic liquids as advanced lubricant fluids, *Molecules*, 14, 2888–2908 (2009).
2. F. Zhou, Y. Liang, and W. Liu, Ionic liquid lubricants: Designed chemistry for engineering applications, *Chem. Soc. Rev.*, 38, 2590–2599 (2009).
3. I. Minami, Ionic liquids in tribology, *Molecules*, 14, 2286–2305 (2009).
4. M. Palacio and B. Bhushan, A review of ionic liquids for green molecular lubrication in nanotechnology, *Tribol. Lett.*, 40, 247–268 (2010).
5. E. Schlücker and P. Waserscheid, Ionic liquids in mechanical engineering, *Chemie Ingeniuer Technik*, 83, 1476–1484 (2011).
6. C.F. Ye, W.M. Liu, Y.X. Chen, and L.G. Yu, Room-temperature ionic liquids: A novel versatile lubricant, *Chem. Commun.*, 21, 2244–2245 (2001).

7. Z. Mu, F. Zhou, S. Zhang, Y. Liang, and W. Liu, Effect of the functional groups in ionic liquid molecules on the friction and wear behavior of aluminum alloy in lubricated aluminum-on-steel contact, *Tribol. Int.*, 38, 725–731 (2005).

8. A.E. Jiménez, M.D. Bermúdez, P. Iglesias, F.J. Carrión, and G. Martínez-Nicolás, 1-n-alkyl-3-methylimidazolium ionic liquids as neat lubricants and lubricant additives in steel-aluminium contacts, *Wear*, 260, 766–782 (2006).

9. J. Qu, P.J. Blau, S. Dai, H. Luo, and H.M. Meyer III, Ionic liquids as novel lubricants and additives for diesel engine applications, *Tribol. Lett.*, 35, 181–189 (2009).

10. J.S. Wikes, A short history of ionic liquids—From molten salts to neoteric solvents, *Green Chem.*, 4, 73–80 (2002).

11. H. Matsumoto, T. Matsuda, and Y. Miyazaki, Room temperature molten salts based on trialkylsulfonium cations and bis (trifluoromethylsulfonyl) imide, *Chem. Lett.*, 81, 1430–1431 (2000).

12. M. Imanari, K. Uchida, K. Miyano, H. Seki, and K. Nishikawa, NMR study on relationships between reorientational dynamics and phase behavior of room-temperature ionic liquids: 1-alkyl-3-methylimidazolium cations, *Phys. Chem. Chem. Phys.*, 12, 2959–2967 (2012).

13. S. Kawada, S. Watanabe, Y. Kondo, R. Tsuboi, and S. Sasaki, Tribochemical reactions of ionic liquids under vacuum conditions, *Tribol. Lett.*, 54, 309–315 (2014).

14. A.E. Jiménez and M.D. Bermudez, Ionic liquids as lubricants for steel-aluminum contacts at low and elevated temperatures, *Tribol. Lett.*, 26, 53–60 (2007).

15. T. Yagi, S. Sasaki, H. Mano, K. Miyake, M. Nakano, and T. Ishida, Lubricity and chemical reactivity of ionic liquids used for sliding metals under high-vacuum conditions, *J. Eng. Tribol.*, 223, 1083–1090 (2009).

16. C.M. Jin, C.F. Ye, B.S. Philips, J.S. Zabinski, X.Q. Liu, W.M. Liu, and J.M. Shreeve, Polyethylene glycol functionalized dicationic ionic liquids with alkyl or polyfluoroalkyl substituents as high temperature lubricants, *J. Mater. Chem.*, 16, 1529–1535 (2006).

17. K.W. Street Jr., W. Morales, V.R. Koch, D.J. Valco, R.M. Richard, and N. Hanks, Evaluation of vapor pressure and ultra-high vacuum tribological properties of ionic liquids, *Tribol. Trans.*, 54, 911–919 (2011).

18. W. Liu, C. Ye, Q. Gong, H. Wang, and P. Wang, Tribological performance of room-temperature ionic liquids as lubricant, *Tribol. Lett.*, 13, 81–85 (2002).

19. A. Suzuki, Y. Shinka, and M. Masuko, Tribological characteristics of imidazolium-based room temperature ionic liquids under high vacuum, *Tribol. Lett.*, 27, 307–313 (2007).

20. Y. Kondo, S. Yagi, T. Koyama, R. Tsuboi, and S. Sasaki, Lubricity and corrosiveness of ionic liquids for steel-on-steel sliding contacts, *J. Eng. Tribol.*, 226, 991–1006 (2011).

21. S. Mori, Adsorption of benzene on the fresh steel surface formed by cutting under high vacuum, *Appl. Surf. Sci.*, 27, 401–410 (1987).

22. R. Lu, S. Mori, K. Kobayashi, and H. Nanao, Study of Tribochemical decomposition of ionic liquids on a nascent steel surface, *Appl. Surf. Sci.*, 255, 8965–8971 (2009).

23. M. Kasrai and D. S. Urch, Electronic structure of iron (II) and (III) fluorides using X-ray emission and X-ray photoelectron spectroscopies, *J. Chem. Soc. Faraday Trans. II*, 75, 1522–1531 (1979).

24. D. Li, M. Cai, D. Feng, F. Zhou, and W. Liu, Excellent lubrication performance and superior corrosion resistance of vinyl functionalized ionic liquid lubricants at elevated temperature, *Tribol. Int.*, 44, 1111–1117 (2011).

25. B.S. Philips, G. Johm, and J.S. Zabinski, Surface chemistry of fluorine containing ionic liquids on steel substrates at elevated temperature using Mossbauer spectroscopy, *Tribol. Lett.*, 26, 85–91 (2007).

26. V. Totolin, N. Ranetcais, V. Hamciuc, N. Shore, N. Dorr, C. Ibanescu, B.C. Simionescu, and V. Harabagiu, Influence of ionic structure and tribological properties of poly (dimethylsiloxane-alkylene oxide) graft copolymers, *Tribol. Int.*, 67, 1–10 (2013).
27. Q. Lu, H. Wang, C. Ye, W. Liu, and Q. Xue, Room temperature ionic liquid 1-eth yl-3-hexylimidazolium-bis(trifluoromethylsulfonyl)-imide as lubricant for steel–steel contact, *Tribol. Int.*, 37, 547–552 (2004).
28. W. Zhao, J. Pu, Q. Yu, Z. Zeng, X. Wu, and Q. Xue, A novel strategy to enhance micro/nano-tribological properties of DLC film by combining micro-pattern and thin ionic liquids film, *Colloids Surf. A*, 428, 70–78 (2013).
29. A.J. Nelson, S. Glenis, and A.J. Frank, XPS and UPS investigation of PF6 doped and undoped poly 3methyl thiophene, *J. Chem. Phys.*, 87, 5002–5006 (1987).
30. Z. Wang, Y. Xia, Z. Liu, and Z. Wen, Conductive lubricating grease synthesized using the ionic liquid, *Tribol. Lett.*, 46, 33–42 (2012).
31. H. Wang, Q. Lu, C. Ye, W. Liu, and Z. Cui, Friction and wear behaviors of ionic liquid of alkylimidazolium hexafluorophosphates as lubricants for steel/steel contact, *Wear*, 256, 44–48 (2004).

2 Friction of Polyethylene-b-Poly(ethylene glycol) Diblock Copolymers against Model Substrates

Sophie Bistac, Diane Fischer, and Maurice Brogly

CONTENTS

2.1 INTRODUCTION

Block copolymers are made by covalent bonding of two or more different monomers forming blocks of repeating units. The simplest block copolymers are A_nB_m diblock copolymers, which consist of an A block with n units and a B block with m units. As in most cases, these sequences are incompatible; thus, block copolymers give rise to a rich variety of microstructures both in bulk and in solution. The length scale of these microstructures is of the order of the magnitude of the copolymer size itself [1] (i.e., 5–50 nm). Advances in synthetic chemistry [2] provide unparalleled control over molecular-scale morphology in this class of polymers. Hence, block copolymers are strong candidates for applications in advanced technologies such as new thermoplastic elastomers, drug release agents, recovery agents for contaminated water, nanopatterning materials for photolithography, and structure-directing agents for hybrid mesostructure used in catalysis or separation technologies [3]. Block copolymers also open new opportunities as responsive materials [4], an intensively growing field dedicated to the development of polymer-based biomaterials, sensors, and smart coatings. Block copolymer–based smart coatings, as other functional nanomaterials, exhibit unexpected properties (e.g., self-cleaning, self-lubricating) due to the fact that their thickness is limited only to a few hundred nanometers or less. The precise role of block copolymer molecular structure (composition, sequence length, etc.) in the microphase and nanophase separation in bulk copolymer materials or in selected solvents is extensively documented [5–8]. However, in thin films, the phase separation is subjected to more complex mechanisms such as self-assembly, lateral ordering, and organization governed by confinement or adsorption onto a substrate [9–11]. If by nature one block of the copolymer is semicrystalline, further competitive processes occur between self-organization, adsorption, and crystallization [12,13]. Very few studies have focused on the specific ordering of block copolymers having two crystalline blocks [14–16]. Double-crystalline block copolymers have received attention because they may lead to unexpected properties resulting from the increase in system complexity. Double-crystalline block copolymers include a combination of a crystalline sequence of polyalkanes (polyethylene [PE], polypropylene [PP], poly(ethyl ethylene) [PEE], polyethers (poly(ethylene oxide) [PEO], poly(propylene oxide) [PPO])), poly(L-lactide) (PLLA), or poly(ε-caprolactone) (PCL), to name the most representative examples.

Poly(ethylene oxide) (PEO) is also called poly(ethylene glycol) (PEG). The two polymers' chemical structures are identical, but they differ in their molecular weight: PEO generally has long chains, while PEG has shorter chains.

Bulk assembly in solid state is described in the literature for these types of double-crystalline block copolymers [14,15]. In these studies, the authors focused on double-crystalline amphiphilic block copolymers of PE and PEO. The synthesis of linear PE with PEO blocks of controlled molecular weights allows to study confined crystallization, new morphologies, and phase separation due to the high chain regularity and strong incompatibility of these blocks. As an example [16], for a polymerization degree of 29 and 20, respectively, for the PE and PEO blocks, the crystallization of the PE blocks was unconfined, while the crystallization of the PEO blocks was confined between preexisting PE crystalline lamellae. Thus, for both PE and PEO crystals, an interdigitated single-crystalline layer morphology was observed. The slower crystallization kinetics of PEO in polyethylene-block-poly(ethylene oxide) (PE-b-PEO) copolymer, relative to PEO homopolymer, is due to nanoconfinement of PEO chains tethered to PE crystals. PE-b-PEO copolymers are also good candidates for applications such as emulsifiers, stabilizers, or surfactants [17]. In this case, adsorption mechanisms as well as surface and interface-driven phase separation or self-assembling mechanisms have to be considered and analyzed to understand the key properties of PE-b-PEO amphiphilic copolymers for their use as surface modifier molecules. Interest has also been recently focused on their several useful biointerfacial properties, which are essential in surface modification of polymers for biomedical applications. PEO appears very attractive for surface modification since it can simultaneously impart biocompatibility and lubricity to materials. Indeed, the role of PEO-based amphiphilic copolymers in reducing nonspecific protein adsorption and cell adhesion, which can improve the biocompatibility of biomaterials and sensitivity of biosensor surfaces, has been investigated [18]. An interesting way to underline the hydrophobic/hydrophilic character of a surface is to measure its friction properties. Many interfacial mechanisms of adsorption or adhesion in a confined or dynamic interface are intimately linked to the sliding behavior of the macromolecular chains at the interface [19]. Gao et al. [20] have shown that the addition of PEO homopolymer to poly(oxy methylene) (POM)/poly(tetrafluoroethylene) (PTFE) fiber composites enhanced the formation of a transfer film during sliding contact and, accordingly, further improved the friction and wear performances. In a recent paper [21], the coefficient of friction of pure PEO film was determined by lateral force microscopy (LFM) experiments as a function of film thickness. The results show a constant friction coefficient value (0.28) for thicknesses in the range of 500–1500 nm and a monotonic decrease in the friction coefficient (from 0.28 to 0.16) with decreasing thickness (from 500 nm down to 20 nm). Shi et al. [22] have demonstrated that in the case of wet lubrication, a PEO-based solution can strongly reduce (by up to 40%) the friction coefficient of prosthetic joints (aluminum–cartilage contact). The friction reduction behavior increases with the PEO molecular weight in the range of 10,000–100,000 g mol^{-1}. Due to its affinity toward water, PEO polymer brushes have been observed [23] to greatly enhance the formation of aqueous lubricant films when grafted onto the surfaces of tribopairs. It has also been shown [24] that poly(L-lysine)-graft-poly(ethylene glycol) caused an improvement in the aqueous lubrication properties when adsorbed on both hydrophobic and hydrophilic thermoplastic surfaces. On the basis of in situ LFM and X-ray photoelectron spectroscopy, Li et al. [25] have investigated the lubrication behavior of an aqueous solution of poly(ethylene oxide)-poly(propylene oxide)-poly(ethylene oxide) (PEO-PPO-PEO) symmetric triblock copolymer on thin thermoplastic films of PP, PE, and cellulose. The proposed lubrication mechanism states that on hydrophobic surfaces (PP, PE), the outer layer is composed of flexible PEO chains, while the inner layer of PPO blocks anchors the copolymer to the surface. The exposed flexible PEO blocks enhance the lubrication of the surface, while, on the hydrophilic cellulose surface, the PEO blocks of the copolymer adhere to the surface, exposing the PPO blocks and hindering lubrication. Nevertheless, neither the friction properties of PEO-PE amphiphilic copolymers nor the influence of their block length nor the influence of substrate chemistry are described in the literature.

The objective of this study was thus to investigate the friction properties of PE-b-PEG diblock copolymers against two model substrates. Several PE-b-PEG copolymers, with different molecular weights and PEG/PE ratios, were used. The two model substrates were a hydrophilic glass plate, on the one hand, and a hydrophobic glass plate (obtained by a methyl-terminated silane grafting), on the other hand. The influences of both the PE-b-PEG copolymer composition and the nature of the substrate on the friction properties are discussed.

2.2   EXPERIMENTAL PART

2.2.1   MATERIALS

2.2.1.1   PE-b-PEG Diblock Copolymers

PE-b-PEG diblock copolymers were purchased from Sigma Aldrich. Polymer balls of 4 mm diameter were used as received (spherical pellets) for the friction test. The general chemical structure of PE-b-PEG copolymer is given in Figure 2.1.

Different PE-b-PEG copolymers, with molecular weight M_n varying from 575 to 2250 g mol^{-1}, were chosen. The molecular weight and block ratio (determined by proton nuclear magnetic resonance [H-NMR]) are presented in Table 2.1.

COP A, COP B, COP C, and COP D have one PE block of constant length and one PEG block of varying length. The different compositions and chain lengths of these copolymers allow for a better understanding of the influence of each block on the friction properties of the copolymer. Smooth polymer balls (diameter equal to 4 mm) were used as received for the friction experiments.

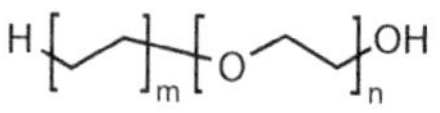

FIGURE 2.1   Structure of PE-b-PEG diblock copolymer.

TABLE 2.1

Chemical Composition of PE-b-PEG Copolymers Investigated

	Chemical Composition of PE-b-PEG Copolymers					
	M_n PE (g mol^{-1})	M_n PEG (g mol^{-1})	PE Block (wt%)	PEG Block (wt%)	PE Block (mol%)	PEG Block (mol%)
COP A	334	101	77	23	84	16
COP B	355	153	70	30	78	22
COP C	326	385	45	55	57	43
COP D	311	1474	17	83	25	75

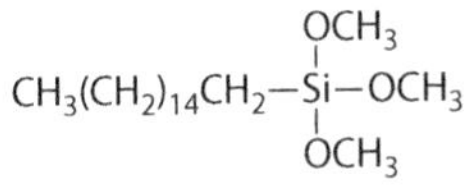

$$OCH_3$$
$$CH_3(CH_2)_{14}CH_2-Si-OCH_3$$
$$OCH_3$$

FIGURE 2.2 Structure of hexadecyltrimethoxysilane.

2.2.1.2 Substrates

In order to better understand the effect of substrate chemistry on the friction properties, smooth PE-b-PEG balls (4 mm diameter, used as received) were investigated on two model substrates: hydrophilic and hydrophobic glass disks (25 mm diameter).

2.2.1.2.1 Hydrophilic Substrate

The hydrophilic substrate was prepared using glass disks cleaned with acetone and ethanol in an ultrasonic bath. The glass disks were then dried at 80°C for 5 min. The surface hydrophilicity was verified by water wettability measurements.

2.2.1.2.2 Hydrophobic Substrate

Cleaned glass disks (cleaning as described previously) were used to prepare the hydrophobic substrate. The surface was activated by immersion of the cleaned glass slides in a piranha solution (70% H_2SO_4 + 30% H_2O_2) at 30°C for 30 min. After water rinse, the glass disks were immersed in a solution of hexadecyltrimethoxysilane (Sigma Aldrich) of millimolar concentration dissolved in toluene. The chemical formula of hexadecyltrimethoxysilane is presented in Figure 2.2.

The time of immersion in the grafting solution was fixed to 6 h. After the functionalization procedure, the substrates were washed with toluene to remove the ungrafted molecules. The complete covering of the substrate surface and the efficiency of the grafting with CH_3 end group at the surface were verified by contact angle measurements.

2.2.2 Methods

2.2.2.1 Pin-on-Disk Tribometer

The friction properties of PE-b-PEG diblock copolymers were evaluated using a CSM–Anton Paar pin-on-disk tribometer (model TRB-S-DE-0000, Peseux, Switzerland) (Figure 2.3) under ambient conditions (in air and at room temperature equal to 20°C).

For the friction properties, a copolymer ball of 4 mm diameter was used as a pin on both substrates, the hydrophilic and hydrophobic glass disks (see Figure 2.4).

All tests were performed under a normal force of 1 N and a sliding speed of 1 mm s^{-1}. The number of revolutions (rotations) varied between one and five: during the first revolution, the polymer ball was always sliding against a "fresh" glass surface, and after this first revolution, the polymer was sliding on the track previously created. The first revolution, then, is more representative of the copolymer–substrate

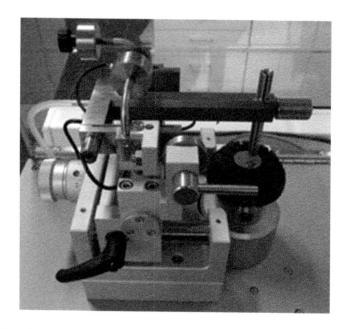

FIGURE 2.3 Pin-on-disk tribometer used in the investigation.

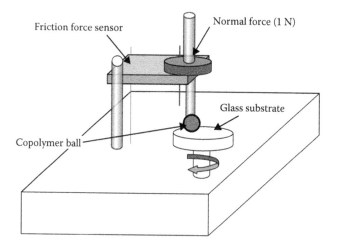

FIGURE 2.4 Schematic of the pin-on-disk tribometer.

interactions, and the tests performed for five revolutions should reflect the influence of the polymer transfer layer on the friction coefficient.

The friction coefficient μ (equal to friction force divided by normal force) was recorded as a function of time. The influence of copolymer composition on the friction properties was studied for both the hydrophilic and hydrophobic substrates.

2.2.2.2 Optical Microscopy

The morphology of PE-b-PEG tracks obtained after friction tests on both the hydrophilic and hydrophobic substrates was observed with an optical microscope, Olympus BX51 (Hamburg, Germany), in the transmission mode with a magnification of 50×. Images were recorded after one revolution or five revolutions.

2.3 RESULTS AND DISCUSSION

The surface hydrophilicity of the glass disk was determined by wettability measurements, and the average water contact angle of the surface was equal to 31° (±0.3). Figure 2.5 shows a waterdrop deposited on the hydrophilic glass.

The complete covering of the substrate surface and the efficiency of the grafting with CH_3 end group at the surface were verified by contact angle measurements. The average water contact angle of the hydrophobic surface was equal to 91° (±1.5). Figure 2.6 shows a waterdrop deposited on the hydrophobic glass.

Wettability results therefore clearly demonstrated the hydrophilic and hydrophobic characters of both substrates.

2.3.1 FRICTION PROPERTIES

2.3.1.1 Friction of Copolymer A (23 wt% PEG) against Hydrophilic and Hydrophobic Substrates

The friction coefficient µ of copolymer A against the hydrophilic and hydrophobic glass was measured as a function of time, as illustrated in Figure 2.7.

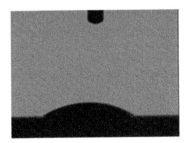

FIGURE 2.5   Contact angle of a waterdrop deposited on the hydrophilic substrate.

FIGURE 2.6   Contact angle of a waterdrop deposited on the hydrophobic substrate.

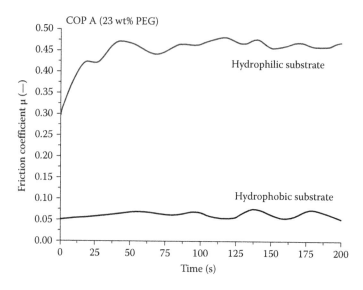

FIGURE 2.7 Friction coefficient as a function of time for COP A (23% w/w PEG) against hydrophilic and hydrophobic glass substrates (sliding speed = 1 mm s^{-1} and normal force = 1 N).

An important effect of the substrate's nature is evidenced, with a higher friction coefficient obtained for the hydrophilic substrate. Indeed, the friction coefficient measured at 200 s (corresponding to five revolutions) was close to 0.47 against the hydrophilic surface and 0.05 against the hydrophobic surface.

An increase in the friction coefficient with time was also observed, mainly for the hydrophilic substrate, during the first revolution (the first revolution was performed between 0 and 60 s). This could be explained by the polymer ball's wear and transfer during the first revolution. After the first revolution, the polymer ball was again sliding on the same track, thus explaining the stabilization of the friction coefficient until 200 s (this time corresponds to the duration of the five revolutions).

The optical microscopy image (see Figure 2.8) of the hydrophilic glass after friction of the copolymer during only one revolution shows indeed a significant transfer of the polymer layer onto the glass substrate.

The transfer of copolymer A was much less on the hydrophobic glass, as illustrated in Figure 2.9, in agreement with a low friction coefficient.

2.3.1.2 Friction of Copolymer B (30 wt% PEG) against Hydrophilic and Hydrophobic Substrates

The evolution of the friction coefficient μ of copolymer B against the hydrophilic and hydrophobic glass as a function of time is presented in Figure 2.10.

The effect of the substrate's nature is also evidenced, with a higher friction coefficient obtained for the hydrophilic substrate. However, the difference between the friction coefficient of the hydrophilic and hydrophobic substrates was less pronounced for copolymer B compared with copolymer A.

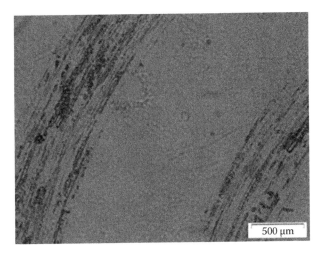

FIGURE 2.8 Optical microscopy image (50× magnification) of COP A (23% w/w PEG) wear tracks on the hydrophilic substrate after one revolution.

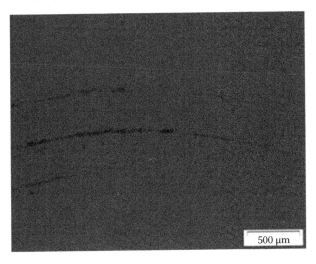

FIGURE 2.9 Optical microscopy image (50× magnification) of COP A (23% w/w PEG) wear tracks on the hydrophobic substrate after one revolution.

As an example, the friction coefficients measured after five revolutions were 0.47 and 0.05 for copolymer A against the hydrophilic and hydrophobic surfaces, respectively, while they were 0.27 and 0.07 for copolymer B against the hydrophilic and hydrophobic surfaces, respectively. The friction coefficients obtained against the hydrophobic substrate were similar for copolymers A and B (0.05 and 0.07, respectively). On the other hand, the friction coefficient of copolymer B against the hydrophilic substrate was half of that of copolymer A. Moreover, the friction coefficients for copolymer B were quite stable with time, contrary to copolymer A, for which an increase in μ was observed during the first revolution against the hydrophilic glass.

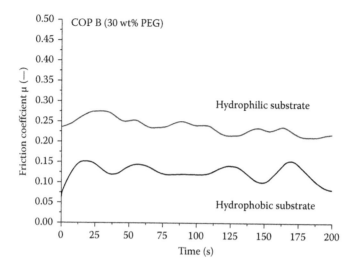

FIGURE 2.10 Friction coefficient as a function of time for COP B (30% w/w PEG) against hydrophilic and hydrophobic glass substrates (sliding speed = 1 mm s^{-1} and normal force = 1 N).

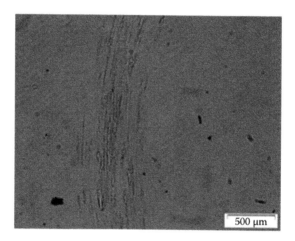

FIGURE 2.11 Optical microscopy image (50× magnification) of COP B (30% w/w PEG) wear tracks on the hydrophilic substrate after five revolutions.

A low transfer was observed for copolymer B after a friction test against the hydrophilic substrate, as shown in Figure 2.11, compared with copolymer A.

The transfer of copolymer B against the hydrophobic substrate was also very low, similar to that observed for copolymer A.

2.3.1.3 Friction of Copolymer C (55 wt% PEG) against Hydrophilic and Hydrophobic Substrates

Figure 2.12 presents the evolution of the friction coefficient μ of copolymer C against the hydrophilic and hydrophobic glass as a function of time.

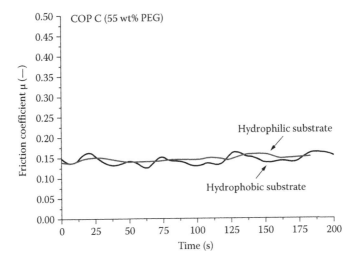

FIGURE 2.12 Friction coefficient as a function of time for COP C (55% w/w PEG) against hydrophilic and hydrophobic glass substrates (sliding speed = 1 mm s⁻¹ and normal force = 1 N).

A surprising result is obtained for copolymer C, for which the effect of the substrate surface properties becomes negligible: quite identical values of μ (about 0.15) were obtained for both substrates and the friction coefficients were stable with time. This value of μ (0.15) was higher than the value obtained for copolymers A and B against the hydrophobic glass (close to 0.07) and was lower than the value obtained for copolymers A and B against the hydrophilic glass (0.47 and 0.27, respectively).

The transfer observed for copolymer C after a friction test against both substrates was very low, as illustrated in Figure 2.13, for the hydrophilic glass slide.

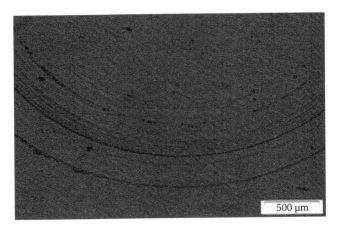

FIGURE 2.13 Optical microscopy image (50× magnification) of COP C (55% w/w PEG) wear tracks on the hydrophilic substrate after five revolutions.

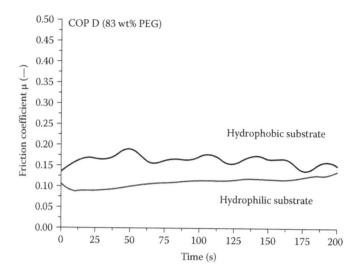

FIGURE 2.14 Friction coefficient as a function of time for COP D (83% w/w PEG) against hydrophilic and hydrophobic glass substrates (sliding speed = 1 mm s^{-1} and normal force = 1 N).

2.3.1.4 Friction of Copolymer D (83 wt% PEG) against Hydrophilic and Hydrophobic Substrates

The friction coefficient μ of copolymer D against the hydrophilic and hydrophobic glass as a function of time is presented in Figure 2.14.

For copolymer D, a unique result was observed: a higher friction coefficient is obtained for the hydrophobic substrate, even though the difference between the values was somehow small sometimes. In both cases, the transfer induced by friction was negligible, as demonstrated in Figure 2.15 for the hydrophobic glass, for which the friction coefficient was higher compared with the hydrophilic glass.

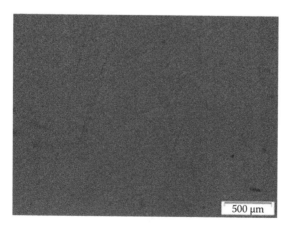

FIGURE 2.15 Optical microscopy image (50× magnification) of COP D (83% w/w PEG) wear tracks on the hydrophobic substrate after one revolution.

2.4 DISCUSSION

2.4.1 INFLUENCE OF PE-b-PEG CHEMICAL COMPOSITION

Friction results show that copolymer A, compared with the other copolymers, exhibits a higher friction coefficient against the hydrophilic glass, as illustrated in Figure 2.16, which compares the friction coefficients of the different copolymers against the hydrophilic glass.

Copolymer A has the lowest PEG content (23%), and consequently, it is the least hydrophilic copolymer. One could then expect copolymer A to possess a lower friction coefficient against the hydrophilic substrate. This surprising result is also associated to a significant copolymer wear during friction, inducing a thick residual copolymer transfer layer onto the substrate after friction. Copolymer A possesses lower chain length, inducing lower mechanical properties and thus favoring wear. Another hypothesis is linked to the crystallinity of the blocks. PE-b-PEG copolymers are indeed double-crystalline block copolymers [14,15], with both blocks being able to crystallize due to the phase separation induced by the strong incompatibility of these blocks. However, the crystallinity of each block will depend on the PEG/PE ratio. For high PE content, PE crystallization will be favored, hindering PEG crystallization, and vice versa. The PEG blocks of copolymer A are therefore not able to crystallize, because the longer PE blocks will crystallize and hinder PEG crystallization [16].

The PEG blocks of copolymer A are thus amorphous and, consequently, more mobile and able to adsorb onto the hydrophilic glass surface. This can explain the observed higher friction coefficient and the higher wear (as seen by optical microscopy).

Moreover, a decrease in the friction coefficient against the hydrophilic glass was observed when the PEG content was increased. This result could also be explained by the increase in PEG crystallinity when the PEG block length is increased [16]. For copolymer D (83% of PEG), the PEG blocks will be highly crystalline (hindering PE

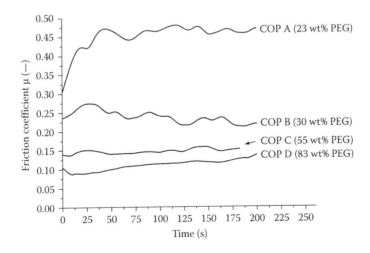

FIGURE 2.16 Friction coefficient as a function of time for different PE-b-PEG copolymers on the hydrophilic glass substrate.

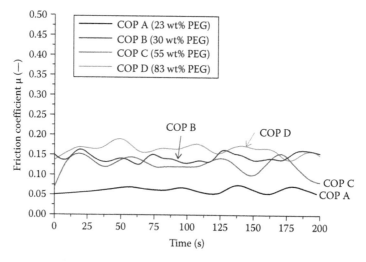

FIGURE 2.17 Friction coefficient as a function of time for different PE-b-PEG copolymers on the hydrophobic glass substrate.

crystallization) and, consequently, will be less mobile and less available to adsorb on the hydrophilic glass.

Friction results also show that copolymer A, compared with the other copolymers, exhibits a lower friction coefficient against the hydrophobic glass, as shown in Figure 2.17, which compares the friction coefficients of the copolymers against the hydrophobic glass.

This result is directly linked to the higher hydrophobicity of copolymer A: the polar PEG blocks avoid the hydrophobic surface, and the high content of copolymer A induces an easier sliding on the glass, thus explaining the low friction coefficient, with only slight wear. Moreover, for copolymer A, the crystallinity of the PE blocks is greater than in the other copolymers. A high PE crystallinity is also able to reduce the friction coefficient [26]. For the other copolymers, the friction coefficient overall depends on the PE content, with a slight increase in the friction coefficient when the PE content is decreased.

To sum up, the influence of copolymer composition on friction against the hydrophilic glass is then original, with a decrease in the friction coefficient when the polar PEG content is increased. This result underlines the fact that the mobility of the PEG blocks has a greater influence than the chemical composition of the copolymers (i.e., the PEG content). Indeed, for low PEG content in the copolymer, the PEG blocks are not able to crystallize and are therefore more mobile (amorphous state) to adsorb on the hydrophilic glass. On the other hand, for high PEG content in the copolymer, the PEG blocks are highly crystalline and, consequently, less mobile to adsorb on the glass substrate. The effect of block mobility and flexibility on lubrication properties was also demonstrated for PEO-PPO-PEO triblock copolymer [25]. In this study, the triblock copolymer was able to show a lubrication effect against hydrophobic surfaces (PE, PP). However, in the case of a hydrophilic surface (cellulose), the flexible PEO blocks adhere to the surface, thus hindering lubrication.

2.4.2   INFLUENCE OF SUBSTRATE SURFACE CHEMISTRY

Another interesting result is the effect of substrate surface chemistry, as illustrated in Figure 2.18.

For copolymer A, a high friction coefficient was obtained against the hydrophilic glass, compared with the hydrophobic substrate, as expected. Hydrophobic surfaces

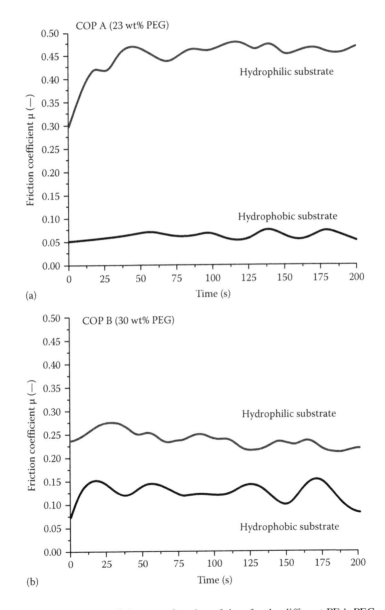

(a)

(b)

FIGURE 2.18   Friction coefficient as a function of time for the different PE-b-PEG copolymers on the hydrophilic and the hydrophobic glass substrate.   (*Continued*)

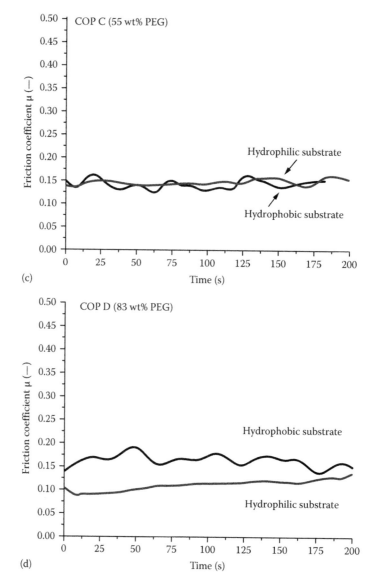

FIGURE 2.18 (*Continued*) Friction coefficient as a function of time for the different PE-b-PEG copolymers on the hydrophilic and the hydrophobic glass substrate.

are indeed able to reduce friction due to their low surface energy [27]. This effect is also present for copolymer B, but the difference between the two friction coefficients was lower. However, for copolymers C and D, a surprising behavior was observed. Identical friction coefficients were observed for copolymer C against both the hydrophobic and hydrophilic substrates. For copolymer D, a higher friction coefficient was observed for the hydrophobic glass compared with the hydrophilic glass.

The PE/PEG ratio of copolymer C was close to 1, meaning that this polymer is as hydrophilic (due to the PEG blocks) as it is hydrophobic (due to the PE blocks). This balance could explain the similar affinities toward both the hydrophilic and hydrophobic substrates and, consequently, the similar friction coefficients.

Copolymer D contains a high proportion of PEG, which is highly crystalline, and has a low proportion of PE, whose crystallinity is prevented by the PEG crystals. The polar PEG blocks, which are not mobile due to their crystalline state, avoid the hydrophobic glass surface. The nonpolar PE blocks then interact with the hydrophobic glass surface but also dissipate more energy during friction due to their greater mobility (amorphous state). This could explain the slightly higher friction coefficient.

The influence of substrate chemistry is complex and strongly depends on the competition between PEG–PEG interactions PE blocks–PE blocks interactions, PE blocks–substrate interactions, and PEG blocks–substrate interactions. Strong PEG blocks–PEG blocks interactions (as in the case of crystals) will decrease the interactions with the substrate. When the crystallinity of a copolymer block (PE or PEG) is decreased, its mobility is increased (amorphous state), as well as its ability to interact with a substrate and energy dissipation during friction is also increased.

2.5 CONCLUSIONS

The objective of this study was to investigate the friction properties of PE-b-PEG copolymers against hydrophilic and hydrophobic surfaces. Experimental results have demonstrated a complex influence of copolymer composition. A decrease in the friction coefficient against the hydrophilic substrate is indeed observed when the PEG content is increased. This result can be explained by the amorphous state of the PEG blocks, whose mobility allows adsorption on the hydrophilic surface. On the contrary, for high PEG content, highly crystalline chains are less available to interact with the polar glass substrate. The influence of the nature of the substrate is also subtle. For copolymers with a PE/PEG ratio greater than 1, a higher friction coefficient is measured for the hydrophilic glass, compared with the hydrophobic substrate.

If the PE/PEG ratio is close to 1, similar friction coefficients are obtained for both hydrophobic and hydrophilic surfaces. And if the PE/PEG ratio is less than 1, a higher friction coefficient is measured for the hydrophobic glass compared with the hydrophilic glass. For a high proportion of PEG, PE crystallinity will be hindered, and the amorphous, nonpolar, and mobile PE blocks will interact with the hydrophobic surface, inducing energy dissipation during friction and, consequently, a higher friction coefficient.

The friction coefficient of PE-b-PEG diblock copolymers against both hydrophilic and hydrophobic surfaces strongly depends on the competition between PE blocks–PE blocks interactions, PEG blocks–PEG blocks interactions, PE blocks–substrate interactions, and PEG blocks–substrate interactions. This subtle effect could be an interesting means to find the best lubricant for a given application, as a function of the surface energy of the surface to be lubricated.

ACKNOWLEDGMENT

The authors would like to thank the Université de Haute Alsace (UHA, Mulhouse, France) for the financial support.

REFERENCES

1. T. P. Lodge, Block copolymers: Past successes and future challenges, *Macromol. Chem. Phys.* 204, 265–273 (2003).
2. N. Hadjichristidis, S. Pispas, and G. A. Floudas, Block copolymers, *Synthetic Strategies, Physical Properties and Applications*, pp. 385–408, John Wiley & Sons, Hoboken, NJ (2003).
3. N. Hadjichristidis, M. Pitsikalis, and H. Iatrou, Synthesis of block copolymers, *Adv. Polym. Sci.* 189, 1–124 (2005).
4. C. Tsitsilianis, Responsive polymer materials, in: *Design and Applications*, S. Minko (ed.), pp. 27–49, Blackwell Publishing, Ames, IA (2006).
5. L. Leibler, The theory of microphase separation in block copolymers, *Macromolecules* 13, 1602–1617 (1980).
6. I. W. Hamley (ed.), *The Physics of Block Copolymers*, John Wiley & Sons, Chichester, U.K. (2004).
7. M. Lazzai, G. Liu, and S. Lecommandoux (eds.), *Block Copolymers in Nanosciences*, Wiley VCH, Weinheim, Germany (2006).
8. J. F. Gohy, Block copolymer micelles, *Adv. Polym. Sci.* 190, 65–136 (2005).
9. I. W. Hamley, Ordering in thin films of copolymers: Fundamentals to potential applications, *Prog. Polym. Sci.* 34, 1161–1210 (2009).
10. J. K. Kim, S. Y. Yang, Y. Lee, and Y. Kim, Functional nanomaterials based on block copolymer self-assembly, *Prog. Polym. Sci.* 35, 1325–1349 (2010).
11. J. N. L. Albert and T. H. Epps, Self-assembly of block copolymers thin films, *Mater. Today* 13, 24–33 (2010).
12. G. Reiter, G. Castelein, P. Hoerner, G. Riess, A. Blumen, and J. U. Sommer, Nanometer-scale surface patterns with long-range order created by crystallization of diblock copolymers, *Phys. Rev. Lett.* 83, 3844–3847 (1999).
13. S. B. Darling, Directing self-assembly of block copolymers, *Prog. Polym. Sci.* 32, 1152–1204 (2007).
14. M. A. Hillmyer and F. S. Bates, Influence of crystallinity on the morphology of poly(ethylene oxide) containing diblock copolymers, *Macromol. Symp.* 117, 121–130 (1997).
15. R. V. Castillo, A. J. Müller, J. M. Raquez, and P. Dubois, Crystallization kinetics and morphology of biodegradable double crystalline PLLA-b-PCL diblock copolymers, *Macromolecules* 43, 4149–4160 (2010).
16. L. Sun, Y. Liu, L. Zhu, B. S. Hsiao, and C. A. Avila-Orta, Self-assembly and crystallization behavior of a double-crystalline polyethylene-block-poly(ethylene oxide) diblock copolymer, *Polymer* 45, 8181–8193 (2004).
17. R. Pons, Polymeric surfactants as emulsions stabilizers, in: *Amphiphilic Block Copolymers—Self-assembly and Applications*, P. Alexandridis and B. Lindman (eds.), pp. 409–421, Elsevier, Amsterdam, the Netherlands (2000).
18. Y. Chang, W. L. Chu, W. Y. Chen, J. Zheng, L. Liu, R. C. Ruaan, and A. Higuchi, A systematic SPR study of human plasma protein adsorption behavior on the controlled surface packing of self-assembled poly(ethylene oxide) triblock copolymer surfaces, *J. Biomed. Mater. Res. Part A* 93, 400–408 (2010).

19. M. Brogly, A. Fahs, and S. Bistac, Modification of biopolymers friction by surfactant molecules, in: *Surfactants in Tribology*, G. Biresaw and K. L. Mittal (eds.), vol. 3, pp. 93–110, CRC Press, Boca Raton, FL (2013).
20. Y. Gao, S. Sun, Y. He, X. Wang, and D. Wu, Effect of poly(ethylene oxide) on tribological performance and impact fracture behavior of polyoxymethylene/polytetrafluoroethylene fiber composite, *Compos. Part B* 42, 1945–1955 (2011).
21. X. Gu and G. Wang, Interfacial morphology and friction properties of thin PEO and PEO/PAA blend films, *Appl. Surf. Sci.* 257, 1952–1959 (2011).
22. L. Shi, V. I. Sikavitsas, and A. Striolo, Experimental friction coefficients for bovine cartilage measured with a pin-on-disk tribometer: Testing configuration and lubricant effects, *Ann. Biomed. Eng.* 39, 132–146 (2011).
23. X. Yan, S. S. Perry, N. D. Spencer, S. Pasche, S. M. De Paul, M. Textor, and M. S. Lim, Reduction of friction at oxide interfaces upon polymer adsorption from aqueous solutions, *Langmuir* 20, 423–428 (2004).
24. S. Lee and N. D. Spencer, Poly(L-lysine)-graft-poly(ethylene glycol): A versatile aqueous lubricant additive for tribosystems involving thermoplastics, *Lubric. Sci.* 20, 21–34 (2008).
25. Y. Li, O. J. Rojas, and J. P. Hinestroza, Boundary lubrication of PEO-PPO-PEO triblock copolymer physisorbed on polypropylene, polyethylene, and cellulose surfaces, *Ind. Eng. Chem. Res.* 51, 2931–2940 (2012).
26. K. S. Kanaga Karuppiah, A. L. Bruck, S. Sundararajan, J. Wang, Z. Lin, Z. H. Xu, and X. Li, Friction and wear behavior of ultra-high molecular weight polyethylene as a function of polymer crystallinity, *Acta Biomater.* 4, 1401–1410 (2008).
27. H. Liu, S. Imad-Uddin Ahmed, and M. Scherge, Microtribological properties of silicon and silicon coated with diamond like carbon, octadecyltrichlorosilane and stearic acid cadmium salt films: A comparative study, *Thin Solid Films* 381, 135–142 (2001).

3 Aqueous Lubrication with Polyelectrolytes
Toward Engineering Applications

Seunghwan Lee

CONTENTS

3.1 INTRODUCTION: AQUEOUS LUBRICATION WITH BRUSH-LIKE POLYMER LAYERS

The last couple of decades have witnessed a rapid progress in research on aqueous lubrication [1–5]. Lubrication with water is an attractive idea because water has outstanding merits as base lubricant. For example, water displays very low fluidic drag once hydrodynamic lubrication is activated because of its low viscosity. The high heat capacity of water allows its use as both lubricant and coolant at the same time. Aqueous lubricants are ideally suited for the recent trend in science and technology of emphasizing the importance of ecological preservation and human health due to their excellent eco-compatibility and biocompatibility. Despite these advantages, the practical use of water as lubricant or base stock for lubricant formulation is very limited because of certain concomitant drawbacks. These include low viscosity, low pressure coefficient of viscosity, limited operating temperature range, and corrosion. Among these, its low viscosity and low pressure coefficient of viscosity are major hurdles since they prevent the onset of the all-important elastohydrodynamic lubrication (EHL) mechanism. For this reason, aqueous lubrication studies to date have mainly focused on the modification of tribopair surfaces using boundary lubricant additives rather than modifying bulk fluid properties of water. Recent studies in the aqueous lubrication of metals and ceramics have employed a variety of water-compatible additives, including nanoparticles [6–8], fullerenes [9], nanotubes [10,11], graphite [12], surfactants [13–17], ionic liquids [18,19], and organic compounds [20–24]. These studies aimed at practical engineering applications, reflecting the rapidly growing interest and progress in the subject. In this context, emerging interest in seawater-compatible lubrication is also notable [25–29].

The most extensively and systematically studied additive materials for aqueous lubrication to date are probably brush-forming polymer chains [1–5]. It is well known that a layer of brush polymer chains is generated when one end of the polymer chains is immobilized on the surface at a high graft density and the grafted polymer chains are exposed to good or theta solvent (see Figure 3.1). According to the scaling theory of Alexander and de Gennes [30,31], if two opposing surfaces are grafted with brush-like polymer coatings, repulsive forces, $F(D)$, develop between them that is a function of several parameters including polymer chain length (L), graft density (s; the spacing between adjacent polymer chains), and the distance between the two substrates (D), as shown in Equation 3.1:

$$F\left(D\right) \cong \frac{k_B T}{s^3} \left[\left(\frac{2L^{9/4}}{D} \right) - \left(\frac{D}{2L} \right)^{3/4} \right]$$

(3.1)

where
  k_B is the Boltzmann constant
  T is the temperature

The first and second terms in square brackets on the right side of Equation 3.1 account for the osmotic repulsion experienced by the polymer brushes as they are compressed and the reduction in free energy due to the compression of the

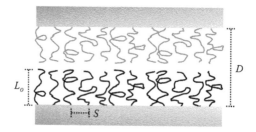

FIGURE 3.1 A schematic illustration of surface-grafted polymer chains in the Alexander-de Gennes model (L: brush height in equilibrium, s: spacing between adjacent polymer chains, D: distance between the two surfaces). (From Lee, S. and Spencer, N.D., Achieving ultralow friction by aqueous, brush-assisted lubrication, in: Erdemir, A. and Martin, J.-M., eds., *Superlubricity*, Elsevier, Amsterdam, The Netherlands, 2007, pp. 365–396.)

overstretched chains, respectively. The resulting repulsive forces, which increase with decreasing distance (D) between the two surfaces, thus provide a cushioning layer and contribute to countering the external load. Additionally, a fluidic layer formed between the two shearing surfaces further contributes to smooth gliding at the tribological interfaces [32,33]. This principle can be applicable to any pair of polymer chains/solvent combination as long as the solubility of the polymer in the solvent is sufficiently high. The polystyrene/toluene pair was employed in the first report on this remarkable lubricating effect [34]. However, this approach is particularly meaningful for fluids with poor pressure response, such as water. The critical problem of these fluids as lubricant can be possibly resolved by combining them with highly soluble polymer chains grafted on the tribopair surfaces.

One of the first requirements for polymers to be employed as aqueous lubricant additives is good water solubility. Among many candidates, poly(ethylene glycol) (PEG)-based (co)polymers (also known as poly(ethylene oxide) (PEO)) have been most extensively studied to date [35–51], primarily because of their excellent and unique water solubility [52,53]. Additionally, prior to tribological considerations, PEG chains have shown a broad range of useful applications, such as in biosurfaces [54–60] and colloidal stabilization [61–63], for which a number of approaches to immobilize them onto a variety of surfaces have been well established and are available [54–63]. Even though the end-group functionalization to graft onto surfaces could be cumbersome because of PEG's inertness, its electrostatic neutrality helps to suppress the interaction with the underlying substrates regardless of their electrostatic characteristics and to maintain the brush-like conformation. Thus, PEG can be readily applicable to a very broad range of materials with positive, negative, or nonpolar surface characteristics in an aqueous environment. Of course, this feature is not restricted to PEG but is common to all neutral polymer chains.

3.2 LUBRICATING PROPERTIES OF CHARGED POLYMER BRUSH LAYERS

Research on charged polymers as lubricious coatings for aqueous lubrication started somewhat later than that on neutral polymers. J. Klein and coworkers were the first who showed the superior lubricity of charged polymer chains relative to neutral ones using surface force apparatus (SFA) [33]. The authors employed an amphiphilic diblock copolymer, poly(methyl methacrylate)-$block$-poly(sulfonated glycidyl methacrylate) (PMMA-b-PSGMA, $(CH_2–C(CH_3)CO_2CH_3)_{41}–b–(CH_2–C(CH_3)CO_2 CH_2CHOHCH_2SO_3^-Na^+)_{115}$, Figure 3.2a), and a dense brush-like layer of PSGMA was spontaneously generated via the hydrophobic interaction of PMMA blocks with the hydrophobized mica surface in an aqueous solution. Extremely low coefficients of friction of ≤ 0.0006 between them were observed against the applied load of up to ca. 0.3 MPa [33]. More importantly, while the neutral brushes (PEG) and the charged polymer chains in adsorbed conformation (chitosan, "flat"-lying) showed a rapid increase in the friction forces at higher than 0.3 volume fractions, a low coefficient of friction on the order of 0.0006–0.001 mediated by the polyelectrolyte brushes remained persistent up to the point of brush removal, that is, volume fraction being near unity. The authors attributed the superior lubricating capabilities of

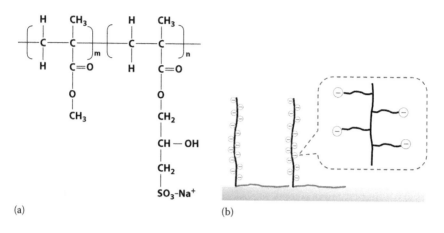

(a)   (b)

FIGURE 3.2 (a) The molecular structure of the diblock copolymer, poly(methyl methacrylate)-*block*-poly(sulfonated glycidyl methacrylate) (PMMA-*b*-PSGMA), employed in the study by Klein and coworkers, (b) a schematic illustration for a possible conformation of PMMA-*b*-PSGMA on a hydrophobized mica surface in an aqueous environment. (From Raviv, U. et al., *Nature*, 425, 163, 2003.)

charged PSGMA chains to enhanced osmotic pressure by the counterions embedded within the polyelectrolyte brushes and also to the atomic-scale bearing-like activity of charged moieties along the polymer chains [33]. Even though this study was carried out under a very well-controlled condition, for example, molecularly smooth mica surface (absence of random local asperities), mild apparent contact pressure (up to 0.3 MPa), and homogeneous polymer coating, the results imply that polyelectrolyte chains can be taken as an effective means to improve the pressure responsiveness of water for engineering applications as well. As will be discussed below in detail, however, similar approaches with charged amphiphilic diblock copolymers on macroscale contacts using pin-on-disk tribometry failed to reveal the superior lubricity of charged copolymers to neutral counterparts. In the same context, it has not been shown if a PMMA-*b*-PSGMA diblock copolymer would function equally effectively at macroscopic tribological contacts under higher contact pressure. Thus, whether the exceedingly superior lubricity of PMMA-*b*-PSGMA system in the above study [33] is related to the mild pressure generated by SFA and/or the uniqueness of the structure of PMMA-*b*-PSGMA is not clear and will be further discussed below. One point to note is that while the superior lubricity of the polyelectrolyte chains, PSGMA, to the neutral polymer chains, PEG, employed in that particular study is obvious, the lubricating capability of PEG chains is also greatly influenced by the structural features, such as surface grafting method or polymer chain length [36,37,42,43]. The particular PEG chains (ca. 3.4 kDa) employed in the study were monofunctionalized with trimethylammonium ($(CH_3)_3N^+$) groups, and thus, the grafting stability of the resulting PEG layer may not be optimal, and it could have a negative influence on the lubricating properties as well.

SFA is an ideal instrument to investigate surface forces due to the extremely smooth morphological features of mica surface and the capability to determine the absolute distance between the two opposing surfaces. This is necessary to assess the interfacial friction forces between opposing polymer brushes with high accuracy. Thus, the experimental results can be analyzed on the basis of theoretical models taking into account the structural features of surface-grafted polymer brushes. Following the first study by the Klein group on PSGMA chains [33], a few more studies with SFA have been reported, such as polystyrene-*b*-poly(acrylic acid) (PS-*b*-PAA) [64] and poly([2-(methacryoyloxy)ethyl]trimethylammonium chloride) (PMETAC) [65]. SFA was employed to investigate the lubricating behavior of various biological polyelectrolyte chains too, including lubricin [66], hyaluronic acid [67], mixtures of hyaluronic acid and lipids [68], porcine gastric mucin (PGM) [69], and bovine submaxillary mucin (BSM) [70]. It should be noted that not all these biological polyelectrolytes displayed an extended brush conformation. Despite its limited accuracy in determining the distance between the two opposing surfaces, AFM has also been employed as an experimental tool to investigate the lubricating effect of brush-like polyelectrolyte chains, such as poly(acrylic acid) (PAA) [71,72], [3-(2-methylpropionamide)propyl]trimethylammonium chloride (MAPTAC) [73], poly(2-(methacryloyloxyethyl phosphorylcholine) (PMPC) [74], and mixed polymer chains of PEO and poly(methacryloxyethyl trimethylammonium chloride) (PMETAC) [51]. Tribological properties of biological polyelectrolyte chains have also been studied with AFM. These include lubricin [75,76], hyaluronic acid [75,76], and glycosaminoglycan (GAG) [77,78]. Finally, macroscopic-scale studies with more conventional tribometers were also reported on several polymer systems, such as poly(2,3-dihydroxypropyl methacrylate) [79], PMMA [80], poly(phosphorylcholine) (PMPC) [81], poly(phosphorylcholine-*co*-2-dimethylaminoethyl methacrylate) P(MPC-*co*-DMAEMA) [81] and poly(MPC-*co*-MTAI)2-methacryloyloxyethyl phosphorylcholine (PMPC) [82], and poly{[2-(methacryloyloxy)ethyl] trimethylammonium chloride} (PMTAC) [83]. Aqueous suspensions of poly(*N*-isopropylacrylamide) (PNIPAAm) microgels grafted with poly(3-sulfopropyl methacrylate potassium salt) (PSPMK) brushes have been shown to display excellent lubricating properties, although the brush polymers were grafted on spherical, small-scale substrates [84].

For macroscale contacts, it is not clear whether a similar repulsion/lubrication mechanism, that is, brush–brush repulsion and osmotic pressure, as in SFA or AFM experiments, is possible because of the irregularly high local contact pressures arising from the local surface asperities. Even with SFA, at contact pressure above ca. 0.2 MPa, a collapse and/or detachment of polymer chains from the surface has been reported [85]. Nevertheless, as shown by Lee and coworkers [35,36,45], even if the brush-like polymers do not function via the brush–brush repulsion mechanism, they can still act as effective boundary lubrication additives in water due to their excellent hydrating capability of the tribological interface. Thus, macroscopic-scale lubrication studies on charged polymer brushes with more conventional tribometers [79–84] are especially encouraging for engineering applications due to their very high relevance in operative conditions, including contact scale, speed, and surface roughness. However, the macroscale lubrication studies available to date have been mostly limited to applied external loads of 0.5 or 1.0 N. Obviously, this is because the polymer layers will start to delaminate under higher loads and lose lubricating function.

A most advanced research area of aqueous lubrication with polyelectrolyte brushes in view of engineering applications is probably for biomedical applications. This is partly related to the fact that in biomedical applications requiring adequate lubrication, such as catheterization, endoscopy, and contact lenses, the contact pressure is generally very low due to the involvement of soft biological tissues as the sliding partner, and thus, the risk of delamination of lubricating films under tribological contacts is significantly reduced. It is interesting to note that Ikada and coworkers [86] have presented a few examples of polymer brushes for the lubrication of catheters and other tubular biomedical devices in 1990. This was a few years ahead of the publications by Klein and coworkers on the lubricating performance of the polystyrene/toluene system [33]. It is also interesting to note that the polymers studied by Ikada and coworkers were polyelectrolytes such as polyacrylamide (PAAm) [86] and poly(dimethylacrylamide) (PDMMAm) [86–90]. Polyelectrolyte chain layers, mostly grafted by UV-irradiation-initiated grafting-from approach, have shown a substantial reduction in the coefficient of friction for catheter materials based on nylon 6, polypropylene (PS), poly(vinyl chloride) (PVC), and ethylene-vinyl acetate (EVA) copolymers. Tribological applications of polyelectrolyte chains at higher contact pressures have emerged as well. For example, Kawaguchi and coworkers [91,92] have shown a great reduction in both friction forces and wear debris generation from orthopedic implants (polyethylene liners) under ca. 280 kg force by surface modification with poly(2-methacryloyloxyethyl phosphorylcholine) (PMPC) chains. A number of different types of hydrophilic polymer chains, including oligo(ethylene glycol) methacrylate (OEGMA) (neutral), poly(2-(dimethylamino) ethyl methacrylate) (PDMAEMA) (cationic), and poly(2-methacryloyloxyethyl phosphate) (PMPA) (anionic) [93] have also been studied in the same context.

3.3 LIMITS OF IRREVERSIBLY IMMOBILIZED POLYELECTROLYTE BRUSH CHAINS

A common feature of the majority of studies mentioned in the previous section is that polyelectrolyte chains are irreversibly immobilized onto the tribopair surfaces. In many nontribological scientific and technical applications, the irreversible immobilization of polymer chains provides high grafting stability for operation. However, it could be insufficient or even problematic for macroscale applications. As mentioned above, the external load in the previous studies was limited to 1 N as nearly all polyelectrolyte brush films are ruptured at higher loads. Moreover, if the lubricating film failure occurs due to high tribological stress, the recovery or replenishment is not feasible under operation. Thus, one is faced with the question whether this approach will lead to a practical solution for engineering tribology. Of course, this question is not limited to only polyelectrolyte chains. However, it may signify that the superior pressure response of charged polyelectrolyte chains discussed above may not be sufficient to resolve the film durability problem under much harsher tribological stresses encountered in engineering applications.

A simple reason for this limit is that polymer brushes, including polyelectrolyte chains, are organic substances and the anchoring stability on the engineering materials is not sufficient to withstand harsh tribostress between, for instance, oxides and metals. Moreover, the irreversible immobilization of polymer chains by chemical (covalent) bonds may provide initially high grafting stability, but once the films are damaged or ruptured, they cannot be reestablished until the film preparation procedure is restarted from the beginning, as mentioned above. For this reason, it could be more convenient to use polymers as additives blended in base lubricant and allow them to adsorb "spontaneously" onto the tribopair surfaces to form lubricious films for long-term applications. The additives can participate in tribochemical reactions in aqueous base lubricant [20–24] very similar to friction-modifying additives in oil-based lubricants [94]. They can also simply be physisorbed via electrostatic or hydrophobic interaction with the substrates and thus continuously reused in the repeating process of adsorption and desorption. Thus, physically grafted polymer chains have an advantage that they can be re-adsorbed onto the surface as long as there are ample polymers in solutions surrounding the contacting area. While the detached polymers themselves can re-adsorb onto the surface, it is more effective if excess polymers are present in solution or surrounding the contact area. In either case, an important property requirement for the additives to restore lubricating film is fast surface adsorption kinetics. The exchange between initially adsorbed and reservoir polymers was directly visualized by fluorescence labeling of the PEG-based copolymers [41]. The prolonged lubrication performance via the continuous supply of additives from the surrounding base lubricant was termed as "self-healing lubrication mechanism." Even though the practical application of this approach is limited to an enclosed tribosystem, such as bearings, the working principle is the same as the classical oil-based lubricants and antifriction modifiers [94].

3.4 REVERSIBLE FORMATION OF POLYELECTROLYTE BRUSH LAYERS

Given that charged polymers can provide superior lubricity compared to neutral ones (Section 3.2) and that physically grafted polymers can provide more persistent lubricity from the reversible adsorption of the lubricant layer when excess polymers are present in the base lubricant (Section 3.3), it is a reasonable proposition that the combination of the two may provide a synergetic improvement in aqueous lubrication. The design of polyelectrolyte chain layers with respect to the charge characteristics of anchoring groups (or moieties), buoyant groups, as well as the substrate is, however, not a trivial issue. For example, a majority of engineering materials, such as oxides and metal/metal oxides, are negatively charged in a neutral pH aqueous environment. In order to graft polymer chains onto these surfaces by means of electrostatic attraction, "anchoring" groups or moieties should be positively charged. This excludes the random choice of "buoyant" polymer chains in terms of their charge characteristics, because they have to avoid the electrostatic attraction with both tribopair surfaces and the anchoring units at the same time. As mentioned above, this is again one of the merits of neutral polymer chains, such as PEG, as "buoyant" polymer chains as mentioned above. In short, among the possible interactions between substrates, buoyant polymer chains, and anchoring moieties, the interaction of substrate–buoyant should be suppressed as much as possible, whereas the interaction of substrate–anchor should be facilitated as much as possible.

Thus, a simple way to test the feasibility of the "self-healing" lubrication of polyelectrolyte brushes is to employ neutral, nonpolar surfaces as the tribopair. Røn et al. [95] have compared a series of diblock copolymer systems composed of hydrophobic anchoring blocks (e.g., polystyrene (PS), poly(2-methoxyethylacrylate) (PMEA), poly(dimethylsiloxane) (PDMS), polycaprolactone (PCL), etc.) and hydrophilic buoyant block of either neutral PEO or negatively charged PAA. Examples of copolymer include X-*b*-PAA vs X-*b*-PEO, where X = PS, PMEA, PDMS, PCL, etc. The molecular weight of each block was targeted to be 5 kDa. PDMS was employed as the tribopair as it is relatively easy to lubricate with water under mild contact pressure. PDMS is also compliant and possible to activate the soft EHL mechanism [96,97] as long as its surface can become wet by the lubricant, water, by inducing hydrophilicity to the surface [39]. The results obtained from the PDMS–PDMS sliding contacts in aqueous solutions of PS-*b*-PAA vs PS-*b*-PEO solutions are shown in Figure 3.3. Contrary to expectations, the neutral PEO-based diblock copolymers were substantially more effective in lowering friction than PAA-based counterparts. The ineffective lubricating properties of PS-*b*-PAA and other PAA-based diblock copolymers [95] were attributed to the retarded formation of the lubricious layers on a nonpolar PDMS surface due

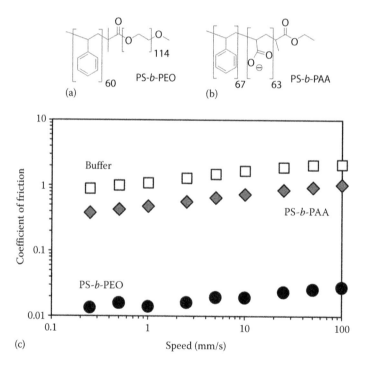

FIGURE 3.3 The molecular structures of the diblock copolymers, (a) polystyrene-*block*-poly(ethylene oxide) (PS-*b*-PEO) and (b) polystyrene-*block*-poly(acrylic acid) (PS-*b*-PAA), and (c) the plot of coefficient of friction vs speed acquired from the sliding contacts between PDMS surfaces in the diblock copolymer solutions (in HEPES buffer 1 mM pH 7.0) with pin-on-disk tribometry. (From Røn, T. et al., *Langmuir*, 29, 7782, 2013.)

to the electrostatic repulsion between charged moieties (polyelectrolytes) on the same side of the PDMS. A comparative characterization showed that the adsorbed masses of the X-*b*-PAA copolymer were generally much smaller than those of X-*b*-PEO [95]. Nevertheless, this result does not entirely exclude the feasibility of the "self-healing" lubrication of polyelectrolyte brushes. It is noted again that in the pioneering study by Klein et al. on charged polymer chains as a lubricious layer in an aqueous environment, PMMA-*b*-PSGMA was grafted to a hydrophobized mica surface via spontaneous adsorption based on hydrophobic interaction with PMMA block, even though "self-healing" behavior was not discussed by Klein et al. in this study [33]. Other studies involving amphiphilic, polyelectrolyte-based copolymers, such as hydrophobically modified polyelectrolytes [98–100] or mucins [101–105], *do* display facile adsorption onto nonpolar surfaces from aqueous solution and often effective lubrication too [101,102]. Thus, the unsuccessful lubrication performance of the diblock copolymers employed in the previously mentioned study [95] can be simply due to nonoptimal structural features of the copolymers. This includes nonoptimal total molecular weight, molecular weight ratio between anchoring and buoyant blocks (neutral and charged blocks), overall charge density, hydrophobicity of the anchoring block, and charge characteristics of hydrophilic block. The surface adsorption of diblock copolymers can be significantly enhanced by screening the electrostatic repulsion between them and the nonpolar surfaces [95]. Thus, it can be proposed that one of the key parameters determining the efficacy of adsorption/lubrication by "free," amphiphilic charged (co)polymers is the charge density along the polymer, as it determines the accumulated charge density on the nonpolar tribopair surface eventually. This approach may become more successful if the charge density along the polymer chains can be optimally controlled as mentioned above. In this context, two synthetic alternatives were further investigated in the following study by Røn et al. [95]. First, instead of block copolymerization, the graft copolymerization of charged polymer chains on a hydrophobic backbone was carried out [106] (Figure 3.4a). Generally, graft

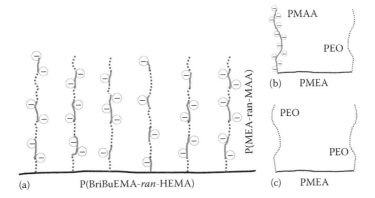

FIGURE 3.4 A schematic illustration of the copolymers (a) P(BriBuEMA-*ran*-HEMA)-*g*-P(MEA-*ran*-tBMA) and (b) PMAA-PMEA-PEO, (c) PEO-PMEA-PEO. (b: From Røn, T. et al., *Polymer*, 55, 4873, 2014; c: Javakhishvili, I. et al., *Macromolecules*, 47, 2019, 2014.)

copolymers are known to provide higher anchoring stability on the surface than block copolymers for multiple interactions of anchoring units with the underlying substrates [107,108]. More importantly, the charged monomer units of "buoyant" polymer chains, poly(methacrylic acid) (PMAA), were pre-copolymerized with nonpolar monomer units, poly(2-methoxyethyl acrylate) (PMEA), in order to reduce the charge density along the chain. The second approach is to polymerize triblock copolymers with a "charged buoyant-*block*-hydrophobic anchoring-*block*-neutral buoyant" topology (or "ABC") (Figure 3.4b) [109]. PMAA was employed as charged block (A) and PEO as nonpolar block (C) and PMEA was employed as the anchoring block (B). This way, the accumulated charge density on the surface is reduced to half compared to the diblock copolymer or triblock copolymer with both charged blocks ("ABA"). The lubricating efficacy of the two copolymers indeed improved [106,109]. In both cases, under proper environment (such as pH and ionic strength of solution) the polyelectrolyte-based copolymers showed friction forces comparable to the reference nonpolar polymers composed of PEO as buoyant chains only. However, it was not proven if the observed lubricity was any superior compared to their neutral counterparts. This raises the question about the necessity of all these complicated polymer synthesis procedure. As mentioned earlier, however, it is still unclear whether the inferior or "nonsuperior" lubricity from charged polymer chains compared to PEO-based polymer chains in these studies [95,106,109] is a generic limit or due to nonoptimal condition of the polymer structure, especially the charge density along the polymer chain. As shown in Figure 3.2b, for example, the negatively charged moieties (SO_3^-) in PMMA-*b*-PSGMA employed in the Klein's study [33] are located at the end of a fairly long repeating entity $(—CO_2CH_2CHOHCH_2SO_3^-Na^+)$ in a vinyl-type monomer unit. This structural feature may provide some flexibility for the negative moieties to avoid each other and could be a contributing factor for the facile adsorption on the hydrophobized mica surface.

3.5 ALTERNATIVE APPROACHES USING POLYELECTROLYTES FOR AQUEOUS LUBRICATION

Despite the problems of polyelectrolytes when they are employed as brush-like polymer layers for engineering applications discussed so far, they are still very attractive for the purpose of aqueous lubrication for a number of reasons. First of all, water solubility of polyelectrolytes is generally superior to neutral polymers. While not all neutral polymers are readily soluble in water, especially at neutral pH, nearly all polyelectrolytes are readily soluble in water. Second, when polyelectrolyte-based coatings are present on both opposing shearing surfaces, charged moieties along the polymer chains can provide electrostatic double layer repulsion, regardless of the exact conformation of polyelectrolytes, although its magnitude may vary from case to case. Third, charged moieties in polyelectrolytes typically have high reactivity so that chemical or physical interactions between themselves or with other moieties can be employed to modify the polymer functional properties. This can lead to the formation of covalent/ionic crosslinking or complexation with other polymers. All these modifications can substantially alter structural/compositional features and thus the lubricity of polyelectrolyte-based polymeric layers. Thus, apart from brush polymers, other polymer film fabrication approaches utilizing the higher chemical/ physical reactivity of polyelectrolytes are readily available. Emerging examples include (1) polyelectrolyte multilayer (PEM) [110–120], (2) polyelectrolyte complexation (PEC) [121–125], and (3) crosslinking and potential formation of hydrogels [126–129]. A common effect of all these approaches is to strengthen the mechanical integrity of the polymer layers by enhancing the polymer chain–chain interaction in one way or another. As a result of the increased interaction between polymer chains, unique high resilience of polymer chains in brush conformation is diminished, and thus higher friction forces are often observed. However, the durability of the films is substantially enhanced that it may be better suited for industrial applications where super-low friction forces are not necessarily required whereas wear resistance and durability of the lubricant layers are important.

3.5.1 POLYELECTROLYTE MULTILAYER (PEM) OR LAYER-BY-LAYER (LBL) SYSTEM

PEM (or termed as "layer-by-layer" [LBL]) is an organic layer system fabricated by alternating deposition of polycations and polyanions on top of each other [110,111]. As the alternating layers of oppositely charged polyelectrolytes can be deposited indefinitely, in principle, very thick layers with macroscale thickness can be formed. A first tribological application of LBL systems was reported by Cohen and coworkers [112], who employed poly(allylamine hydrochloride) (PAH) and PAA to fabricate multilayer coatings on stainless steel, silicon oxide, and glass substrates. The applied external load was up to 3.5 N. Following the studies with pristine PEMs, composites of nanoparticles and PEMs emerged as advanced coating materials. A number of nanoparticles were used to enhance the mechanical stability and tribological performance of the films. These include multiwall carbon nanotubes [113], silver nanoparticles [113], C_{60}-ethylenediamine adduct [114], ZnS [115], copper hydroxide [116], Cu nanoparticles [117], CuS [118], and CuS/ZnS [119]. In these studies, composites with inorganic nanoparticles showed improvements in antiwear and/or antifriction properties. The PEM and PEM/nano-objects composites were, however, targeted for applications at ambient environment, such as in microelectromechanical systems (MEMS) [116]. Thus, electrostatic attraction between polycations and polyanions in this approach was employed as a means to fabricate the film, which has little to do with aqueous lubrication. Exceptions to this so far include the studies by Cohen and coworkers [112,113], in which a PEM coating fabricated from PAH and PAA was applied to orthopedic implant material, in particular metal/ultrahigh molecular weight polyethylene (UHMWPE) pair, and the tribological test was conducted in calf serum. Tribological tests with pin-on-disk tribometer for up to 5×10^5 cycles showed a 33% reduction in wear by the PEM coating compared to an uncoated control. While this study had a particular objective to improve the lubrication and antiwear properties of orthopedic implants, it is still necessary to establish the fundamental relationship between on the structure of the PEM before and after tribological stress applied to the system in an aqueous environment. Another example of application of PEM for aqueous lubrication is an LBL multilayer formed from hyaluronic acid and chitosan on hydrophilized PDMS [120]. Under a mild contact pressure (about 0.21 MPa), the coatings reduced the coefficient of friction by about two orders of magnitude, that is, as effectively as other amphiphilic copolymers for the self-mated tribological contacts of PDMS in an aqueous environment.

3.5.2 POLYELECTROLYTE COMPLEXATION (PEC)

Polyelectrolyte complex or complexation (PEC) is, in a way, similar to the PEM system discussed above in a sense that it is a complex system generated by the electrostatic attraction between polycations and polyanions. The distinction between PEC and PEM, at least in this review, is that PEC is not necessarily achieved via controlled deposition of alternating oppositely charged polyelectrolytes. It can be simply mixed in an aqueous solution, and yet display spontaneous interaction between polycations and polyanions, as well as enhanced lubrication. Spencer and coworkers have recently shown [121] that surface-tethered poly(methacrylic acid) (PMMA) brushes (polyanions) interact with poly(L-lysine)-*graft*-poly(ethylene glycol) (PLL-*g*-PEG) (polycation) to give a significant enhancement in mechanical properties such as elasticity but also a slight degradation in friction properties (i.e., higher friction forces). The net effect is basically twofold: First, the aforementioned charge accumulation on the surface is suppressed by complex formation between polyanions and polycations. Second, as a result, the film layer became much thicker and tougher. Eventually, this led to an enhancement of mechanical properties of the film, increased friction forces, but at the same time improved antiwear properties. Nikogeorgos et al. [122] reported complex formation of porcine gastric mucin (PGM) and chitosan, which resulted in significant improvements in the lubrication properties [122], as shown in Figure 3.5. Unlike PMMA and PLL-*g*-PEG in the study by Spencer and coworkers [121], this study employed two "free polymers" mixed at varying ratios in aqueous solvent.

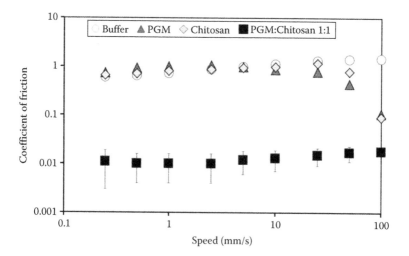

FIGURE 3.5 Coefficient of friction vs speed plots of porcine gastric mucin (PGM), chitosan, and the mixture solution of PGM:chitosan (1:1) in PBS as studied with pin-on-disk tribometry. (From Nikogeorgos, N. et al., *Soft Matter*, 11, 489, 2015.)

Thus, from an engineering application point of view, blending of two components into base lubricant, for example, water is highly preferred and sufficient for practical use without any need for the surface modification of materials in advance. A major driving force for the synergistically improved lubricating properties is the charge neutralization, which facilitated the adsorption of the polymers (a polycation [chitosan] and a polyanion [PGM]) onto the nonpolar surface, PDMS, as supported by the increased adsorbed mass upon mixing [122]. Other contributing factors, such as crosslinking or cohesion between the two component polymers, are likely to be active as well, since not only friction but also antiwear properties were significantly improved for the PGM–chitosan complex compared to either component alone [122].

The concept and scope of polyelectrolyte complex (PEC) can, in fact, be extended from two macromolecules with opposite charges to one macromolecule and a second chemical moiety with opposite charge, such as a small surfactant or even salt. Liu and coworkers [123] have recently shown that the friction forces of grafted polyelectrolyte chain layers can be modulated by more than two orders of magnitude, such as a change in coefficient of friction from 0.01 to 1, by exposing it to different monovalent salts. The response of cationic polymer chains was controlled by exposing it to different anions whose size increased in the order $Cl^- < ClO_4^- < PF_6^- <$ bis(trifluoromethanesulfonimide) (TFSI$^-$). This change in coefficient of friction was attributed to the dehydration and collapse of polyelectrolyte chains induced by ion-pairing interactions. On the other hand, polyanion polymer chains were exposed to the same tetra-alkylammonium cation but with varying lengths of hydrophobic tails (C4 to C16). For anionic brushes, the friction increase was attributed to the hydrophobicity induced by electrostatic interactions between surfactants and polymer chains. Polyanionic brushes with the carboxylate and sulfonate side groups revealed different friction responses. This was attributed to the carboxylate groups producing stronger specific interactions with the quaternary ammonium and with the multivalent metal ions as well. This particular study was performed using the polyelectrolyte brushes that were irreversibly immobilized on the surface. It is of interest how the principles in this study can be applied to "free polymers." Dedinaite and coworkers [124] have studied the effect of exposing hyaluronic acid to the contact between opposing phospholipid bilayers using AFM and observed an increase of adhesion and friction forces. Israelachvili and coworkers [125] have reported a synergistic interaction between hyaluronic acid and lubricin using SFA. In this study, a hyaluronic acid layer was chemically grafted on a mica surface and crosslinked. While the friction forces were higher, wear-resistant properties were significantly improved upon chemical grafting and crosslinking. This film showed no visible effect upon exposure to a lipid solution, 1,2-dioleoyl-sn-glycero-3-phosphocholine. These studies are particularly interesting because both hyaluronic acid and lipids are integral components of synovial fluids. The investigation into the properties of the major individual components of synovial fluids, such as hyaluronic acids (HAs), phospholipids, proteoglycan, and sucrose, is a long-standing topic of study to understand the excellent lubrication mechanism of articular joints by synovial fluids [130,131]. Many studies have confirmed that the fluid mixtures of these components reveal synergetic and superior lubricity compared to the individual components [132,133]. Yet, detailed mechanism behind it has not been clarified to date. In particular, "free" hyaluronic acid, despite its high hygroscopic characteristic, has been long known to fail in the effective lubrication of various engineering materials [67]. The contrasting synergistic effects of hyaluronic acid and lipids in these studies [124,125] appear to be related to the interaction mechanism of the involved molecules with the underlying substrates. This synergy can also be viewed in terms of PEC.

3.5.3 CROSSLINKED POLYELECTROLYTES AND HYDROGELS

The last category of studies is crosslinking between polyelectrolytes, often leading to the formation of hydrogels in aqueous solutions. The crosslinking bond can be covalent or ionic. In the latter case, a distinction from PEC systems discussed above is, in fact, blurred. In an early study by Klein and coworkers [126], chitosan was crosslinked using a polyanionic crosslinking agent, namely, sodium hexametaphosphate (SHMP) (also known as Graham's salt, $(NaPO_3)_6$). Other examples include poly(acrylamide) (PAAm) brushes covalently crosslinked with bis(acrylamide) [127,128], poly(MPC), and poly(3-sulfopropyl methacrylate potassium salt-co-2-(methacryloyloxy)ethyltrimethylammonium chloride) (poly(SPMK-co-MTAC)) crosslinked via ionic interactions [81]. Similar to PEM or PEC discussed above, the net effect of crosslinking is that the friction properties degrade (higher friction forces), whereas wear-resistant properties improve. This is attributed to the reduced flexibility of polymer chains, which leads to a mechanically tougher film. The fabrication of these films usually starts with immobilized polyelectrolyte chains in "brush-like" conformation, which shows excellent lubricating properties. Thus, the film preparation typically requires an immobilized polyelectrolyte brush layer in the first place. On the other hand, a recent study by Giasson et al. [129] has shown that "free" poly(L-lysine)-b-poly(acrylic acid)-b-poly(L-lysine) crosslinked with PAA displayed both improved frictional and antiwear properties. While this study was performed with SFA under mild contact pressure, it is of interest whether a similar beneficial lubricating effect can be achieved with tribological contacts of engineering scale.

3.6 SUMMARY AND CONCLUSIONS

Brush-like polymer chains constructed from polyelectrolytes are known to provide extra repulsion between opposing shearing surfaces, arising mainly from increased osmotic pressure within the layer. Thus, it can be considered as an effective method to enhance the lubricity of water as base lubricant fluid. The question addressed in this review article was, simply speaking, whether such superior aqueous lubricating properties observed from charged polymer chains could be applicable to macroscale engineering tribology. Presently, this effect seems to be observed only when dense, brush-like polyelectrolyte chains are prepared via an irreversible immobilization method on the substrates and, more importantly, under mild tribostress conditions. The lubrication of charged surfaces with "free" polyelectrolyte chains, while keeping brush-like conformation, is challenged by the complexity of electrostatic attraction/repulsion with the target surfaces. The adsorption and lubrication of free amphiphilic polyelectrolyte chains on nonpolar surfaces suffer from electrostatic repulsion between neighboring polyelectrolytes on the same substrate and thus limit the efficacy. Overall, attempts to utilize polyelectrolyte chains for aqueous lubrication without disrupting the brush-like conformation appear to be very difficult. Alternative approaches, such as PEM, PEC, and crosslinking, to using polyelectrolyte chains for aqueous lubrication are emerging. The common features of all these approaches are that, first, the lubricating efficacy strengthens due to interactions with other moieties rather than relying upon the genuine lubricity of polyelectrolytes alone and, second, the interactions with other moieties or between polyelectrolyte chains generally lead to a reduction in the flexibility of polymer brushes, higher friction forces, and improved antiwear properties. As long as the fabrication process can be simplified, these features may be more suitable for the desired lubricating effects in engineering tribology.

ACKNOWLEDGMENT

The author appreciates the financial support from the European Research Council (ERC, Starting Grant 2010, Project number 261152).

REFERENCES

1. S. Lee and N.D. Spencer, Achieving ultralow friction by aqueous, brush-assisted lubrication, in: *Superlubricity*, A. Erdemir and J.-M. Martin (eds.), pp. 365–396, Elsevier, Amsterdam, The Netherlands (2007).
2. N.D. Spencer (ed.), *Aqueous Lubrication: Natural and Biomimetic Approaches*, World Scientific Publishing, Singapore (2014).
3. J. Klein, Hydration lubrication, *Friction*, 1, 1–23 (2013).
4. S. Jahn and J. Klein, Hydration lubrication: The macromolecular domain, *Macromolecules*, 48, 5059–5075 (2015).
5. A. Dedinaite, Biomimetic lubrication, *Soft Matter*, 8, 273–284 (2012).
6. J.E. St. Dennis, K. Jin, V.T. John, and N.S. Pesika, Carbon microspheres as ball bearings in aqueous-based lubrication, *ACS Appl. Mater. Interfaces*, 11, 2215–2218 (2011).
7. S. Radice and S. Mischler, Effect of electrochemical and mechanical parameters on the lubrication behaviour of Al_2O_3 nanoparticles in aqueous suspensions, *Wear*, 261, 1032–1041 (2006).
8. L. Gara and Q. Zou, Friction and wear characteristics of water-based ZnO and Al_2O_3 nanofluids, *Tribol. Trans.*, 55, 345–350 (2012).
9. Y.H. Liu, P.X. Liu, L. Che, C.Y. Shu, and X.C. Lu, Tunable tribological properties in water-based lubrication of water-soluble fullerene derivatives via varying terminal groups, *Chin. Sci. Bull.*, 57, 4641–4645 (2012).
10. K. Kristiansen, H. Zeng, P. Wang, and J.N. Israelachvili, Microtribology of aqueous carbon nanotube dispersions, *Adv. Funct. Mater.*, 21, 4555–4564 (2011).
11. Y. Peng, Y. Hu, and H. Wang, Tribological behaviors of surfactant-functionalized carbon nanotubes as lubricant additive in water, *Tribol. Lett.*, 25, 247–253 (2006).
12. Q. Chen, X. Wang, Z. Wang, Y. Liu, and T. You, Preparation of water-soluble nanographite and its application in water-based cutting fluid, *Nano Res. Lett.*, 8, 52 (2013).
13. M.W. Sulek and T. Wasilewski, Tribological properties of aqueous lubrication of alkyl polyglucosides, *Wear*, 260, 193–204 (2006).
14. M.W. Sulek, T. Wasilewski, and K.J. Kurzydlowsik, The effect of concentration on lubricating properties of aqueous solutions of sodium lauryl sulfate and ethoxylated sodium lauryl sulfate, *Tribol. Lett.*, 40, 337–345 (2010).

15. L. Serreau, M. Beauvais, C. Heitz, and E. Barthel, Adsorption and onset of lubrication by a double-chained cationic surfactant on silica surfaces, *J. Colloid Interface Sci.*, 332, 382–388 (2009).
16. W.H. Briscoe and J. Klein, Friction and adhesion hysteresis between surfactant mono-layers in water, *J. Adhesion*, 83, 705–722 (2007).
17. I.U. Vakarelski, S.C. Brown, Y.I. Rabinovich, and B.M. Moudgil, Lateral force micros-copy investigation of surfactant-mediated lubrication from aqueous solution, *Langmuir*, 20, 1724–1731 (2004).
18. B.S. Phillips and J.S. Zabinski, Ionic liquid lubrication effects on ceramics in a water environment, *Tribol. Lett.*, 17, 533–541 (2004).
19. S. Stolte, S. Steudte, O. Areitioaurtena, F. Pagano, J. Thöming, P. Stepnowski, and A. Igartua, Ionic liquids as lubricants or lubrication additives: An ecotoxicity and biode-gradability assessment, *Chemosphere*, 29, 1135–1141 (2012).
20. S. Liu, D. Guo, G. Li, and H. Lei, Lubricating properties of organic phosphate ester aqueous solutions, *Tribol. Lett.*, 37, 573–580 (2010).
21. B. Wang, J. Sun, and Y. Wu, Lubricating performances of nano organic-molybdenum as additives in water-based liquid during cold rolling, *Adv. Mater. Res.*, 337, 550–555 (2011).
22. F. Chinas-Catillo, J. Lara-Romero, and G. Alonso-Nunez, Tribology of aqueous thiomo-lybdate and thiotungstate additives in low-pressure contacts, *Tribol. Trans.*, 56, 366–373 (2013).
23. C. Zhang, J. Liu, C. Zhang, and S. Liu, Friction reducing and anti-wear property of metallic friction pairs under lubrication of aqueous solutions with polyether added, *Wear*, 292–293, 11–16 (2012).
24. M.W. Sulek, W. Sas, T. Wasilewski, A. Bak-Sowinska, and U. Piortrwska, Polymers (polyvinylpyrrolidones) as active additives modifying the lubricating properties of water, *Ind. Eng. Chem. Res.*, 51, 14700–14707 (2012).
25. Z. Wang and D. Gao, Comparative investigation on the tribological behavior of rein-forced plastic composite under natural sea water lubrication, *Mater. Des.*, 51, 983–988 (2013).
26. B. Chen, J. Wang, and F. Yan, Synergism of carbon fiber and polyimide in polytetraflu-oroethylene-based composites: Friction and wear behavior under sea water lubrication, *Mater. Des.*, 36, 366–371 (2012).
27. N. Liu, J. Wang, B. Chen, and F. Yan, Tribochemical aspects of silicon nitride ceramic sliding against stainless steel under the lubrication of sea water, *Tribol. Int.*, 61, 205–213 (2013).
28. G. Cui, Q. Bi, S. Zhu, L. Fu, J. Yang, Z. Qiao, and W. Liu, Synergistic effect of alumina and graphite on bronze matrix composites: Tribological behaviors in sea water, *Wear*, 303, 216–224 (2013).
29. G. Cui, Q. Bi, J. Yang, and W. Liu, The bronze–silver self-lubricating composite under sea water condition, *Tribol. Int.*, 60, 83–92 (2013).
30. S. Alexander, Adsorption of chain molecules with a polar head: A scaling description, *J. de Physique*, 38, 983–987 (1977).
31. P.-G. de Gennes, Conformations of polymers attached to an interface, *Macromolecules*, 13, 1069–1075 (1980).
32. U. Raviv and J. Klein, Fluidity of bound hydration layers, *Science*, 297, 1540–1543 (2002).
33. U. Raviv, S. Giasson, N. Kampf, J.-F. Gohy, R. Jerome, and J. Klein, Lubrication by charged polymers, *Nature*, 425, 163–165 (2003).
34. J. Klein, E. Kumacheva, D. Mahalu, D. Perahia, and L.J. Fetters, Reduction of friction forces between solid surfaces bearing polymer brushes, *Nature*, 370, 634–636 (1994).

35. S. Lee, M. Müller, M. Ratoi-Salagean, J. Vörös, S. Pasche, S.M. De Paul, H.A. Spikes, M. Textor, and N.D. Spencer, Boundary lubrication of oxide surfaces by poly(L-lysine)-g-poly(ethylene glycol) (PLL-g-PEG) in aqueous media, *Tribol. Lett.*, 15, 231–239 (2003).

36. M. Müller, S. Lee, H.A. Spikes, and N.D. Spencer, The influence of molecular architecture on the macroscopic lubrication properties of the brush-like co-polyelectrolyte poly(L-lysine)-g-Poly(ethylene glycol) (PLL-g-PEG) adsorbed on oxide surfaces, *Tribol. Lett.*, 15, 395–405 (2003).

37. X. Yan, S.S. Perry, N.D. Spencer, S. Pasche, S.M. De Paul, M. Textor, and M.S. Lim, Reduction of friction at oxide interfaces upon polymer adsorption from aqueous solutions, *Langmuir*, 20, 423–428 (2004).

38. M. Müller, X. Yan, S. Lee, S.S. Perry, and N.D. Spencer, Lubrication properties of a brushlike copolymer as a function of the amount of solvent absorbed within the brush, *Macromolecules*, 38, 5706–5713 (2005).

39. S. Lee and N.D. Spencer, Aqueous lubrication of polymers: Influence of surface modification, *Tribol. Int.*, 38, 922–930 (2005).

40. M. Müller, X. Yan, S. Lee, S.S. Perry, and N.D. Spencer, Preferential solvation and its effect on the lubrication properties of a surface-bound, brushlike copolymer, *Macromolecules*, 38, 3861–3866 (2005).

41. S. Lee, M. Müller, R. Heeb, S. Zürcher, S. Tosatti, M. Heinrich, F. Amstad, S. Pechmann, and N.D. Spencer, Self-healing behavior of a polyelectrolyte-based lubricant additive for aqueous lubrication of oxide materials, *Tribol. Lett.*, 24, 217–223 (2006).

42. S. Lee and N.D. Spencer, Adsorption properties of poly(L-lysine)-*graft*-poly(ethylene glycol) (PLL-g-PEG) at a hydrophobic interface: Influence of tribological stress, pH, salt concentration, and polymer molecular weight, *Langmuir*, 24, 9479–9488 (2008).

43. W. Hartung, T. Drobek, S. Lee, S. Zürcher, and N.D. Spencer, The influence of anchoring-group structure on the lubricating properties of brush-forming graft copolymers in an aqueous medium, *Tribol. Lett.*, 31, 119–128 (2008).

44. W. Hartung, Aqueous lubrication of ceramics by means of brush-forming graft copolymers, Doctoral thesis, No. 18428, ETH, Zurich, Switzerland (2009).

45. S. Lee and N.D. Spencer, Poly(L-lysine)-*graft*-poly(ethylene glycol): A versatile aqueous lubricant additive for tribosystems involving thermoplastics, *Lubric. Sci.*, 20, 21–34 (2008).

46. T. Drobek and N.D. Spencer, Nanotribology of surface-grafted PEG layers in an aqueous environment, *Langmuir*, 24, 1484–1488 (2008).

47. S. Lee, S. Zürcher, A. Dorcier, G.S. Luengo, and N.D. Spencer, Adsorption and lubricating properties of poly(L-lysine)-*graft*-poly(ethylene glycol) on human-hair surfaces, *ACS Appl. Mater. Interfaces*, 1, 1938–1945 (2009).

48. S.S. Perry, X. Yan, F.T. Limpoco, S. Lee, M. Müller, and N.D. Spencer, Tribological properties of poly(L-lysine)-*graft*-poly(ethylene glycol) films: Influence of polymer architecture and adsorbed conformation, *ACS Appl. Mater. Interfaces*, 1, 1224–1230 (2009).

49. R. Heeb, S. Lee, N.V. Venkataraman, and N.D. Spencer, Influence of salt on the aqueous lubrication properties of end-grafted, ethylene glycol-based self-assembled monolayers, *ACS Appl. Mater. Interfaces*, 1, 1105–1112 (2009).

50. U. Raviv, J. Frey, R. Sak, P. Laurat, R. Tadmor, and J. Klein, Properties and interactions of physigrafted end-functionalized poly(ethylene glycol) layers, *Langmuir*, 18, 7482–7495 (2002).

51. T. Pettersson, A. Naderi, R. Makuska, and P.M. Claesson, Lubrication properties of bottle-brush polyelectrolytes: An AFM study on the effect of side chain and charge density, *Langmuir*, 24, 3336–3347 (2008).

52. F.E. Bailey Jr. and R.W. Callard, Some properties of poly(ethylene oxide) in aqueous solution, *J. Appl. Polym. Sci.*, 1, 56–62 (1959).

53. M. Ataman, Properties of aqueous salt solutions of poly(ethylene oxide). Cloud points, θ temperatures, *Colloid Polym. Sci*, 265, 19–25 (1987).

54. M. Zhang, T. Desai, and M. Ferrari, Proteins and cells on PEG-immobilized silicon surfaces, *Biomaterials*, 19, 953–960 (1988).

55. S.J. Sofia, V. Premnath, and E.W. Merrill, Poly(ethylene oxide) grafted to silicon surfaces: Grafting density and protein adsorption, *Macromolecules*, 31, 5059–5070 (1998).

56. S. Jo and K. Park, Surface modification using silanated poly(ethylene glycol)s, *Biomaterials*, 21, 605–616 (2000).

57. E. Ostuni, R.G. Chapman, R.E. Holmlin, S. Takayama, and G.M. Whitesides, A survey of structure-property relationships of surfaces that resist the adsorption of protein, *Langmuir*, 17, 5605–5620 (2001).

58. S. Pasche, S.M. De Paul, J. Voeroes, N.D. Spencer, and M. Textor, Poly(L-lysine)-*graft*-poly(ethylene glycol) assembled monolayers on niobium oxide surfaces: A quantitative study of the influence of polymer interfacial architecture on resistance to protein adsorption by ToF-SIMS and in situ OWLS, *Langmuir*, 19, 9216–9225 (2003).

59. L.D. Unsworth, H. Sheardown, and J.L. Brash, Protein resistance of surfaces prepared by sorption of end-thiolated poly(ethylene glycol) to gold: Effect of surface chain density, *Langmuir*, 21, 1036–1041 (2005).

60. X. Fan, L. Lin, and P.B. Messersmith, Cell fouling resistance of polymer brushes grafted from Ti substrates by surface-initiated polymerization: Effect of ethylene glycol side chain length, *Biomacromolecules*, 7, 2443–2448 (2006).

61. M.C. Woodle and D.D. Lasic, Sterically stabilized liposome, *Biochim. Biophys. Acta (BBA)*, 1113, 171–199 (1992).

62. K.-L. Gosa and V. Uricanu, Emulsions stabilized with PEO-PPO-PEO block copolymers and silica, *Colloids Surf. A*, 197, 257–269 (2002).

63. A.R. Studart, E. Amstad, and L.J. Gauckler, Colloidal stabilization of nanoparticles in concentrated suspensions, *Langmuir*, 23, 1081–1090 (2007).

64. B. Liberelle and S. Giasson, Friction and normal interaction forces between irreversibly attached weakly charged polymer brushes, *Langmuir*, 24, 1550–1559 (2008).

65. I.E. Dunlop, W.H. Briscoe, S. Titmuss, R.M.J. Jacobs, V.L. Osborne, S. Edmondson, W.T.S. Huck, and J. Klein, Direct measurement of normal and shear forces between surface-grown polyelectrolyte layers, *J. Phys. Chem. B*, 113, 3947–3956 (2009).

66. B. Zappone, M. Ruths, G.W. Greene, G.D. Jay, and J.N. Israelachvili, Adsorption, lubrication, and wear of lubricin on model surfaces: Polymer brush-like behavior of a glycoprotein, *Biophys. J.*, 92, 1693–1708 (2007).

67. M. Benz, N. Chen, and J.N. Israelachvili, Lubrication and wear properties of grafted polyelectrolytes, hyaluronan and hylan, measured in the surface forces apparatus, *J. Biomed. Mater. Res. Part A*, 71A, 6–15 (2004).

68. J. Yu, X. Banquy, G.W. Greene, D.D. Lowrey, and J.N. Israelachvili, The boundary lubrication of chemically grafted and cross-linked hyaluronic acid in phosphate buffered saline and lipid solutions measured by the surface forces apparatus, *Langmuir*, 28, 2244–2250 (2012).

69. N.M. Harvey, G.E. Yakubov, J. Stokes, and J. Klein, Normal and shear forces between surfaces bearing porcine gastric mucin, a high-molecular-weight glycoprotein, *Biomacromolecules*, 11, 1041–1050 (2011).

70. B. Zappone, N.J. Patil, J.B. Madsen, K.I. Pakkanen, and S. Lee, Molecular structure and equilibrium forces of bovine submaxillary mucin adsorbed at a solid–liquid interface, *Langmuir*, 31, 4524–4533 (2015).

71. G. Duner, E. Thormann, O. Ramstrom, and A. Dedinaite, Friction between surfaces—Polyacrylic acid brush and silica—Mediated by calcium ions, *J. Dispers. Sci. Technol.*, 31, 1285–1287 (2010).

72. B. Lego, W.G. Skene, and S. Giasson, Swelling study of responsive polyelectrolyte brushes grafted from mica substrates: Effect of pH, salt, and grafting density, *Macromolecules*, 43, 4384–4393 (2010).

73. M.A. Plunkett, A. Feiler, and M.W. Rutland, Atomic force microscopy measurements of adsorbed polyelectrolyte layers. 2. Effect of composition and substrate on structure, forces, and friction, *Langmuir*, 19, 4180–4187 (2003).

74. Z. Zhang, A.J. Morse, S.P. Armes, A.L. Lewis, M. Geoghegan, and G.J. Leggett, Effect of brush thickness and solvent composition on the friction force response of poly(2-(methacryloyloxy)ethylphosphorylcholine), *Langmuir*, 27, 2514–2521 (2011).

75. D.P. Chang, N.I. Abu-Lail, J.M. Coles, F. Guilak, D.J. Gregory, and S. Zauscher, Friction force microscopy of lubricin and hyaluronic acid between hydrophobic and hydrophilic surfaces, *Soft Matter*, 5, 3438–3445 (2009).

76. J.M. Coles, D.P. Chang, and S. Zauscher, Molecular mechanisms of aqueous boundary lubrication by mucinous glycoproteins, *Curr. Opin. Colloid Interface Sci.*, 15, 406–416 (2010).

77. L. Han, L.D. Dean, D.C. Ortiz, and A.J. Grodzinsky, Lateral nanomechanics of cartilage aggrecan macromolecules, *Biophys. J.*, 92, 1384–1398 (2007).

78. L. Han, L.D. Dean, P. Mao, C. Ortiz, and A.J. Grodzinsky, Nanoscale shear deformation mechanisms of opposing cartilage aggrecan macromolecules, *Biophys. J.*, 93, L23–L25 (2007).

79. M. Kobayashi and A. Takahara, Synthesis and frictional properties of poly(2,3-dihydroxypropyl methacrylate) brush prepared by surface-initiated atom transfer radical polymerization, *Chem. Lett.*, 34, 1582–1583 (2005).

80. H. Sakata, M. Kobayashi, H. Otsuka, and A. Takahara, Tribological properties of poly(methyl methacrylate) brushes prepared by surface-initiated atom transfer radical polymerization, *Polym. J.*, 37, 767–775 (2005).

81. M. Kobayashi, M. Terada, and A. Takahara, Polyelectrolyte brushes: A novel stable lubrication system in aqueous conditions, *Faraday Discuss.*, 156, 403–412 (2012).

82. M. Kobayashi, Y. Terayama, N. Hosaka, M. Kaido, A. Suzuki, N. Yamada, N. Torikai, K. Ishiharae, and A. Takahara, Friction behavior of high-density poly(2-methacryloyloxyethyl phosphorylcholine) brush in aqueous media, *Soft Matter*, 3, 740–746 (2007).

83. M. Kobayashi, H. Tanaka, M. Minn, J. Sigimura, and A. Takahara, Interferometry study of aqueous lubrication on the surface of polyelectrolyte brush, *ACS Appl. Mater. Interfaces*, 6, 20365–20371 (2014).

84. G. Liu, Z. Liu, N. Li, X. Wang, F. Zhou, and W. Liu, Hairy polyelectrolyte brushes-grafted thermosensitive microgels as artificial synovial fluid for simultaneous biomimetic lubrication and arthritis treatment, *ACS Appl. Mater. Interfaces*, 6, 20452–20463 (2014).

85. N. Kampf, J.-F. Gohy, R. Jerome, and J. Klein, Normal and shear forces between a polyelectrolyte brush and a solid surface, *J. Polym. Sci. Part B*, 43, 193–204 (2005).

86. Y. Uyama, H. Tadokoro, and Y. Ikada, Surface lubrication of polymer films by photo induced graft polymerization, *J. Appl. Polym. Sci.*, 39, 489–498 (1990).

87. Y. Uyama, H. Tadokoro, and Y. Ikada, Low-frictional catheter materials by photo induced graft polymerization, *Biomaterials*, 12, 71–75 (1991).

88. K. Ikeuchi, T. Takii, H. Norikane, N. Tomita, T. Ohsumi, X. Uyama, and Y. Ikada, Water lubrication of polyurethane grafted with dimethylacrylamide for medical use, *Wear*, 161, 179–185 (1993).

89. K. Ikeuchi, M. Kouchiyama, N. Tomita, Y. Uyama, and Y. Ikada, Friction control with a graft layer of a thermo-sensing polymer, *Wear*, 199, 197–201 (1996).

90. Y. Ikada, Surface modification of polymers for medical applications, *Biomaterials*, 15, 725–735 (1994).

91. T. Moro, Y. Takatori, K. Ishihara, T. Konno, Y. Takigawa, T. Matsushita, U.-I. Chung, K. Nakamura, and H. Kawaguchi, Surface grafting of artificial joints with a biocompatible polymer for preventing periprosthetic osteolysis, *Nat. Mater.*, 3, 829–836 (2004).

92. M. Kyomoto, T. Moro, F. Miyaji, T. Konno, M. Hashimoto, H. Kawaguchi, Y. Takatori, K. Nakamura, and K. Ishihara, Enhanced wear resistance of orthopaedic bearing due to the cross-linking of poly(MPC) graft chains induced by gamma-ray irradiation, *J. Biomed. Mater. Res. Part B*, 84B, 320–327 (2008).

93. M. Kyomoto, T. Moro, K. Saiga, M. Hashimoto, H. Ito, H. Kawaguchi, Y. Takatori, and K. Ishihara, Biomimetic hydration lubrication with various polyelectrolyte layers on cross-linked polyethylene orthopaedic bearing materials, *Biomaterials*, VV, 4451–4459 (2012).

94. R. McDonald, Zinc dithiophosphates, in: *Lubricant Additives: Chemistry and Applications*, L.R. Rudnick (ed.), pp. 51–62, CRC Press, Boca Raton, FL (2009).

95. T. Røn, I. Javakhishvili, K. Jankova, S. Hvilsted, and S. Lee, Adsorption and aqueous lubricating properties of charged and neutral amphiphilic diblock copolymers at a compliant, hydrophobic interface, *Langmuir*, 29, 7782–7792 (2013).

96. B.J. Hamrock and D. Dowson, Minimum film thickness in elliptical contacts for different regimes of fluid-film lubrication, Leads, U.K., *Proceedings of the Fifth Leeds-Lyon Symposium on Tribology*, Mechanical Engineering Publication, Bury St. Edmunds, U.K., pp. 22–27 (1979).

97. M. Esfahanian and B.J. Hamrock, Fluid-film lubrication regimes revisited, *Tribol. Trans.*, 34, 628–632 (1991).

98. C. Poncet, F. Tiberg, and R. Audebert, Ellipsometric study of the adsorption of hydrophobically modified polyacrylates at hydrophobic surfaces, *Langmuir*, 14, 1697–1704 (1998).

99. Y. Zhang, M. Tirrell, and J.W. Mays, Effect of ionic strength and counterions valency on adsorption of hydrophobically modified polyelectrolytes, *Macromolecules*, 29, 7299–7301 (1996).

100. J.G. Göbel, N.A.M. Besseling, M.A. Cohen Stuart, and C. Poncet, Adsorption of hydrophobically modified polyacrylic acid on a hydrophobic surface: Hysteresis caused by an electrostatic adsorption barrier, *J. Colloid Interface Sci.*, 209, 129–136 (1999).

101. N. Nikogeorgos, J.B. Madsen, and S. Lee, Influence of impurities and contact scale on the lubricating properties of bovine submaxillary mucin (BSM) films on a hydrophobic surface, *Colloids Surf. B*, 122, 760–766 (2014).

102. J.B. Madsen, B. Svensson, M. Abou Hachem, and S. Lee, Proteolytic degradation of bovine submaxillary mucin (BSM) and its impact on adsorption and lubrication at a hydrophobic surface, *Langmuir*, 31, 8303–8309 (2015).

103. K.I. Pakkanen, J.B. Madsen, and S. Lee, Conformation of bovine submaxillary mucin layers on hydrophobic surface as studied by biomolecular probes, *Int. J. Biol. Macromol.*, 72, 790–796 (2015).

104. G.E. Yakubov, J. McColl, J.H.H. Bongaerts, and J.J. Ramsden, Viscous boundary lubrication of hydrophobic surfaces by mucin, *Langmuir*, 25, 2313–2321 (2009).

105. L. Shi and K.D. Caldwell, Mucin adsorption to hydrophobic surfaces, *J. Colloid Interface Sci.*, 224, 372–381 (2000).

106. I. Javakhishvili, T. Røn, K. Jankova, S. Hvilsted, and S. Lee, Synthesis, characterization, and aqueous lubricating properties of amphiphilic graft copolymers comprising 2-methoxyethyl acrylate, *Macromolecules*, 47, 2019–2029 (2014).

107. H.D. Bijsterbosch, M.A. Cohen Stuart, and G.J. Fleer, Effect of block and graft copolymers on the stability of colloidal silica, *J. Colloid Interface Sci.*, 210, 37–42 (1999).

108. J. Rieger, C. Passirani, J.-P. Benoit, K. Van Butsele, R. Jérôme, and C. Jérôme, Copolymers of poly(ethylene oxide) and poly(epsilon-caprolactone) with different architectures, and their role in the preparation of stealthy nanoparticles, *Adv. Funct. Mater.*, 16, 1506–1514 (2006).

109. T. Røn, I. Javakhishvilli, N.J. Patil, K. Jankov, B. Zappone, S. Hvilsted, and S. Lee, Aqueous lubricating properties of charged (ABC) and neutral (ABA) triblock copolymer chains, *Polymer*, 55, 4873–4883 (2014).

110. A.R. Esker, C. Mengel, and G. Wegner, Ultrathin films of a polyelectrolyte with layered architecture, *Science*, 280, 892–895 (1998).

111. D. Yoo, S.S. Shiratori, and M.F. Rubner, Controlling bilayer composition and surface wettability of sequentially adsorbed multilayer of weak polyelectrolytes, *Macromolecules*, 31, 4309–4318 (1998).

112. P.V. Pavoor, B.P. Gearing, A. Bellare, and R.E. Cohen, Tribological characteristics of polyelectrolyte multilayers, *Wear*, 256, 1196–1207 (2004).

113. P.V. Pavoor, B.P. Gearing, R.E. Gorga, A. Bellare, and R.E. Cohen, Engineering the friction-and-wear behavior of polyelectrolyte multilayer nanoassemblies through block copolymer surface capping, metallic nanoparticles, and multiwall carbon nanotubes, *J. Appl. Polym. Sci.*, 92, 439–448 (2004).

114. X. Dai, Y. Zhang, Y. Guan, S. Yang, and J. Xu, Mechanical properties of polyelectrolyte multilayer self-assembled films, *Thin Solid Films*, 474, 159–164 (2005).

115. G. Yang, H. Ma, Z. Wu, and P. Zhang, Tribological behavior of ZnS-filled polyelectrolyte multilayers, *Wear*, 262, 471–476 (2007).

116. G. Yang, Z. Wu, and P. Zhang, Study on the tribological behaviors of polyelectrolyte multilayers containing copper hydroxide nanoparticles, *Tribol. Lett.*, 25, 55–60 (2006).

117. G. Yang, Z.-G. Geng, H. Ma, Z. Wu, and P. Zhang, Preparation and tribological behavior of Cu-nanoparticle polyelectrolyte multilayers obtained by spin-assisted layer-by-layer assembly, *Thin Solid Films*, 517, 1778–1783 (2009).

118. Y.-B. Guo, D.-G. Wang, and S.-W. Zhang, Adhesion and friction of nanoparticles/polyelectrolyte multilayer films by AFM and micro-tribometer, *Tribol. Int.*, 44, 906–915 (2011).

119. G. Yang, X. Chen, K. Ma, L. Yu, and P. Zhang, The tribological behavior of spin-assisted layer-by-layer assembled Cu nanoparticles-doped hydrophobic polyelectrolyte multilayers modified with fluoroalkylsilane, *Surf. Coat. Technol.*, 205, 3365–3371 (2011).

120. J.H.H. Bongaerts, J.J. Cooper-White, and J.R. Stokes, Low biofouling chitosan-hyaluronic acid multilayers with ultra-low friction coefficients, *Biomacromolecules*, 10, 1287–1294 (2009).

121. A. Li, S.N. Ramkrishna, T. Schwarz, E.M. Benetti, and N.D. Spencer, Tuning surface mechanical properties by amplified polyelectrolyte self-assembly: Where grafting-from meets grafting-to, *ACS Appl. Mater. Interfaces*, 5, 4913–4920 (2013).

122. N. Nikogeorgos, P. Efler, A.B. Kayitmazer, and S. Lee, Bio-glues to enhance slipperiness of mucins: Improved lubricity and wear resistance of porcine gastric mucin (PGM) layers assisted by mucoadhesion with chitosan, *Soft Matter*, 11, 489–498 (2015).

123. Q. Wei, M. Cai, F. Zhou, and W. Liu, Dramatically tuning friction using responsive polyelectrolyte brushes, *Macromolecules*, 46, 9368–9379 (2013).

124. C. Liu, M. Wang, J. An, E. Thormann, and A. Dedinaite, Hyaluronan and phospholipids in boundary lubrication, *Soft Matter*, 8, 10241–10244 (2012).

125. S. Das, X. Banquy, B. Zappone, G.W. Greene, G.D. Jay, and J.N. Israelachvili, Synergistic interactions between grafted hyaluronic acid and lubricin provide enhanced wear protection and lubrication, *Biomacromolecules*, 14, 1669–1677 (2013).

126. N. Kampf, U. Raviv, and J. Klein, Normal and shear forces between adsorbed and gelled layers of chitosan, a naturally occurring cationic polyelectrolyte, *Macromolecules*, 37, 1134–1142 (2004).

127. A. Li, E.M. Benetti, D. Tranchida, J.N. Clasohm, H. Schönherr, and N.D. Spencer, Surface-grafted, covalently cross-linked hydrogel brushes with tunable interfacial and bulk properties, *Macromolecules*, 44, 5344–5351 (2011).

128. A. Li, S.N. Ramakrishna, E.S. Kooij, R.M. Espinosa-Marzal, and N.D. Spencer, Poly(acrylamide) films at the solvent-induced glass transition: Adhesion, tribology, and the influence of crosslinking, *Soft Matter*, 8, 9092–9100 (2012).
129. S. Giasson, J.-M. Lagleize, J. Rodriguez-Hernandez, and C. Drummond, Boundary lubricant polymer films: Effect of cross-linking, *Langmuir*, 29, 12936–12949 (2013).
130. G.D. Jay, Characterization of a bovine synovial fluid lubricating factor. I. Chemical, surface activity and lubricating properties, *Connect. Tissue Res.*, 28, 71–88 (1992).
131. G.D. Jay, K. Haberstroh, and C.-J. Cha, Comparison of boundary lubricating ability of bovine submaxillary fluid, lubricin, and healon, *J. Biomed. Mater. Res.*, 40, 414–418 (1998).
132. T.A. Schmidt, N.S. Gastelum, Q.T. Nguyen, B.L. Schumacher, and R.L. Sah, Boundary lubrication of articular cartilage: Role of synovial fluid constituents, *Arthritis Rheum.*, 56, 882–891 (2007).
133. D.A. Swann, R.B. Hendren, R. Student, E.L. Radin, and S.L. Sotman, The lubricating activity of synovial fluid glycoproteins, *Arthritis Rheum.*, 24, 22–30 (1981).

4 Use of Polymers in Viscosity Index Modification of Mineral Oils and Pour Point Depression of Vegetable Oils

Dogan Grunberg, Mert Arca, Dan Vargo,
Sevim Z. Erhan, and Brajendra K. Sharma

CONTENTS

4.1 INTRODUCTION

4.1.1 LUBRICANTS AND THEIR FUNCTIONS

Surface-to-surface contact results in friction and wear. This is one of the fundamental laws of nature. One of the main functions of lubricants is to reduce friction and wear of surfaces in contact. Lubricants may also assume many other functions such as preventing corrosion, heat dissipation, emissions reduction, washing away metal chips, and cleaning. The result of the various roles of lubricants is a reduction in energy use. The common applications of lubricants are in internal combustion engines, gearboxes, hydraulic systems, compressors, turbines, and metalworking applications (cutting, rolling, and forming metal).

In general, lubricants are composed of a base oil and an additive package. The lubricants can be mixed with water, using an appropriate emulsifier, and used as emulsion for metalworking and other applications. The base oil in most cases is mineral oil, which is obtained by vacuum distillation of petroleum crude oil followed by a number of petroleum refining processes. Other synthetic base stocks including polyalphaolefins (PAOs), gas-to-liquid base oils (GTL), polyolesters, and polyglycols and naturally occurring base stocks including vegetable oils are also used. In this chapter, two fundamental properties of lubricant base stocks, the viscosity index (VI) and pour point (PP), will be discussed, and polymeric chemicals (viscosity index improvers [VII] and pour point depressants [PPDs]) that are used to improve these two properties of base stocks will be analyzed and discussed in depth. The analysis will be limited to VI and VI improvement for mineral oil base stocks and PP depression for vegetable oils.

Kramer et al. [1] classified mineral base oils into three general groups: paraffinic, naphthenic, and aromatic oils. Paraffinic base oils are manufactured using crude oils that have relatively high alkane contents. After vacuum distillation, aromatics are removed by solvent extraction and then the dearomatized base is dewaxed. Paraffinic oils are generally classified as solvent neutrals (SNs), bright stocks, and spindle oils. Solvent neutral means the base oil has been solvent refined and any acid or base in the base oil has been neutralized [1,2]. Spindle oil refers to low-viscosity paraffinic oils, and bright stock refers to high-viscosity paraffinic base oils. Naphthenic base oils contain a high proportion of cycloalkanes (naphthenes) and less than 55%–60% paraffinic carbon content. These oils are characterized by their excellent low-temperature properties and are suitable for applications that require low PPs, high solvency, and high compatibility with certain resins. Table 4.1 shows the main differences between naphthenic and paraffinic base stocks. They have similar viscosity at 40°C, but there are differences in VI and PP that alter their suitability for use in certain applications [3]. Aromatic oils are extraction by-products obtained during the production of paraffinic oils. These oils have aromatic content, resulting in low aniline points, which makes them useful in applications where a high degree of solvency is desired to dissolve aromatic components or for use as plasticizers in plastics. These oils are dark colored and have relatively high flash points.

The American Petroleum Institute (API) has classified lubricant base stocks into five categories. Mineral oils are placed into three groups (group I, group II, and group III) based on the amount of saturation and sulfur content present and their VI.

TABLE 4.1
Lubricant Base Stocks

Crude Type	Naphthenic	Paraffinic
Viscosity (cSt at 38°C)	20.53	20.53
Pour point (°C)	−42.5	−18
Viscosity index	15*	100
Flash point (°C)	171	199
Gravity, API	24.4	32.7
Color (ASTM)	1.5	0.5

Source: Pirro, D.M. and Wessol, A.A., *Lubrication Fundamentals*, Marcel Dekker, New York, 2001.
* Recent Hydrocal naphthenic base oils from Calumet have VI in the range of 27–81.

PAOs are classified as group IV, and other base oils such as esters and vegetable oils form group V. Categories of the five groups are shown in Table 4.2 [3].

Examples of additives used in finished lubricants include antioxidants, antiwear agents, extreme pressure agents, friction modifiers, antifoam, corrosion inhibitors, dispersants, emulsifiers, demulsifiers, biocides, anti-mist additives, thickeners, VI modifiers, and PPDs [4,5]. The additive package can make up as low as 0.1%–1% of the lubricant as in turbine oils and as high as 15%–20% in engine oils. Global additive demand in 2015 was 4.2 million metric tons, which is slightly up from 2014 (4.05 million metric tons). Around 60% of these additives are consumed in making heavy-duty and passenger car engine oils, and so dispersants (26%), VII (22%), and detergents (19%) account for the major portion by additive functionality.

Based on their application, lubricants can be classified into two groups: automotive lubricants and industrial lubricants. Worldwide demand for finished lubricants went down from 37.4 million metric tons in 2004 [6] to 35.6 million metric tons in 2015 [7], of which 56% is automotive lubricants, 26% industrial oils (hydraulic fluids, industrial gear lubricants, and others), 10% process oils (rubber process oils, white oils, transformer oils, and others), 5% metalworking fluids and corrosion preventives, and 3% grease, as shown in Figure 4.1.

TABLE 4.2
API Base Stock Categories

	Sulfur (%)	Saturates (%)	VI
Group I	>0.03	≤90	80–120
Group II	≤0.03	≥90	80–120
Group III	≤0.03	≥90	≥120
Group IV	Polyalphaolefins (PAOs)		
Group V	All other base stocks not included in I–IV		

Source: Pirro, D.M. and Wessol, A.A., *Lubrication Fundamentals*,
Marcel Dekker, New York, 2001.

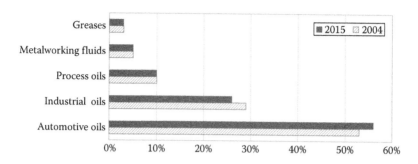

FIGURE 4.1 Worldwide percentage consumption of lubricants in different market segments in the year 2004 and 2015. (From Mang, T. and Dresel, W., *Lubricants and Lubrication*, Wiley-VCH, Weinheim, Germany, 2007; Jacobs, C., 2016–2017, Lubricants industry factbook: A supplement to *Lubes'N'Greases*, in: *Lubes N Greases*, 22, LNG Publishing Company, Inc., Falls Church, VA, 2016.)

4.1.2 Environmentally Friendly Lubricants

The lubricant industry has experienced a slow but continuous move toward biodegradable (environmentally friendly) lubricants in the last 15 years. The terms biodegradable and environmentally friendly were not widely used in the lubricant industry 15 years ago. The environmental friendliness of a base stock or lubricant has become an important consideration for current lubricant manufacturers. However, the meaning of biodegradability is not always clear. The term biodegradable is applied to fluids that, using standard methods and assays, are converted over time from the lubricating fluids to lower molecular weight (MW) components that have essentially no environmental impact [8].

There are two main biodegradability tests for lubricants. The first method measures the evolution of CO_2 over a period of 28 days. In this method, oil is mixed with mineral salts and the amount of CO_2 evolved during biodegradation is measured. Biodegradation of modified vegetable oils is routinely studied with this method, and it has been shown that some chemically modified vegetable oils are not biodegradable [9]. ASTM D5988 and U.S. Environmental Protection Agency shake flask test and other similar methods are based on CO_2 decomposition [3,10].

Another method to determine biodegradability is Coordinating European Council (CEC) L-33-T-82 test method [3]. The main difference between CEC L-33-T-82 and ASTM D5988 is that in CEC L-33-T-82, IR spectroscopy method is used to follow the intensity of the CH_2 band at 2928 cm^{-1} and the CH_3 band at 2958 cm^{-1} after 0, 7, and 21 days. In order for an oil to be considered biodegradable, it must show more than 60% conversion to CO_2 following the method standardized in ASTM D5988 or 90% conversion in CEC L-33-T-82 [3,11]. Table 4.3 shows biodegradability of different base stocks using these two methods. Mineral oil and polyglycol are not biodegradable in either of these tests [12].

Mobil EAL 224H (environmentally aware lubricant) is one of the vegetable-based hydraulic oils on the market. Figure 4.2 shows comparative biodegradability of Mobil EAL 224H and mineral and synthetic oils [3].

TABLE 4.3

Carbon Dioxide Evolution Using the EPA Shake Flask Test and CEC Methods

	Biodegradability (%)	
Base Stock	Shake Flask [3]	CEC L-33-T-82 [12]
Mineral oil	42–48	15–35
Vegetable oil	72–80	70–100
Synthetic ester	55–84	5–30
Polyglycol	6–38	0–25

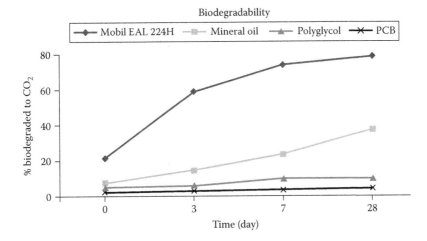

FIGURE 4.2 Comparative biodegradability of various oils. PCB, polychlorinated biphenyls.

As a result of local and national regulations and, to some degree, environmental awareness of consumers, the demand for environmentally friendly lubricants is growing. In the year 2002, the total Western European market for all lubricants was 5020 kilotons/year, of which 50 kilotons/year were based on vegetable oils. The same year, the U.S. market for all lubricants was 8250 kilotons/year, of which only 25 kilotons/year were based on vegetable oils [13]. The demand for biolubricants has been increasing since then, and in 2011, it was 505.6 kilotons and is expected to reach 785 kilotons in 2018, growing at a compounded annual growth rate (CAGR) of 6.6% from 2013 to 2018 [14]. Worldwide, biolubricants market is expected to reach $3.15 billion by 2021, growing at a CAGR of 6.3% between 2016 and 2021 [15]. Hydraulic fluid application accounts for the largest share in terms of value in 2015 and is also expected to witness the higher CAGR from 2016 to 2021. Commercial transport, mainly marine industry, is expected to be the fastest growing end-use segment of biolubricants, because of concerns about safe disposal of lubricants into the marine ecosystem. The market of biolubricants in North America is projected to grow at the highest CAGR during 2016–2021 due to the Vessel General Permit

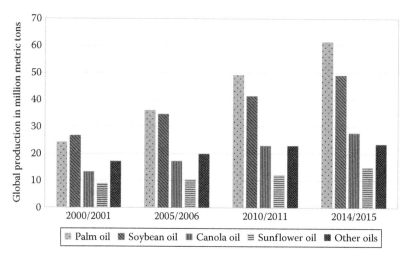

FIGURE 4.3   Global production of vegetable oils (in million metric tons) from 2000–2001 to 2014–2015. Total production in 2014–2015 was 176.73 million metric tons and expected to reach 179.56 million metric tons in 2015–2016. (From World vegetable oil production, 2016, Statista, https://www.statista.com/statistics/263978/global-vegetable-oil-production-since-2000-2001/, accessed September 20, 2016.)

imposed by the U.S. EPA and abundance of feedstock, such as soybean in the United States and canola in Canada. Vegetable oil–based lubricants can be produced from soybean, rapeseed/canola, sunflower, palm, cottonseed, and other vegetable oils. The availability of the vegetable oil is as important as the performance characteristics. Because of the high oxidative stability and good low-temperature properties, rapeseed/canola oil is the best choice in terms of performance. However, rapeseed/canola oil is not widely available in the United States, and as a result soybean oil is the preferred choice because of its large domestic production. World production of different vegetable oils from 2000/2001 to 2014/2015 is given in Figure 4.3 [16].

It has been reported that more than half of the lubricants that are sold worldwide pollute the environment due to total-loss lubrication and spillage and through evaporation [13]. Because of their high biodegradability, renewable character, and superior performance in wear reduction, vegetable oils are used to formulate hydraulic fluids, two-cycle engine oils, chain-saw oils, mold-release oils, metalworking fluids, and railway greases. Even with the advantages of vegetable oil–based lubricants including excellent antiwear properties, high biodegradability, low toxicity, and high VI, their use is still limited to a small number of applications due to their high cost, poor thermo-oxidative stability, and poor low-temperature flow properties compared to traditional petroleum-based lubricants [17]. Recent developments in genetic modification led to the development of high oleic acid content vegetable oils such as high-oleic sunflower oil. These high oleic acid grades exhibit a significantly improved oxidation stability compared to conventional vegetable oils [18]. Other ways to improve oxidation stability and low-temperature flow properties of vegetable oil are structural modifications either chemically or using heat bodying [19], heat modification, and the use of diluents [20,21] and additives, such as PPDs. More than 90% of chemical modifications have been reported on the fatty acid carboxyl groups, while the remaining 10% involve reactions at the double bonds present in the fatty acid part of triacylglycerol [22]. Some examples of these reactions at double-bond sites are polymerization by heat modification [23–26]; alkylation, arylation, cyclization, hydrogenation, and epoxidation [27,28]; ring opening of epoxy [29–32]; and other reactions [33–36].

4.1.3 Viscosity Index and Pour Point

Viscosity can be defined as the resistance of a liquid to flow. VI is a measure of the change in viscosity with temperature. A high VI shows that the viscosity changes less drastically with temperature. PP is the lowest temperature at which a sample of liquid can be made to flow by gravity alone. Mineral oils respond well to PPDs, and PPs can be decreased by almost 30°C [37,38]. As a result, this chapter will not discuss crystallization of mineral oils and depression of their PPs. However, a major barrier to widespread adoption of vegetable oil–based lubricants is their poor low-temperature properties, exacerbated by a poor response to current PPDs. Another challenge is developing biodegradable PPDs. In general, VIs of mineral oils are low, and this is a major deficiency in mineral oils. Conversely, VIs of vegetable oils are very high and will not require the use of VI modifiers. Thus, in this chapter, only the VI improvement for mineral oils and the PP depression for vegetable oils will be discussed.

Without VI modifier additives, one would have to use different types of engine oils in winter and in summer because the viscosity changes drastically with the season. This problem is obvious in Canada where the temperature in a year can vary between −30°C and 30°C. The PPs of soybean oil and rapeseed oil are −9°C and −18°C, respectively [39]. Without PPDs, the commercialization of soybean oil–based lubricants in the United States or rapeseed oil–based lubricants in Europe would not have been possible.

4.2 DISCUSSION

4.2.1 Viscosity and VI Improvement in Mineral Oils

It is crucial to use a lubricant with the correct viscosity for any given application in order to achieve satisfactory lubrication. The film thickness and film-forming ability of a lubricant are highly dependent on its viscosity. Absolute viscosity, also called dynamic viscosity, is defined as

$$\tau = \frac{F}{A} = \eta \frac{\partial u}{\partial y} \tag{4.1}$$

where
    the term $\partial u / \partial y$ is the shear rate
    u is the unit velocity (m/s)
    y is the unit distance between layers (m)
    η is the dynamic viscosity in Poise (Pa · s or N · s/m^2)
    τ is the shear stress (N/m^2), which is given by the force applied (F) and the area
        (A) over which it is applied.

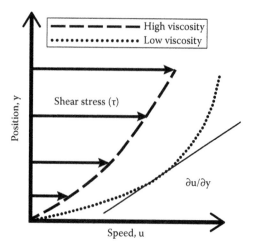

FIGURE 4.4 A general parallel flow in a no-slip boundary. Higher viscosity would have a steeper slope.

Figure 4.4 shows the relationships between these parameters. As the viscosity decreases, $\partial u/\partial y$ decreases.

The unit of absolute viscosity is Pa · s. The viscosity that is often used in the lubricant industry is kinematic viscosity, which measures the rate of flow of the liquid through a capillary under the influence of gravity [40]. Kinematic viscosity (m^2/s), v, is defined as

$$v = \frac{\eta}{Q} \tag{4.2}$$

where
  η is the dynamic viscosity (N · s/m^2)
  Q is the density of the fluid (kg/m^3)

The standard unit of kinematic viscosity is m^2/s, but the most common unit used in industry is centistokes (cSt) where 1 cSt equals 1 mm^2/s or 10^{-6} m^2/s.

In the case of a car engine, if the viscosity of the oil is too low, the oil will not be able to maintain a film under high shear, and this will lead to piston wear and ultimately failure. If the viscosity of the oil is too high, then the film will be too thick and moving the piston will require an excessive amount of force, which will reduce fuel efficiency. Having the optimum viscosity would result in minimum friction and wear as well as minimum energy consumption.

4.2.1.1 Viscosity Index

As discussed earlier, VI is a measure of the change in viscosity with temperature. As the temperature increases, the average kinetic energy of the molecules in the liquid increases and liquid flows more easily and its viscosity decreases. The opposite is observed with a decrease in temperature. VI is a dimensionless number that measures the change in internal friction with temperature [6,41,42]. Kinematic viscosity of a sample is measured at

40°C and 100°C, and viscosity change is compared with an empirical reference scale. The scale that is used in this chapter ranges from 80 to 400 [6,43].

Automobile engine oil requires a very high VI for optimal performance. If an engine oil is prepared without VII and used in both warm and cold climates, the following would occur: In winter, the viscosity would increase sharply and moving the pistons would become very hard and would require high energy/fuel consumption. In summer, the viscosity would decrease sharply, and at low enough viscosity, it would not be possible to maintain the lubricant film between the piston and the cylinder. This would lead to high wear and, if not handled promptly, to engine failure. One way to avoid this situation is to use a low-viscosity lubricant in winter and a high-viscosity lubricant in summer. By incorporating VI modifiers in lubricant, a single motor oil can be used in all seasons. Lubricants of this type are known as multigrade motor oils; the viscosity of base stock and multigrade motor oils are shown in Table 4.4 [18,44–47].

4.2.1.2 Viscosity Index Modifiers

Two of the earliest VI modifiers (also called improvers) that are still in use today are polymethacrylates (PMAs) and polyisobutylene (PIB). The use of PMAs as VII dates back to the research by Rohm and Haas in 1937, whereas the use of PIB dates back to the research by I.G. Farbenindustrie AG in 1938 [44]. The commercialization of VII occurred around 1950. These types of products were used for the development of multigrade engine oils.

Today, the main types of VII in use are polyalkylmethacrylates (PAMAs), olefin copolymers (OCPs), and hydrogenated styrene–isoprene copolymers. Various other VII that are manufactured in small volumes are not covered in this chapter. The largest consumption of VII is still in engine oil formulation, but they are also used in industrial lubricants. Some VII, such as PMAs, can also function as PPDs. The chemical structures of these VII are shown in Figures 4.5 through 4.8.

4.2.1.3 How Viscosity Improvement Works

Solubility in a given base stock is crucial for the selection of appropriate VII. Solubility in a given base stock depends on the polymer type (e.g., homopolymer, random, block, star), chain length, molecular weight (MW), chemical type (dependent on monomers), and temperature of the medium.

In general, the MW of VII varies between 10,000 and 250,000 g/mol [6]. The VII are generally used at 1%–10% (w/w) in the base fluids. The operating mechanism of VII can be explained by the following model. (Note: The model and calculations are only approximations.) If a VI improver in the MW range of 10,000–250,000 g/mol has been added into a paraffinic oil at room temperature, the spontaneity of the dissolution process is described by the Gibbs energy of mixing (ΔG_{MIX}), which is a function of temperature in kelvin (T), mixing enthalpy in joules (ΔH_{MIX}), and mixing entropy in joules per kelvin (ΔS_{MIX}):

$$\Delta G_{MIX} = \Delta H_{MIX} - T\Delta S_{MIX} \qquad (4.3)$$

In Equation 4.3, the most problematic is the entropy term because it is usually very complex for polymers. Enthalpy of mixing is in most cases positive for polymers.

TABLE 4.4

Kinematic Viscosities at 40°C and 100°C, VI and Pour Points of Various Mineral and Vegetable Oils and ExxonMobil's Mobil Clean 5000 (SAE 5W-30) Motor Oil Blended Using Paraffinic Base Stocks

Base Stock	Mineral Oils				Vegetable Oils			Multigrade Motor Oil
	Spindle [44]	150 SN [44]	500 SN [44]	Bright Stock [44]	Soybean Oil [8,18,45]	Rapeseed Oil [18,45]	Sunflower Oil [18,46]	SAE 5W-30 [47]
Viscosity at 40°C (cSt)	12.7	27.3	95.5	550	32.93	34.8	32.3	63.1
Viscosity at 100°C (cSt)	3.1	5.0	10.8	33	8.08	7.8	7.7	11.1
Viscosity index (VI)	100	103	97	92	233	208	212	170
Pour point (PP) (°C)	−15	−12	−9	−9	−9	−18	−12	−45

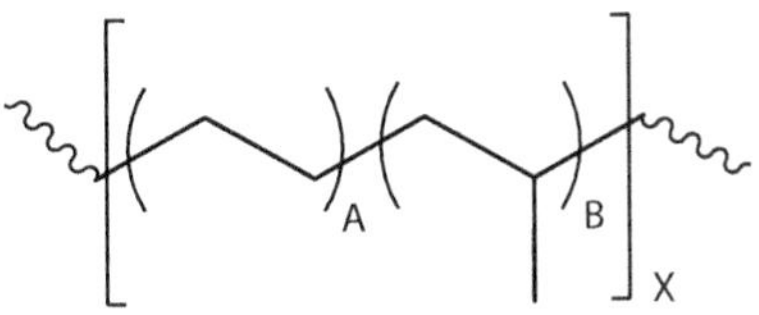

FIGURE 4.5 The generic structure of an olefin copolymer (OCP), used mainly in engine and hydraulic oils.

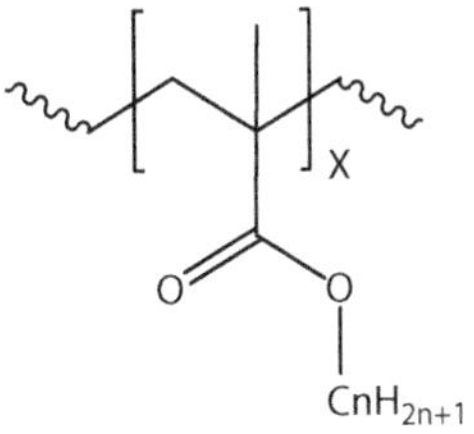

FIGURE 4.6 Polyalkylmethacrylates (PAMAs) used mainly in hydraulic, gear oils and environmentally friendly lubricants.

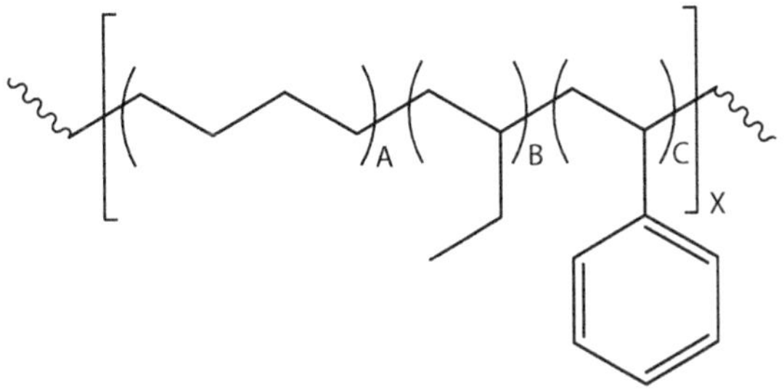

FIGURE 4.7 Hydrogenated styrene-butadiene copolymers (HSDs) used mainly in engine oils.

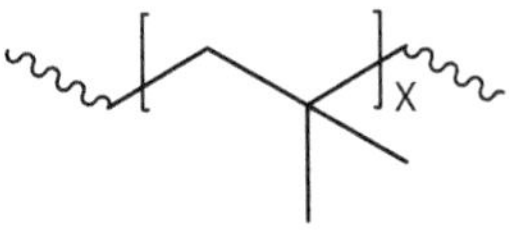

FIGURE 4.8 Polyisobutylene (PIB) used mainly in gear oils. Polyisobutylene consumption as VI improvers is not as significant as other VI improvers shown in Figures 4.5 through 4.7.

Thus, the dissolution process is generally dependent on entropy. Even though dissolution may be enthalpically unfavorable, entropically it is favorable as the number of states increases drastically after dissolution. In the case of the collapsed coil, there are only a few available states. The simplest equation for entropy of mixing is given by Equation 4.4, where k is the Boltzmann constant (J/K) and W is the number of possible states:

$$\Delta S_{MIX} = k \ln W \qquad (4.4)$$

The overall enthalpy of mixing of components 1 and 2, ΔH_{MIX} (cal), can be predicted by the Hildebrand [48] Equation 4.5:

$$\frac{\Delta H_{MIX}}{V_m} = \left[\left(\frac{\Delta E_1}{V_1} \right)^{1/2} - \left(\frac{\Delta E_2}{V_2} \right)^{1/2} \right]^2 \psi_1 \psi_2 \tag{4.5}$$

where

V_m is the total molar volume of the mixture (cm³)
ΔE_1 is the energy of vaporization (cal)
V_1 is the molar volume of the solvent (cm³)
ψ_1 is the volume fraction of the solvent

The parameters with the subscript 2 refer to the polymer.

$\Delta E/V$ is generally referred to as the cohesive energy density and its square root as the solubility parameter, δ (cal$^{1/2}$/cm$^{3/2}$), given by Equation 4.6:

$$\left(\frac{\Delta E}{V} \right)^{0.5} = \delta \tag{4.6}$$

After rearranging Equation 4.5, the heat of mixing per cubic centimeter at a given concentration is given by Equation 4.7:

$$\frac{\Delta H_{MIX}}{V_m} \psi_1 \psi_2 = \left(\delta_1 - \delta_2 \right)^2 \tag{4.7}$$

If $(\delta_1 - \delta_2)^2 = 0$, then solution is assured, and if δ values of oil and VI improver are nearly equal, then these will be mutually miscible. This also suggests that VI improver and oil will interact strongly with each other than among themselves [49]. The solubility parameter for polymethylmethacrylates (δ_{PMMA}) $\approx$ 9.5 cal$^{1/2}$/cm$^{3/2}$, and the solubility parameter for paraffinic oils (δ_{oil}) $\approx$ 7.5 cal$^{1/2}$/cm$^{3/2}$ [50], so this results in the heat of mixing per cubic centimeter at a given concentration ($\Delta H_{MIX}/V_m$ $\psi_1 \psi_2$) $= (7.5 - 9.5)^2 = 4$.

The solubility parameters of paraffinic oil (solvent) and VI improver (PMMA) are not very similar; thus, ΔH_{MIX} is positive. This approximate result shows that the spontaneity of the dissolution process depends on the entropy of mixing.

The solubility of VII is dependent on temperature. This can be seen clearly if, in the ΔG_{MIX} Equation 4.3, ΔH_{MIX} is positive. To allow a negative ΔG_{MIX} and spontaneous dissolution, the entropy term ($-T\Delta S_{MIX}$) must be negative. As explained earlier, S_{MIX} is positive. Then, as temperature increases, the magnitude of the term $-T\Delta S_{MIX}$ increases and the dissolution process becomes more spontaneous and the solubility of the VI improver polymer in the paraffinic oil increases. Conversely, as temperature decreases, the solubility of the VI improver decreases.

The VI modifier additives have a thickening effect on base oil. The thickening efficiency and increase in viscosity depends on the molecular weight and the concentration of the VI improver polymer [51,52]. Thus, as temperature decreases, the solubility of the VI improver decreases and the increase in oil viscosity is compensated by the reduced thickening effect of the VI improver. Conversely, as temperature goes up and oil viscosity decreases, the VI improver becomes more soluble in the oil and its increased thickening effect compensates for the decrease in oil viscosity.

Covitch and Trickett [53] proposed that it is more appropriate to consider the effect of the polymer compared to the viscous behavior of the base oil without polymer, when considering physicochemical reasons on why polymers alter the viscosity–temperature relationships of fluids.

A polymer's contribution to the viscosity of a solution is measured by its intrinsic viscosity, η, (dl/g). To determine intrinsic viscosity according to Huggins [54], specific viscosity (η_{sp}) is measured over a range of polymer concentration c (g/dl), and η_{sp}/c is extrapolated to infinite dilution as given by Equation 4.8:

$$\frac{\eta_{sp}}{c} = [\eta] + K_1 [\eta]^2 \tag{4.8}$$

where
  K_1 is the Huggins constant
  $\eta_{sp} = (\eta/\eta_0) - 1$
  η_0 is the viscosity of pure solvent

Alternatively, the Kraemer relationship [54], shown in Equation 4.9, can also be used to determine intrinsic viscosity by extrapolation to c = 0:

$$\ln\left(\frac{\eta_r}{c}\right) = [\eta] - K_2 [\eta]^2 \tag{4.9}$$

where relative viscosity $\eta_r = (\eta/\eta_0)$.

Intrinsic viscosity (dl/g) is proportional to the hydrodynamic volume of spherical polymer coils in solutions as shown by Equation 4.10:

$$V_e = \frac{M_w [\eta]}{2.5N} \tag{4.10}$$

where
  V_e is the hydrodynamic volume (dl)
  M_w is the weight-average molecular weight (g/mol)
  N is Avogadro's number (mol^{-1})

This relationship is based upon Einstein's viscosity relation, Equation 4.11, which models the polymers as equivalent hydrodynamic solid spheres [55]. According to

Einstein, the viscosity increase caused by adding spherical particles to a liquid only depends on the initial viscosity of the solvent and the total volume fraction of the spheres:

$$\frac{\eta}{\eta_0} = 1 + 2.5\psi \qquad (4.11)$$

where

η is the viscosity of sphere containing solutions

η_0 is the viscosity of pure solvents

ψ is the volume fraction of spheres

OCPs are primarily copolymers of ethylene and propylene. They can also be, however, ethylene/propylene/diene copolymers, which are diene-modified OCPs. When the ethylene/propylene ratio is around 45/55 to 55/45, the OCP is amorphous and flows at room temperature. When the ethylene/propylene ratio is more than 60/40, the OCP is semicrystalline at room temperature and does not flow. When the ethylene/propylene ratio is high, thickening efficiency is high, but low-temperature properties are poor since at higher ethylene ratios, the OCP is in semicrystalline form. When the ethylene–propylene ratio is decreased resulting in an amorphous OCP, thickening is reduced to some extent and low-temperature properties are improved [56–58]. OCPs are generally manufactured by solution Ziegler–Natta polymerization [59]. Metallocene polymerization is also used where a higher level of control over stereoregularity, molecular weight distribution, and composition is required [60–62].

PAMAs are linear polymers that are composed of short-, intermediate-, and long-chain alkylmethacrylates. Short-chain alkylmethacrylates affect the polymer coil size, especially at low temperatures, and thus influence the VI of the solution. Intermediate-chain alkylmethacrylates are the reason for the solubility of the polymer in the mineral base oils. Longer-chain alkylmethacrylates interact with wax and thus demonstrate PP reduction effects. When PAMAs are used as VII, the longer-chain alkylmethacrylates are not a major part of the copolymer as PP reduction is not the intended function of the additive [63,64]. The thermal coil expansion behavior of PAMAs viscosity modifiers is thus highly dependent upon monomer composition [53]. PAMAs are manufactured through anionic polymerization followed by hydrogenation [44].

Hydrogenated styrene–butadiene copolymers (HSDs) are generally random copolymers of styrene and butadiene. HSDs can be random, block, or star-shaped polymers. In general, their composition is 50%–60% styrene and 40%–50% butadiene. Optimum thickening is achieved when butadiene polymerizes in 1,4 configuration. However, this results in semicrystalline regions in the polymer, which negatively affect the low-temperature properties of the lubricant. In order to prevent semicrystalline regions, 30%–40% of butadiene is polymerized in 1,2 configuration. Molecular weight is generally between 75,000 and 200,000. HSDs are manufactured through anionic solution polymerization followed by hydrogenation [44].

Lariviere et al. studied the viscosities of dilute polymer solutions over the temperature range from −10°C to 150°C using a capillary viscometer [65]. The solvent used was a mineral oil with 95% saturates. The VII studied were OCPs (M_w 304,100 and M_w/M_n 1.13), hydrogenated diene polymers, HDP (M_w 469,000 and M_w/M_n 1.10), which is related to HSD, and PMAs (M_w 169,600 and M_w/M_n 1.95). The results of this study are shown in Figures 4.9 through 4.11. The thickening effect of the OCP and the HDP was found to be higher at 40°C than at 100°C. The thickening effect was found to decrease with increasing temperature. The authors claimed that this observation was due to stronger intermolecular hydrodynamic interactions at low temperatures. It was also observed that the thickening effect increased with increasing temperature for the PMA VI improver. The authors concluded that at temperatures between 10°C and 150°C the polymer coil dimension remains constant for the OCP and the HDP VII, but increases with the increasing temperature for the PMA VI improver [65]. The increasing coil dimension of PMA with increasing temperature (increasing solubility) explains why its thickening effect increases with increasing temperature.

In Figure 4.9, the curve for 0.7% between −10°C and 30°C shows that the OCP has good VI improvement property. In Figure 4.10, the VI improvement for 0.7% PMA is much larger and spans a wide range of temperature from 10°C to 150°C. Figure 4.11 shows that VI improvement properties of HDP are the worst among the three VI improver families.

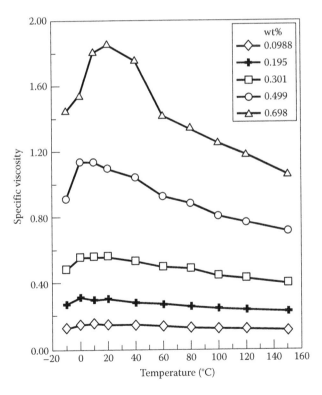

FIGURE 4.9 Specific viscosity of OCP at different concentrations as a function of temperature. Specific viscosity, $\eta_{sp} = (\eta - \eta_0)/\eta_0$, where η_0 is solvent viscosity and η is solution viscosity. (From LaRiviere, D. et al., *Lubr. Sci.*, 12, 133, 2000.)

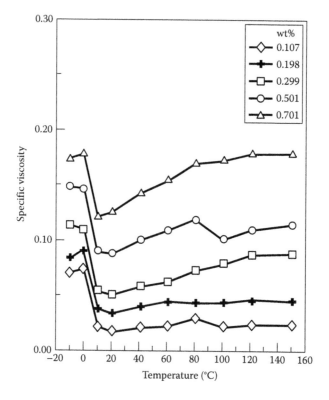

FIGURE 4.10 Specific viscosity of PMA at different concentrations as a function of temperature. Specific viscosity, $\eta_{sp} = (\eta - \eta_0)/\eta_0$, where η_0 is solvent viscosity and η is solution viscosity. (From LaRiviere, D. et al., *Lubr. Sci.*, 12, 133, 2000.)

Although VI is a practical and useful metric for lubricating oils, it is not as useful to understand the mechanism behind their functioning. The coil expansion mechanism associated with VI improvement is still used widely, but Covitch and Trickett [53] have shown that consideration of the viscosity–temperature profile of base oil containing polymer compared to the polymer-free base oil is more appropriate to explain the mechanism by which polymers increase VI. They also concluded that coil size expansion with temperature is not necessary to achieve a significant elevation of VI, but polymers that do expand with temperature have higher VI contributions than those that do not. Authors reported that OSP increases the viscosity of base oil by about the same proportion, regardless of temperature. The degree of viscosity increase from 100°C to 40°C will necessarily be less for a polymer-containing low-viscosity base oil (4 cSt) than a polymer-free high-viscosity base oil (6 cSt) of equal KV viscosity at 100°C.

Overall, PMA-type VII offer the best performance in most cases. However, OCPs have adequate performance, and they are manufactured in very large quantities and thus are cheaper than PMAs. Thus, with the exception of highly demanding applications (wide VI range requirement), OCPs are preferred over PMAs.

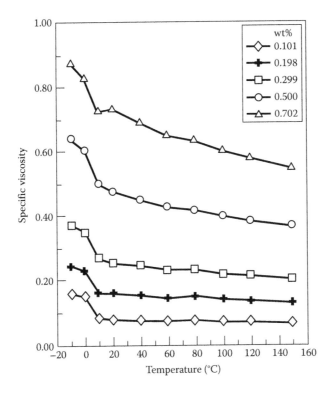

FIGURE 4.11 Specific viscosity of HDP at different concentrations as a function of temperature. Specific viscosity, $\eta_{sp} = (\eta - \eta_0)/\eta_0$, where η_0 is solvent viscosity and η is solution viscosity. (From LaRiviere, D. et al., *Lubr. Sci.*, 12, 133, 2000.)

4.2.2 CRYSTALLIZATION, POUR POINT, AND POUR POINT DEPRESSION IN VEGETABLE OILS

At low temperatures, vegetable oils solidify due to crystallization. This is one of the biggest disadvantages of vegetable oils as lubricants. Crystallization results in a sharp increase in the viscosity of vegetable oils, leading to poor pumpability, lubrication, and rheological behavior [16].

4.2.2.1 Crystallization

Crystallization consists of two major processes: nucleation and crystal growth [66]. Nucleation, or the formation of a crystalline cluster, or nuclei, from the liquid state, is the most important factor in controlling crystallization [67,68]. In order for nucleation to continue, molecules must organize into a crystal lattice. Entropically, this process is not favored, as it is a transition from the relatively disordered liquid phase to an ordered crystal lattice. However, after nucleation, energy is released in the form of the latent heat of fusion. Thus, an energy barrier must be overcome for nucleation to occur. Since entropy change for this process is negative, crystallization is more favored at low temperature. As temperature decreases, the positive contribution of the $-T\Delta S$ terms to ΔG decreases and the process eventually becomes spontaneous. The nuclei become stable when they reach a critical size; below this critical size, the nucleation process is reversible. The critical size depends on the environmental conditions. Crystal growth only occurs around the nuclei that succeed in achieving the critical cluster size [66]. The requirement of critical cluster size reflects kinetic control in crystallization. Thus, both thermodynamic and kinetic processes are active in crystallization. Even if crystallization is thermodynamically favored at a given temperature, critical cluster size is required for crystallization to continue. Before discussing crystallization and low-temperature properties of vegetable oils, their composition will be discussed.

Vegetable oils are made up of triacylglycerols and less than 2% of other compounds such as free fatty acids and phospholipids. However, in the case of refined vegetable oils, the compounds other than vegetable oils are present in insignificant amounts since free fatty acids and phospholipids are removed in the neutralization and degumming stages of refining, respectively.

Triacylglycerols are esters made of one glycerol molecule and three fatty acid molecules. Glycerol is the common molecule in all triacylglycerol molecules; the fatty acids can vary in length and degree of unsaturation. Biological fatty acids are mainly even-numbered hydrocarbon chains. In terms of the degree of unsaturation, three kinds are present. The first one is saturated fatty acids in which the hydrocarbon chain is fully reduced with no double bond. Monounsaturated fatty acids have a single double bond, and polyunsaturated fatty acids are the ones with more than one double bond. Single carbon bonds rotate freely, while double bonds in hydrocarbon chains are restricted. Fatty acids with double bond exist as two isomers: *cis* form with double-bond hydrogens on the same side of the molecule forming a bent chain structure and *trans* form with double-bond hydrogens on the opposite side of the molecule, forming a straight chain structure. Fatty acids with *cis* double bonds are found predominantly. The *trans* configuration is more stable and is thus formed during heating, hydrogenation, and other chemical treatments [69]. Physical properties of saturated, *cis*, and *trans* fatty acids are different; for example, oleic acid (18:1 *cis*) has a melting point 56°C lower than stearic acid (18:0), whereas elaidic acid (18:1 *trans*) has a lower melting point by just 25°C [70]. The melting points of fatty acids decrease with each addition of double bonds; for example, the melting points of the C18 fatty acids decrease progressively from stearic acid (18:0, 69°C) to oleic acid (18:1 *cis*, 13°C), linoleic acid (18:2, −5°C), and linolenic acid (18:3, −11°C). The melting points of even-numbered saturated fatty acids increase progressively with increasing chain length. Short-chain saturated fatty acids up to C8 are liquid at room temperature, saturated fatty acids with C10 and higher are solid at room temperature, and the very long-chain saturated fatty acids have a waxy consistency. Due to these variations in the triacylglycerol molecule, the crystal lattice in vegetable oils can be composed of several relatively stable and different lattice structures. These different chain packings or polymorphic forms have quite discrete energies, lattice arrangements, and crystal habits.

Polymorphism is the ability of a molecule to exist in more than one crystalline form depending on the molecular arrangement within the crystal lattice [71].

Polymorphism affects crystallization because the crystallization process is dependent on the crystalline forms in the lattice. Polymorphism depends strongly on the position of the fatty acid on the glycerol unit and the chemical properties of the fatty acids, such as the chain length, parity (odd or even), and unsaturation (*cis* or *trans*) [72]. Triacylglycerols are oriented in a chair or tuning fork configuration in the crystalline lattice. The triacylglycerol can take either a double- or triple-chain-length structure. Triacylglycerols with varied chain lengths and degrees of unsaturation (almost all vegetable oils) obtain triple-chain-length structure [71]. It is widely recognized that vegetable oils retain some degree of ordering in the pure liquid phase and have been suggested to associate in a basic structure in which alternating chair-like conformations align in a dimeric unit, with temperatures well above the melting point needed to fully dissociate this ordering. While these are not crystalline, this natural ordering of the liquid phase provides a framework for crystal formation following cooling at low temperatures [71]. The triacylglycerol molecules are large structures, and in tuning fork configuration, they exhibit a relatively high degree of multimolecular order even in the liquid state [70]. This ordered liquid has been suggested to reflect a dynamic structured liquid with "icebergs" of organized structures forming and dissolving continuously, as shown in Figure 4.12. This proposed order suggests a lamellar structure for most triacylglycerols in the liquid state. As the temperature drops during cooling, the mean size of the "icebergs" is thought to increase (Figure 4.12).

In vegetable oil crystallization, crystalline growth occurs by the addition of more triacylglycerol molecules to the nuclei. The incorporation of a new triacylglycerol molecule into an existing crystal lattice depends on the probability of it having the correct configuration at the correct site on a crystal surface [71]. Growth continues as long as there is a driving force for crystallization. For growth to occur, molecules from the liquid phase must migrate to the surface of the crystal, where rearrangement and orientation takes place [71].

4.2.2.2 Pour Point

Crystallization of vegetable oils is a complex process and cannot be directly measured. One indirect method for analyzing crystallinity in vegetable oils is by measuring its PP. As defined before, PP is the lowest temperature at which a sample of liquid can be made to flow by gravity alone [44]. PP is widely used in the lubricant industry for characterizing low-temperature properties.

Three factors must be considered while measuring PP. First, low temperature is the main factor driving wax crystal growth, so a low-temperature end point should be defined. Second, the rate of cooling has an effect on the size and number of wax crystals formed. Finally, while measuring the low-temperature properties, low shear rate is necessary as wax crystals can be interrupted under high shear conditions. In order to measure wax formation, PP tests have to be conducted under conditions of low shear stress [73]. There are several standard methods that are used to measure PP.

ASTM D97 is the most common method due to its quickness. Oil is cooled to different temperatures until no flow is observed while holding the test jar horizontally

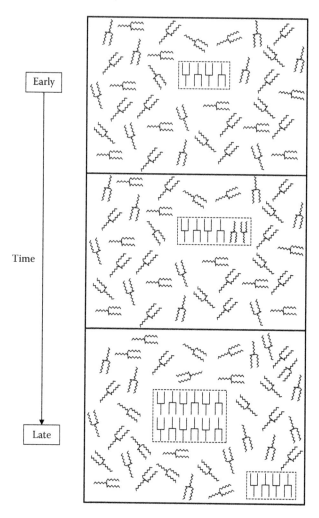

FIGURE 4.12 Pictorial representation of mechanism for nucleation of triacylglycerols as a function of time. Straight chains indicate crystallized triacylglycerols, wriggly chains indicate triacylglycerols in the fluid phase.

for 5 s. This method takes 2 h using a high cooling rate, which does not allow wax crystals to fully form [74]. This incomplete formation of wax crystals may lead to results that are not an accurate representation of the true PP.

ASTM D3829 standard is used for engine oils and measures the lowest temperature where oil can be pumped. This method takes 16 h and measures the viscosity at end temperature [75].

ASTM D5133 [76] and ASTM D2983 [77] measure apparent viscosity at low temperatures with a Brookfield viscometer. Both methods use a low shear rate. The ASTM D5133 standard requires a constant cooling rate of 1°C/h, while ASTM D2983 requires a low-temperature cooling bath.

The methods described earlier are based on the bulk properties of fluid and, as such, require a relatively long time for analysis. By using a differential scanning calorimetry (DSC), these problems could be solved. Moreover, the reproducibility of DSC testing is better than that of the ASTM D97 method [78]. Crystallization is an exothermic process that can be monitored using DSC. The temperature at which the crystallization process is complete can be determined. However, this method is not an ASTM standard, but it generally correlates with ASTM D97 [79]. Examining the PP of vegetable oils using DSC is more complex than that of pure molecules due to the complex mixture of triacylglycerol molecules containing different fatty acid with varying chain lengths and abundances in vegetable oils. In short, special attention must be given while obtaining a PP for vegetable oils using the DSC method [80].

4.2.2.3 Pour Point Depressants

PPDs are used to lower the PPs of lubricant formulations [38,81,82]. PPDs are crucial for manufacturing vegetable oil–based lubricants due to the fact that vegetable oils generally have higher PPs than mineral oils. The biggest problem is poor additive response in vegetable oils compared to mineral oils. Most PPDs can lower the PPs of mineral oils by almost 30°C, but these are not as effective in vegetable oils. PPDs for mineral oils with long, waxy side chains are not sufficient to decrease the PP of vegetable oils due to the presence of longer fatty acid chains in triacylglycerols. This results in poor additive response in vegetable oils. For vegetable oil applications, only two types of PPDs are commercially viable and show adequate performance: PMAs and poly(styrene-maleic anhydride) copolymers (PSMAs), also referred to as malan–styrene copolymers. In the mid-1930s, PMAs entered the market as an oil thickener and VI improver for mineral oils. PMAs showed good VI improvement characteristics, that is, good performance in a wide range of temperatures, which led to the first multigrade oils during World War II [83]. Even though PMAs were not the first PPDs, today they are widely used for both VI improvement and PP depression. The chemical structure of PMA is shown in Figure 4.6.

As described earlier, PMAs can be categorized into three types based on their alkyl chain. A higher chain length with 14 or more carbons in the alkyl group is desired for PP depression. As molecular weight is proportional to viscosity during the polymerization of methacrylic acid, viscosity is controlled. The M_w ranges from 7,000 to 750,000 Da, and in higher molecular weights, the viscosity of PMAs becomes difficult to control. In order to overcome this problem, PMAs are mixed with light mineral oil or vegetable oil depending on the end-use application [4].

Vegetable oils do not respond linearly to PPD concentration; the response will plateau at a certain concentration. This is shown in Figure 4.13, where after using 2% PMA in three types of vegetable oils the PP does not decrease, but remains constant [39]. Overtreatment with PPD can also reverse the benefits. The structure of malan–styrene copolymers can be seen in Figure 4.14.

Biodegradability and low toxicity are other important criteria for selecting a PPD for vegetable oil–based applications. If the additives blended into vegetable oils are

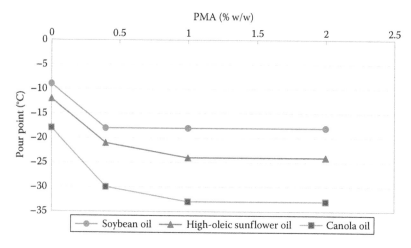

FIGURE 4.13 Effect of PMA copolymer of ~8000 Da on pour point in different vegetable oils. (From Asadauskas, S. and Erhan, S.Z., *J. Am. Oil. Chem. Soc.*, 76, 313, 1999.)

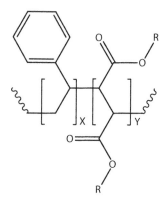

FIGURE 4.14 Malan-styrene copolymers (PSMAs), mainly used as PPD in automatic transmission fluids and environmentally friendly lubricants.

toxic or nonbiodegradable, the benefits of using a biodegradable lubricant are diminished. In order for a lubricant to be fully environmentally friendly, both the base stock and the additives have to be biodegradable and nontoxic. One of the major problems is that PMAs and PSMAs are typically sold as solutions in mineral oil, which is used as the solvent in polymerization reactions of these materials.

4.2.2.4 How Pour Point Depression Works

Even when crystallization is thermodynamically favorable at a particular temperature, crystallization growth can be suppressed if the formed nuclei cannot reach a critical size. If the critical cluster size is not reached, the cluster is not stable, and thus crystallization and redissolution into liquid phase are reversible processes at a given temperature. Therefore, controlling the low-temperature properties of vegetable oils means controlling the crystallization kinetics. "Pour point depressants cannot eliminate crystals but instead they have been created to suppress formation of large crystals during solidification" [84]. They do this by co-crystallizing with the triacylglycerol molecules.

Similar chemistry is used in PMAs used as a PPD and in PMAs used as a VII. They both use methacrylate monomers and this is where the similarity ends. The difference is in MW and monomer building blocks. VII are composed of short-, medium-, and long-chain alkylmethacrylates, while the monomer building blocks in PPD contain long-chain (C12–C18) alkylmethacrylates with branching [6]. "PMAs for VI improvement applications have MW in the range of 250,000 and 750,000. In contrast PMAs for pour point depression applications have MW in the range of 7,000–10,000" [44]. Also when PMAs are produced to be used in environmentally friendly lubricants, the polymerization reaction is carried out in a suitable vegetable oil, and the product is sold to lubricant blenders as a solution in vegetable oil. One example of such environmentally friendly PMA PPDs is the VISCOPLEX 10 series manufactured by Evonik RohMax Corporation [85].

In Figure 4.15, the inclusion of a PMA PPD into the growing triacylglycerol crystal lattice can be seen. Figure 4.15 is only a cartoon representation since crystallites are on the order of μm and polymers are on the order of nm.

PSMAs are manufactured by the esterification of an approximately 1:1 styrene–maleic anhydride copolymer. The copolymer of styrene–maleic anhydride is dissolved in a suitable solvent (usually mineral oil) and esterified to about 70% with a C8–18 alcohol mixture using an acid catalyst at 150°C–160°C [44]. The mechanism of PP depression by PSMA copolymer is very similar to that of PMAs. When Figures 4.6 and 4.14 are compared, it can be seen that the structures of PMAs and PSMAs are similar. Like PMAs, PSMAs may also be used as VII.

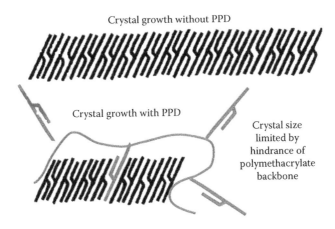

Crystal growth without PPD

Crystal growth with PPD

Crystal size limited by hindrance of polymethacrylate backbone

FIGURE 4.15 Pour point depression (crystal growth suppression) in a triacylglycerol crystal lattice by a PMA pour point depressant molecule (cartoon representation only). (From Erhan, S.Z., Vegetable oils as lubricants, hydraulic fluids, and inks, in: *Bailey's Industrial Oil and Fat Products, Industrial and Nonedible Products from Oils and Fats*, 6th edn., Vol. 6, Bailey, A.E. and Shahidi, F., eds., John Wiley & Sons, Inc., Indianapolis, IN, 2005.)

TABLE 4.5

Effect of PMA and PSMA on Pour Points (Temperature in °C) of Vegetable Oils

Vegetable Oil	Neat Vegetable Oil	1% (w/w) PMA	1% (w/w) PSMA (7671A)
Soybean oil	−9	−18	−18
Sunflower oil	−12	−24	−24
Canola oil	−18	−33	−36

Source: Asadauskas, S. and Erhan, S.Z., *J. Am. Oil. Chem. Soc.*, 76, 313, 1999.

Both PMA and PSMA have molecular weights in the range of ~7000–8000 Da and are 50% solutions in canola oil.

The worldwide production volume of PSMAs for PPD and VII applications is much smaller compared to PMAs. Lubrizol Corporation is the only large-scale manufacturer. When PSMAs are produced to be used in environmentally friendly lubricant formulations, the polymerization reaction is carried out in a suitable vegetable oil, and the product is sold to blenders as a solution in vegetable oil. One example of an environmentally friendly PSMA PPD is the 7671A series manufactured by Lubrizol Corporation. A comparison of PMA and PSMA performance in different vegetable oils is given in Table 4.5.

The data in Table 4.5 show that both PMA and PSMA have similar performances with just a slightly better performance by PSMA. Since PMAs are produced in larger quantities and have lower costs, therefore, PMAs remain a more common choice than PSMAs. It can be seen that even at 1% w/w, the additive response is still poor for both PMA and PSMA. PPs reduced by only 9°C and 12°C using either PMA or PSMA in soybean oil and sunflower oil, respectively, while in canola oil, PP reduction is 15°C with PMA and 18°C with PSMA. PPDs in general can achieve PP reductions of up to 30°C in mineral oil at concentrations of only 0.1%–0.4% w/w. Another problem with vegetable oils treated with PPDs

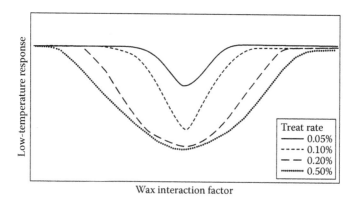

FIGURE 4.16 Low temperature response as a function of treat rate and Wax Interaction Factor (WIF). (From Souchik, J., Pour point depressants, in: *Lubricant Additives: Chemistry and Applications*, 2nd edn., Rudnick, L.R., ed., CRC Press, Boca Raton, FL, 2009.)

is that the lowered PPs are not maintained for prolonged cooling periods. After 1 week of cooling, vegetable oil and PPD mixtures stop flowing at higher temperatures than their measured PPs [39].

PPDs can be ranked using the wax interaction factor (WIF), which defines interactions between the alkyl side of a PPD and wax. Each oil has different wax content, and by using the PPD with a suitable WIF, the optimum result can be achieved [73]. Since vegetable oils have a waxy structure, using a PPD with high WIF would result in lower PP temperatures. Figure 4.16 shows the low-temperature response of base oil as a function of WIF and treat rate (PPD concentration). Increasing the treat rate increases the peak area, which results in a less specific response. Specific response could be achieved by using lowest treat rate (0.05%) at a specific WIF.

4.3 CONCLUSIONS

Low VI is a major problem for mineral oils, and low-temperature flow properties are a major problem for vegetable oils. VI improvement in mineral oils is a well-established area of research. PMAs display the highest VII performance over a wide range of temperatures. However, in many applications, performance requirements are less strenuous and OCPs can deliver satisfactory performance. Cost is a major factor for additive selection resulting in OCPs retaining the largest volume as VII. Lubricants are expected to last longer and operate over a wider range of temperatures as machinery technology develops. Thus, the future demand for high-performance VII like PMA is likely to increase. Vegetable oils have high VI even in their neat form; thus, VI improvement is not a major need for vegetable oil–based formulations.

Currently, only two types of PPDs, PMAs and PSMAs, display adequate performance in vegetable oils. As the performance of both polymers is similar, and PMAs are manufactured in much larger quantities, PMAs are cheaper and used widely. However, PP depression in vegetable oils provided by PMAs and PSMAs is still far from the desired goal of −30°C as is the case with mineral oils. Since additive response in mineral oils is excellent, not much progress is expected in the area of mineral oil PP depression. Modified versions of PMAs and PSMAs and, if possible, entirely new chemistries designed specifically for vegetable oils are needed to reach the desired level of PP depression in vegetable oils. There is still a lot of room for development of better performing VI modifiers for mineral oils and PPDs for vegetable oils.

REFERENCES

1. D. C. Kramer, B. K. Lok, and R. R. Krug, The evolution of base oil technology, in: *Turbine Lubrication in the 21st Century*, W. R. Herguth and T. M. Warne (eds.), ASTM International, West Conshohocken, PA (2001).
2. J.-P. Favennec, *Petroleum Refining: Refinery Operation and Management*, Vol. 5, Editions Technip & Ophrys, Paris, France (2001).
3. D. M. Pirro and A. A. Wessol, *Lubrication Fundamentals*, Marcel Dekker, New York (2001).

4. S. P. Srivastava, *Advances in Lubricant Additives and Tribology*, CRC Press, Taylor & Francis Group, Boca Raton, FL (2009).
5. P. J. Blau (ed.), Glossary of terms, in: *ASM Handbook, Vol. 18: Friction Lubrication and Wear Technology*, Materials Information Society, Metals Park, OH, 1–24 (1992).
6. T. Mang and W. Dresel, *Lubricants and Lubrication*, Wiley-VCH, Weinheim, Germany (2007).
7. C. Jacobs, 2016–2017 Lubricants industry factbook: A supplement to *Lubes'N'Greases*, in: *Lubes N Greases*, Vol. 22, LNG Publishing Company, Inc., Falls Church, VA (2016).
8. A. Adhvaryu and S. Z. Erhan, Epoxidized soybean oil as a potential source of high-temperature lubricants, *Ind. Crops Prod.* 15, 247–254 (2002).
9. R. L. Shogren, Z. Petrovic, Z. Liu, and S. Z. Erhan, Biodegradation behavior of some vegetable oil-based polymers, *J. Polym. Environ.* 12, 173–178 (2004).
10. ASTM Committee, ASTM D5988-12 standard test method for determining aerobic biodegradation of plastic materials in soil, in: *ASTM Annual Book Standards*, ASTM International, West Conshohocken, PA (2012).
11. M. P. Schneider, Plant-oil-based lubricants and hydraulic fluids, *J. Sci. Food Agric.* 86, 1769–1780 (2006).
12. S. Erhan and B. K. Sharma, Vegetable oil-based biodegradable industrial lubricants, in: *Biocatalysis and Biotechnology for Functional Foods and Industrial Products*, C. T. Hou and J.-F. Shaw (eds.), CRC Press, Taylor & Francis Group, Boca Raton, FL (2006).
13. D. Hörner, Recent trends in environmentally friendly lubricants, *J. Synth. Lubr.* 18, 327–347 (2002).
14. Synthetic & bio-based lubricants market—Global industry analysis, market size, share, trends, analysis, growth and forecast 2012–2018 (January 23, 2013), http://www.transparencymarketresearch.com/global-synthetic-and-bio-based-lubricants-market.html, accessed September 20, 2016.
15. Biolubricants Market by Type (Vegetable Oil, Animal Fat), Application (Hydraulic Fluids, Metalworking Fluids, Chainsaw Oils, Mold Release Agents), End Use (Industrial, Commercial Transport, Consumer Automobile)—Global Forecasts to 2021 (August 2016), http://www.researchandmarkets.com/reports/3830547/biolubricants-market-by-type-vegetable-oil#rela9, accessed September 20, 2016).
16. US Department of Agriculture, and USDA Foreign Agricultural Service. Global production of vegetable oils from 2000/01 to 2014/15 (in million metric tons) (2016). In Statista - The Statistics Portal. https://www.statista.com/statistics/263978/global-vegetable-oil-production-since-2000-2001/, accessed September 20, 2016.
17. B. K. Sharma, U. Rashid, F. Anwar, and S. Erhan, Lubricant properties of Moringa oil using thermal and tribological techniques, *J. Therm. Anal. Calorim.* 96, 999–1008 (2009).
18. S. Z. Erhana, B. K. Sharma, and J. M. Perez, Oxidation and low temperature stability of vegetable oil-based lubricants, *Ind. Crops Prod.* 24, 292–299 (2006).
19. S. Z. Erhan, A. Adhvaryu, and B. K. Sharma, Chemically functionalized vegetable oils, in: *Synthetics, Mineral Oils, and Bio-Based Lubricants: Chemistry and Technology*, L. R. Rudnick (ed.), CRC Press, Boca Raton, FL (2006).
20. K. M. Doll and B. K. Sharma, Physical properties study on partially bio-based lubricant blends: Thermally modified soybean oil with popular commercial esters, *Int. J. Sustain. Eng.* 5, 33–37 (2012).
21. S. N. Shah, B. R. Moser, and B. K. Sharma, Glycerol tri-ester derivatives as diluent to improve low temperature properties of vegetable oils, *ASTM Spec. Tech. Publ.* 1477, 450–463 (2011).
22. K. M. Doll, B. R. Moser, B. K. Sharma, and S. Z. Erhan, Current uses of vegetable oil in the surfactant, fuel, and lubrication industries, *Chim. Oggi Chem. Today* 24, 41–44 (2006).

23. Z. Liu, B. K. Sharma, S. Z. Erhan, A. Biswas, R. Wang, and T. P. Schuman, Oxidation and low temperature stability of polymerized soybean oil-based lubricants, *Thermochim. Acta* 601, 9–16 (2015).

24. M. Arca, B. K. Sharma, N. P. J. Price, J. M. Perez, and K. M. Doll, Evidence contrary to the accepted Diels-Alder mechanism in the thermal modification of vegetable oil, *JAOCS J. Am. Oil Chem. Soc.* 89, 987–994 (2012).

25. Z. S. Liu, B. K. Sharma, and S. Z. Erhan, From oligomers to molecular giants of soybean oil in supercritical carbon dioxide medium: 1. Preparation of polymers with lower molecular weight from soybean oil, *Biomacromolecules* 8, 233–239 (2007).

26. A. Biswas, A. Adhvaryu, D. G. Stevenson, B. K. Sharma, J. L. Willet, and S. Z. Erhan, Microwave irradiation effects on the structure, viscosity, thermal properties and lubricity of soybean oil, *Ind. Crops Prod.* 25, 1–7 (2007).

27. P. A. Z. Suarez, M. S. C. Pereira, K. M. Doll, B. K. Sharma, and S. Z. Erhan, Epoxidation of methyl oleate using heterogeneous catalyst, *Ind. Eng. Chem. Res.* 48, 3268–3270 (2009).

28. B. K. Sharma, K. M. Doll, and S. Z. Erhan, Oxidation, friction reducing, and low temperature properties of epoxy fatty acid methyl esters, *Green Chem.* 9, 469–474 (2007).

29. B. K. Sharma, K. M. Doll, and S. Z. Erhan, Ester hydroxy derivatives of methyl oleate: Tribological, oxidation and low temperature properties, *Bioresour. Technol.* 99, 7333–7340 (2008).

30. B. R. Moser, B. K. Sharma, K. M. Doll, and S. Z. Erhan, Diesters from oleic acid: Synthesis, low temperature properties, and oxidation stability, *JAOCS J. Am. Oil Chem. Soc.* 84, 675–680 (2007).

31. K. M. Doll, B. K. Sharma, and S. Z. Erhan, Synthesis of branched methyl hydroxy stearates including an ester from bio-based levulinic acid, *Ind. Eng. Chem. Res.* 46, 3513–3519 (2007).

32. B. K. Sharma, K. M. Doll, and S. Z. Erhan, Chemically modified fatty acid methyl esters: Their Potential use as lubrication fluids and surfactants, in: *Surfactants in Tribology*, Vol. 2, G. Biresaw and K. L. Mittal (eds.), CRC Press, Boca Raton, FL, 387–408 (2011).

33. H. L. Ngo, R. O. Dunn, B. K. Sharma, and T. A. Foglia, Synthesis and physical properties of isostearic acids and their esters, *Eur. J. Lipid Sci. Technol.* 113, 180–188 (2011).

34. S. Z. Erhan, B. K. Sharma, Z. Liu, and A. Adhvaryu, Lubricant base stock potential of chemically modified vegetable oils, *J. Agric. Food Chem.* 56, 8919–8925 (2008).

35. B. K. Sharma, Z. Liu, A. Adhvaryu, and S. Z. Erhan, One-pot synthesis of chemically modified vegetable oils, *J. Agric. Food Chem.* 56, 3049–3056 (2008).

36. B. K. Sharma, A. Adhvaryu, Z. Liu, and S. Z. Erhan, Chemical modification of vegetable oils for lubricant applications, *JAOCS J. Am. Oil Chem. Soc.* 83, 129–136 (2006).

37. L.-J. Wu, F. Zhang, Z.-Y. Guan, and S.-F. Guo, Synthesis of new EVA graft copolymer and its pour point depressant performance evaluation for Daqing crude oil, *J. Cent. South Univ. Technol. Engl. Ed.* 15, 488–491 (2008).

38. J. Balzer, M. Feustel, M. Krull, and W. Reimann, Graft polymers, their preparation and use as pour point depressants and flow improvers for crude oils, residual oils and middle distillates, U.S. Patent 5439981 (1995).

39. S. Asadauskas and S. Z. Erhan, Depression of pour points of vegetable oils by blending with diluents used for biodegradable lubricants, *J. Am. Oil Chem. Soc.* 76, 313–316 (1999).

40. ASTM Committee, ASTM D445-15a standard test method for kinematic viscosity of transparent and opaque liquids (and calculation of dynamic viscosity), in: *ASTM Annual Book of Standards*, ASTM International, West Conshohocken, PA (2015).

41. R. M. Gresham, The basics of viscosity index, *Tribol. Lubr. Technol.* 63, 27–28 (2007).

42. J. Wright, A simple explanation of viscosity index improvers, *Mach. Lubr.* 3/2008, (2008). http://www.machinerylubrication.com/Read/1327/viscosity-index-improvers.

43. J. Igarashi, M. Kagaya, T. Satoh, and T. Nagashima, High viscosity index petroleum base stocks—The high potential base stocks for fuel economy automotive lubricants, SAE Technical Paper Series, SAE Paper No. 920659, International Congress & Exposition, Detroit, MI, February 24–28 (1992).

44. R. M. Mortier and S. T. Orszulik, *Chemistry and Technology of Lubricants*, Blackie Academic, London, U.K. (1997).

45. T. Di-Hua and Y. Bin, Modification of the chemical structure of an environmentally-friendly castor oil lubricant, *J. Synth. Lubr.* 21, 59–64 (2004).

46. E. Durak, M. Çetinkaya, M. Yeinigün, and F. Karaosmanolu, Effects of sunflower oil added to base oil on the friction coefficient of statically loaded journal bearings, *J. Synth. Lubr.* 21, 207–222 (2004).

47. Mobil 1™ 5W-30 Mobil Passenger Vehicle Lube, United States, https://www.mobil.com/english-US/Passenger-Vehicle-Lube/pds/GLXXMobil-1-5W30, (2016) accessed September 20, 2016.

48. J. H. Hildebrand and R. L. Scott, *Solubility of Nonelectrolytes*, 3rd edn., Reinhold, New York (1950).

49. P. Somasundaran, S. Krishnakumar, and S. C. Mehta, A new model to describe the sorption of surfactants on solids in non-aqueous media, *J. Colloid Interface Sci.* 292, 373–380 (2005).

50. C. E. J. Carraher, *Introduction to Polymer Chemistry*, CRC/Taylor & Francis, Boca Raton, FL (2010).

51. U. F. Schodel, Automatic transmission fluids, *International Tribology Colloquium*, TAE, Esslingen, Germany, Vol. 3, p. 22.7-1 (1992).

52. H. C. Evans and D. W. Young, Polymers and viscosity index, *Ind. Eng. Chem.* 39, 1676–1681 (1947).

53. M. J. Covitch and K. J. Trickett, How polymers behave as viscosity index improvers in lubricating oils, *Adv. Chem. Eng. Sci.* 5, 134–151 (2015).

54. F. W. Billmeyer, *Textbook of Polymer Science*, 3rd edn., John Wiley & Sons, Inc., New York (1984).

55. A. Einstein, Berichtigung zu meiner Arbeit: Eine neue Bestimmung der Moleküldimensionen, *Ann. Phys.* 339, 591–592 (1911).

56. G. Verstrate and M. J. Struglinski, Polymers as lubricating-oil viscosity modifiers, *ACS Symp. Ser.* 462, 256–272 (1991).

57. G. T. Spiess, J. E. Johnston, and G. Verstrate, Ethylene propylene copolymers as lube oil viscosity modifiers, in: *Fifth International Colloquium, Additives for Lubricants and Operational Fluids*, Vol. 11, W. J. Bartz (ed.), Technische Akademie Esslingen, Ostfildern, Germany, pp. 8.10-1–8.10-11 (1986).

58. K. Marsden, Literature review of OCP viscosity modifiers, *Lubr. Sci.* 1, 265–280 (1989).

59. J. Boor, *Ziegler-Natta Catalysts and Polymerizations*, Academic Press, New York (1979).

60. V. K. Gupta, S. Satish, and I. S. Bhardwaj, Metallocene complexes of group 4 elements in the polymerization of monoolefins, *J. Macromol. Sci. Rev. Macromol. Chem. Phys.* C34, 439–514 (1994).

61. M. Hackmann and B. Rieger, Metallocenes: Versatile tools for catalytic polymerization reactions and enantioselective organic transformations, *Cat. Tech.* 79 (December) (1997).

62. J. C. W. Chien and D. He, Olefin copolymerization with metallocene catalysts. I. Comparison of catalysts, *J. Polym. Sci. Part Polym. Chem.* 29, 1585–1593 (1991).

63. J. P. Arlie, J. Dennis, and G. Parc, *Viscosity Index Improvers 1. Mechanical and Thermal Stabilities of Polymethacrylates and Polyolefins*, Institute of Petroleum, London, U.K. (1975).

64. P. Neudoerfl, State of the art in the use of polymethacrylates in lubricating oils, *Int. Colloq. Addit. Lubr. Oper. Fluids*, 11, 8.2-1 (1986).

65. D. Lariviere, A.-F. A. Asfour, A. Hage, and J. Z. Gao, Viscometric properties of viscosity index improvers in lubricant base oil over a wide temperature range. Part I: Group II base oil, *Lubr. Sci.* 12, 133–143 (2000).

66. P. D. Glynn and E. J. Reardon, Solid-solution aqueous-solution equilibria—Thermodynamic theory and representation, *Am. J. Sci.* 290, 164–201 (1990).
67. J. W. Mullin, *Crystallization*, Butterworth-Heinemann, Oxford, U.K. (2001).
68. A. G. Walton, Nucleation in liquids and solutions, in: *Nucleation*, A. C. Zettlemoyer (ed.), Marcel Dekker, New York (1969).
69. R. R. Allen, Hydrogenation, in: *Bailey's Industrial Oil and Fat Products*, D. Swern (ed.), John Wiley & Sons, Inc., New York (1982).
70. J. B. German and C. Simoneau, Phase transitions of edible fats and triglycerides: Theory and applications, in: *Phase/State Transitions in Foods: Chemical, Structural, and Rheological Changes*, M. A. Rao and R. W. Hartel (eds.), Marcel Dekker, New York (1998).
71. L. Hernqvist, Crystal structures of fats and fatty acids, in: *Crystallization and Polymorphism of Fats and Fatty Acids*, N. Garti and K. Sato (eds.), Marcel Dekker, New York (1988).
72. K. Sato, Solidification and phase transformation behaviour of food fats—A review, *Lipid Fett.* 101, 467–474 (1999).
73. J. Souchik, Pour point depressants, in: *Lubricant Additives: Chemistry and Applications*, 2nd edn., L. R. Rudnick (ed.), CRC Press, Boca Raton, FL (2009).
74. ASTM Committee, ASTM D97-16 standard test method for pour point of petroleum products, in: *ASTM Annual Book of Standards*, ASTM International, West Conshohocken, PA (2016).
75. ASTM Committee, ASTM D3829-14 standard test method for predicting the borderline pumping temperature of engine oil, in: *ASTM Annual Book of Standards*, ASTM International, West Conshohocken, PA (2014).
76. ASTM Committee, ASTM D5133-15 standard test method for low temperature, low shear rate, viscosity/temperature dependence of lubricating oils using a temperature-scanning technique, in: *ASTM Annual Book of Standards*, ASTM International, West Conshohocken, PA (2015).
77. ASTM Committee, ASTM D2983-15 standard test method for low-temperature viscosity of lubricants measured by brookfield viscometer, in: *ASTM Annual Book of Standards*, ASTM International, West Conshohocken, PA (2015).
78. A. Adhvaryu, S. Z. Erhan, and J. M. Perez, Wax appearance temperatures of vegetable oils determined by differential scanning calorimetry: Effect of triacylglycerol structure and its modification, *Thermochim. Acta* 395, 191–200 (2003).
79. P. Claudy, J.-M. Letoffe, B. Neff, and B. Damin, Diesel fuels: Determination of onset crystallization temperature, pour point and filter plugging point by differential scanning calorimetry. Correlation with standard test methods, *Fuel* 65, 861–864 (1986).
80. A. Govindapillai, N. H. Jayadas, and M. Bhasi, Analysis of the pour point of coconut oil as a lubricant base stock using differential scanning calorimetry, *Lubr. Sci.* 21, 13–26 (2009).
81. S. M. Samoilov and V. N. Monastyrskii, New polymeric pour-point depressant additives (A review), *Chem. Technol. Fuels Oils* 9, 161–164 (1973).
82. S. Deshmukh and D. P. Bharambe, Synthesis of polymeric pour point depressants for Nada crude oil (Gujarat, India) and its impact on oil rheology, *Fuel Process. Technol.* 89, 227–233 (2008).
83. B. G. Kinker, Polymethacrylate viscosity modifiers and pour point depressants, in: *Lubricant Additives: Chemistry and Applications*, 2nd edn., L. R. Rudnick (ed.), CRC Press, Boca Raton, FL (2009).
84. S. Z. Erhan, Vegetable oils as lubricants, hydraulic fluids, and inks, in: *Bailey's Industrial Oil and Fat Products, Industrial and Nonedible Products from Oils and Fats*, 6th edn., Vol. 6, A. E. Bailey and F. Shahidi (eds.), John Wiley & Sons, Inc., Indianapolis, IN (2005).
85. VISCOPLEX® pour point depressants, http://oil-additives.evonik.com/product/oil-additives/en/products/viscoplex-pour-point-depressants/pages/default.aspx, accessed September 21, 2016.

Section II

Nanomaterials in Tribology

5 Polyethylene Nanotubes Formed by Mechanochemical Fragments and Displaying Positive Charge

Fernando Galembeck, Douglas Soares da Silva, Lia Beraldo da Silveira Balestrin, and Thiago Augusto de Lima Burgo

CONTENTS

5.1 INTRODUCTION

Triboelectricity is an important but challenging research topic that still does not benefit fully from well-established theories or models [1]. There is no consensus on the mechanisms for electrostatic charging, more than 25 centuries after the first reports on charge generation by friction, although many authors revisited this topic during the last 200 years [2]. Recent results on the triboelectricity of common organic polymers show that this is the outcome of mechanochemical reactions [3,4] producing ionic polymer fragments [5]. The formation of these fragments is explained following the

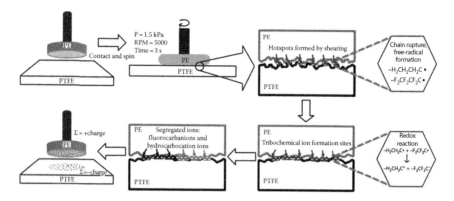

FIGURE 5.1 Schematic representation of the sequence of mechanochemical events triggered by mutual friction of PTFE and PE: chain rupture leads to free-radical formation that is followed by electron transfer from fluorocarbon to hydrocarbon fragments. These two types of polymer chains are immiscible, and they are thus segregated, forming adjacent macroscopic domains with excess positive or negative charge. (Reprinted with permission from Burgo, T.A.L. et al., *Langmuir*, 28, 7407, 2012.)

sequence of events shown in Figure 5.1, where the main steps are chain breaking under mechanical stress followed by electron transfer between fragments, driven by differences in electronegativity. In the case of mutually tribocharged polyethylene (PE) and polytetrafluoroethylene (PTFE), PE fragments are cationic while PTFE fragments are anionic. They form separated domains on tribocharged surfaces due to the immiscibility of the polymer chains [6]. Moreover, many solvents extract the charged fragments [7], allowing charge elimination from polymer surfaces by using a simple rinsing step. These recent findings are now opening new avenues for explaining and controlling hitherto challenging electrostatic phenomena in insulators, including tribocurrent developed at metal–insulator interfaces under relative motion [8].

Ionic fragments of PE and PTFE contain a hydrophobic chain and a charged chain end. They are thus amphiphilic, like most surfactants and polyelectrolytes. Amphiphilic substances often present a rich self-assembly behavior, forming many different structures [9]. Knowledge accumulated on this topic helped to explain challenging phenomena as, for instance, the insolubility of cellulose [10]. We may then expect the ionic fragments from hydrophobic polymers to display interesting self-assembly properties, too.

One expected behavior is burying the charged chain ends in between polymer hydrophobic chains and chain orientation at solid surfaces, avoiding exposure of the ionic chain ends to the external atmosphere and thus minimizing the solid–air interfacial tension. Charge occlusion or trapping is thus consistent with the long experimental half-lives of polymer electrostatic charge produced by friction, even under high relative humidity. In another context [11], it can also contribute to the rather long half-lives of electrostatic potential imparted by atmospheric ions deposited on polymers, following corona discharge [12,13]. A related behavior is observed with peroxy radicals [14] from hydrophobic polymers that are also amphiphilic polymer fragments showing surprisingly long half-lives.

Ethanol is an excellent solvent for the ions derived from PTFE and PE, and dry residues from ethanol extracts were observed by analytical transmission electron microscopy (TEM). Indeed, this property played an important role in the identification of PTFE fragments as the charge carriers in the negative domains formed when PTFE was tribocharged by mutual friction with PE.

Early results for PTFE fragments showed the kind of images usually obtained for disordered macromolecular solutes, resembling featureless liquid drops partially wetting the substrate [3].

Positive domains on PTFE were previously identified as formed by PE, using different analytical techniques: infrared and Raman spectroscopies together with pyrolysis [3]. For this reason, the use of analytical TEM was not necessary, recalling that this technique is rather complex and time consuming.

More recently, we decided to examine the substances extracted with ethanol from positive tribocharged domains by TEM. Surprisingly, the micrographs show self-arrayed material, completely different from the material extracted from the negative domains. The observed structures do not show similarity to any published images for PE systems, even though this polymer has been extensively examined by TEM, for many reasons. This article reports and discusses these results.

5.2 MATERIALS AND METHODS

5.2.1 MATERIALS

PTFE thick film, low density polyethylene (LDPE) foam, and high density polyethylene (HDPE) stubs were cut from sheets and polymer cylinders acquired from a local supplier (O Seringueiro, Campinas, Brazil) of materials for technical applications. Polymer identity was verified using infrared spectroscopy in reflectance mode.

5.2.2 TRIBOELECTRIFICATION

The sample used for X-ray photoelectron spectroscopy (XPS) measurements was prepared as follows: a square polymer slab of PTFE was fixed with a double-sided adhesive tape to an aluminum holder that was, in turn, mounted on a tabletop balance AM 5500 Automarte 10 mg resolution (Marte, Sao Paulo, Brazil), as shown in Figure 5.2. PTFE was rubbed with the lower face of a spinning HDPE stub, previously fixed on the chuck of a drilling tool spun at 5000 rpm for 3 s. Pressure applied by the spinning stub over the slab was 12 kPa, and it was calculated by dividing the vertical force (measured using the balance) by the area of the lower face of the stub.

Tribocharging for TEM was done by fixing a square polymer slab of PTFE onto an aluminum holder and rubbing it with LDPE foam held on the chuck of the drilling tool spun, using the same speed and time as earlier but at a lower pressure, 1.5 kPa.

5.2.3 XPS MEASUREMENTS

XPS measurements were done with a K-Alpha (Thermo Fisher, Waltham, MA) instrument using a 200 μm X-ray beam spot size and treating data with the Avantage 5.921 software from Thermo Fisher.

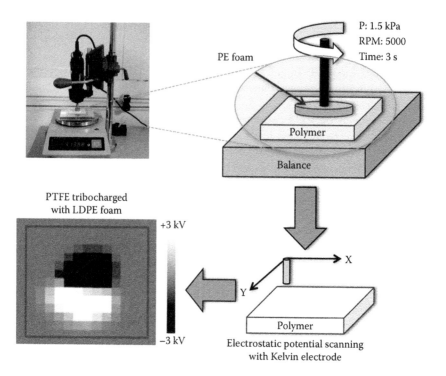

FIGURE 5.2 Tribocharging experiments are done by fixing a square sheet of one polymer to an aluminum holder mounted on a tabletop balance and rubbing the sheet with a disk of another polymer, fixed on the chuck of a drilling machine spun at 5000 rpm for 3 s. The pressure exerted by the spinning disk on the sample was defined by adjusting the vertical position of the drilling machine and measuring the force applied to the balance. (From Burgo, T.A.L. et al., *Langmuir*, 28, 7407, 2012.)

5.2.4 TRANSMISSION ELECTRON MICROSCOPY

Samples for bright-field micrographs were obtained by extracting the positive areas of PTFE sheared with PE foam, with ethanol. The extract was obtained by placing a 10 μL ethanol droplet over a positive PTFE area, followed by sucking the liquid with a micropipette and placing it on a coated microscope grid. The grid coating was a parlodion thin film that was in turn coated with an evaporated carbon film. Following evaporation under air, the samples were stored in a desiccator. TEM examination was done using a Libra 120 instrument (Carl Zeiss NTS GmbH, Oberkochen, Germany) operating at 120 kV.

5.3 RESULTS

The mutual abrasion of PTFE and PE produces an electrostatic charge whose distribution can be obtained by mapping the surfaces with a scanning Kelvin electrode. Typical electrostatic potential maps are shown in Figure 5.3.

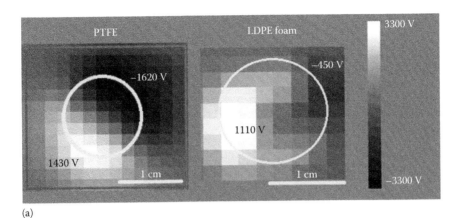

(a)

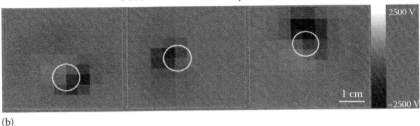

(b)

FIGURE 5.3 Electrostatic potential maps of (a) PTFE plate (left) rubbed with LDPE foam (right) and (b) three different PTFE plates rubbed with HDPE cylindrical stubs. Rectangles show the borders of each PTFE plate, while the circles show the rubbed areas. The three maps in (b) are from independent tribocharging experiments made using pristine samples. In every case, the charged area extends beyond the rubbed area.

5.3.1 XPS

XPS examination of the tribocharged HDPE allows detection of F and O compounds on the rubbed areas in HDPE. These areas are also identified due to damage visible in optical microscopy image, as shown in Figure 5.4.

The XPS maps are dominated by carbon, as expected for PE. The presence of fluorine residues evidences fluorocarbon transfer from PTFE to PE, during mutual shearing, while the presence of oxygen is assigned to the oxidation of PE surface layers and to the mechanochemical formation of peroxide and other compounds, under air. Most importantly, the areas containing higher amounts of F are those where visual change is most evident in the PE surface.

5.3.2 TRANSMISSION ELECTRON MICROSCOPY

TEM was also used to examine the contents of the material extracted from areas on tribocharged PTFE carrying excess positive charge. This was done on samples previously mapped for electrostatic potential distribution, choosing positive areas and extracting its contents with ethanol, as described in Section 5.2.

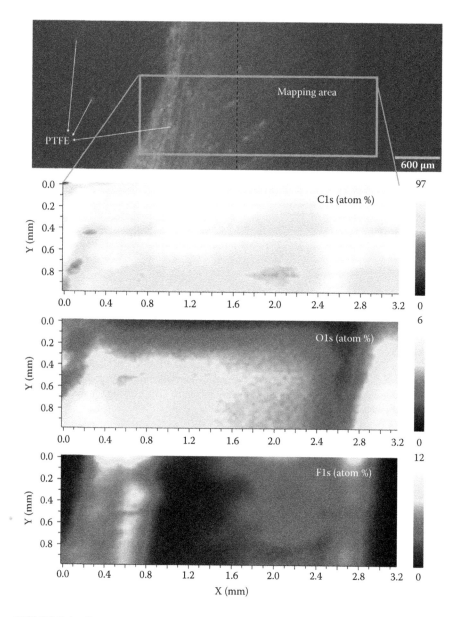

FIGURE 5.4 From top to bottom: Picture of the HDPE sample following rubbing with PTFE under well-defined conditions. Mechanical damage to the surface is visually observed; C, O, and F maps from the rubbed area. The grayscale bars to the right show the concentration of the mapped element in atom %. Note that XPS does not detect H atoms.

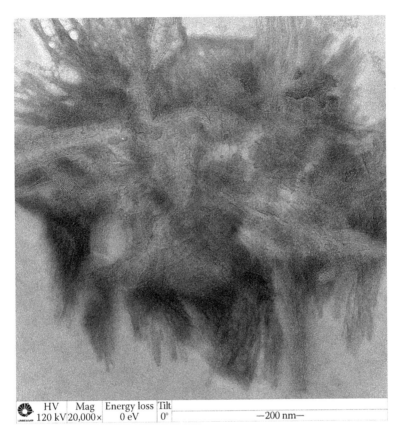

HV | Mag | Energy loss | Tilt
120 kV | 20,000× | 0 eV | 0° | —200 nm—

FIGURE 5.5 Bright-field transmission electron micrograph from the dry extract of a tribocharged positive area on PTFE.

The experimental procedure is possible due to the high stability of charge distribution on the tribocharged surfaces, which allows sufficient time for extracting charge carriers from the tribocharged materials.

A representative transmission bright-field micrograph is shown in Figure 5.5, and it shows broad areas covered with material similar to what is frequently seen when examining dry solutes.

This micrograph shows two interesting features: (1) bundles of parallel lines with *ca.* 4 nm spacing and *ca.* 2 nm line thickness and (2) agglomerated discs with *ca.* 4 nm outer diameter and *ca.* 2 nm wall thickness. These dimensions do not correspond to crystallographic features of PE, evidenced in wide-angle X-ray diffractograms. The 2 nm dimension is below the usual range for PE lamellae thickness that are usually within the 5–20 nm range, while 2 nm is the height of the noncrystalline part in otherwise ideal PE nanocrystals [15]. This is more evident in Figure 5.6, under higher magnification.

The examination of these samples under dark-field conditions shows that the areas showing the parallel lines do not diffract light; thus they are not part of

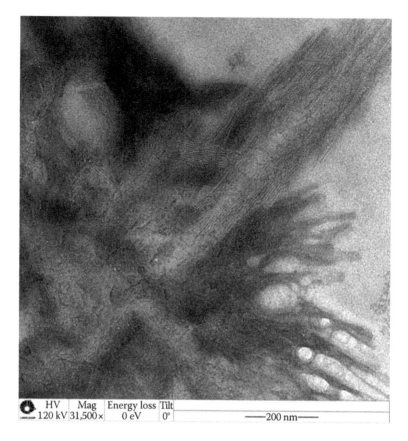

HV	Mag	Energy loss	Tilt
120 kV	31,500×	0 eV	0°

————200 nm————

FIGURE 5.6 Another bright-field transmission electron micrograph from the dry extract of a tribocharged positive area on PTFE but under higher magnification than in Figure 5.5.

crystalline domains. The bundled material shows some degree of cohesion, and it is typically separated from the adjacent dark areas devoid of morphology features. In many cases, the discs appear at the end of the bundled lines, raising the possibility that these well-organized structures are nanosized tubes made out of PE fragments.

There is no previous literature report on the formation of PE nanotubes, but in the present case, this may be understood considering the amphiphilic nature of the PE fragments. We propose that PE fragment chains acquire the usual trans conformation and they are thus prone to form the usual lamellar structure. However, positive chain ends prevent crystallization leaving the fragments in a noncrystalline state. Moreover, positive charges at chain ends repel each other, and this can be minimized by placing them well apart, along the nanotubes. The charged PE chain ends are expected to make a positive excess contribution to the polymer film surface energy, if they are in contact with the atmosphere. For this reason, the charged chain ends should be occluded by nonmodified chain fragments that protect them from contact with atmospheric molecules and ions. Based on these arguments, we propose that nanotubes are formed by PE chains arranged along the nanotube walls and they are extended to minimize electrostatic repulsion among their ends. The rigid nanotubes pack forming bundles due to geometric constraints. We may also hypothesize that a few negative-charge groups are located in between them, probably derived from atmospheric ions, either from water vapor or from gas ionization triggered by micro-discharges and triboplasma [16] formed during friction.

The formation of PE nanotubes is thus the outcome of three factors: the van der Waals attraction between the hydrophobic neighboring chains causing their close packing, the electrostatic repulsion between charged chain ends, and the tendency of the latter to avoid the outmost surface layer, locating themselves beneath hydrophobic polymer fragments.

If the structures observed in this work can be produced in significant amounts and if all the assumptions and explanations presented here are verified, this could provide a way to make cylindrical nanoelectrets that currently have no precedent and could prove useful for many applications, including for making charged tips for Kelvin microscopy and other useful devices.

5.4 CONCLUSION

The mutual friction of PTFE and LDPE produces PE nanotubes up to 400 nm long, soluble in ethanol and transferable to a microscope grid for examination by transmission electron microscope, where they appear to form bundles. The nanotube walls are noncrystalline, and the appearance of this unprecedented morphology arises from the amphiphilic character of the mechanochemical PE fragments that carry positive charge at the chain ends, in line with previously published results.

REFERENCES

1. M.W. Williams, What creates static electricity? *Am. Sci.* 100, 316–323 (2012).
2. C.A. Rezende, R.F. Gouveia, M.A. da Silva, and F. Galembeck, Detection of charge distributions in insulator surfaces, *J. Phys. Condens. Matter* 21(263002), 1–19 (2009).
3. T.A.L. Burgo, T.R.D. Ducati, K.R. Francisco, K.J. Clinckspoor, F. Galembeck, and S.E. Galembeck, Triboelectricity: Macroscopic charge patterns formed by self-arraying ions on polymer surfaces, *Langmuir* 28, 7407–7416 (2012).
4. L.B.S. Balestrin, D.D. Duque, D.S. Silva, and F. Galembeck, Triboelectricity in insulating polymers: Evidence for a mechanochemical mechanism, *Farad. Discuss.* 170, 369–383 (2014).
5. F. Galembeck, T.A.L. Burgo, L.B.S. Balestrin, R.F. Gouveia, C.A. Silva, and A. Galembeck, Friction, tribochemistry and triboelectricity: Recent progress and perspectives, *RSC Adv.* 4, 64280–64298 (2014).
6. H.G. Elias, *Macromolecules*, Plenum Press, New York, pp. 213–216 (1984).
7. K.R. Francisco, T.A.L. Burgo, and F. Galembeck, Tribocharged polymer surfaces: Solvent effect on pattern formation and modification, *Chem. Lett.* 41, 1256–1258 (2012).
8. T.A.L. Burgo and A. Erdemir, Bipolar tribocharging signal during friction force fluctuations at metal–insulator interfaces, *Angew. Chem. Int. Ed.* 53, 12101–12105 (2014).
9. J. Israelachvilli, Physical principles of surfactant self-association into micelles, bilayers, vesicles and microemulsion droplets, in: *Surfactants in Solution*, Vol. 4, K.L. Mittal (ed.), pp. 3–33, Springer, New York (1986).

10. M. Kihlman, B.F. Medronho, A.L. Romano, U. Germgård, and B. Lindman, Cellulose dissolution in an alkali based solvent: Influence of additives and pretreatments, *J. Braz. Chem. Soc.* 24, 295–303 (2013).

11. T.A.L. Burgo, C.A. Rezende, S. Bertazzo, A. Galembeck, and F. Galembeck, Electric potential decay on polyethylene: Role of atmospheric water on electric charge build-up and dissipation, *J. Electrostat.* 69, 401–409 (2011).

12. J.A. Giacometti, S. Fedosov, and M.M. Costa, Corona charging of polymers: Recent advances on constant current charging, *Braz. J. Phys.* 29, 269 (1999).

13. T.A.L. Burgo, L.B.S. Balestrin, and F. Galembeck, Corona charging and potential decay on oxidized polyethylene surfaces, *Polym. Degrad. Stabil.* 104, 11–17 (2014).

14. G.G. De Barros and F. Galembeck, Polytetrafluoroethylene–iron oxide composite. Interactions with acrylic acid and vinyl acetate, *J. Polym. Sci., Part A, Polym. Chem.* 25, 2369–2383 (1987).

15. A. Osichow, C. Rabe, K. Vogtt, T. Narayanan, L. Harnau, M. Drechsler, M. Ballauff, and M. Mecking, Ideal polyethylene nanocrystals, *J. Am. Chem. Soc.* 135, 11645–11650 (2013).

16. G. Heinicke, *Tribochemistry*, Carl Hanser, Berlin, Germany (1984).

6 Nanotechnology and Performance Development of Cutting Fluids

The Enhanced Heat Transfer Capabilities of Nanofluids

Nadia G. Kandile and David R. K. Harding

CONTENTS

6.1 INTRODUCTION

Lubrication primarily involves three basic tribological processes—friction, wear, and adhesion. Many types of lubricants exist, including ionic liquids [1] to reduce friction, wear, and adhesion down to the nanoscale level in equipment with moving parts and/or used in the machining of materials. Of critical importance, however, is also reducing friction, wear, and adhesion during the manufacture of equipment. In other words, friction, wear, and adhesion must be kept to a minimum during equipment manufacture. Heat removal is an integral part of manufacturing. The use of lubricating cutting fluids is an ideal way to achieve these goals. Nanocutting fluids have achieved a definite place in this area of nanotechnology.

6.1.1 NANOTECHNOLOGY

Nanotechnology is defined by the National Nanotechnology Initiative as *the manipulation of matter with at least one dimension sized from 1 to 100 nanometers* [2]. Nanotechnology involves the manipulation of matter at the atomic and molecular levels ranging from the research level to the manufacturing level. Hence, the evolution of nanocutting fluids has produced solutions to the health and safety needs through the use of minimum quantity lubrication (MQL) in cutting procedures [3]. Huge investments have been made in Europe, the United States, and Japan alone in nanotechnology [2]. Nanotechnology involves surface science, organic chemistry, molecular biology, semiconductor physics, and microfabrication [2]. Medical applications stand to gain immensely from nanotechnological research efforts.

Nanotechnology has given us the ability to consider the smaller-to-larger (bottom-up) approach as opposed to the larger-to-smaller (top-down) modeling approach to produce smaller nanoparticles. The range of current applications for nanotechnologies is vast. Analytical nanotechniques used include atomic force microscopy (AFM), scanning tunneling microscopy (STM), scanning confocal microscopy (SCM), and scanning acoustic microscopy (SAM)—all are scanning probe techniques [4]. Nanolithography, a branch of nanotechnology, is concerned with the study and application of fabricating nanometer-scale structures, meaning patterns with at least one lateral dimension between 1 and 100 nm. Lithography also supports nanotechnology with the following techniques: optical lithography, X-ray lithography, dip pen nanolithography, electron beam lithography, nanoimprint lithography, deep ultraviolet lithography, focused ion beam machining, atomic layer deposition, and molecular vapor deposition. All these techniques support the production of nanotubes, nanowires, and other nanomaterials. Of course, nanotechnology brings with it toxicity and safety concerns related to the manufacturing and application of nanodevices.

6.1.2 NANOTECHNOLOGY, NANOFLUIDS, AND NANOCUTTING FLUIDS

Due to nanotechnology, highly efficient thermal conducting fluids called "nanofluids" have emerged. Nanofluids are engineered colloidal suspensions of nanoparticles (10–100 nm) in base fluids [5]. The applicability of these fluids to act as coolants is mainly due to the enhanced thermophysical properties resulting from nanoparticle inclusion [6,7]. Nanofluids are not without controversy, and this is possibly related to the underestimated system complexity and the presence of the solid–liquid interface. The large surface area of the nanoparticles produces boundary layers between nanoparticles and the liquid and contributes significantly to the fluid properties, resulting in a three-phase system. The approach to nanofluids as three-phase systems (solid, liquid, and interface) instead of the traditional two-phase systems (solid and liquid) allows for a deeper understanding of correlations between the nanofluid parameters, properties, and cooling performance [8].

The use of nanofluids as coolants will allow for smaller size and better positioning of radiators. Owing to the fact that there would be less fluid due to the higher efficiency, coolant pumps could be shrunk and truck engines could be operated at higher temperatures, allowing for more horsepower while still meeting stringent emission standards.

The use of high thermally conductive nanofluids in radiators can lead to a reduction in the frontal area of the radiator by up to 10%. This reduction in aerodynamic drag can lead to fuel savings of up to 5%. The application of nanofluids has also contributed to the reduction of friction and wear, parasitic losses, and the operation of components such as pumps and compressors and subsequently leads to more than 6% fuel savings. It is conceivable that greater fuel savings could be obtained in the future [9].

Nanofluids can be classified into two categories [5]:

1. Metallic nanofluids that contain elemental metallic nanoparticles for example copper (Cu), iron (Fe), gold (Au), and silver (Ag)
2. Nonmetallic nanofluids, that is, those that contain nonmetallic nanoparticle compounds such as aluminium oxide (Al_2O_3), copper oxide (CuO), silicon carbide (SiC), and carbon nanotubes (CNTs)

Currently, the most prevalent cooling fluids are water, ethylene glycol (EG), and engine oil, which have much lower thermal conductivities than many solids, including metals, such as silver, copper, and iron, or nonmetallic materials, such as aluminium CuO, SiC, and CNTs. This fact was the starting point of an idea, which led to the creation of mixtures of solid and fluid in order to improve thermal conductivity and, consequently, improves heat transfer performance [10].

Maxwell was the first to initiate the use of small solid particles in fluids to increase thermal conductivity. His idea was based on suspending milli- or microsized solid particles in the fluids [11]. However, the larger milli- or even microsized particles caused several technical problems such as [12]

1. Faster settling time
2. Clogging of microchannels of devices
3. Abrasion of the surface
4. Erosion of pipelines
5. Large pressure drops

The advantages of nanosized particles over microparticles are as follows [13]:

1. Longer suspension time (greater stability), which leads to lower erosion and clogging of channels, especially in microdevices
2. Much greater surface areas due to the number of surface atoms versus internal atoms; larger surface area/volume ratios (e.g., 1000 times larger), which allow for increased thermal conductivity of the particle
3. Lower demand for pumping power, one form of energy saving
4. Reduction in the machining shop's inventory of heat transfer fluids

From a heat transfer point of view, various results with great disparities have been reported in recent years. For instance, it has been claimed that improving the thermal transport properties of nanofluids would have several advantages, and the most important ones are summarized as follows [14]:

1. Improvement in the efficiency of heat exchange
2. Reduction in the size of the system
3. Provision of much greater safety margins
4. Reduction in costs

Machining is one of the most critical processes in manufacturing that involves the controlled removal of material from the substrate during action with a cutting tool. Since machining involves the plastic deformation of the workpiece material and friction between the tool-and-chip and tool-and-workpiece interfaces, a large fraction of the energy supplied is converted into heat. During the machining of low-strength alloys, heat generation is low, but when ferrous and other high-strength alloys are machined, a lot of heat is generated, which increases proportionally with the cutting speed. Many techniques have evolved for the effective removal of heat from the vicinity of the machining area. One of the techniques is the application of a coolant in the form of a fluid during the machining process. Although high-speed machining is desirable, in many cases for higher productivity, the consequences of heat generation need to be minimized.

For many years, coolants, popularly known as metalworking fluids (MWFs), were successfully employed for the removal of heat during machining. However, it was realized that some MWFs can cause serious damage to the environment and to the health of the operator. It was shown by Klocke and Eisenblatter [15] that cutting fluids also created a waste disposal problem and added to the cost of manufacturing. These negative consequences of cutting fluids prompted researchers to switch to technologies that involved minimum use of cutting fluids [15].

Many alternatives have been explored to minimize the quantity of cutting fluid used during machining. Examples of such techniques include

1. Dry machining
2. Cryogenic cooling
3. Coated tools
4. Minimum quantity lubrication

Cutting fluids are the best choice to serve as both a cooling medium and a lubricant to improve the quality of the operations involved in machining and thereby increase productivity [16].

Cutting fluids are used in machining to improve the tribological characteristics of the machining process and also to dissipate the heat generated. Several types of cutting fluids have been explored. Some of these employ petroleum-based metal cutting fluids and vegetable oil–based metal cutting fluids. Conventional cutting fluids are mostly petroleum based, the continued application of which creates problems, such as (1) environmental pollution, for example, greenhouse gas production and cutting fluid disposal, and (2) negative impacts on the health of operators.

The main functions of cutting fluids can be summarized as follows:

1. Cooling the extremely hot cutting zone by taking away some of the heat generated
2. Reduction of the coefficient of friction between the tool-and-chip and tool-and-workpiece interfaces so as to lower the heat generated due to frictional effects, thus lubricating the surfaces
3. Reduction of the thermal distortion of the workpiece by lowering the thermal gradient
4. Removal of chips from the machining zone, thereby facilitating the disposal of chips
5. Protection of the finished workpiece surface from corrosion and rust formation

Some highly desirable properties for ideal cutting fluids are as follows:

1. Good lubrication
2. High cooling capacity
3. Low viscosity to provide free flow of the cutting fluid
4. Chemical stability
5. Noncorrosive
6. High flash point to reduce fire risks
7. Allergy-free and nontoxic to operators
8. Lower evaporation rate to reduce emissions
9. Low cost

The type of cutting fluid to be used depends on its application. The choice of cutting fluid can comprise the machining operation. Choices depend on one or more of the following: water, oils, oil–water emulsions, pastes, gels, sprays, and air or other gases such as carbon dioxide. Cutting fluids can contain petroleum distillates, animal fats, plant oils, and/or other raw ingredients. While cast iron and brass are machined dry, other metals require cutting fluids in order to reduce temperatures, reduce wear on the machining contact points (blade, tip), reduce the health risk of dry powders resulting from dry machining, and reduce rust formation on all parts in the machining operation.

Cutting fluids can be classified into four primary categories based on their thermophysical properties, the cutting process, and the method of application:

1. Cutting oils are made up entirely of mineral or vegetable oils and are used primarily for operations where lubrication is required [17]. Chemical additives like sulfur improve oil lubricant capabilities. Areas of application depend on the properties of the particular oil that is used, particularly for heavy cutting operations on tough steels.
2. Soluble oils are mixtures of oil and water and have increased cooling capabilities over straight oils and offer some rust protection. They are the most common, cheapest, and effective form of cutting fluids consisting of oil droplets suspended in water with a typical water-to-oil ratio of 30:1. Emulsifying agents are also added to promote the stability of the emulsion. For heavy-duty work, extreme-pressure additives are used. Oil emulsions are typically used for aluminium and copper alloys.
3. Chemical cutting fluids possess good flushing and cooling abilities. They tend to form more stable emulsions but may have harmful effects on the skin. They can be classified as follows:
   a. Semisynthetics, which are similar to soluble oils in performance characteristics but differ in composition because 30% or less of the total volume of the concentrate contains inorganic or other compounds that dissolve in water. Semisynthetics have better maintenance characteristics than soluble oils but do contaminate easily when exposed to other machine fluids and may pose a dermatitis risk to workers.
   b. Synthetics are chemical fluids that contain inorganic or other chemicals dissolved in large amounts of water and offer superior cooling performance.
   c. Maintenance is also not a major issue with synthetics; however, cases of dermatitis are more prevalent in workers, and the lubrication functionality is weaker than with semisynthetics [18].
4. The selection of the type of nanocutting fluid is highly dependent on the end use. Nanocutting fluids are mixtures of a conventional cutting fluid and nanoparticles [19]. In addition to thermal control in the machining process, cutting fluids are used to remove particulates from the machining area.

Nanocutting fluids and their performance in thermal control allow for MQL. In other words, nanocutting fluids have led us to MQL and presented a viable way to maintain liquefied lubrication without using vast amounts of fluid.

Various strategies have been reported to improve MWF performance sustainability [20,21]. Biocide-free metal cutting fluids when compared to conventional mineral oil–based fluids were found to reduce machining forces and workpiece surface roughness. MQL applications of MWFs have been applied to reduce environmental and health impacts [20] and enhance penetration into the cutting zone [21–28]. Studies have shown that nanoparticle additives decrease friction and wear and reduce cutting forces while improving thermal conductivity and machining performance compared to nanoparticle-free MWFs [25–27]. Thus, advantages of using such nanocutting fluids include the reduction of energy and fluid consumption, reduction of related economic and environmental impacts, and enhancement of cutting tool life [29].

In summary, nanocutting fluids have an important role to play in the machining of solid materials. Many studies have shown that adding nanoparticles to cutting fluids can improve their efficiency in terms of friction and wear and thermal conductivity [30]. This role will be expanded in the following sections of this chapter.

The present chapter will discuss only the enhancement in the heat transfer of conventional cutting fluids by incorporating nanomaterials.

6.2 EXPERIMENTAL

6.2.1 NANOPARTICLES, THEIR PROPERTIES AND THEIR USE IN THE PREPARATION OF NANOFLUIDS

Nanoparticles have attracted considerable interest in recent years because of their excellent physical and chemical properties. Recently, inorganic nanoparticles have been extensively investigated in the field of tribology [31–34]. Due to the unique nature of nanoparticles such as small size effect, macroquantum tunnel effect, large surface area, and high activity, as well as isotropy, low melting point, shear strength, and good thermal stability, nanoparticles as lubricant additives have been found to possess an excellent tribological performance with self-repairing properties and environment friendliness [35–38].

Another important characteristic of nanoparticles is their fast Brownian motion at the nanoscale, which is due to the fact that the random motion of nanoparticles transports energy directly from nanoparticles to the suspending fluid. The particles provide a large surface area for molecular collisions from the higher momentum of particles when they are used at higher mass concentrations. The transfer of thermal energy is at a greater distance inside the base fluid before it is released on the colder side [39]. In addition, the microconvection effect, which is due to fluid mixing around these particles, plays a significant role in heat transfer enhancement [40]. It has been shown that the thermal conductivity of fluids can be enhanced through the addition of nanoparticles due to this feature [41].

Nanoparticles can be produced using several methods [42,43], which can be categorized as follows:

1. Transition metal salt reduction [44,45]
2. Thermal decomposition and photochemical methods [46–48]
3. Ligand reduction and displacement from organometallics
4. Metal vapor synthesis
5. Electrochemical synthesis [49,50]

Methods to separate particles by size from a colloidal suspension, which contains nanoparticles of various sizes, are as follows:

1. Precipitation, which is suitable for a wide distribution of colloid nanoparticles in the suspension.

2. Centrifugation and gel filtration, which are well suited for suspensions of colloidal nanoparticles with a narrow size distribution.
3. Gel electrophoresis, which is suitable for separating nanoparticles, takes advantage of the difference in the charge density of the particles and is also suitable for separating particles with a small cluster size.
4. A combination of these various methods might also prove advantageous. However, a problem with sorting the various-sized nanoparticles using these methods is that only a fraction of the nanoparticles of a given size may be collected and then only in small quantities. The digestive ripening method and high-temperature melting technique have been proposed to resolve this problem [51].

The actual physical and chemical composition of the nanoparticles is, of course, of paramount importance. This then leads to the following issues that need to be considered when preparing nanocutting fluids [52]:

Application: The widest use of nanoparticles is in cutting fluids [53]. The particular application in turn defines the issues that need to be addressed. The choices include metals, metal oxides, metal–organic composites, and organic-only nanoparticles. The application of nanoparticles in cutting fluids defines the criteria that need to be addressed.

Size and shape: The size range of nanoparticles can be varied. For example, Au nanospheres have been prepared with diameters ranging from 9 to 99 nm [52].

Heat transfer improvements by the addition of nanoparticles to cutting fluids have been known for some time [54–58]. The review by Saidur et al. [59] concluded the following:

1. Nanorefrigerants had higher heat transfer capabilities than the base fluid alone.
2. Refrigerants containing CNTs had greater heat transfer capabilities than the fluid alone. Improved heat transfer capabilities ranged from 21% to 276% above that of the fluid alone.
3. TiO_2 gave a 21.6% better heat transfer than a hydrofluorocarbon (HFC-134a) refrigerant or polyester (POE) oil.
4. In another study, 6% Al_2O_3 increased the heat transfer efficiency of an ethylene glycol–water system from 78.1% to 81.11%.
5. In any refrigeration cycle, there are pressure drops. Studies indicate a greater pressure drop for the nanorefrigerants over the base fluid based on the increase in viscosity.
6. They also identified inconsistencies between reports from different research groups. They concluded with the note that much more needs to be done in the area of nanofluids to achieve a better understanding of heat transfer mechanisms in order to fully benefit from the coolant advantages of suspended nanoparticles.

The long-term stability of nanoparticle suspensions is important for all applications, although any particular application may allow for some flexibility. Stability will also depend on the carrier fluid and/or the environment of the nanoparticle. Surfactants can be added to the nanofluid to maintain an even dispersion of suspended nanoparticles. The surface modification of the nanoparticle can also be employed to prevent clustering, also called aggregation, of the nanoparticle into larger particles. Lazzarri et al. [60] used dynamic light scattering to compare the stability of spherical 100–200 nm nanoparticles of polylactic acid (PLA) or polymethyl methacrylate (PMMA) in salt solutions, biological fluids, and homogenates. They found that the PMMA nanoparticles were stable in all fluids, whereas the PLA nanoparticles aggregated in gastric and spleen homogenates [60]. Aggregation is one major result of the lack of nanoparticle stability.

Aggregation, the collecting or self-assembly of individual nanoparticles into a distinct larger mass, can result from the increased surface area for a given weight/volume of nanomaterial. Aggregation, especially if leading to sedimentation, will reduce the effective surface area. This will have a significant effect on nanocutting fluidics. Philip et al. conducted experiments with a Fe_3O_4 nanofluid where they manipulated particle aggregation structures with variable strengths of external magnetic fields. They found that a stronger magnetic field yielded longer particle structures, which caused more aggregation in the nanofluids. This in turn results in a large thermal conductivity enhancement (up to 300% at a volume fraction of 0.82%) [61]. This enhancement provides evidence that increasing particle aggregation can play a significant role in increasing thermal conductivity enhancement [62].

The aggregation of nanoparticles can be reduced by using a capping agent. Capping is a very common procedure in many aspects of the modification of materials, especially silver nanoparticles and gold nanoparticles. The procedure is based on the concept of blocking (capping) some or all chemically/physically reactive surface sites. Capping is carried out in order to improve performance characteristics or to allow selected surface modifications. Capping can be induced during preparation or can happen circumstantially, that is, resulting from impurities in the preparation of the nanoparticles. In the wider world of nanoparticle chemistry, capping is used for a number of different reasons. These reasons range from catalysis to poisoning (the direct opposite of catalysis), in addition to preventing aggregation, especially in nanocutting fluids [63].

Figure 6.1 shows an example of a nanoparticle spectrum [63]. This spectrum ranges from the naked reactive nanoparticle (*activity promotion*) through to a heavily

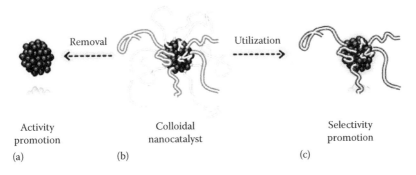

Removal   Utilization

Activity   Colloidal   Selectivity
promotion   nanocatalyst   promotion

(a)   (b)   (c)

FIGURE 6.1   A schematic showing (a) naked nanocatalyst, (b) heavily capped nanocatalyst, and (c) purpose-designed capped nanocatalyst. (From Niu, Z. and Li, Y., *Chem. Mater.*, 26, 72, 2013.)

capped nanoparticle (*colloidal nanocatalyst*). Either can be modified to produce a nanoparticle (*selective promotion*) with selected purpose(s) in mind. Figure 6.2 [63] presents a selection of common capping agents that have been used for the modification of nanoparticles. The structures clearly illustrate the range of options available for nanoparticle surface modification. These options vary from highly polar (e.g., PEI, PAMAM) to very hydrophobic (e.g., OAM, OA).

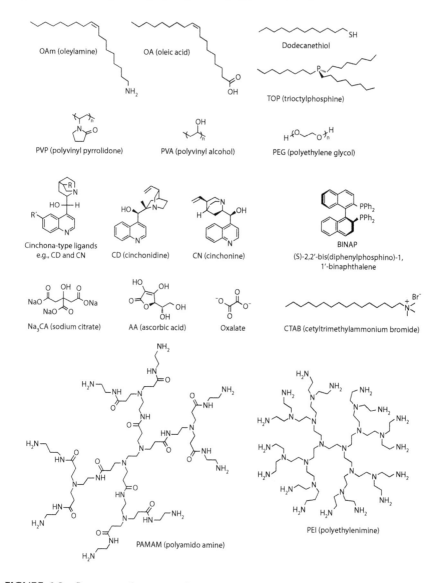

FIGURE 6.2 Structures of representative capping agents in nanoparticle synthesis, including long-chain hydrocarbons, polymers, chiral ligands, polycarboxylic acids, polyhydroxy compounds, cationic surfactant, and dendrimers. The capping agents shown illustrate the range of polarities available for capping nanoparticles. (From Niu, Z. and Li, Y., *Chem. Mater.*, 26, 72, 2013.)

Heat transfer is the exchange of thermal energy between physical systems, depending on the temperature and pressure, by dissipating heat. The fundamental modes of heat transfer are conduction or diffusion, convection, and radiation. Many industrial processes involve facilitated heat transfer, which is accomplished using heat transfer fluids such as water, EG, and engine oil. Thermal properties of these fluids determine the thermal efficiency as well as the size of the equipment needed. Hence, many different techniques are being employed to improve the thermal properties of these fluids, especially thermal conductivity [64].

Heat transfer studies with nanofluids are not just confined to cutting fluids. For the time being, nanofluids play an important role in heat pipes to increase the heat transfer compared to conventional fluids.

There are several methods to improve the heat transfer efficiency. Some methods are as follows:

1. Utilization of extended surfaces
2. Application of vibration to the heat transfer surfaces
3. Usage of microchannels

Heat transfer efficiency can also be improved by increasing the thermal conductivity of the working fluid. Commonly used heat transfer fluids such as water, EG, and engine oil have relatively low thermal conductivities when compared to the thermal conductivity of solids. The high thermal conductivity of solids can be used to increase the thermal conductivity of a fluid by adding small solid particles to that fluid. The feasibility of the usage of such suspensions of solid particles with sizes of the order of 2 mm or micrometers was previously investigated by several researchers, and the following significant drawbacks were observed [64]:

1. The particles settle rapidly, forming a layer on the surface and reducing the heat transfer capacity of the fluid.
2. If the circulation rate of the fluid is increased, sedimentation is reduced, but the erosion of the heat transfer devices, pipelines, etc., increases rapidly.
3. The larger-size particles tend to clog the flow channels, particularly if the cooling channels are narrow.
4. The pressure drop in the fluid increases considerably.
5. Finally, conductivity enhancement based on particle concentration is achieved (i.e., the greater the particle volume fraction is, the greater the enhancement and greater the problems, as indicated earlier).

Saidur et al. reported on the use of nanoparticles as refrigerants [59]. They reviewed the heat transfer or heat conduction capability of a number of nanoparticles and presented summaries as show in Figure 6.3.

Figure 6.3 illustrates that the heat conductivity (transfer) capabilities of the metals and nonmetallic oxides far exceed that of water and organic materials. They also compared metallic, nonmetallic, and CNT nanoparticles in various base fluids and noted that thermal conductivity/transfer improvements ranged from 15% to 150%.

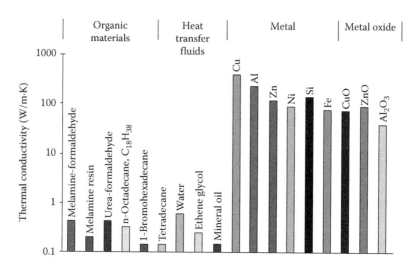

FIGURE 6.3 Common organic, fluid, and inorganic materials used in nanofluid heat control. (From Saidur, R. et al., *Renew. Sustain. Energy Rev.*, 15, 310, 2011.)

Several mechanisms of heat transfer augmentation have been proposed by various researchers. Sergis and Hardalupas [65] evaluated 131 published references based on the statistical analysis of mechanisms for heat transfer enhancement. They found that most of the researchers (33%) attribute the augmentation to Brownian motion, followed by the interfacial layer theory (23%), the shear thinning behavior of flows (5%), aggregation and diffusion (2%), and other mechanisms. However, some of the proposed theories are physically inconsistent. Prasher et al. [66] proposed an internally consistent physical theory to explain heat transfer augmentation of nanofluids. Basically, they employed very basic principles and explain that most nanofluid behavior is not strange or anomalous at all. While their approach may or may not be correct, it seems more plausible and physically consistent when compared with many other explanations and theories.

Nanoparticles with magnetic characteristics typically contain cobalt, nickel, or iron as stand-alone ions or inorganic metallic complexes. Applications of magnetic nanoparticles include medical diagnostics, medical therapies, immunoassays, wastewater treatment, organometallic complexes, free and immobilized catalysts, biomedical imaging, information storage, and genetic engineering.

The Argonne National Laboratory (United States) has pioneered the concept of nanofluids by applying nanotechnology to thermal engineering. Nanofluids are a new class of solid–liquid composite materials consisting of solid nanoparticles (in the range of 1–100 nm) or CNTs, dispersed in a heat transfer fluid such as EG, water, or oil [67,68].

Nanofluids have been considered for applications as advanced heat transfer fluids for almost two decades. However, due to the wide variety and the complexity of the nanofluid systems, no agreement has been achieved on the magnitude of potential benefits of using nanofluids for heat transfer applications. Compared to conventional solid–liquid suspensions for heat transfer intensifications, nanofluids having properly dispersed nanoparticles possess the following advantages:

1. High specific surface area and therefore produce greater heat surface transfer between particles and fluids
2. High dispersion stability with the predominant Brownian motion of particles
3. Reduced pumping power as compared to pure liquid to achieve equivalent heat transfer intensification
4. Reduced particle clogging as compared to conventional slurries, thus promoting system miniaturization
5. Adjustable properties, including thermal conductivity and surface wettability, by varying particle concentrations to suit different applications

There are two methods for the preparation of nanofluids using nanoparticles: (1) the single-step and (2) the two-step method to enhance heat transfer, control the thermophysical properties, and ensure the stability of the suspension.

The top-down (breakdown) approach of preparing nanoparticles involves reducing the size of existing large particles (the *buildings*) to nanosized particles (the *brick dust*) in a similar fashion as destroying an existing building. Hence, the nano-building materials (nano-building bricks) are generated by drilling, grinding, and crushing and may also involve chemical treatment as well prior to their assembly into nanoparticles. The bottom-up approach involves preparing the nanobricks, starting with a liquid- or solid-phase approach, prior to assembly into nanoparticles.

The single-step preparation process means the synthesis of nanofluids in one step. Several single-step methods have been developed for nanofluid preparation. Akoh et al. [69] developed a single-step direct evaporation method, known as Vacuum Evaporation onto a Running Oil Substrate (VEROS), but it was difficult to isolate nanoparticles from the fluids. Eastman et al. [70] developed a modified VEROS technique, in which Cu vapor was directly condensed into nanoparticles by contact with flowing low-vapor-pressure EG. Zhu et al. [71] presented a single-step chemical process for the preparation of Cu nanofluids by reducing $CuSO_4 \cdot 5H_2O$ with $NaH_2PO_2 \cdot H_2O$ in EG under microwave irradiation. This method also proved to be a good way to produce mineral oil–based silver nanofluids. Lo et al. [72] developed a vacuum-based submerged nanoparticle synthetic method to prepare CuO, Cu_2O, and other Cu-based nanofluids with different dielectric liquids. A suitable power source was required to produce an electric arc between 6,000 °C and 120,000 °C, which melts and vaporizes the metal rod in the region where the arc was created. The vaporized metal was condensed and then dispersed with deionized water to produce nanofluids. An advantage of the one-step synthesis method is that nanoparticle agglomeration is minimized. However, a prime problem of the one-step process is that only low-vapor-pressure fluids can be prepared with such a process.

The two-step preparation of nanofluids begins with the preparation of the nanomaterial in the first step, followed by the direct mixing of the base fluid with the nanomaterials in the second. The two-step preparation process is widely used in the synthesis of nanofluids by mixing base fluids with commercially available nanopowders obtained from different mechanical, physical, and chemical routes such as milling, grinding, sol-gel, and vapor phase methods. An ultrasonic vibrator or a higher-shear mixing device is generally used to stir nanopowders with host fluids. The frequent use of ultrasonication or stirring is required to reduce particle agglomeration. Eastman et al. [70], Lee et al. [73], and Wang et al. [74] used the two-step method to produce aluminium nanofluids. Murshed et al. [75] prepared a TiO_2–water nanosuspension by the same method. Xuan and Li [76] used commercially available Cu nanoparticles to prepare nanofluids of both water and transformer oil. Kwak and Kim [77] used a two-step method to prepare CuO-dispersed EG nanofluids by sonication without stabilizers. The two-step method can also be used for the synthesis of CNT-based nanofluids. Single-walled carbon nanotube (SWCNT) and multiwalled carbon nanotube (MWCNT) were first produced by a pyrolysis method and then suspended in base fluids with or without the use of surfactants [67,68,78]. Some authors suggested that the two-step process is more suitable for preparing nanofluids containing oxide nanoparticles than those containing metallic nanoparticles [70]. Stability is a big issue that is inherently related to this operation as the powders easily aggregate due to strong van der Waals forces among nanoparticles. In spite of such disadvantages, this process is still popular as it is the most economical process for nanofluid production.

The two-step process is commonly used for the synthesis of CNT-based nanofluids. SWCNTs and MWCNTs are cylindrical allotropes of carbon. SWCNTs consist of a single cylinder of graphene, while MWCNTs contain multiple graphene cylinders nesting within each other [79–81]. The CNTs are usually produced by a pyrolysis method and then suspended in a base fluid with or without the use of a surfactant.

Some authors suggest that the two-step process works well only for nanofluids containing oxide nanoparticles dispersed in deionized water as opposed to those containing heavier metallic nanoparticles [67,71,79]. Since nanopowders can be obtained commercially in large quantities, some economic advantage exists in using two-step methods that rely on the use of such powders.

Several studies, including the earliest investigations of nanofluids, used a two-step method in which nanoparticles or nanotubes were first produced as a dry powder and then dispersed into a fluid in a second processing step. In contrast, the one-step method entails the synthesis of nanoparticles directly in the heat transfer fluid.

Nanofluids represent the cutting-edge technology of liquid coolants where the heat transfer properties of conventional base fluids are enhanced by the addition of nanoparticles in the form of stable dispersions. The higher thermal conductivity of heat transfer nanofluids allows for higher efficiency, better performance, and reduced costs. The single-step synthesis method overcomes the drawbacks of two-step synthesis such as the agglomeration of particles during storage, transportation, or redispersion, which results in poor thermal conductivity. Single-step methods could be physical methods like the direct evaporation technique and submerged nanosynthesis or chemical methods like solution phase reduction, the polyol method, and the microwave method [68,71,82]. A review of the literature shows there are only a few examples of the single-step chemical method of nanofluid synthesis.

The following nanofluids have been successfully prepared by the one-step method:

1. Cu/EG nanofluids prepared by a vapor condensation method [78].
2. Dielectric nanofluids via the vacuum submerged arc nanoparticle synthesis system [72,82]. This method reportedly avoids aggregation quite well.
3. $CuSO_4 \cdot 5H_2O$ reduced in EG with $NaPO_2$ using microwave energy [71].
4. Ag/ethanol nanofluids using polyvinylpyrrolidone stabilization using microwaves [83,84].
5. Graphene oxide nanosheets can be phase transferred with the cationic oleyldiamine [85] (Figure 6.4). This phase transfer may well allow the removal of at least some impurities.

Yu and Xie [86] discussed the following nanofluid production methods:

1. A continuous system for producing CuO nanofluids [85] using ultrasound and microwaves [87]
2. Au and Ag nanoparticle production in a liquid phase transfer system that delivers a nanofluid of the metal complexed with dodecylamine in cyclohexane [88,89]
3. Kerosene/Fe_3O_4 nanofluids with oleic acid [104] using a phase transfer process [90]

Nanofluids have been considered for applications as advanced heat transfer fluids for almost two decades. However, due to the wide variety and the complexity of the nanofluid systems, no agreement has been achieved on the magnitude of potential benefits of using nanofluids for heat transfer applications. While many nanocutting

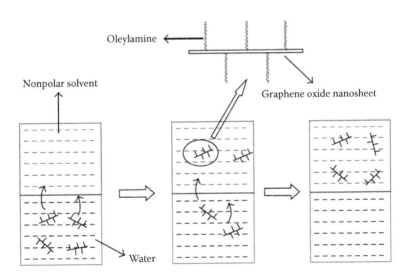

FIGURE 6.4 Use of oleyldiamine for graphene oxide nanosheet phase transfer. (From Yu, W. et al., *Nanoscale Res. Lett.*, 6, 47, 2011.)

fluids may be prepared using standard nanofluid preparative procedures, two examples are presented here.

Khandekar et al. [19] prepared a nanocutting fluid by the two-step method. They added 1% of in-house-prepared Al_2O_3 to Servocut S followed by 2 h ultrasonication. Servocut S is a commercial rich milky water-based emulsion cutting oil with "a special emulsifier" to produce the complete dispersion of the Al_2O_3 that "does not split" during use [91]. From this study [19], it was reported that adding of 1% Al_2O_3 nanoparticles (by volume) to the conventional cutting fluid greatly enhances its wettability characteristics compared to pure water and other conventional cutting fluids. Also, the great reduction of crater and flank wear was attributed to enhanced thermal properties, the improvement in wettability, the lubricating characteristics of the nanocutting fluid, and a reduction of 50% and 30% in the cutting force while machining with nanocutting fluids [19]. This stable preparation did not result in any sedimentation during processing. They compared their results with dry cutting and conventional cutting fluids. Their study covered wettability, tool wear, cutting force, chip morphology, chip thickness, and surface roughness. Figure 6.5 compares the surface roughness results.

Their overall results can be summarized as follows:

1. 1% Al_2O_3 added to Servocut S enhanced wettability.
2. 1% Al_2O_3 added to Servocut S reduced crater and flank wear.
3. 1% Al_2O_3 added to Servocut S produced 50% reductions in cutting force over dry cutting and about 30% reduction over Servocut S used alone.
4. Similar reductions were seen for the respective surface roughness study.

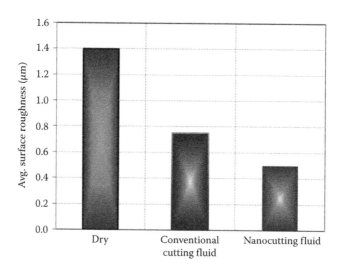

FIGURE 6.5 Comparison of surface roughness after use of a dry abrasive (no fluid), a conventional cutting fluid and nanocutting fluid. (From Khandekar, S. et al., *Mater. Manuf. Proc.*, 27, 963, 2012.)

Gu et al. [92] prepared TiO_2 nanoparticles from $TiOSO_4$ and a linear alkylbenzene sulfonate (LAS, sodium dodecyl benzene sulfonate) in solution, followed by precipitation with urea. The dried powders were then dual coated with a silane coupling agent (KH-570, gamma-methacryloxypropytrimethoxysilane) and OP-10 (the 10-mol ethoxylate of p-octylphenol, also called an octylphenol polyoxyethylene {10} ether) in sequence in order for them to be dispersed stably in water as lubricant additives. The resulting molecule contains active groups that can interact chemically with both the inorganic and organic substances, can couple organic substances and inorganic substances, and can greatly improve electrical properties, resistance to water, acid/alkali, and weathering. It is mainly used as a surface treatment agent of glass fibers. It is also widely used in the surface treatment of micro glass beads, silica hydrated white carbon black, talcum, mica, clay, and fly ash. It can also enhance the overall properties of polyesters, polyacrylates, PNC, and organosilicons. KH-270 is used as an adhesion promoter at organic-inorganic interfaces, a surface modifier or a crosslinker of polymers. OP-10 acts as a nonionic surfactant/emulsifier. The structures of KH-570 and OP-10 are shown in Figure 6.6 [92].

The results showed that the use of the surface-modified TiO_2 nanoparticles produced a considerable improvement in the load carrying capacity, the friction reducing, and antiwear abilities of pure water. The wear scar diameter and the coefficient of friction of the water-based lubricating fluids with coated TiO_2 nanoparticles decreased. In addition, the thick deep furrows on the surface wear scar decreased with the increased coated TiO_2 concentration. The power consumption in drilling process was lower and the cutting surface was smoother using the water-based lubricating fluids with added coated TiO_2 nanoparticles compared to the base fluid. The reason for modified TiO_2 nanoparticles improving tribological properties of water-based lubricating fluid was proposed to be due to the formation of the dynamic deposition film. An analysis of the worn surface during the rubbing process supported this contention [105]. The quality of their particles was tested on a four-ball tribometer. The water-based nanofluid with a 2%–10% particle concentration was tested in a drilling application on Q235 steel. Q235 is a carbon steel that also contains small amounts of Si, Mn, Cr, Ni, Mo, V, S, P, and Cu. Their study showed that double-coated TiO_2 particles can

1. Maintain dispersity and stability in water
2. Present improved antiwear, adhesion, and friction performance over water
3. Reduce friction and power consumption over water alone of 6%

In another study, Ikeda et al. [93] prepared hydrophilic nanocarbon particles and suspended them in water to produce non-oily cutting fluids.

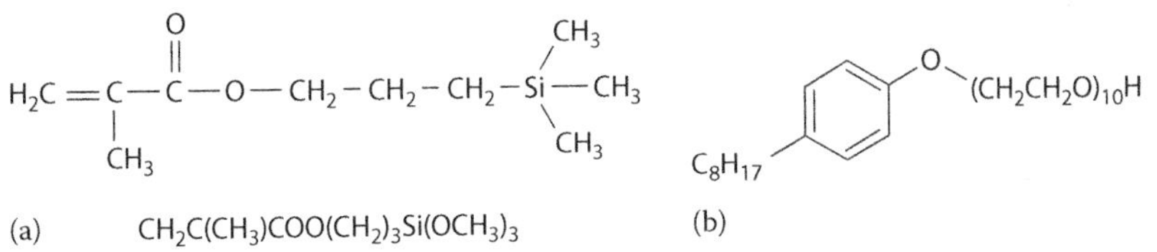

(a) $CH_2C(CH_3)COO(CH_2)_3Si(OCH_3)_3$ (b)

FIGURE 6.6   Structures of (a) KH-570 and (b) PO-10. (From Gu, Y. et al., *J. Nanomater.*, 785680, 2014.)

6.2.2 Minimum Quantity Lubrication

MQL is also called, near dry machining (ND), microlubrication or microlubrification, or microdosing. Cutting fluids still appear to remain the best choice for heat removal in machining [3]. Classically, cutting fluids have been used in a "flood" lubrication and coolant capacity. MQL is one of the important choices for classic cutting fluid use in machining. Nanofluids hold much potential as MQL candidates. MQL faces competition from cryogenic cooling, water vapor cooling, heat pipe cooling, and solid lubricants. MQL also addresses, in part, the move toward dry machining. Nanofluids with their enhanced heat removal capability and viscosity play an important role in the use of MQL. Nanofluids contribute to reductions in costs, health risks, and environmental risks.

MQL technology involves spraying or dripping the (nano) cutting fluid into the cutting zone such that the tool tip and the workpiece are coated with the fluid. An aerosol or drip system is used as opposed to constantly flooding the cutting site with copious amounts of fluid. Considerable heat is removed from the particle waste in this process. The waste can be in the form of chips, turnings, filings, or shavings. This waste is also called swarf and can be metal, wood, plastic, or ceramic. The waste is generally very clean with the MQL process and is readily recycled. If oil is used, it is generally vaporized at the cutting site such that the wasted swarf is clean. The vapor must be captured to prevent air pollution. Srikant et al. [3] and Park et al. [94] have recently reviewed MQL with respect to nanofluids and ball milling.

Although MQL has been used as an alternative solution for flood cooling as well as dry machining, its benefit has only been realized in mild machining conditions. The heat generation during more aggressive machining conditions cannot be effectively removed by the small amount of oil mist applied by the MQL process.

To extend the applicability of MQL to more aggressive machining conditions, Park et al. developed a potential additive to MQL lubricant [94]. After the preliminary wetting angle measurement of the various lubricants was established, one commercially available MQL vegetable oil was chosen. This was then mixed in a high-speed mixer with exfoliated nanographene particles. The resulting nanoenhanced MQL lubricant was evaluated for its tribological and machining behavior together with the suspension stability of the mixture. Friction coefficients of the new nanoenhanced MQL oil were also measured in terms of load, speed, and lubricant. Finally, MQL ball milling tests with the nanographene-enhanced lubricant were performed to show a remarkable performance improvement in reducing both central wear and flank wear as well as chipping at the cutting edge [94].

6.2.3 Measurement of Thermal Conductivity for Nanofluids

There are two basic methods for the measurement of the thermal properties of a fluid: the steady-state method and the transient method.

1. Although steady-state methods are simple, theoretically, they involve rather elaborate techniques in practice. These techniques include a thermal guard to eliminate lateral heat flow and an electronic control system to enable stable conditions during the test.

2. Transient methods provide fast measurement and reduce unwanted modes of heat transfer. Most thermal property measurements of nanofluids have been done using the transient method of measurement.

Thermal conductivity (k) is reported as SI units in watts per meter kelvin (W/m·K), that is, at a given temperature. The dimensions of thermal conductivity are $M^1L^1T^{-3}\Theta^{-1}$. These variables of conductivity are mass (M), length (L), time (T), and temperature (Θ) [95].

In the case of nanofluids, the conductivity is often reported as a conductivity ratio of knf/kf (knf, nanofluid conductivity/kf, base fluid conductivity).

In their review, Paul et al. [96] outlined the following most commonly used techniques for the measurement of the thermal conductivity of a nanofluid (Figure 6.7):

1. Transient hot-wire (THW) method
2. Thermal constants analyzer method
3. 3ω (3-omega) method
4. Steady-state parallel plate method
5. Temperature oscillation method
6. Cylindrical cell method
7. Thermal comparator method

They also discussed their advancements made with the Powell thermal comparator device.

The popularity of these methods is shown in Figure 6.8, and these methods are discussed in the following text.

6.2.3.1 Transient Hot-Wire Method

The most common method is the THW method. Horrocks and McLaughlin [97] report that the THW method was propsed 1931 by Stahlane and Pyk for the measurement of absolute thermal conductivity of solids (as powders). Modifications since then have resulted in a greater accuracy. Advantages of the THW method include

1. Elimination of errors due to convection
2. Higher speed compared to other techniques
3. Simple design of the hot-wire apparatus

The basis for the THW technique is shown in Figure 6.9 [75].

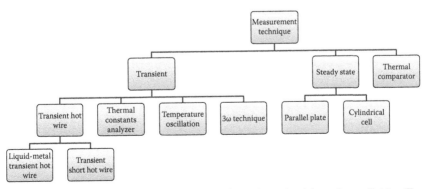

FIGURE 6.7 Techniques for measuring the thermal conductivity of nanofluids. (From Paul, G., *Renew. Sustain. Energy Rev.*, 14, 1913, 2010.)

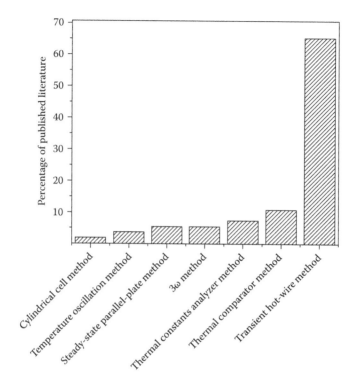

FIGURE 6.8 Comparison of the thermal conductivity measurement techniques for nanofluids. (From Paul, G., *Renew. Sustain. Energy Rev.*, 14, 1913, 2010.)

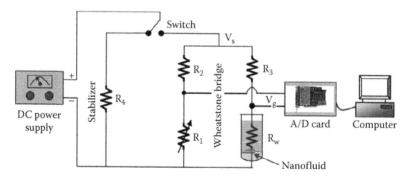

FIGURE 6.9 Schematic of the transient hot-wire experimental setup, where V_s is the voltage supplied. R_1–R_4 and R_w are resisters. (From Chen, Y. and Wang, X., *Mater. Lett.*, 62, 2215, 2008.)

The principle of this method involves a platinum wire that is used as a heater and as a thermometer for conductivity measurements. This method measures both the temperature and the time the wire takes to produce an abrupt response as an electrical pulse. In general, a THW instrument consists of a probe, which is inserted into the nanofluid for the conductivity measurement. Many THW instruments have

FIGURE 6.10   A cylindrical transient hot-wire apparatus. (From Merckx, B. et al., *Adv. Civil Engin.*, 2012, 635395, 2012.)

been manufactured over the years. An example clearly presenting this principle is shown in Figure 6.10 [98], although this apparatus was for larger-scale conductivity measurements.

The hot-wire probe functions as a thermometer as well as a heat source. The temperature is controlled by a current supplied to the wire. As the heat dissipates from the wire, the temperature of the nanofluid and the wire both increase. This temperature and hence the thermal conductivity measurement are dependent on the nanofluid. The liquid metal THW technique and the transient short hot-wire (TSHW) technique are variations of the basic THW technique.

An example of a nanoscale thermal conductivity study was reported by Sahooli and Sabbaghi who studied the thermal characteristics of CuO nanoparticles in EG [99]. They used the THW method to study the dispersion of the CuO nanoparticles at an optimum pH of 7.8.

The liquid metal THW method can also be used for electrically conducting liquids where mercury forms the "hot wire" [100]. The thermal conductivity of highly corrosive liquids, such as molten carbonates, can also be measured by the THW method [101,102].

In each example, the hot wire forms one resistor in a Wheatstone bridge circuit. A constant voltage is applied to the hot wire to raise its temperature. The temperature rise of the wire is calculated from the change in the resistance of the wire over time measured from the voltage offset of the initially balanced Wheatstone bridge.

6.2.3.2   Thermal Constants Analyzer (TCA) Method

The TCA method, which relies on the transient plane source (TPS) theory, is used to calculate the thermal conductivity of nanofluid using a thermal constants analyzer. Similar to the THW method, the plate used in the TCA acts as a temperature sensor

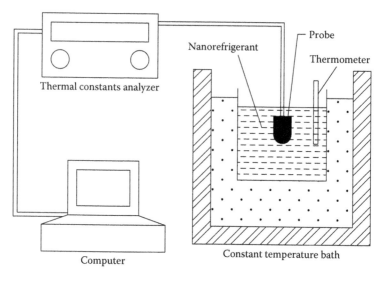

FIGURE 6.11 Schematic of the basic scheme for a thermal constants analyzer used to evaluate CNTs as nanorefrigerants. (From Jiang, W. et al., *Int. J. Thermal Sci.*, 48, 1108, 2009.)

and the heat source. The Fourier law of heat conduction, like the THW method, is used as the principle for measuring the thermal conductivity. The advantages for this method include

1. Speed of measurements.
2. Measurement of a wide range of thermal conductivities (from 0.02 to 200 W/m · K).
3. No sample preparation is needed.
4. Flexibility of sample size [103].

A schematic of a thermal circuit is shown in Figure 6.11 [104]. The analyzer is composed of a constant temperature bath with a vessel containing the nanofluid and a probe placed in the nanofluid—in this case a refrigerant. A thermometer is placed in the nanofluid to measure the temperature of the nanofluid.

Zhu et al. [103] used the TCA method to evaluate the dispersion of Al_2O_3 nanoparticles in water.

6.2.3.3 3ω (3-Omega) Method

The 3ω method is used for measuring the thermophysical properties of thin films and liquids, yet appears less popular for measuring the thermal conductivity of nanofluids. It was originally proposed in 1912 as a method for studying the thermal behavior of light bulb filaments [105]. In 1987, it was used for measuring thermal conductivity in solids [106], and its use has expanded since then. The principle of the 3ω approach for measuring heat transfer in nanofluids is depicted in Figure 6.12 [107].

In this 3ω technique, a microsized metal strip is deposited on the surface to be studied. The strip acts both as a heater and a thermometer. A current is

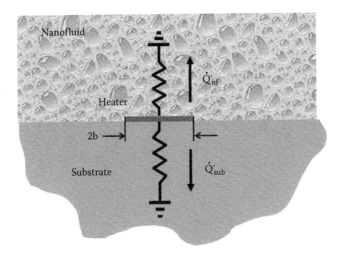

FIGURE 6.12 The principle of the 3ω (3-omega) method for measuring the heat conductivity of nanofluids. (From Oh, D. et al., *Int. J. Heat Fluid Flow*, 29, 1456, 2008.)

passed through the strip at an angle ω. This generates heat at a second harmonic frequency, which results in a temperature change. The ensuing temperature oscillations are measured by measuring the 3ω voltage. Advantages of the 3ω method include

1. Accuracy in measuring thermal conductivity.
2. Speed, as equilibration time takes only a few seconds.
3. Reduction of heater dimensions, which produces a reduction in blackbody infrared reducing errors.
4. The total heat generated in heater $\dot{Q}'_{total}$ passes through either the nanofluid $\dot{Q}'_{nf}$ or the substrate $\dot{Q}'_{sub}$.

6.2.3.4 Steady-State Parallel Plate Method

In the steady-state parallel plate method, the fluid is placed between two copper plates. This method is designed to transfer the heat in one direction as shown in Figure 6.13 [74].

The apparatus was originally designed by Challoner and Powell [108]. Wang et al. [74] used this method for measuring the thermal conductivity of aluminium and copper oxide nanofluids. Two parameters need to be controlled: (1) the accurate measurement of the temperature increase in each thermocouple and (2) the difference in temperature that needs to be minimized when the thermocouples are at the same temperature.

6.2.3.5 Temperature Oscillation Method

The temperature oscillation method involves a temperature oscillation or heat flux that imposes a temperature response from a nanofluid. This response results from

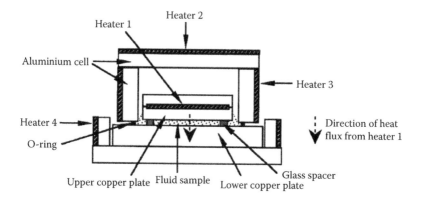

FIGURE 6.13 Experimental setup for the steady-state parallel plate method. (From Wang, X. et al., *J. Thermophys. Heat Transf.*, 13, 474, 1999.)

averaged or localized thermal conductivity in the direction of the nanofluid chamber height. The method used here is based on the oscillation method proposed and developed by Roetzel et al. [96,109,110]. Das et al. [111] used this technique to measure the thermal conductivity of nanofluids containing Al_2O_3 and CuO nanoparticles dispersed in water.

The testing system, Figure 6.14, requires as follows:

1. A custom-designed test cell
2. Cooling water on both ends of the test cell
3. Thermostatic bath
4. Electrical power to the Peltier element
5. Temperatures recorded through a number of thermocouples
6. Temperature data is amplified and filtered
7. Computer

6.2.3.6 Cylindrical Cell Method

The cylindrical cell method is a steady-state method used for measuring thermal conductivity in fluids. The equipment consists of coaxial inner and outer cylinders (Figure 6.15). The cell is also cylindrical and in the study by Kurt and Kayfeci [112] was filled with various concentrations of EG in water whose thermal conductivity was studied at various temperatures.

An electrical heater is contained inside the inner cylinder. The equipment is insulated to reduce heat loss, and as such, heat loss is seen as negligible. Heat is applied such that it flows through the test liquid to the cooling water. Two calibrated thermocouples are used to measure the outer surface temperature of the glass tube (T_i) and the inner cylinder (T_o). The thermocouples are positioned in the middle of the test section and connected to a multichannel digital readout with an accuracy of 0.1 °C. The required measurements for the calculation of the thermal conductivity are the T_i and T_o temperatures, adjusted voltage, and current of the heater.

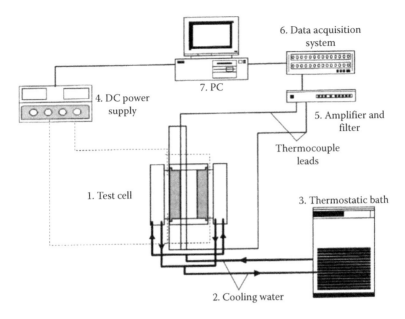

FIGURE 6.14   The basic system for the temperature oscillation method. (From Das, S.K. et al., *J. Heat Transf.*, 125, 567, 2003.)

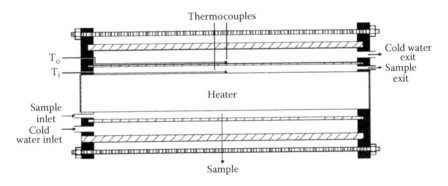

FIGURE 6.15   Cross section of the cylindrical cell equipment. (From Kurt, H. and Kayfeci, M., *Appl. Energy*, 10, 1016, 2009, see also http://www.sciencedirect.com/.)

6.2.3.7   Thermal Comparator Method

Paul et al. [96] have taken Powell's [113] thermal comparator concept and designed an effective apparatus for measuring the thermal conductivity of nanofluids (Figure 6.16).

A comparator is a device that compares two voltages or currents and produces a signal indicating which is larger. The concept is based on a single point of contact with the sample being assessed for its thermal conductivity. The transfer of heat from the hot material to the cold is very quickly measured by such a device. The temperature difference between the probe tip and a reference in the heated probe is measured with a thermocouple. The probe (Figure 6.16a) is copper. The heater

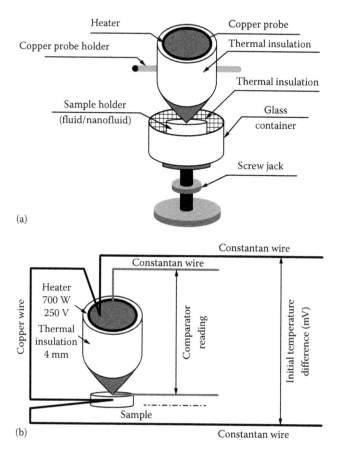

FIGURE 6.16 Thermal conductivity measurements based on (b) the thermal comparator method and (a) the principle for recording the differential thermo-emf. (From Paul, G., *Renew. Sustain. Energy Rev.*, 14, 1913, 2010.)

compensates for heat loss from the probe and allows for a constant temperature difference between sample and probe. The probe and sample material are connected by a constantan thermocouple. The constantan wire is made from a copper–nickel alloy usually consisting of 55% copper and 45% nickel and has constant resistivity over a wide range of temperatures.

Paul et al. [96] reported an effective use of their thermal comparator in a study of EG and water nanofluids containing ZnO_2 or TiO_2. Nano-ZnO_2 showed the larger increase in thermal conductivity over the base fluid, EG. They concluded their review of the seven techniques used to measure thermal conductivity in nanofluids with the following points [96]:

1. THW is the most popular due to its speed, accuracy, and reproducibility.
2. The use of two thermal conductivity techniques would go a long way to validating data, especially where anomalies are observed to occur between base fluid and nanofluid thermal conductivities with a single technique.

3. Much more study of thermal conductivity needs to be carried out at temperatures other than ambient. They particularly targeted subzero temperatures as needing study.
4. Nanofluid aging studies also need to be carried out.
5. Their thermal comparator development bodes well as an alternative device for measuring thermal conductivity for nanofluids.

The thermal comparator technique used to measure thermal conductivity in nanofluids needs only a point contact with the sample whose conductivity is to be measured. Further, the measurement needs to be almost instantaneous. These features render this technique very suitable for the measurement of thermal conductivity of different liquids. Following the original concept of Powell [113], a thermal comparator setup was developed to measure the thermal conductivity of nanofluids. It is well known that when two materials at different temperatures are brought in contact over a small area, heat transfer takes place from the hotter to the colder body. As a result, an intermediate temperature is very quickly attained at the point of contact. The contact temperature depends on the thermal conductivity of the two materials. Thermocouples are used to measure the voltage proportional to the temperature difference between the thermocouple probe tip and a reference located within the heated probe is measured. Using samples of known thermal conductivity, a calibration curve is prepared. Using the calibration curve, the thermal conductivity of the unknown samples can be estimated readily. The setup consists of a metallic copper probe, a temperature-controlled heating coil, a direct current microvoltmeter, and a voltage stabilizer (shown in Figure 6.16). The probe is the most important part of the setup as the success of the method depends on the heat flow from the probe to test material through a very small area of contact.

The THW method for estimating the thermal conductivity of solids and fluids is found to be the most accurate and reliable technique among the methods discussed earlier. Most of the thermal conductivity measurements in nanofluids reported in the literature have been conducted using the THW method. The temperature oscillation method helps in estimating the temperature-dependent thermal conductivity of nanofluids. The steady-state method has the difficulty that steady-state conditions have to be attained while performing the measurements. A comparison of the thermal conductivity values of nanofluids obtained by various measurement methods and reported in literature is shown in Table 6.1 [114].

6.3 DISCUSSION

One of the biggest contributions of nanoparticles to the advancement of cutting science has been in the area of heat transfer. As discussed earlier, the range of nanoparticles available for nanocutting fluids is broad. In many cases, the transfer of heat is much greater/more efficient for nanofluids than the basic, unmodified cutting fluids or totally dry cutting procedures. The heat transfer capabilities of nanofluids are important to other processes such as refrigeration. The nature of the cutting fluid also influences wetting, the removal of solid machined particulate matter, and, of

TABLE 6.1

Comparison of Thermal Conductivity Values Obtained Using Transient and Steady-State Measurement Techniques

Sl. No.	Base Fluid	Nanoparticle	Avg Particle Size (nm)	Conc. (vol %)	Sonication Time (h)	Temp. (°C)	Enhancement	Method of Measurement	Uncertainty %
							Thermal Conductivity Values		
1	Distilled water	Al_2O_3	36	10	3	27.5–34.7	1.3 times	Steady state	2.5
2	Distilled water	CuO	29	6	3	34	1.52 times	Steady state	2.5
3	Distilled water	Al_2O_3	28.6	1	12	21–51	2%–10.8%	Temperature oscillation	2.7
4	Distilled water	Al_2O_3	28.6	4	12	21–51	9.4%–24.3%	Temperature oscillation	2.7
5	Distilled water	CuO	38.4	1	12	21–51	6.5%–29%	Temperature oscillation	2.7
6	Distilled water	CuO	38.4	1	12	21–51	14%–36%	Temperature oscillation	2.7
7	Distilled water	Al_2O_3	20	1	NA	5–50	10%	SHW method	1
8	Distilled water	Al_2O_3	45	1	15	NA	4.4%	3ω method	NA

Source: Thomas, S. and Sobhan, C.B.P., *Nanoscale Res. Lett.*, 6, 77, 2011.

course, lubrication. The results of the MQL capabilities of nanocutting fluids have made advances in these issues.

Any cutting process will produce particulates both from machine (tool tip) wear and tear and from the surface being cut. These particulates present a health risk as an immediate dry dust or as a liquid mist. The disposal of the particulates presents another problem. In the literature reviewed, many researchers have reported that the addition of nanoparticles into base fluid significantly reduces the friction and increases the load bearing capacity of friction parts.

A new class of cutting fluids can be formulated by mixing nanoparticles (metallic, nonmetallic, ceramics, or carbon) in a conventional cutting fluid. This is because relative to suspended milli- or microsized particles, nanofluids show better stability, rheological properties, extremely good thermal conductivity, and no negative effect on pressure drop [115].

6.3.1 CLASSIFICATION OF NANOFLUID SYSTEMS

The classification of nanofluid systems is based on the type of nanoparticulate material and its thermal conductivity enhancing ability that contributes to its heat transfer efficiency [7]. In general, there are four types of nanomaterials that contribute to the heat efficiency of a nanofluid: ceramic (oxides, carbides, nitrides), metallic, carbon based (graphite, graphene, CNTs), and nanodroplet/nanoemulsions.

1. *Ceramic nanofluids* are the most investigated nanocutting fluids due to their wide availability, low cost, and stability. They show increased thermal conductivity, or values slightly above the predicted effective medium theory (EMT) corrected for the contribution of interfacial thermal resistance and/ or elongated nanoparticle shape [116–120]. The thermal conductivity of solid–liquid suspensions linearly increases with the volume fraction of the solid particles [7]. EMT predicts the thermal conductivity of two component heterogeneous mixtures and is a function of the conductivity of the two materials [119].

2. *Metal-based nanofluids* have been less investigated due to the limited oxidative stability of many metals and the high cost of precious metal. Metallic nanofluids show thermal conductivity increases well above the EMT prediction [68,121–128]. Timofeeva [7] summarized published data on the thermal conductivities of nanofluids containing silver, copper, aluminium, and gold in base fluids EG (EG), water, and other solvents. In all cases, the thermal conductivity enhancement did not increase until the particle volume fraction rose above 0.1%, regardless of temperature. Of the data summarized, silver nanoparticles in water at 90°C gave the best enhancement at 0.8% particle volume fraction %, whereas silver citrate in water or gold in toluene gave the worst at greater than 1.5%.

The data obtained from different research groups is due to variations in preparation methods, particle size, material type, base fluids, and surfactants and uncertainties in the measurements of particle concentration and thermal conductivity [116]. It was suggested [129] that metallic nanoparticles possess geometries dependent on local-ized plasmon resonances (collective oscillations of the metals' free electrons upon optical or other excitation), which are responsible for their abnormal thermal conduc-tivity increases in metallic nanofluids. The use of dry metal nanopowders fabricated in the gas phase has been limited to precious metals (Au, Pt) that are resistant to surface oxidation. The generation of nanoparticles directly in the base fluid produces more homogeneous nanofluids with fewer agglomerates and also provides better control over the surface state [7].

1. *Carbon-based nanomaterials* have been shown to exhibit a wide range of thermal conductivity increases, from very insignificant in amorphous car-bon black, to 2–3 in some suspensions of CNTs [130–133] and graphene oxides [134,135].
2. *Nanodroplets/nanoemulsions* are attractive due to their long-term stability, although the potential of nanodroplets in enhancing thermal conductivity is limited. The development of nanoemulsions may open a new direction for thermal fluid studies [136–138].

6.3.2 NANOPARTICLES USED IN NANOFLUIDS

Singh [139] summarized data on the thermal conductivity of nanofluids (Table 6.2). Particle size is preferably determined with transmission electron microscopy over scanning electron microscopy (SEM). While steady-state and transient methods exist for measuring the thermal conductivity of nanofluids, the THW is preferred by many. A modified AFM with a heat-sensing tip is another technique known as scanning thermal microscopy. Thermal conductivity (k), the ability of a material to conduct heat, is measured in watts per meter kelvin (W/(m·K)). The thermal conductivity ratio is the thermal conductivity of the nanofluid, knf, divided by the thermal con-ductivity, kf, of base fluid.

The thermal conductivity of a heat transfer fluid is widely recognized to be the main factor influencing the heat transfer efficiency. It was reported that the low thermal conductivity of conventional fluids (i.e., 0.1–0.6 W/m·K at 25 °C) is improved when solid particles with significantly higher thermal conductivity values (i.e., 10–430 W/m·K for pure elements) are added. Therefore, the addition of small solid particles to liquids improves the mechanism of the fluid. The magnitudes of the effects reported in the literature are scattered from a few percent, as predicted by the EMT [11,116,140], to hundreds of percent per nanoparticle volume concentration [141,142].

Most of the studies reported that the thermal conductivity increased as the particle volume fraction parameter increased [10].

1. Data from many studies show that particle type is an important param-eter that affects the thermal conductivity of nanofluids [73,120]. For example, the results of thermal conductivity studies with nanofluids containing Al_2O_3 and CuO nanoparticles showed that nanofluids with CuO nanoparticles showed higher enhancement when compared to the

TABLE 6.2
Nanofluids with Their Thermal Conductivity Increase in Nanofluid Thermal Conductivity over Base Fluid Thermal Conductivity and Synthesis Procedure Used as Reported in the Literature

Base Fluid with Conductivity	Nanoparticles, Average Diameter, and Concentration	Method Used for Synthesis	Max. Thermal Conductivity Ratio
Water 0.613	Al_2O_3, <50 nm, up to 4.3 vol%	2-step	1.08
Water 0.613	CuO_3, <50 nm, up to 3.4 vol%	2-step	1.10
Water 0.613	C-MWNT 50 nm, 5 µm, 3 µm, 0.6 vol%	2-step	1.38
EG 0.252	Fe, <10 nm, 6.0 vol%	2-step	1.18
Water 0.613	TiO_2, 15 nm, <5.0 vol%	2-step	1.30
Water 0.613	Cu, 18 nm, up to 5.0 vol%	1-step	1.60
Thiolate	Au, 10–20 nm, 0.1 vol%	2-step	1.09
Citrate	Ag, 6–80 nm, 0.1 vol%	2-step	1.85
α-Olefin	CNT, 25 × 50,000 nm, 1.0 vol%	2-step	2.50
EG 0.252	Al_2O_3, <50 nm, up to 5.0 vol%	2-step	1.18
EG 0.252	CuO, 35 nm, up to 4 vol%	2-step	1.21
EG 0.252	Cu, 10 nm, up to 0.5 vol%	1-step	1.41
Oil (Trans) 0.145	Cu, up to 100 nm, up to 7.6 vol%	2-step	1.43
Water 0.613	Cu, 75–100 nm, 1.0 vol%	1-step	1.23

Source: Singh, A.K., Defence Sci. J., 58, 600, 2008.

nanofluids prepared using Al_2O_3 nanoparticles. Thus, it is seen that Al_2O_3, as a solid material, has a higher thermal conductivity than CuO. However, the thermal conductivity of particle material may not be the dominant parameter that determines the thermal conductivity of the nanofluid [143,144].

2. One of the most important factors that affects the thermal conductivity enhancement is the size of the nanoparticles. This is contrary to the predictions of conventional models such as the Hamilton and Crosser model, which does not take into account the effect of particle size on thermal conductivity [144]. The Hamilton and Crosser model is an adaptation of an earlier Maxwell model. Both models deal with the conduction of heat in solid–liquid suspensions.

3. Two nanofluids were prepared using 26 nm diameter and 600 nm average cylindrical diameter particles respectively. It was found that 4.2 vol% water-based nanofluids with spherical particles had a thermal conductivity enhancement of 15.8%, whereas 4 vol% nanofluids with cylindrical particles had a thermal conductivity enhancement of 22.9%. These results are in line with the fact that nanofluids with CNTs that are cylindrical generally showed greater thermal conductivity enhancement than the spherical particles. One possible reason is the rapid heat transport along the relatively larger distances of the cylindrical particles, which usually have lengths of the order of micrometers [7,143].

4. With regard to the base fluid used in nanofluid suspensions, the fluids used in the preparation of nanofluids are the same fluids used for any heat transfer applications, such as water, EG, and engine oil. According to conventional thermal conductivity models, such as the Maxwell model, as the base fluid thermal conductivity of a mixture decreases, the thermal conductivity ratio (thermal conductivity of nanofluid knf divided by the thermal conductivity of base fluid kf) increases. Thus, poorly conductive fluids are better than highly conductive ones for nanofluid applications. Hence, water is generally avoided, when it comes to nanofluids [143].

5. Changes in temperature affect the Brownian motion of nanoparticles, which results in dramatic changes in the thermal conductivity of nanofluids with a temperature clustering of nanoparticles [143]. In conventional suspensions of larger particles, thermal conductivity depends on temperature alone. Base fluid and liquid and particles separately depend on temperature.

6. Some studies [145] reported that there is a decrease in the thermal conductivity ratio with increasing pH. It was also observed [145] that the rate of change of thermal conductivity with the particle volume fraction was dependent on the pH value. The thermal conductivity enhancement of a 5 vol% Al_2O_3/water nanofluid was 23% at pH 2.0 and decreased to 19% at 11.5. The reported optimum pH for maximum thermal conductivity enhancement is approximately 8.0 for Al_2O_3/water and 9.5 for Cu/water nanofluids [146]. At optimum pH, the surface charge of nanoparticles increases, which creates repulsive forces between nanoparticles, and the severe clustering of nanoparticles is prevented. Excessive clustering could result in sedimentation, which decreases thermal conductivity enhancement [146].

The transport properties of nanofluids are another issue that needs to be considered. Dynamic thermal conductivity and viscosity are not only dependent on nanoparticle volume but on parameters such as particle shape, size, mixture combinations and slip mechanisms, and surfactant. Studies have shown that thermal conductivity and viscosity are higher for nanofluids than base fluids [70]. Various theoretical and experimental studies have been conducted and various correlations have been proposed for the thermal conductivity and dynamic viscosity of nanofluids. However, no general correlations have been established due to the lack of a common understanding of the mechanism of nanofluid behavior [147].

6.3.3 MECHANISMS FOR ENHANCED THERMAL CONDUCTIVITIES

Many mechanisms have been proposed to explain the experimental data of the anomalous increase of the thermal properties of nanofluids. Examples of these include

1. The interaction between nanoparticles and liquids in the form of interfacial thermal resistance [12,117,148–150]
2. Formation of condensed nanolayers around the particles [151–153]
3. The particle size effects [154], agglomeration of nanoparticles [12,118,154,155]
4. The microconvection mechanism due to Brownian motion of nanoparticles in the liquid [40,41,156]
5. Surface plasmon resonance [129,157,158]
6. Near field radiation [159,160]

None of these mechanisms by themselves seems to have the capacity of explaining the various experimental thermal conductivity enhancements in nanofluids. However, it appears that combinations of these mechanisms could explain the majority of the experimental results.

Wei Yu et al. [135] reported in their review that nanofluids show many interesting properties and have found applications such as those in the energy, mechanical, and biomedical fields. There are many reviews that present overviews of various aspects of nanofluids [8,96,162–168], including preparation, characterization, measurements of thermal conductivity, theory and modeling, thermophysical properties, and convective heat transfer.

Xie et al. [161] studied the thermal conductivity of EG nanofluids with MgO, TiO_2, ZnO, Al_2O_3, and SiO_2 nanoparticles [161]. Their results are given in Table 6.3 [161], which shows that the MgO–EG nanofluid was found to have superior features with the highest thermal conductivity and lowest viscosity.

TABLE 6.3
Properties of Oxides and Their Nanofluids

	Thermal Conductivity[a] (W/m·K)	Density (g/cm³)	Crystalline	Viscosity (cP) with 5.0 vol% 30 °C	Thermal Conductivity Enhancement of Nanofluids (%) with 5.0 vol%
MgO	48.4	2.9	Cubic	17.4	40.6
TiO_2	8.4	4.1	Anatase	31.2	27.2
ZnO	13.0	5.6	Wurtzite	129.2	26.8
Al_2O_3	36.0	3.6	γ	28.2	28.2
SiO_2	10.4	2.6	Noncrystalline	31.5	25.3

Source: Xie, H. et al., *J. Exp. Nanosci.*, 5, 463, 2010.

cP (centipoise) is a unit of dynamic viscosity. 1 centipoise = 0.01 gram per centimeter-second in the centimeter (1 cP = 0.01 g/cm/s).

[a] Thermal conductivities of the oxides are for the corresponding bulk materials. Al_2O_3 has three crystalline forms—α, β, and γ.

6.3.4 Mechanical Applications

Some nanoparticles show excellent lubricating properties [162]. This is due to the fact that they have an elastic modulus and can form a soft protective film on the surface being modified [162].

Magnetic fluids form a special group of nanofluids [163]. Magnetic liquid rotary seals operate with no maintenance and extremely low leakage in a very wide range of applications, due to the magnetic properties of the magnetic nanoparticles in the liquid [164].

6.3.5 Friction Reduction

Zhou et al. [165] evaluated the tribological behavior of Cu nanoparticles additives in oil in a four-ball tribometer. The results showed that Cu nanoparticles had better friction and antiwear properties than zinc dithiophosphate, especially at high applied loads. The nanoparticles showed a striking improvement in the load carrying capacity of the base oil [165]. The dispersion of solid particles was found to play an important role, especially when a slurry layer was formed. Water-based Al_2O_3 and diamond nanofluids were applied in an MQL grinding process of cast iron. During grinding, a dense and hard slurry layer was formed on the wheel that benefited the grinding performance. Nanofluids reduced grinding forces, improved surface roughness, and prevented workpiece burnishing. Compared to dry grinding, MQL grinding was seen to significantly reduce grinding temperatures [166].

Wear and friction properties of surface-modified Cu nanoparticles as additives in 50CC oil were studied [37]. Higher oil temperatures resulted in a better tribological performance of the Cu nanoparticles. It was proposed that a thin copper protective film with a lower elastic modulus and hardness was formed on the worn surface, which resulted in good tribological performances for the Cu nanoparticles, especially when the oil temperature was high [167]. Wang et al. [168] reported that room temperature ionic liquid MWCNT composites were evaluated as lubricant additives. These composites exhibited excellent dispersibility, good friction reduction, and antiwear properties [168]. Wang et al. studied the tribological properties of ionic liquid–based nanofluids containing functionalized MWNTs at 200–800 N loads [168]. The nanofluids exhibited friction reduction properties below the 800 N load and remarkable antiwear properties [169]. $Mn_{0.78}Zn_{0.22}Fe_2O_4$ magnetic nanoparticles were also shown to be efficient lubricant additives in 46 turbine oil [170]. It improved wear resistance, load carrying capacity, and the antifriction capability of the base oil. Thus, the wear scar diameter decreased by 25% compared to the base oil [170].

Chen and Mao reported on the dispersion stability enhancement and self-repair principles of ultrafine tungsten disulfide in green lubricating oil [171]. Ultrafine tungsten disulfide particulates were able to fill and level up the furrows on surfaces, thus repairing the abrasive surface well. In addition, the particulates formed a WS_2 film with low shear stress by adsorbing and depositing in the pores of the abrasive surface, making the abraded surface smoother. The FeS film formed during tribochemical reactions protected the abrasive surface further, promoting the self-repair of the abrasive surface. The tribological properties of liquid paraffin with SiO_2 nanoparticles that were made by a sol-gel method were investigated by Peng et al. [172]. At optimal concentrations of SiO_2 nanoparticles, better tribological properties than pure paraffin oil were observed. The antiwear properties were a function of particle size. Oleic acid surface–modified SiO_2 nanoparticles with an average diameter of 58 nm provided a better load carrying capacity, antiwear, and friction reduction than pure liquid paraffin [172]. Nanoparticles can easily penetrate between the rubbing surfaces because of their small size. During the frictional process, a thin physical tribofilm of the nanoparticles forms between rubbing surfaces. This film not only bears the load; it also separates the rubbing surfaces [172]. The spherical SiO_2 nanoparticles are able to roll between the rubbing surfaces in a sliding friction mode and the original pure sliding friction becomes a mixed sliding and rolling friction [172]. Therefore, the friction coefficient declines markedly and then remains constant [172].

6.3.6 SOME EXAMPLES OF HEAT TRANSFER WITH NANOFLUIDS

Recent heat transfer studies involve CNTs suspended in aluminium [173], CuO and ZnO in xanthan gum [174], CNTs [175], CuO oil flow, [176], $CaCO_3$ [177], graphite/oil mixtures [178], and ceramic composites [179].

CNTs [180] play an important role in the arena of nanocutting fluids. Sonication is used to reduce agglomeration and improve heat transfer. The preparation of the CNTs involves dissolving gum arabic (0.25%) in EG and adding 0.5% MWCNTs. The MTCNTs were imaged by SEM. The mixture was then sonicated. The agglomerates were sized with an optical microscope and imaged by TEM.

The effect of sonication on heat transfer and viscosity can be summarized as follows:

1. Increased sonication energy input led to increased (nonlinear) thermal conductivity.
2. Increased shear rate resulted in an increase in viscosity.
3. At a fixed shear rate, viscosity increased then decreased.
4. Sonication reduces the size of the CNTs.

In another study [178], the thermal and viscosity characteristics of graphite in oil were investigated. Colloidal graphite, a heavy alkylbenzene heat transfer oil (LD320), and a dispersant (CH-5) were mixed in a ball mill (aluminium balls:oil, ratio 1:1) to generate the nanofluids. The heat transfer capabilities of the nanofluids in this study were improved with the addition of the CH-5 dispersant. It was proposed that the dispersion of the colloidal graphite was essential to improve the thermal conductivity even though the graphite concentration is much more important than temperature. This was an indication of clustering being the prime component for 36% improved heat transfer.

Wu et al. [181] examined the tribological properties of lubricating oils: API-SF engine oil and a base oil with CuO, TiO_2, and nanodiamond nanoparticles as additives. Friction and wear experiments were performed using a reciprocating tribometer. CuO added in standard oil exhibits good friction reduction and antiwear properties. CuO nanoparticles in the API-SF engine oil and the base oil decreased the friction coefficient by 18.4% and 5.8%, respectively, and reduced wear by 16.7% and 78.8%, respectively, as compared to the oils without CuO nanoparticles. The antiwear mechanism is attributed to the deposition of CuO nanoparticles on the worn surface, which may decrease the shearing stress, thus improving the tribological properties.

Peng et al. [182] discussed the tribological properties of diamond and SiO_2 nanoparticles that were prepared by a surface modification method, which involved added oleic acid. The results of the dispersion ability and the dispersion stability of oleic acid–modified diamond and SiO_2 nanoparticles are shown in Figure 6.17. The tribological tests were conducted on a ball-on-ring tribometer. The results showed that both nanoparticles in liquid paraffin, even at a tiny concentration have better antiwear and antifriction properties than the pure paraffin oil. SEM was used to examine the plowing of nanoscale grooves of worn surfaces from the diamond and SiO_2 nanoparticles. The optimal concentration of diamond particles that minimizes the wear scar diameter was 0.2–0.5 wt% and that of SiO_2 nanoparticles is 0.1–1 wt%.

Zhang et al. [183] examined the tribological properties of blank polyalphaolefin (PAO) and PAO-containing $CaCO_3$ nanoparticles. The test was conducted on a

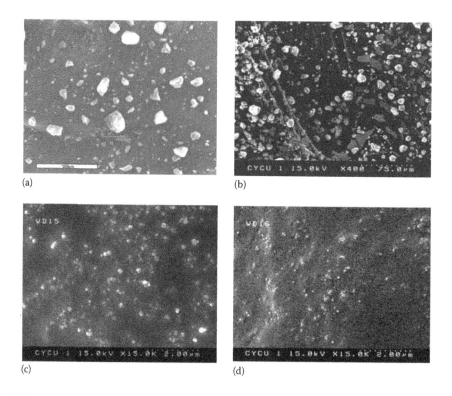

(a)

(b)

(c)

(d)

FIGURE 6.17 SEM images of diamond and SiO_2 nanoparticles: (a) nonmodified diamond nanoparticles, (b) nonmodified SiO_2 nanoparticles, (c) oleic acid surface–modified diamond particles with diameter 110 nm, and (d) oleic acid surface–modified SiO_2 particles with diameter 92 nm. (From Peng, D.X. et al., *Tribol. Int.*, 42, 911, 2009.)

reciprocating ball-on-ring tribometer. The results showed that $CaCO_3$ nanoparticles can dramatically improve the load carrying capacity, as well as the antiwear and friction reduction properties of PAO base oil. In addition, higher applied loads, moderate frequencies, longer duration times, and lower surface temperatures were seen as beneficial to the deposition of $CaCO_3$ nanoparticles accumulating on rubbing surfaces. X-ray photoelectron spectroscopy revealed a boundary film composed of $CaCO_3$, CaO, iron oxide, and some organic compounds on the worn surfaces.

6.4   SUMMARY

Current heat transfer fluids have limited thermal conductivity and therefore greatly lower their heat exchange efficiency. Many studies have reported attempts to improve the thermal transport properties of the fluids. Therefore, there is a need for the development of advanced heat transfer fluids with higher thermal conductivity and improved heat transfer. In recent years, many studies have been carried out on applications of nanoparticles in the field of lubrication. The reductions of friction and wear by nanolubricants are dependent on the characteristics of nanoparticles, such as size, shape, and concentration. Their nanometer size allows them to enter into the contact zone to reduce erosion and clogging. Many studies have shown that the heat transfer rate was found to be higher with the use of nanofluids than with their corresponding base fluids alone. The thermal conductivity enhancement of nanofluids depends on various factors such as the particle type and volume fraction, type of base fluid, size and shape of nanoparticles, and operating temperature. The properties of nanofluids make them attractive for many reasons for use by industries with serious heat transfer needs.

REFERENCES

1. N. G. Kandile and D. R. K. Harding, Nanotribology. Progress toward improved lubrication for the control of friction using ionic liquid lubricants, in: *Surfactants in Tribology*, Vol. 4, G. Biresaw and K. L. Mittal (eds.), pp. 183–213, CRC Press, Boca Raton, FL (2014).
2. National Nanotechnology Initiative (NNI), Supplement to the President's FY 2015 Budget. Office of Science and Technology Policy, Washington, DC (2014).
3. R. R. Srikant, M. M. S. Prasad, M. Amrita, A. V. Sitaramaraju, A. V. Krishna, and P. Vansi, Nanofluids as a potential solution for minimum quantity lubrication: A review, *Proc. Inst. Mech. Eng. B J. Eng. Manuf.*, 228, 3–20 (2014).
4. R. V. Lapshin, Feature-oriented scanning methodology for probe microscopy and nanotechnology, *Nanotechnology*, 15, 1135–1151 (2004); R. V. Lapshin, Feature-oriented scanning probe microscopy, in: *Encyclopedia of Nanoscience and Nanotechnology*, Vol. 14, H. S. Nalwa (ed.), pp. 105–115, American Scientific Publishers, Valencia, CA (2004).
5. J. A. Eastman, S. R. Phillpot, S. U. S. Choi, and P. Keblinski, Thermal transport in nanofluids, *Annu. Rev. Mater. Res.*, 34, 219–246 (2004).
6. V. Vasu, K. R. Krishna, and A. C. S. Kumar, Analytical prediction of forced convective heat transfer of fluids embedded with nanostructured materials (nanofluids), *Pramana*, 69, 411–421 (2007).

7. E. V. Timofeeva, Nanofluids for heat transfer—Potential and engineering strategies, Chapter 19, in: *Two Phase Flow, Phase Change and Numerical Modeling*, A. Ahsan (ed.), pp. 435–449, InTech Europe, Rijeka, Croatia (2011). Available from http://www.intechopen.com/books/two-phase-flow-phase-change-and-numericalmodeling/nanofluids-for-heat-transfer-potential-and-engineering-strategies.

8. K. V. Wong and O. De Leon, Applications of nanofluids: Current and future, *Adv. Mech. Eng.*, 2010, 1–11, Article ID 519659 (2010).

9. D. Singh and J. Routbort, Heavy vehicle systems optimization merit review and peer evaluation: Annual report, Argonne National Laboratory, Department of Energy, Washington, DC (2006).

10. S. Özerinç, S. Kakaç, and A. G. Yazıcıoğlu, Enhanced thermal conductivity of nanofluids: A state-of-the-art review, *Microfluid. Nanofluid.*, 8, 145–170 (2010).

11. J. C. Maxwell, *A Treatise on Electricity and Magnetism*, Dover Publications, Oxford Clarendon Press (1873). Available from http://en.wikipedia.org/wiki/A_Treatise_on_Electricity_and_Magnetism (2015).

12. B.-X. Wang, L.-P. Zhou, and X.-F. Peng, A fractal model for predicting the effective thermal conductivity of liquid with suspension of nanoparticles, *Int. J. Heat Mass Transf.*, 46, 2665–2672 (2003).

13. S. K. Das, S. U. S. Choi, W. Yu, and T. Pradeep, *Nanofluids: Science and Technology*, John Wiley & Sons, Hoboken, NJ (2008).

14. Z. Han, Nanofluids with enhanced thermal transport properties, PhD thesis, Department of Mechanical Engineering, University of Maryland, College Park, MA (2008).

15. F. Klocke and G. Eisenblatter, Dry cutting, *Ann. CIRP*, 46, 519–526 (1997).

16. J. B. Zimmerman, A. F. Clarens, K. F. Hayes, and S. J. Skerlos, Design of hard water stable emulsifier systems for petroleum- and bio-based semi-synthetic metalworking fluids, *Environ. Sci. Technol.*, 37, 5278–5288 (2003).

17. S. Kalpakjian and S. Schmid (eds.), Manufacturing engineering and technology, in: *Manufacturing Engineering and Technology*, SI 6th edn., pp. 585–590, Prentice Hall, Upper Saddle River, NJ (2001).

18. Cutting Fluid Management for Small Machining Operations, *A Practical Pollution Prevention Guide*, Iowa Waste Reduction Center, University of Northern Iowa, Cedar Falls, IA. Creation of this manual was funded by the U.S. Environmental Protection Agency, Risk Reduction Engineering Lab under Cooperative Agreement CR 821492-01-2 (1996).

19. S. Khandekar, M. Ravi Sankar, V. Agnihotri, and J. Ramkumar, Nano-cutting fluid for enhancement of metal cutting performance, *Mater. Manuf. Process.*, 27, 963–967 (2012).

20. S. H. Seyedmahmoudi, S. Harper, M. C. Weismiller, and K. R. Haapala, Evaluating the use of zinc oxide and titanium dioxide nanoparticles in a metalworking fluid from a toxicological perspective, *J. Nanopart. Res.*, 17, 104 (2015).

21. M. M. A. Khan and N. R. Dhar, Performance evaluation of minimum quantity lubrication by vegetable oil in terms of cutting force, cutting zone temperature, tool wear, job dimension and surface finish in turning AISI-1060 steel, J. Zhejiang University, *Science A* 7(11) 1790–1799 (2006).

22. M. Winter, R. Bock, C. Herrmann, H. Stache, H. Wichmann, and M. Bahadir, Technological evaluation of a novel glycerol based biocide-free metalworking fluid, *J. Clean. Prod.*, 35, 176–182 (2012).

23. P. H. Lee, T. S. Nam, C. Li, and S. W. Lee, Environmentally friendly nano-fluid minimum quantity lubrication (MQL) meso-scale grinding process using nano-diamond particles, in: *Proceedings of Sixth International Conference on Manufacturing Automation (ICMA)*, pp. 44–49 (2010).

24. S. Wang and A. F. Clarens, Analytical model of metalworking fluid penetration into the flank contact zone in orthogonal cutting, *J. Manuf. Process.*, 15, 41–50 (2013).

25. K. Weinert, I. Inasaki, J. W. Sutherland, and T. Wakabayashi, Dry machining and minimum quantity lubrication, *CIRP Ann.*, 53, 511–537 (2004).

26. A. Kotnarowski, Influence of nanoadditives on lubricants tribological properties, *Mater. Sci. Forum*, 14, 366–370 (2008).

27. M. Mosleh, N. D. Atnafu, J. H. Belk, and O. M. Nobles, Modification of sheet metal forming fluids with dispersed nanoparticles for improved lubrication, *Wear*, 267, 1220–1225 (2009).

28. L. Rapoport, N. Fleischer, and R. Tenne, Applications of WS2(MoS2) inorganic nanotubes and fullerene-like nanoparticles for solid lubrication and for structural nanocomposites, *J. Mater. Chem.*, 15, 1782–1788 (2005).

29. P. Krajnik, F. Pusavec, and A. Rashid, Nanofluids: Properties, applications and sustainability aspects in materials processing technologies, *Adv. Sustain. Manuf.*, 3, 107–113 (2011).

30. S. J. Skerlos, K. F. Hayes, A. F. Clarens, and F. Zhao, Current advances in sustainable metalworking fluids research, *Int. J. Sustain. Manuf.*, 1, 180–202 (2008).

31. Y. Gao, G.-X. Chen, Y. Oli, Z.-J. Zhang, and Q.-J. Qun-ji, Study on tribological properties of oleic acid-modified TiO$_2$ nanoparticle in water, *Wear*, 252, 454–458 (2002).

32. J.-F. Zhou, Z.-S. Wu, Z. Zhang, W.-M. Liu, and H.-X. Dang, Study on an antiwear and extreme pressure additive of surface coated LaF$_3$ nanoparticles in liquid paraffin, *Wear*, 249, 333–337 (2001).

33. S. Chen and W.-M. Liu, Characterization and anti-wear ability of non-coated ZnS nanoparticles and DDP-coated ZnS nanoparticles, *Mater. Res. Bull.*, 36, 137–143 (2001).

34. L. Rapoport, N. Fleischer, and R. Tenner, Fullerene-like WS$_2$ nanoparticles: Superior lubricants for harsh conditions, *Adv. Mater.*, 15, 651–655 (2003).

35. Y. Zhang, J. Yan, L. Suni, G. Yang, Z. Zhang, and P. Zhang, Friction reducing anti-wear and self-repairing properties of nano-Cu additive in lubricating oil, *J. Mech. Eng.*, 46, 74–79 (2010).

36. M. Zhang, X. Wang, W. Liu, and X. Fu, Performance and anti-wear mechanism of Cu nanoparticles as lubricating oil additives, *Ind. Lubr. Tribol.*, 61, 311–318 (2009).

37. H. Yu, Y. Xu, P. Shi, B. Xu, X. Wang, and Q. Liu. Tribological properties and lubricating mechanisms of Cu nanoparticles in lubricant, *Trans. Nonferrous Met. Soc. China*, 18, 636–641 (2008).

38. Y. Choi, C. Lee, Y. Hwang, M. Park, J. Lee, C. Choi, and M. Jung. Tribological behavior of copper nanoparticles as additives in oil, *Curr. Appl. Phys.*, 9, 124–127 (2009).

39. A. P. Sasmit, S. A. Khan, and S. Arun, Nanofluids heat transfer: Preparation, Characterization and theoretical aspects, in: *eBook Collection (EBSCOhost) Nanofluids: Research, Development and Applications*, Y. Zhang (ed.), pp. 55–90, Nova, New York (2013).

40. R. Prasher, P. Bhattacharys, and P. E. Phelan, Thermal conductivity of nanoscale colloidal solutions (nanofluids), *Phys. Rev. Lett.*, 94, 025901 (2005).

41. S. P. Jang and S. U. S. Choi, Role of Brownian motion in the enhanced thermal conductivity of nanofluids, *Appl. Phys. Lett.*, 84, 4316–4318 (2004).

42. O. Masala and R. Seshadri, Synthesis routes for large volumes of nanoparticles, *Annu. Rev. Mater. Res.*, 34, 41–81 (2004).

43. A. S. Edlestein and R. C. Cammarata (eds.), *Nanomaterials: Synthesis, Properties and Applications*, Taylor & Francis, Oxford, U.K. (1996).

44. D. L. Van Hyning, W. G. Klemperer, and C. F. Zukoski, Silver nanoparticle formation: Predictions and verification of the aggregative growth model, *Langmuir*, 17, 3128–3135 (2001).

45. I. Sondi, D. V. Goia, and E. J. Matijević, Preparation of highly concentrated stable dispersions of uniform silver nanoparticles, *J. Colloid Interface Sci.*, 260, 75–81 (2003).

46. D. K. Lee and Y. S. Kang, Synthesis of silver nanocrystallites by a new thermal decomposition method and their characterization, *ETRI J.*, 26, 252–256 (2004).

47. S. U. Son, I. K. Park, J. Park, and T. Hyeon, Synthesis of Cu_2O coated Cu nanoparticles and their successful applications to Ullmann-type amination coupling reactions of aryl chlorides, *Chem. Commun.*, 7, 778–779 (2004).

48. K. Mallick, M. J. Witcomb, and M. S. Scurrell, Polymer stabilized silver nanoparticles: A photochemical synthesis route, *J. Mater. Sci.*, 39, 4459–4463 (2004).

49. H. Ma, B. Yin, S. Wang, Y. Jiao, W. Pan, S. Huang, S. Chen, and F. Meng, Synthesis of silver and gold nanoparticles by a novel electrochemical method, *ChemPhysChem*, 5, 66–75 (2004).

50. B. Nikoobakht, Z. L. Wang, and M. A. El-Sayed, Self-assembly of gold nanorods, *J. Phys. Chem. B*, 104, 8635–8640 (2000).

51. S. Stoeva, K. J. Klabunde, C. M. Sorensen, and I. Dragieva, Gram-scale synthesis of monodisperse gold colloids by the solvated metal atom dispersion method and digestive ripening and their organization into two- and three-dimensional structures, *J. Am. Chem. Soc.*, 124, 2305–2311 (2002).

52. S. Horikoshi and N. Serpone, Introduction to nanoparticles, in: *Microwaves in Nanoparticle Synthesis*, 1st edn., S. Horikoshi and N. Serpone (eds.), pp. 1–24, Wiley-VCH Verlag GmbH & Co. KGaA, New York (2013).

53. K. Sharma, A. K. Tiwari, and A. R. Dixit, Progress of nanofluid application in Machining: A review, *Mater. Manuf. Process.*, 30, 813–828 (2015).

54. S. K. Das, N. Putra, and W. Roetzel, Pool boiling characteristics of nano-fluids, *Int. J. Heat Mass Transf.*, 46, 51–62 (2003).

55. S. K. Das, N. Putra, and W. Roetzel, Pool boiling of nano-fluids on horizontal narrow tubes, *Int. J. Multiphase Flow*, 29, 1237–1247 (2003).

56. I. C. Bang and S. H. Chang, Boiling heat transfer performance and phenomena of Al_2O_3–water nanofluids from a plain surface in a pool, *Int. J. Heat Mass Transf.*, 48, 2407–2424, 2005.

57. C. B. Sobhan and G. P. Peterson, *Microscale and Nanoscale Heat Transfer: Fundamentals and Engineering Applications*, CRC Press, Taylor & Francis Group, Oxford, U.K. (2008).

58. P. Hao, D. Guoliang, H. Haitao, J. Weiting, Z. Dawei, and W. Kaijiang, Nucleate pool boiling heat transfer characteristics of refrigerant/oil mixture with diamond nanoparticles, *Int. J. Refrig.*, 33, 347–358 (2010).

59. R. Saidur, S. N. Kazi, M. S. Hossain, M. M. Rahman, and H. A. Mohammed, A review on the performance of nanoparticles suspended with refrigerants and lubricating oils in refrigeration systems, *Renew. Sustain. Energy Rev.*, 15, 310–323 (2011).

60. S. Lazzarri, D. Moscatelli, F. Codari, M. Salmona, M. Morbidelli, and L. Diomede, Colloidal stability of polymeric nanoparticles in biological fluids, *J. Nanopart. Res.*, 14, 920 (2012).

61. A. J. Philip, P. D. Shima, and B. Raj, Evidence for thermal conduction through percolating structures of nanofluids, *Nanotechnology*, 19, 305706 (2008).

62. J. Liao, Y. Zhang, W. Yu, L. Xu, C. Ge, J. Liu, and N. Gu, Linear aggregation of gold nanoparticles in ethanol, *Colloids Surf. A Physiochem. Eng. Asp.*, 223, 177–183 (2003).

63. Z. Niu and Y. Li, Removal and utilization of capping agents in nanocatalysis, *Chem. Mater.*, 26, 72–83 (2013).

64. S. K. Das, S. U. S. Choi, and H. E. Patel, Heat transfer in nanofluids—A review, *Heat Transf. Eng.* 27, 3–19 (2006).

65. A. Sergis and Y. Hardalupas, Anomalous heat transfer modes of nanofluids: A review based on statistical analysis, *Nanoscale Res. Lett.*, 6, 391 (2011).

66. R. Prasher, P. E. Phelan, and P. Bhattacharya, Effect of aggregation kinetics on the thermal conductivity of nanoscale colloidal solutions (nanofluid), *Nano Lett.*, 6, 1529–1534 (2006).

67. P. Keblinski, J. A. Eastman, and D. G. Cahill, Nanofluids for thermal transport, *Mater. Today*, 8, 36–44 (2005).

68. J. A. Eastman, S. U. S. Choi, S. Li, W. Yu, and L. J. Thompson, Anomalously increased effective thermal conductivities of ethylene glycol-based nanofluids containing copper nanoparticles, *Appl. Phys. Lett.*, 78, 718–720 (2001).

69. H. Akoh, Y. Tsukasaki, S. Yatsuya, and A. Tasaki, Magnetic properties of ferromagnetic ultrafine particles prepared by vacuum evaporation on running oil substrate, *J. Cryst. Growth*, 45, 495–500 (1978).

70. J. A. Eastman, U. S. Choi, S. Li, L. J. Thompson, and S. Lee, Enhanced thermal conductivity through the development of nanofluid, *Mater. Res. Soc. Symp. Proc.*, 457, 3–11 (1997).

71. H. Zhu, Y. Lin, and Y. Yin, A novel one-step chemical method for preparation of copper nanofluids, *J. Colloid Interface Sci.*, 277, 100–103 (2004).

72. C. Lo, T. Tsung, and L. Chen, Shape-controlled synthesis of Cu-based nanofluid using submerged arc nanoparticle synthesis system (SANSS), *J. Cryst. Growth*, 277, 636–642 (2005).

73. S. Lee, S. U. S. Choi, S. Li, and J. A. Eastman, Measuring thermal conductivity of fluids containing oxide nanoparticles, *J. Heat Transf.*, 121, 280–289 (1999).

74. X. Wang, X. Xu, and S. U. S. Choi, Thermal conductivity of nanoparticle-fluid mixture, *J. Thermophys. Heat Transf.*, 13, 474–480 (1999).

75. S. M. S. Murshed, K. C. Leong, and C. Yang, Enhanced thermal conductivity of TiO_2—Water Based Nanofluids, *Int. J. Thermal Sci.*, 44, 367–373 (2005).

76. Y. Xuan and Q. Li, Heat transfer enhancement of nanofluids, *Int. J. Heat Fluid Flow*, 21, 58–64 (2000).

77. K. Kwak and C. Kim, Viscosity and thermal conductivity of copper oxide nanofluid dispersed in ethylene glycol, *Korea-Aust. Rheol. J.*, 17, 35–40 (2005).

78. Q. Yu, Y. J. Kim, and H. Ma, Nanofluids with plasma treated diamond nanoparticles, *Appl. Phys. Lett.*, 92, 103111 (2008).

79. M. S. Liu, M. C. C. Lin, I. T. Huang, and C.-C. Wang, Enhancement of thermal conductivity with carbon nanotube for nanofluids, *Int. Commun. Heat Mass Transf.*, 32, 1202–1210 (2005).

80. D. Bom, R. Andrews, D. Jacques, J. Anthony, B. Chen, M. S. Meier, and J. P. Seleque, Thermogravimetric analysis of the oxidation of multiwalled carbon nanotubes: Evidence for the role of defect sites in carbon nanotube chemistry, *Nano Lett.*, 2, 615–619 (2002).

81. H. Xie, H. Lee, W. Youn, and M. Choi, Nanofluids containing multiwalled carbon nanotubes and their enhanced thermal conductivities, *J. Appl. Phys.*, 94, 4967–4971 (2003).

82. H. Lo, T. T. Tsung, L. C. Chen, C. H. Su, and H. M. Lin, Fabrication of copper oxide nanofluid using the submerged arc nanoparticle synthesis system (SANSS), *J. Nanopart. Res.*, 7, 313–320 (2005).

83. A. K. Singh and V. S. Raykar, Microwave synthesis of silver nanofluids with polyvinylpyrrolidone (PVP) and their transport properties, *Colloid Polym. Sci.*, 286, 1667–1673 (2008).

84. A. Kumar, H. Joshi, R. Pasricha, A. B. Mandale, and M. Sastry, Phase transfer of silver nanoparticles from aqueous to organic solutions using fatty amine molecules, *J. Colloid Interface Sci.*, 264, 396–401 (2003).

85. W. Yu, H. Xie, X. Wang, and X. Wang, Highly efficient method for preparing homogeneous and stable colloids containing graphene oxide, *Nanoscale Res. Lett.*, 6, 47 (2011).

86. W. Yu and H. Xie, A review on nanofluids: Preparation, stability mechanisms, and applications, *J. Nanomater.*, 2012, 1–17, Article ID 435873 (2012).

87. H. T. Zhu, C. Y. Zhang, Y. M. Tang, and J. X. Wang, Novel synthesis and thermal conductivity of CuO nanofluid, *J. Phys. Chem. C*, 111, 1646–1650 (2007).

88. Y. Chen and X. Wang, Novel phase-transfer preparation of monodisperse silver and gold nanoparticles at room temperature, *Mater. Lett.*, 62, 2215–2218 (2008).

89. X. Feng, H. Ma, S. Huang, W. Pan, X. Zhang, F. Tian, C. Gao, Y. Cheng, and J. Luo, Aqueous-organic phase transfer of highly stable gold, silver, and platinum nanoparticles and new route for fabrication of gold nanofilms at the oil/water interface and on solid supports, *J. Phys. Chem. B*, 110, 12311–12317 (2006).

90. W. Yu, H. Xie, L. Chen, and Y. Li, Enhancement of thermal conductivity of kerosene-based Fe_3O_4 nanofluids prepared via phase-transfer method, *Colloid Surf. A Physiochem. Eng.*, 355, 109–113 (2010).

91. Servocut S., Product data sheet—SERVOCUT 335 & 345. http://www.ioclindustrialoil. co.in/speciality_lubricants/pdf/servo%20cut.pdf (2017).

92. Y. Gu, X. Zhao, Y. Liu, and Y. Lv, Preparation and tribological properties of dual-coated TiO_2 nanoparticles as water-based lubricant additives, *J. Nanomater.*, 2014, 1–8, Article ID 785680 (2014).

93. S. Ikeda, S. Kawasaki, A. Nobumoto, H. Ono, S. Ono, M. Rusop, H. Muhazli, and B. H. Muhamad, Preparation and applications of hydrophilic nano-carbon particles, *Adv. Mater. Res.*, 832, 767–772 (2014).

94. K. Park, B. Ewald, and P. Y. Kwon, Effect of nano-enhanced lubricant in minimum quantity lubrication balling milling, *J. Tribol.*, 133, 031803 (2011).

95. Thermal conductivity: Theory, properties, and applications, T. Tritt, Kluwer (ed.), Academic/Plenum Publishers, New York (2004). *Thermal conductivity*, https:// en.wikipedia.org/wiki/Thermal_conductivity.

96. G. Paul, M. Chopkar, I. Manna, and P. K. Das, Techniques for measuring the thermal conductivity of nanofluids: A review, *Renew. Sustain. Energy Rev.*, 14, 1913–1924 (2010).

97. J. K. Horrocks and E. McLaughlin, Non-steady state measurements of thermal conductivities of liquids polyphenyls. *Proc. R. Soc. Lond.*, 273(A), 259–274 (1963).

98. B. Merckx, P. Dudoignon, J. P. Garnier, and D. Marchand, Simplified transient hot-wire method for effective thermal conductivity measurement in geo materials: Microstructure and saturation effect, *Adv. Civil Eng.*, 2012, 635395 (2012).

99. M. Sahooli and S. Sabbaghi, Investigation of thermal properties of CuO nanoparticles on the ethylene glycol-water mixture, *Mater. Lett.*, 93, 254–257 (2013).

100. S. E. Gustafsson, Transient plane source techniques for thermal conductivity and thermal diffusivity measurements of solid materials. *Rev. Sci. Instrum.*, 62, 797–804 (1991).

101. H. Xie, H. Gu, M. Fujii, and X. Zhang, Short hot wire technique for measuring thermal diffusivity of various materials, *Meas. Sci. Technol.*, 17, 208–214 (2006).

102. Y. Xuan and Q. Li, Heat transfer enhancement of nanofluids. *Int. J. Heat Mass Transf.*, 21, 58–64 (2000).

103. D. S. Zhu, X. F. Li, N. Wang, X. J. Wang, J. W. Gao, and H. Li, Dispersion behavior and thermal conductivity characteristics of Al_2O_3-H_2O nanofluids, *Curr. Appl. Phys.*, 9, 131–139 (2009).

104. W. Jiang, D. Ding, and H. Peng, Measurement and model on thermal conductivities of carbon nanotube nanorefrigerants, *Int. J. Thermal Sci.*, 48, 1108–1115 (2009).

105. F. Faghani, Thermal conductivity measurement of PEDOT:PSS by the 3ω technique, PhD thesis, Linköping University, Norrköping, Sweden.

106. D. G. Cahill, Thermal conductivity measurement from 30 to 750 K: The 3ω method, *Rev. Sci. Instrum.*, 61, 802 (1990).

107. D. Oh, A. Jain, J. K. Eaton, K. E. Goodson, and J. S. Lee, Thermal conductivity measurement and sedimentation detection of aluminum oxide nanofluids by using the 3 omega method, *Int. J. Heat Fluid Flow*, 29, 1456–1461 (2008).

108. A. R. Challoner and R. W. Powell, Thermal conductivity of liquids: New determinations for seven liquids and appraisal of existing values. *Proc. R. Soc. Lond. Ser. A*, 238, 90–106 (1956).

109. W. Roetzel, S. Prinzen, and Y. Xuan, Measurement of thermal diffusivity using temperature oscillations, in: *Thermal Conductivity*, Vol. 21, C. J. Cremers and H. A. Fine (eds.), pp. 201–207, Plenum Press, New York (1990).

110. W. Czarnetzki and W. Roetzel, Temperature oscillation techniques for simultaneous measurement of thermal diffusivity and conductivity, *Int. J. Thermophys.*, 16, 413–422 (1995).

111. S. K. Das, N. Putra, P. Thiesen, and W. Roetzel, Temperature dependence on thermal conductivity enhancement for nanofluids, *J. Heat Transf.*, 125, 567–574 (2003).

112. H. Kurt and M. Kayfeci, Prediction of thermal conductivity of ethylene glycol-water solutions by using artificial neural networks, *Appl. Energy*, 10, 1016 (2009), see also http://www.sciencedirect.com/.

113. R. W. Powell, Experiments using a simple thermal comparator for measurement of thermal conductivity, surface roughness and thickness of foils or of surface deposits, *J. Sci. Instrum.*, 34, 485–492 (1957).

114. S. Thomas and C. B. P. Sobhan, A review of experimental investigations on thermal phenomena in nanofluids, *Nanoscale Res. Lett.*, 6, 77 (2011).

115. W. Daungthongsuk and S. Wongwises, A critical review of convective heat transfer of nanofluid, *Renew. Sustain. Energy Rev.*, 11, 797–817 (2007).

116. J. Buongiorno, D. C. Venerus, N. Prabhat, T. McKrell, J. Townsend, R. Christianson, Y. V. Tolmachev et al., A bench mark study on the thermal conductivity of nanofluids, *J. Appl. Phys.*, 106, 1–14, Article ID 094312 (2009).

117. C.-W. Nan, R. Birringer, D. R. Clarke, R. David, and H. Gleiter, Effective thermal conductivity of particulate composites with interfacial thermal resistance, *J. Appl. Phys.*, 81, 6692–6699 (1997).

118. W. Evans, R. Prasher, J. Fish, P. Meakin, P. Phelan, and P. Keblinski, Effect of aggregation and interfacial thermal resistance on thermal conductivity of nanocomposites and colloidal nanofluids, *Int. J. Heat Mass Transf.*, 51, 1431–1438 (2008).

119. E. V. Timofeeva, J. L. Routbort, and D. Singh, Particle shape effects on thermophysical properties of alumina nanofluids, *J. Appl. Phys.*, 106, 1–10, Article ID 014304 (2009).

120. E. V. Timofeeva, W. Yu, D. M. France, D. Singh, and J. L. Routbort, Base fluid and temperature effects on the heat transfer characteristics of SiC in EG/H_2O and H_2O nanofluids, *J. Appl. Phys.*, 109, 014914, 2011.

121. T. Cho, I. Baek, J. Lee, and S. Park, Preparation of nanofluids containing suspended silver particles for enhancing fluid thermal conductivity of fluids, *J. Ind. Eng. Chem.*, 11, 400–406 (2005).

122. T. K. Hong, H. S. Yang, and C. J. Choi, Study of the enhanced thermal conductivity of Fe nanofluids, *J. Appl. Phys.*, 97, 064311/1–064311/4 (2005).

123. M. Chopkar, S. Kumar, D. R. Bhandari, P. K. Das, and I. Manna, Development and characterization of Al_2Cu and Ag_2Al nanoparticle dispersed water and ethylene glycol based nanofluid, *Mater. Sci. Eng. B Solid State Mat. Adv. Tech.*, 139, 141–148 (2007).

124. H. U. Kang, S. H. Kim, and J. M. Oh, Estimation of thermal conductivity of nanofluid using experimental effective particle volume, *Exp. Heat Transf.*, 19, 181–191 (2006).

125. Q. Li and Y. Xuan, Experimental investigation on transport properties of nanofluids, *Heat Transf. Sci. Technol.*, 2000, 757–762 (2000).

126. H. E. Patel, S. K. Das, T. Sundararajan, A. S. Nair, B. George, and T. P. Patel, Thermal conductivities of naked and monolayer protected metal nanoparticle based nanofluids: Manifestation of anomalous enhancement and chemical effects, *Appl. Phys. Lett.*, 83, 2931–2933 (2003).

127. D. C. Venerus, M. Kabadi, S. Lee, and V. Perez-Luna, Study of thermal transport in nanoparticle suspensions using forced Rayleigh scattering, *J. Appl. Phys.*, 100, 094310 (2006).

128. L. Godson, D. M. Lal, and S. Wongwises, Measurement of thermo physical properties of metallic nanofluids for high temperature applications, *Nanoscale Microscale Thermophys. Eng.*, 14, 152–173 (2010).

129. K. H. Lee, S. L. Low, G. K. Lim, and C. C. Wong, Surface plasmon enhanced thermal properties of noble metallic nanofluids, *Adv. Sci. Lett.*, 3, 149–153 (2010).

130. N. N. V. Sastry, A. Bhunia, T. Sundararajan, and S. K. Das, Predicting the effective thermal conductivity of carbon nanotube based nanofluids, *Nanotechnology*, 19, 1–8, Article ID 055704 (2008).

131. S. U. S. Choi, Z. G. Zhang, W. Yu, F. E. Lockwood, and E. A. Grulke, Anomalous thermal conductivity enhancement in nanotube suspensions, *Appl. Phys. Lett.*, 79, 2252–2254 (2001).

132. X. Zhang, H. Gu, and M. Fujii, Effective thermal conductivity and thermal diffusivity of nanofluids containing spherical and cylindrical nanoparticles, *J. Appl. Phys.*, 100, 044325 (2006).

133. Y. Hwang, J. K. Lee, C. H. Lee, Y. M. Jung, S. I. Cheong, C. G. Lee, B. C. Ku, and S. P. Jang, Stability and thermal conductivity characteristics of nanofluids, *Thermochim. Acta*, 455, 70–74 (2007).

134. W. Yu, H. Xie, and D. Bao, Enhanced thermal conductivities of nanofluids containing graphene oxide nanosheets, *Nanotechnology*, 21, 055705 (2010).

135. W. Yu, H. Xie, X. Wang, and X. Wang, Significant thermal conductivity enhancement for nanofluids containing graphene nanosheets, *Phys. Lett. A*, 375, 1323–1328 (2011).

136. B. Yang and Z. H. Han, Thermal conductivity enhancement in water-in-FC72 nanoemulsion fluids, *Appl. Phys. Lett.*, 88, 261914 (2006).

137. Z. H. Han and B. Yang, Thermophysical characteristics of water-in-FC72 nanoemulsion fluids, *Appl. Phys. Lett.*, 92, 013118 (2008).

138. Z. H. Han, F. Y. Cao, and B. Yang, Synthesis and thermal characterization of phase changeable indium/polyalphaolefin nanofluids, *Appl. Phys. Lett.*, 92, 243104 (2008).

139. A. K. Singh, Thermal conductivity of nanofluids, *Defence Sci. J.*, 58, 600–607 (2008).

140. R. L. Hamilton and O. K. Crosser, Thermal conductivity of heterogeneous two-component systems, *Ind. Eng. Chem. Fund.*, 1, 187–191 (1962).

141. S. Kabelac and J. F. Kuhnke, Heat transfer mechanisms in nanofluids—Experiments and theory, *Annals of the Assembly for International Heat Transfer Conference*, Vol. 13, p. KN-11, 2006.

142. W. Yu, D. M. France, J. L. Routbort, and S. U. S. Choi, Review and comparison of nanofluid thermal conductivity and heat transfer enhancements, *Heat Trans. Eng.*, 29, 432–460, 2008.

143. S. R. Babu, P. R. Babu, and D. V. Rambabu, Effects of some parameters on thermal conductivity of nanofluids and mechanisms of heat transfer improvement, *Int. J. Eng. Res. Appl. (IJERA)*, 3, 2136–2140 (2013).

144. S. M. S. Murshed, K. C. Leong, and C. Yang, Thermophysical and electro kinetic properties of nano fluids—A critical review, *Appl. Therm. Eng.*, 28, 2109–2125 (2008).

145. S. A. Adio, M. Sharifur, and J. P. Meyer, Investigation into effective viscosity, electrical conductivity, and pH of γ-Al_2O_3-glycerol nanofluids in Einstein concentration regime, *Heat Transf. Eng.*, 36, 1241–1251 (2015).

146. E. Abu-Nada, Effects of variable viscosity and thermal conductivity of Al_2O_3-water nanofluid on heat transfer enhancement in natural convection, *Int. J. Heat Fluid Flow*, 30, 679–690 (2009).

147. L. Xue, Effect of liquid layering at the liquid/solid interface on thermal transport, *Int. J. Heat Mass Transf.*, 47, 4277–4284 (2004).

148. M. Chandrasekar and S. Suresh, A review on the mechanisms of heat transport in nanofluids, *Heat Trans. Eng.*, 30, 1136–1150 (2009).

149. O. M. Wilson, X. Hu, D. G. Cahill, and P. V. Braun, Colloidal metal particles as probes of nanoscale thermal transport in fluids, *Phys. Rev. B*, 66, 224301 (2002).

150. J. L. Barrat and F. Chiaruttini, Kapitza resistance at the liquid-solid interface, *Mol. Phys.*, 101, 1605–1610 (2003).

151. W. Yu and S. U. S. Choi, The role of interfacial layers in the enhanced thermal conductivity of nanofluids: A renovated Maxwell model, *J. Nanopart. Res.*, 5, 167–171 (2003).

152. H. Q. Xie, M. Fujii, and X. Zhang, Effect of interfacial nanolayer on the effective thermal conductivity of nanoparticle-fluid mixture, *Int. J. Heat Mass Transf.*, 48, 2926–2932 (2005).

153. K. C. Leong, C. Yang, and S. M. S. Murshed, A model for the thermal conductivity of nanofluids—The effect of interfacial layer, *J. Nanopart. Res.*, 8, 245–254 (2006).

154. H. Xie, J. Wang, T. Xi, Y. Liu, F. Ai, and Q. Wu, Thermal conductivity enhancement of suspensions containing nanosized alumina particles, *J. Appl. Phys.*, 91, 4568–4572 (2002).

155. R. Prasher, W. Evans, P. Meakin, J. Fish, P. Phelan, and P. Keblinski, Effect of aggregation on thermal conduction in colloidal nanofluids, *Appl. Phys. Lett.*, 89, 143119 (2006).

156. R. P. Prasher, P. Bhattacharya, and P. E. Phelan, Brownian-motion-based convective conductive model for the effective thermal conductivity of nanofluids, *J. Heat Transf.—Trans. ASME*, 128, 588–595 (2006).

157. G. V. Hartland, Measurements of the material properties of metal nanoparticles by time resolved spectroscopy, *Phys. Chem. Chem. Phys.*, 6, 5263–5274 (2004).

158. I. Kim and K. D. Kihm, Measuring near-field nanoparticle concentration profiles by correlating surface plasmon resonance reflectance with effective refractive index of nanofluids, *Opt. Lett.*, 35, 393–395 (2010).

159. P. M. Rad and C. Aghanajafi, The effect of thermal radiation on nanofluid cooled microchannels, *J. Fusion Energy*, 28, 91–100 (2009).

160. L. Mu, Q. Zhu, and L. Si, Radiative properties of nanofluids and performance of a direct solar absorber using nanofluids, *ASME*, 1, 18–21 (2009).

161. H. Xie, W. Yu, and W. Chen, MgO nanofluids: Higher thermal conductivity and lower viscosity among ethylene glycol-based nanofluids containing oxide nanoparticles, *J. Exp. Nanosci.*, 5, 463–472 (2010).

162. V. Trisaksri and S. Wongwises, Critical review of heat transfer characteristics of nanofluids, *Renew. Sustain. Energy Rev.*, 11, 512–523 (2007).

163. X. Q. Wang and A. S. Mujumdar, Heat transfer characteristics of nanofluids: A review, *Int. J. Therm. Sci.*, 46, 1–19 (2007).

164. X. Q. Wang and A. S. Mujumdar, A review on nanofluids—Part I: Theoretical and numerical investigations, *Brazilian J. Chem. Eng.*, 25, 613–630 (2008).

165. J. Zhou, Z. Wu, Z. Zhang, W. Liu, and Q. Xue, Tribological behavior and lubricating mechanism of Cu nanoparticles in oil, *Tribol. Lett.*, 8, 213–218 (2000).

166. Y. Li, J. Zhou, S. Tung, E. Schneider, and S. Xi, A review on development of nanofluid preparation and characterization, *Powder Technol.*, 196, 89–101 (2009).

167. S. Kakac and A. Pramuanjaroenkij, Review of convective heat transfer enhancement with nanofluids, *Int. Heat Mass Transf.*, 52, 3187–3196 (2009).

168. B. Wang, X. Wang, W. Lou, and J. Hao, Rheological and tribological properties of ionic liquid-based nanofluids containing functionalized multi-walled carbon nanotubes, *J. Phys. Chem. C*, 114, 8749–8754 (2010).

169. L. Wang, C. Guo, and R. Yamane, Experimental research on tribological properties of $Mn_{0.78} Zn_{0.22} Fe_2O_4$ magnetic fluids, *J. Tribol.*, 130, 031801 (2008).

170. L. Wang, C. Guo, R. Yamane, and Y. Wu, Tribological properties of Mn-Zn-Fe magnetic fluids under magnetic field, *Tribol. Int.*, 42, 792–797 (2009).

171. S. Chen and D. Mao, Study on dispersion stability and self-repair principle of ultrafine-tungsten disulfide particulates, in: *Advanced Tribology, Proceedings of CIST2008 &ITS-IFToMM2008*, J. Luo, Y. Meng, T. Shao, and Q. Zhao (eds.), pp. 995–999, Tsinghua University Press, Beijing, China (2009).

172. D. X. Peng, C. H. Chen, Y. Kang, Y. P. Chang, and S. Y. Chang, Size effects of SiO_2 nanoparticles as oil additives on tribology of lubricant, *Ind. Lubr. Tribol.*, 62, 111–120 (2010).

173. B. L. Dehkordi, S. N. Kazi, M. Hamdi, A. Ghadimi, E. Sadeghinezhad, and H. S. C. Metselaar, Investigation of viscosity and thermal conductivity of alumina nanofluids with addition of SDBS, *Heat Mass Transf.*, 49, 1109–1115 (2013).

174. J. K. M. William, S. Ponmani, R. Samuel, R. Nagarajan, and J. S. Sangwai, Effect of CuO and ZnO nanofluids in xanthan gum on thermal, electrical and high pressure rheology of water-based drilling fluids, *J. Petrol. Sci. Eng.*, 117, 15–27 (2014).

175. B. Ruan and A. M. Jacobi, Ultrasonication effects on thermal and rheological properties of carbon nanotube suspensions, *Nanoscale Res. Lett.*, 7, 127 (2012).

176. M. Saeedinia, M. A. Akhavan-Behabadi, and P. Razi, Thermal and rheological characteristics of CuO-base oil nanofluid flow inside a circular tube, *Int. Commun. Heat Mass Transf.*, 39, 152–159 (2012).

177. Z. HaiTao, L. ChangJiang, W. DaXiong, Z. CanYing, and Y. YanSheng, Preparation, characterization, viscosity and thermal conductivity of $CaCO_3$ aqueous nanofluids, *Sci. China Tech. Sci.*, 53, 360–368 (2010).

178. B. Wang, X. Wang, W. Lou, and J. Hao, Thermal conductivity and rheological properties of graphite/oil nanofluids, *Colloids Surf. A Physicochem. Eng. Asp.*, 414, 125–131 (2012).

179. I. Tavman, A. Turgut, M. Chirtoc, H. P. Schuchmann, and S. Tavman, Experimental investigation of viscosity and thermal conductivity of suspensions containing nanosized ceramic particles, *Arch. Mat. Sci. Eng.*, 34, 99–104 (2008).

180. B. Yu, Z. Liu, F. Zhou, W. Liu, and Y. Liang, A novel lubricant additive based on carbon nanotubes for ionic liquids, *Mater. Lett.*, 62, 2967–2969 (2008).

181. Y. Y. Wu and W. C. Tsui, Experimental analysis of tribological properties of lubricating oils with nanoparticles additives, *Wear*, 262, 819–825 (2007).

182. D. X. Peng, Y. Kang, R. M. Hwang, S. S. Shyr, and Y. P. Chang, Tribological properties of diamond and SiO_2 nanoparticles added in paraffin, *Tribol. Int.*, 42, 911–917 (2009).

183. M. Zhang, X. Wang, X. Fu, and Y. Xia, Performance and anti-wear mechanism of $CaCO_3$ nanoparticles as a green additive in poly-alpha-olefin, *Tribol. Int.*, 42, 1029–1039 (2009).

7 Silver Nanoparticles Colloidal Dispersions
Synthesis and Antimicrobial Activity

Ali A. Abd-Elaal and Nabel A. Negm

CONTENTS

7.1 INTRODUCTION

The technological attributes of process fluids (PF), especially their tribological and antiadhesion behavior during machining, are very important. The aim of tribology is to ensure that the relative motion of two surfaces happens with the least energy and material loss. The contaminations of metalworking fluids during metal processing have strongly inappropriate actions on the fluid during the technological process, and also increase the economic costs. The quality of the process fluid is also influenced by a bacterial infection. Metalworking fluid (MWF) is used as lubricant, coolant, and/or metal removing agent in machining operations. The modern synthetic MWFs contain organic and inorganic salts, hydrocarbons, organic esters, and lubricating fluids. These ingredients are excellent sources of nutrition for microorganisms [1]. Microbial contamination of MWF has been frequently associated with occupational health problems, such as hypersensitivity pneumonitis due to inhalation of aerosolized MWF bacteria [2,3] and dermatitis due to MWF microflora exposure [4], which occur in metalworkers. The occurrence of mycobacteria in MWF and their potential occupational health significance has been highlighted [3,5,6]. *Mycobacterium immunogenum*, a nontuberculous *Mycobacterium* species, has been reported in MWF and implicated as possible causative agent of hypersensitivity pneumonitis in machine workers who have been exposed to it [7–9]. Among other MWF microbial communities, *pseudomonads* often constitute the major fraction of gram-negative organisms responsible for endotoxin release and accumulation in the fluids, resulting in occupational health hazards in metalworkers [10]. In modern machine industries, the most commonly used method to control microbial contamination is the use of chemical biocides. However, little information is available on the evaluation of the relative efficacy of commercial biocides against microbial genera or species associated with MWFs.

Nanotechnology is the study of the control of matter at an atomic and molecular scale. Generally, nanotechnology deals with structures of the 100 nm size or smaller and involves development of materials or devices within this size range. Nanotechnology is a very diverse field, ranging from novel extensions of conventional device physics, to completely new approaches based on molecular self-assembly, to developing new materials with dimensions on the nanoscale.

There has been much debate on the future implications of nanotechnology. Nanotechnology has the potential to create many new materials and devices with wide-ranging applications in medicine, electronics, and energy production. On the other hand, nanotechnology has raised many of the same issues as with introduction of a new technology. This includes concerns about the toxicity and environmental impact of nanomaterials [11], and their potential effects on global economics. These concerns have led to a debate among advocacy groups and governments on whether special regulation of nanotechnology is warranted.

7.1.1 METAL NANOPARTICLES

Colloidal particles are receiving attention as important starting points for the generation of micro- and nanostructures [12]. These particles are under active research because they possess interesting physical properties that differ considerably from that of the bulk phase. These arise from their small sizes and high surface/volume ratio [13]. The high surface/volume ratio along with size effects (quantum effects) gives nanoparticles distinctively different (chemical, electronic, optical, magnetic, and mechanical) properties from those of the bulk material. For instance, nanoparticle-based semiconductor sensors exhibit higher sensitivities to air pollutants and have lower detection thresholds as well as lower operating temperatures [14]. Nanoparticles are used in many applications, such as electronic, magnetic, optical [15], bioanalysis [16], and environmental remediation [17].

In recent years [18–21], noble metal nanoparticles have been extensively studied and various approaches have been employed for the preparation of metal nanoparticles. Research on metal colloids is greatly stimulated due to the unique properties of nanomaterials (in optical properties, catalytic activity, and magnetic properties that are different from these in the bulk metals). Among the noble metal nanoparticles, silver nanoparticles have attracted more attention [22,23] for their advantage in various studies such as photosensitive components [24], catalysts [25,26], and surface-enhanced Raman spectroscopy.

7.2   SYNTHESIS OF SILVER NANOPARTICLES

Silver nanoparticles can be synthesized by physical and chemical methods. In the physical methods, particles are prepared by such physical processes as laser ablation, evaporation and condensation, and ion sputtering. In the chemical methods, nanoparticles are prepared by chemical oxidation, reduction, and so forth of the precursor material under controlled reaction conditions. The chemical methods of synthesis are preferred over the physical methods due to their energy efficiency.

7.2.1   PHYSICAL METHODS

The most important physical approaches include evaporation–condensation and laser ablation. Various metal nanoparticles such as silver, gold, lead sulfide, cadmium sulfide, and fullerene have been synthesized using the evaporation–condensation method. The absence of solvent contamination in the prepared thin films and the uniformity of the nanoparticle distribution are the advantages of physical approaches compared to the chemical processes. Physical synthesis of silver nanoparticles using a tube furnace at atmospheric pressure has many disadvantages. For example, a tube furnace occupies a large space, consumes a great amount of energy for raising the temperature around the source material, and requires a long time to achieve thermal stability. Moreover, a typical tube furnace requires power consumption of more than several kilowatts and a preheating time of 10–40 min to reach a stable operating temperature [27]. It was demonstrated that silver nanoparticles could be synthesized via a small ceramic heater with a local heating source. The evaporated vapor can cool at a rapid rate, because the temperature gradient in the vicinity of the heater surface is very steep in comparison with the tube furnace. This makes it possible to produce small nanoparticles in high concentration. This physical method can be useful as a nanoparticle generator for long-term experiments such as inhalation toxicity studies, and as a calibration device for nanoparticle measurement equipment [28].

A silver nanoparticle could be synthesized by laser ablation of metallic bulk materials in solution [29]. The ablation efficiency and the characteristics of nanosilver particles produced depend upon many factors such as the wavelength of the laser impinging on the metallic target, the duration of the laser pulses (femto-, pico-, or nanosecond), the laser fluence defined as the optical energy delivered per unit area, the ablation time duration, solvent type, with or without presence of surfactants [30–33]. One important advantage of laser ablation over other methods of production of metal colloids is the absence of chemical reagents in the solutions. As a result, pure and uncontaminated metal colloids can be prepared by this technique for further applications [34]. Silver nanospheroids (20–50 nm) were prepared by laser ablation in water with femtosecond laser pulses at 800 nm wavelength [35]. The formation efficiency and the size of colloidal particles were compared with those of colloidal particles prepared by nanosecond laser pulses. The results revealed that the formation efficiency for femtosecond pulses was significantly lower than that for nanosecond pulses. The size distribution of colloids prepared by femtosecond pulses was narrower than that of colloids prepared by nanosecond pulses. Furthermore, it was found that the efficiency for femtosecond ablation in water was lower than that in air, while, in the case of nanosecond pulses, the ablation efficiency was similar in both water and air.

7.2.2 CHEMICAL METHODS

The most common approach for synthesis of silver nanoparticles is chemical reduction by organic and inorganic reducing agents. In general, sodium citrate, ascorbate, sodium borohydride (NaBH$_4$), elemental hydrogen, polyols, Tollens reagent, N,N-dimethylformamide, and poly(ethylene glycol)-block copolymers are used for the reduction of silver ions (Ag$^+$) in aqueous and nonaqueous solvents. The above agents reduce silver ions (Ag$^+$) and lead to elemental metallic silver (Ag^0), which is followed by agglomeration into oligomeric clusters. These clusters eventually lead to formation of metallic colloidal silver particles [36–38]. It is important to use protective agents to stabilize the dispersed nanoparticles during the metal nanoparticle preparation, and protect the nanoparticles from agglomeration [39]. The presence of surfactants with functionalities (e.g., thiols, amines, acids, and alcohols) for interactions with the particle surface can prevent particle growth and protect particles from sedimentation, agglomeration, or losing their surface properties. Polymeric compounds such as poly(vinyl alcohol), poly(vinyl pyrrolidone), poly(ethylene glycol), poly(methacrylic acid), and poly(methyl methacrylate) have been reported to be effective protective agents to stabilize nanoparticles. Abd-Elaal et al. [40] prepared silver nanoparticles using trisodium citrate as a reducing agent. They also prepared the nanostructure of three nonionic thiol surfactants based on poly(ethylene glycol) with silver nanoparticles. Negm et al. [41] prepared silver nanoparticles loaded by three Gemini cationic surfactants having different spacer chain length. The resultant silver nanoparticles were characterized using ultraviolet–visible (UV–Vis) and transmission electron microscopy (TEM) spectroscopy. TEM results showed the homogeneity and stability of the formed silver nanoparticles in the presence of the synthesized surfactants. Oliveira et al. [39] prepared dodecanethiol-capped silver nanoparticles using Brust procedure [42], based on phase transfer of an Au^{3+} complex from aqueous to organic phase in a two-phase liquid system, followed by a reduction with sodium borohydride in the presence of dodecanethiol as a stabilizing agent. They reported that small changes in synthesis procedure lead to dramatic modifications in nanoparticle structure, average size, size distribution, stability, and self-assembly patterns. Zhang et al. [43] used a hyperbranched poly(methylene-bis-acrylamide aminoethyl piperazine) with terminal dimethyl amine groups to produce silver colloids. The amide moieties, piperazine rings, tertiary amine groups, and the hyperbranched structure were found to be important to its effective stability and reducing ability.

Uniform and size-controllable silver nanoparticles can be synthesized using microemulsion techniques. The nanoparticle preparation is a two-phase aqueous–organic system based on the initial spatial separation of reactants (metal precursor and reducing agent) in two immiscible phases. The interface between the two liquids and the intensity of interphase transport between the phases, which is mediated by a quaternary alkyl ammonium salt, affect the rate of reaction between metal precursor and reducing agent. Metal clusters formed at the interface are stabilized because their surface is coated with stabilizer molecules in the nonpolar aqueous medium. They are transferred to the organic medium by the interphase transporter [44].

One of the major disadvantages of this method is the use of highly deleterious organic solvents. Thus, large amounts of surfactant and organic solvent must be separated and removed from the final product. For instance, Zhang et al. [45] used dodecane as the oil phase (a less deleterious and even nontoxic solvent), and there was no need to separate the prepared silver solution from the reaction mixture. On the other hand, colloidal nanoparticles prepared in nonaqueous media for conductive inks are well-dispersed in a low vapor pressure organic solvent. These advantages can also be found in the application of metal nanoparticles as catalysts in many organic reactions, which are conducted in nonpolar solvents. It is very important to transfer nanometal particles to different physicochemical environments for most practical applications [46].

A simple and effective method, UV-initiated photoreduction, has been reported for synthesis of silver nanoparticles in the presence of citrate, poly(vinyl pyrrolidone), poly(acrylic acid), and collagen. Huang and Yang [47] produced silver nanoparticles via the photoreduction of silver nitrate in layered inorganic laponite clay suspension as a stabilizing agent. The properties of the produced nanoparticles were studied as a function of UV irradiation time. Bimodal size distribution and relatively large silver nanoparticles were obtained when irradiated under UV for 3 h. Further irradiation disintegrated the silver nanoparticles into smaller sizes with a single distribution mode until a relatively stable size and size distribution were obtained. Silver nanoparticles (nanospheres, nanowires, and dendrites) have been prepared by ultraviolet irradiation photoreduction technique at room temperature using poly(vinyl alcohol) as a protecting and stabilizing agent. The concentrations of both poly(vinyl alcohol) and silver nitrate played a significant role in the growth of nanorods and dendrites [48]. Sonoelectrochemistry technique utilizes ultrasonic power primarily to manipulate the material mechanically. The sonoelectrochemical synthesis method involves alternating sonic and electric pulses, and electrolyte composition plays a crucial role in particle shape formation. It was reported [49] that silver nanospheres could be prepared by sonoelectrochemical reduction using a complexing agent, nitrilotriacetate, to avoid aggregation. Nanosized silver particles with an average size of 8 nm were prepared by photoinduced reduction using poly(styrene sulfonate)/poly(allylamine hydrochloride) polyelectrolyte capsules as microreactors [50]. Moreover, it was demonstrated [51] that the photoinduced method could be used for converting silver nanospheres into triangular silver nanocrystals with desired edge lengths in the range of 30–120 nm. The particle growth process was controlled using dual-beam illumination of nanoparticles. Citrate and poly(styrene sulfonate) were used as stabilizing agents. In the study by Malval et al. [52], silver nanoparticles were prepared through a very fast reduction of Ag^+ by α-aminoalkyl radicals generated from hydrogen abstraction to an aliphatic amine by the excited triplet state of two-substituted thioxanthone series ($TX-O-CH_2-COO^-$ and $TX-S-CH_2-COO^-$) (TX: thioxanthone moiety). The quantum yield of this prior reaction was tuned by a substituent effect on the thioxanthones and led to a kinetic control of the conversion of Ag^+ to Ag^0.

Electrochemical synthesis method can be used to synthesize silver nanoparticles [53]. It is possible to control particle size by adjusting the electrolysis parameters and to improve homogeneity of silver nanoparticles by varying the composition of the electrolytic solution. Polyphenylpyrrole-coated silver nanospheroids (3–20 nm) were synthesized by electrochemical reduction at the liquid/liquid interface. This nanocompound was prepared by transferring the silver metal ions from the aqueous phase to the organic phase, where these reacted with pyrrole monomer [53]. In another study [54], monodisperse silver nanospheroids (1–18 nm) were synthesized by electrochemical reduction inside or outside zeolite crystals depending on the degree of silver exchange of the compact zeolite film–modified electrodes. The addition of sodium dodecyl benzene sulfonate to the electrolyte varied the particle size and narrowed the particle size distribution of the silver nanoparticles [55].

Silver nanoparticles can be synthesized using a variety of irradiation methods. Laser irradiation of an aqueous solution of silver salt and surfactant can produce silver nanoparticles with a well-defined shape and size distribution [56]. Laser irradiation was used in a photosensitization synthetic method of making silver nanoparticles using benzophenone. Low laser power at short irradiation time produced silver nanoparticle of approximately 20 nm, while an increased irradiation power produced nanoparticle of approximately 5 nm. Laser and mercury lamps can be used as light sources for the production of silver nanoparticles [57]. In visible light irradiation studies, the photosensitized growth of silver nanoparticles using thiophene (sensitizing dye) and silver nanoparticle formation by illumination of $Ag(NH_3)^+$ in ethanol have been accomplished [17,58].

Microwave-assisted synthesis is a promising method for the synthesis of silver nanoparticles. It was reported that silver nanoparticles could be synthesized by a microwave-assisted synthesis method employing sodium carboxymethyl cellulose as a reducing and stabilizing agent. The size of the resulting particles depended on the concentration of sodium carboxymethyl cellulose and silver nitrate. The resulting nanoparticles were uniform and stable, and were stable at room temperature for 2 months without any visible changes [59]. The production of silver nanoparticles in the presence of Pt seeds, polyvinylpyrrolidone, and ethylene glycol was also reported [60]. Starch has also been employed as a template and reducing agent for the synthesis of silver nanoparticles with an average size of 12 nm, using a microwave-assisted synthesis method. Starch functions as a template, preventing the aggregation of the produced silver nanoparticles [61]. Microwaves in combination with the polyol process were utilized in the synthesis of silver nanospheres using ethylene glycol and poly N-vinyl pyrrolidone as reducing and stabilizing agents, respectively [62]. In a typical polyol process, an inorganic salt is reduced by the polyol (e.g., ethylene glycol which serves as both a solvent and a reducing agent) at a high temperature.

Yin et al. [63] reported the synthesis of size-controlled silver nanoparticles under microwave irradiation of silver nitrate and trisodium citrate solution in the presence of formaldehyde as a reducing agent. The size distribution of the produced silver nanoparticles was strongly dependent on the states of silver cations in the initial reaction medium. Silver nanoparticles with different shapes can be synthesized by microwave irradiation of a silver nitrate ethyleneglycol-$H_2[PtCl_6]$-poly(vinyl pyrrolidone) solution within 3 min [64]. Moreover, the use of microwave irradiation to produce monodispersed silver nanoparticles using amino acids as reducing agents and soluble starch as a protecting agent was reported [65]. Radiolysis of silver ions in ethylene glycol was reported as a method for preparing silver nanoparticles [66]. Moreover, silver nanoparticles supported on silica aerogel were produced using gamma radiolysis. The produced silver clusters were stable in the pH range of 2–9 and started agglomerating at pH > 9 [67].

Oligochitosan as a stabilizer can be used in a preparation of silver nanoparticles by gamma radiation. It was reported that stable silver nanoparticles (5–15 nm) were synthesized at a pH range of 1.8–9.0 using this method [68]. Silver nanoparticles of 4–5 nm were synthesized by γ-ray irradiation of acetic acid solutions containing silver nitrate and chitosan [69]. In another study, silver nanospheroids of size 1–4 nm were produced by γ-ray irradiation of a silver solution in optically transparent inorganic mesoporous silica. Reduction of silver ions within the matrix is brought about by hydrated electrons and hydroalkyl radicals generated during the radiolysis of the 2-propanol solution. The nanoparticles produced within the silica matrix were stable in the presence of oxygen for several months [70]. Moreover, silver nanoparticles in the range of 60–200 nm have been produced by irradiating a solution of silver nitrate and polyvinyl alcohol using 6 MeV electrons [71].

A pulse radiolysis technique was applied to study the factors controlling the shape and size of the nanoparticles of inorganic and organic species in silver nanoparticle synthesis [72]. Dihydroxy benzene was used to synthesize stable silver nanoparticles with an average size of 30 nm in air-saturated solutions [73]. In the polysaccharide method, silver nanoparticles were prepared using water as an environmentally friendly solvent and polysaccharides as capping/reducing agents. A capping agent is a strongly absorbed monolayer of usually organic molecules used to aid stabilization of nanoparticles. For instance, the synthesis of starch–silver nanoparticles was carried out with starch as a capping agent and β-D-glucose as a reducing agent in a gently heated system [20]. The binding interactions between starch and the produced silver nanoparticles were weak and could be reversible at higher temperatures, allowing for the separation of the synthesized nanoparticles. Silver nanoparticles were synthesized by the reduction of silver ions inside nanoscopic starch templates [20]. The extensive network of hydrogen bonds in starch templates provided surface passivation or protection against nanoparticle aggregation. Green synthesis of silver nanoparticles using negatively charged heparin as a reducing, stabilizing agent and nucleation controller was reported. The method involves heating a solution of silver nitrate and heparin to 70°C for approximately 8 h [74]. TEM demonstrated an increase in size of silver nanoparticles with increasing concentrations of silver nitrate and heparin. Moreover, changes in the heparin concentration influenced the morphology and size of silver nanoparticles. The synthesized silver nanoparticles were highly stable and showed no signs of aggregation even after 2 months. In another study [75], stable silver nanoparticles of size 10–34 nm were synthesized using an autoclave at 15 psi and 121°C for 5 min. The synthesized nanoparticles were stable in the solution for 3 months at approximately 25°C. Smaller silver nanoparticles (≤10 nm) were synthesized by mixing silver nitrate solution with starch and solution of NaOH with glucose in a spinning disk reactor for less than 10 min [76].

Recently, a simple one-step process has been used for the synthesis of silver nanoparticles with a controlled size using the Tollens method. Tollens method involves the reduction of $Ag(NH_3)^+$ with an aldehyde [77]. In the modified Tollens procedure, silver ions were reduced by saccharides in the presence of ammonia. The resulting silver nanoparticle films contained particles of size 50–200 nm, silver hydrosols of size 20–50 nm, and silver nanoparticles of different shapes. In this method, the concentration of ammonia of Tollens reagent and the nature of the reducing agent play an important role in controlling the size and morphology of the silver nanoparticles. It was revealed that the smallest particles were formed at the lowest ammonia concentration. Glucose and the lowest ammonia concentration of 5 mM resulted in the smallest average particle size of 57 nm with an intense maximum surface plasmon absorbance at 420 nm. Moreover, an increase in NH_3 concentration from 5 to 200 mM resulted in simultaneous increases of particle size and polydispersity [78]. Silver nanoparticles with controllable sizes were synthesized by reduction of $[Ag(NH_3)_2]^+$ with glucose, galactose, maltose, and lactose [79]. The nanoparticle synthesis was carried out at various ammonia concentrations (0.005–0.20 M) and pH (11.5–13.0), resulting in average particle sizes of 25–450 nm. The particle size was influenced by NH_3 concentration, the structure of the reducing agent, and pH. Sodium dodecyl sulfate, polyoxyethylene sorbitan monooleate (Tween 80), and polyvinylpyrrolidone were used to stabilize the silver nanoparticles [80,81].

Polyoxometalates (POMs) can be defined as molecular metal oxides formed through condensation reactions of early transition metal–oxygen anions, typically with metals such as W, Mo, and V, in their highest oxidation state with different structures, as shown in Figure 7.1 [82].

Silver, gold, palladium, and platinum nanoparticles can be produced at room temperature by mixing the aqueous solution of the metal ions with reduced polyoxometalates

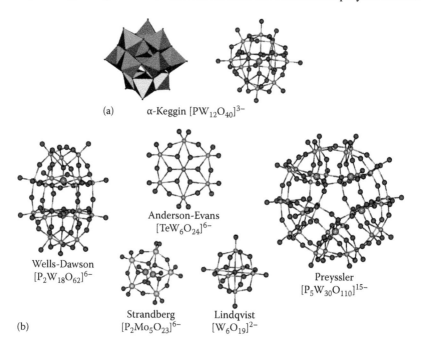

(a)   α-Keggin $[PW_{12}O_{40}]^{3-}$

Wells-Dawson $[P_2W_{18}O_{62}]^{6-}$

Anderson-Evans $[TeW_6O_{24}]^{6-}$

Preyssler $[P_5W_{30}O_{110}]^{15-}$

(b)   Strandberg $[P_2Mo_5O_{23}]^{6-}$ Lindqvist $[W_6O_{19}]^{2-}$

FIGURE 7.1   (a): Polyhedra (left) and ball and stick (right) representation of α-Keggin polyoxotungstate $[PW_{12}O_{40}]^{3-}$; (b) ball and stick structures of other representative polyoxometalates.

as reducing and stabilizing agents [82]. Polyoxometalates are soluble in water and have the ability to undergo stepwise, multielectron redox reactions without disturbing their structure. It was demonstrated that silver nanoparticles can be produced by illuminating a deaerated solution of polyoxometalate/S/Ag^+ (polyoxometalate: $[PW_{12}O_{40}]$, $[SiW_{12}O_{40}]$; S: 2-propanol or 2,4-dichlorophenol). Furthermore, green-chemistry-type one-step synthesis and stabilization of silver nanostructures with Mo^{5+}–Mo^{6+} mixed-valence polyoxometalates in water at room temperature has been reported [82].

7.3 ANTIMICROBIAL EFFECT OF NANOPARTICLES

Nanoparticles, because of their size, can easily pass through the cell boundaries of living cells. The cellular uptake of nanoparticles is several times higher than bulk particles. However, the rate of accumulation of nanoparticles exceeds the cells' ability to remove them. As a result, the cells' structural and functional integrity crumbles, as particles begin to interject and obstruct their physiological processes. This is accomplished by simply "blocking the way" and also by interacting randomly to biomolecules, which were meant to interact differently, for natural metabolism. Because of their extremely small size, the nanoparticles may interact directly with macromolecules such as DNA. Moreover, nanoparticles are easily translocated to other tissues, using the circulatory system. So, nanoparticles are toxic to a cell because they can blend in with any structural or functional component of it and, thereby, disrupt the normal cellular functions.

Several studies [83,84] have been conducted about the nature of interactions between biological components and nanoparticles, and their impact on the first barrier to their entry into the biomembranes. Disruption patterns of polycationic organic nanoparticles on model biological membranes and living cell membranes have been shown to occur even at nanomolar concentrations. The degree of disruption depends on the size and charge of the nanoparticle, and the phase (fluid or liquid crystalline) of the biological membrane.

Not all nanoparticles produce these adverse health effects. The toxicity of nanoparticles depends on various factors including size, aggregation, composition, crystallinity, surface functionalization, and so on. It is also very important to recognize that not all nanoparticles are toxic. Some have been found to be nontoxic, and others were found to be beneficial to the body [85].

Silver nanoparticles are used as antibacterial/antifungal agents in a diverse range of applications. Examples include air sanitizer sprays, face masks, wet wipes, detergents, shampoos, toothpastes, air filters, coatings of refrigerators, washing machines, and food storage containers.

7.3.1 SILVER NANOPARTICLES: SYNTHESIS AND BIOLOGICAL ACTIVITY

Silver nanoparticles have been found to have strong antimicrobial activity and are used in several applications:

- Wound dressings [86]
- Contraceptive devices [87]
- Surgical instruments and bone prostheses [88]

- Coatings on the ocular lenses for the prevention of microbial activity [89]
- Antifungal activity; anti-inflammatory effect [90]
- Antiviral activity [91]

7.3.2 EFFECTS OF VARIOUS PARAMETERS ON STABILIZATION OF SILVER NANOPARTICLES

Silver nanoparticles have been intensively investigated in recent years due to their unique physical and chemical properties. However, the low concentration and poor stability of silver nanoparticles in aqueous dispersion limit their wide applications.

Patil et al. [92] investigated the synthesis of silver nanoparticles in microreactor using surfactants. This investigation studied the effect of process parameters in microreactor such as concentration and flow rate of a precursor. Reduction reaction was carried out using silver nitrate and sodium borohydride. Two surfactants namely sodium dodecyl sulfate (SDS) and cetyl trimethyl ammonium bromide (CTAB) were evaluated for their effect on particle size by controlling nucleation and growth mechanism. Optimum parameters such as flow rate and concentrations of reactants/surfactants to obtain nanosized silver colloidal particles in continuous microreactor were determined. Results showed that use of (0.02 g/mL) SDS with flow rate 1 mL/min (0.001 M) $AgNO_3$ and (0.003 M) $NaBH_4$ 3 mL/min flow rate produced minimum particle size of 4.8 nm. Increasing in the particle size with an increase in the CTAB concentration was due to the agglomeration effect of the cationic CTAB, which has a positive charged head. SDS, on the other hand, favors the reduction in particle size of silver nanoparticles. It was found that the optimum flow conditions and concentration of SDS surfactant (0.02 g/mL) will help to generate monodispersed nanoparticles. It was successfully demonstrated that using a microreactor, monodispersed, nanosized (3–7 nm) silver nanoparticles were synthesized at optimum process conditions. It was also observed that nanoparticles with controlled size could be synthesized continuously using a microreactor.

He et al. [93] employed dimethylene-1,2-bis(dodecyl dimethyl ammonium bromide) (12-2-12) and dodecyl trimethyl ammonium bromide (DTAB) (Figure 7.2) to prepare and stabilize silver nanoparticles. It was found that when using 12-2-12 as a stabilizer, the concentration of silver nanoparticles in aqueous dispersion reached 44.6 mM, and the average size was 11 ± 3 nm: by contrast, using DTAB as a

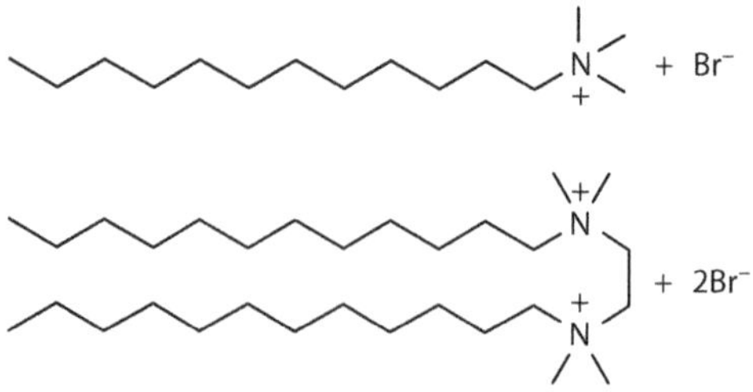

FIGURE 7.2 Chemical structures of dimethylene-1,2-bis(dodecyl dimethyl ammonium bromide) (12-2-12) and dodecyl trimethyl ammonium bromide (DTAB).

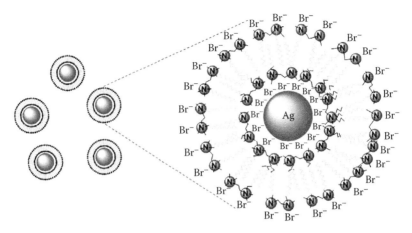

FIGURE 7.3 Schematic illustration of silver nanoparticles stabilized by dimethylene-1,2-bis(dodecyl dimethyl ammonium bromide) (12-2-12).

stabilizer gave the concentration of silver nanoparticle in aqueous dispersion reached to 35.7 mM and 15 ± 10 nm size distribution [83]. Silver nanoparticles stabilized by 12-2-12 also show better long-term stability than those stabilized by DTAB. Negative staining TEM observation showed that the 12-2-12 molecule, due to its double-head architecture and high charge density (Figure 7.2), had more stable and more efficient capping ability to stabilize silver nanoparticles than DTAB (Figure 7.3).

Sarkar et al. [94] synthesized 2D and 3D Ag nanostructures with disk and spherical morphology. They carried out simple silver mirror reaction in the presence of an anionic surfactant. The aldehyde-mediated reduction of silver nitrate occurred at relatively low temperatures (80°C–86°C), which is easy to reproduce, making the process facile and faster. Scanning electron microscopy (SEM) study revealed the temperature-dependent morphological variation of the nanoparticles. At 80°C, only globular morphology was observed; at 82°C–84°C, disk morphology proportionately increased; and at 86°C, they appeared in equal proportion. Nanodisks of uniform thickness could be grown up to micrometer size. The antibacterial study showed the highest activity to date for Ag nanostructures against a series of very devastating bacterial strains. It was found to be more powerful than penicillin for two bacteria (*Escherichia coli* and *Staphylococcus aureus*). The adopted synthesis method was very simple and can be easily implemented for any kind of scientific as well as industrial application due to its cost-effectiveness. An important characteristic of the Ag nanoparticles obtained is its homogeneous colloidal dispersion in water, which is one of the essential requirements for biologically relevant nanodevice applications.

Janardhanan et al. [95] reported a simple wet chemical route for synthesis of silver nanoparticles and their surface properties. Silver nanoparticles of size 40–80 nm were obtained by oxidation process of glucose to gluconic acid with amine in the presence of silver nitrate, and the gluconic acid caps the nanosilver particle. The presence of gluconic acid on the surface of nanosilver particles was confirmed by X-ray photoelectron spectroscopy (XPS) and Fourier transform infrared (FTIR) spectroscopy.

Encapsulation of silver nanoparticle by gluconic acid did not result in surface oxidation, as confirmed by XPS. The silver nanoparticles have also been studied for their formation, structure, morphology, and size using UV–Vis spectroscopy, X-ray diffraction (XRD), and SEM. Further, the antibacterial properties of these nanoparticles showed promising results for *E. coli*. The influence of the alkaline medium toward the particle size and yield was studied by varying the pH of the reaction medium using DEA, NaOH, and Na_2CO_3.

Iqbal et al. [96] synthesized stable silver substituted hydroxyapatite (Ag-HA) nanoparticles with controlled morphology and particle size. CTAB was included as a surfactant in the microwave refluxing process. The nanoparticles produced with different silver ion concentrations (0.05, 0.1, and 0.2 wt.%) were characterized using XRD, FTIR, field emission scanning electron microscope (FESEM), energy dispersive X-ray (EDX), and Brunauer–Emmett–Teller (BET) analysis. FESEM images showed that the resulting nanoparticles were in the size range of 58–72 nm and had uniform elongated spheroid morphology. The dielectric investigation showed an increase in dielectric constant and dissipation factor values with increasing Ag-concentrations. Antibacterial evaluation of the Ag-HA samples using disk diffusion technique and minimum inhibitory concentration demonstrated antibacterial activity against *S. aureus*, *Bacillus subtilis*, *Pseudomonas aeruginosa*, and *E. coli*. This effect was dose dependent and was more pronounced against gram-negative bacteria than gram-positive organisms.

Silver nanoparticles have been used extensively as antimicrobial agents in the health industry, food storage, textile coatings, and a number of environmental applications. Antimicrobial properties of silver nanoparticles allowed the use of these nanoparticles in different fields of medicine, various industries, animal husbandry, packaging, accessories, cosmetics, health, and military. Silver nanoparticles show potential antimicrobial effects against infectious organisms, including *E. coli*, *B. subtilis*, *Vibrio cholerae*, *P. aeruginosa*, *Syphilis typhus*, and *S. aureus* [97,98]. For instance, it was shown that silver nanoparticles mainly in the range of 1–10 nm attached to the surface of *E. coli* cell membrane; disturb its proper function such as respiration and permeability [99]. It was observed that silver nanoparticles had a higher antibacterial effect on *B. subtilis* than on *E. coli*, suggesting a selective antimicrobial effect, possibly related to the structure of the bacterial membrane [100]. As antimicrobial agents, silver nanoparticles were applied in a wide range of applications from disinfecting medical devices and home appliances to water treatment [101].

7.3.3 EFFECT OF SOLVENT ON STABILIZATION OF SILVER NANOPARTICLES

Khan et al. [102] reported the effects of polar-protic and polar-aprotic solvents on absorption spectra and particle size of surfactant-stabilized silver nanoparticles. Cysteine and cetyl trimethyl ammonium bromide, CTAB, were used as the reducing and stabilizing agents, respectively. The resulting orange color silver sols possessed an unusually narrow plasmon absorption shoulder at 450 nm. The absorbance and shape of this shoulder are affected by the protic (methanol, ethanol) and aprotic (acetonitrile, dimethylformamide, dimethyl sulfoxide, and 1,4-dioxane) nature of solvents. These observations are interpreted in terms of the dielectric constant, boiling point, hydrogen bonding, solubilization, and electron donation of the silver particles to the solvents. Absorbance increased with increasing dielectric constants of the reaction mixture. The particle size decreased with decreasing hydrophobic character of the protic solvents. Thus, particle size of 39, 58, and 70 nm was obtained without solvent, with ethanol and methanol, respectively.

The following conclusions were deduced from this study: (1) the particle size and plasmon absorption band change by varying the organic solvent; the absorbance decreased with increasing dielectric constant of the medium and surface-adsorbed species; (2) addition of methanol and ethanol resulted in an increase in the particle size; (3) addition of acetonitrile, dimethylformamide, dioxane, and dimethyl sulfoxide also decreased the absorbance by decreasing the growth rate; (4) higher volume of solvents did not bring any significant change in the absorbance. Various factors such as hydrophobicity, polarity, induction, steric hindrance, resonance, and nucleophilicity were responsible for the optical changes.

7.4   HISTORY OF SILVER AS ANTIMICROBIAL AGENT

For centuries, silver has been used for the treatment of burns and chronic wounds. As early as 1000 BC, silver was used to make water potable [103,104]. Silver nitrate was used in its solid form and was known by different terms such as "Lunar caustic" in English, "Lapis infernale" in Latin, and "Pierre infernale" in French [105]. In 1700, silver nitrate was used for the treatment of venereal diseases, fistulae from salivary glands, and bone and perianal abscesses [106]. In the nineteenth century, granulation tissues were removed using silver nitrate to allow epithelization and to promote crust formation on the surface of wounds. Varying concentrations of silver nitrate were used to treat fresh burns [104,105]. In 1881, Carl S.F. Creed cured ophthalmia neonatorum using silver nitrate eye drops. Creed's son, B. Creed designed silver-impregnated dressings for skin grafting [105,106]. In the 1940s, after penicillin was introduced, the use of silver for the treatment of bacterial infections decreased [107–109]. The use of silver reappeared in the 1960s when Moyer introduced the use of 0.5% silver nitrate for the treatment of burns. He proposed that silver nitrate solution does not interfere with epidermal proliferation and possesses antibacterial property against *S. aureus*, *P. aeruginosa*, and *E. coli* [110,111]. In 1968, silver nitrate was combined with sulfonamide to form silver sulfadiazine cream, which served as a broad-spectrum antibacterial agent and was used for the treatment of burns. Silver sulfadiazine is effective against bacteria like *E. coli*, *S. aureus*, *Klebsiella* sp., and *Pseudomonas* sp. It also possesses some antifungal and antiviral activities [112]. Recently, the emergence of antibiotic-resistant bacteria and limitations on the use of existing antibiotics have made clinicians use silver wound dressings containing varying levels of silver [109,113].

Panacek et al. [79] reported a one-step synthesis protocol for silver colloid nanoparticles. They found high antimicrobial and bactericidal activities of silver nanoparticles on gram-positive and gram-negative bacteria including multiresistant strains such as methicillin-resistant *S. aureus*. The antibacterial activity of silver nanoparticles was found to be size dependent, with 25 nm nanoparticles having the highest antibacterial activity. The nanoparticles were toxic to bacterial cells at concentrations of 1.69 µg/mL Ag.

Shahverdi et al. [114] investigated the combined effect of silver nanoparticles with antibiotics. The silver nanoparticles were synthesized using *Klebsiella pneumoniae* and were evaluated for their antimicrobial activity against *S. aureus* and *E. coli*. The results showed the antibacterial activity of antibiotics like penicillin G, amoxicillin, erythromycin, clindamycin, and vancomycin increased in the presence of silver nanoparticles against *E. coli* and *S. aureus*. The highest synergistic activity was observed with erythromycin against *S. aureus*. Shrivastava et al. [115] reported the synthesis of silver nanoparticles in the size range of 10–15 nm and effect of dose on the gram-negative and gram-positive microorganisms. They concluded that dose-dependent silver nanoparticles had marked activity against gram-negative organisms more than gram-positive organisms.

Pal et al. [116] investigated the antibacterial properties of silver nanoparticles of different shapes and found that efficacy was shape dependent. The silver nanoparticles were prepared by the seeded growth method for the synthesis of spherical nanoparticles. Solution phase method was used for the synthesis of rod-shaped and truncated triangular nanoparticles. The synthesized nanoparticles were purified by centrifugation at 2100 rpm for 10 min and suspended in water. To investigate efficacy kinetics of silver nanoparticles, *E. coli* was inoculated to nutrient broth containing different concentrations of silver nanoparticles. The mixture was incubated at 37°C and placed horizontally on an orbital shaker platform and agitated at 225 rpm. Nutrient agar plates inoculated with 100 µL of bacterial suspension were treated with different concentrations of silver (1, 6, 12, 12.5, 50, or 100 µg/mL). The plates were incubated overnight at 37°C and the characterization of nanoparticles was conducted using UV–Vis spectroscopy and energy-filtering TEM (EFTEM). UV–Vis spectroscopy of nanoparticles synthesized by seeded growth method showed absorption band at 420 nm, demonstrating the presence of spherical nanoparticles, which was also confirmed using TEM images. The nanoparticles synthesized by solution phase method depicted two absorption bands at 418 and 514 nm. The synthesis of rod-shaped nanoparticles was confirmed by EFTEM, while the synthesis of truncated triangular nanoparticles was confirmed by XRD. The inhibition of bacterial growth by spherical nanoparticles was observed at silver concentration of 12.5 µg/mL. Truncated triangular nanoparticles bacterial growth inhibition was observed at 1 µg/mL of silver concentration. On nutrient agar plates, the spherical nanoparticles inhibited bacterial growth at a silver nanoparticle concentration of 6 µg/mL. In the case of truncated triangular nanoparticles, 10 µg/mL concentration of silver led to inhibition of bacterial growth. These findings confirmed that antibacterial activity of silver nanoparticles was shape dependent [117].

Silver has been known to possess strong antimicrobial properties both in its metallic and nanoparticle forms. Hence, it has found a variety of applications in different fields:

- The Fe_3O_4-attached Ag nanoparticles can be used for water treatment and easily removed using a magnetic field to avoid environmental contamination [118].
- Silver sulfadiazine showed good healing of burn wounds due to its slow and steady reaction with serum and other body fluids [112].

- Nanocrystalline silver dressings, creams, and gel effectively reduce bacterial infections in chronic wounds [103,119,120].
- Silver nanoparticle containing polyvinyl nanofibers also show efficient antibacterial property as wound dressing [121].
- Silver nanoparticles are reported to show good wound-healing capacity, effective cosmetic appearance, and scarless healing when tested using an animal model [122].
- Silver-impregnated medical devices such as surgical masks and implantable devices show significant antimicrobial efficacy [123].
- Environmentally friendly antimicrobial nanopaint can be developed with silver nanoparticles [124].
- Inorganic composites are used as antimicrobial and preservatives for various products [125].
- Silica gel microspheres mixed with silica thiosulfate are used as long-lasting antibacterial agent [125].
- Silver nanoparticles have been used for the treatment of burns and various infections [126].
- Silver zeolite is used in food preservation, disinfection, and decontamination of products [127,128].
- Silver nanoparticles can be used for disinfection of water [101].

7.5 SILVER NANOPARTICLE APPLICATION IN TRIBOLOGY

Cutting fluids play a significant role in machining operations and impact shop productivity, tool life, and quality of work. The reduction in the consumption rate of the cutting fluid leads to the minimization of production cost and environmental hazards. This could be achieved by the enhancement of its thermal and tribology properties with the inclusion of suitable additives in the cutting fluid. In recent years various nanoparticles were used as additives in the conventional cutting fluid to enhance its properties.

Saravanakumar et al. [129] prepared silver nanoparticles by chemical reduction using sodium borohydride ($NaBH_4$) as a reducing agent in the presence of cetyl trimethyl ammonium bromide (CTAB) as a capping agent. The prepared silver nanoparticles were dispersed in the cutting fluid using ultrasonic bath and were characterized by TEM and UV–Vis absorption spectra. Experiments were conducted in turning operation to evaluate the enhancement in the properties of the modified cutting fluid and the results were compared with the conventional cutting fluid. Based on the experimental results, the temperature values measured during turning process for minimum cutting fluid is observed with nanoparticle additive. This shows the enhancement in the heat transfer rate at the cutting zone. Authors confirmed that the cutting forces were reduced to an extent of 8.8% for the cutting fluids with the inclusion of nanoparticles. The surface roughness of the workpiece decreased to an extent of 7.5% with the use of nanoparticles cutting fluid. They confirmed the enhancement in the thermal and frictional properties for the cutting fluids dispersed with silver nanoparticles. This leads to reduction in the consumption of cutting fluid, increased cutting tool life, and better surface finish of the machined workpiece. However, the properties of the cutting fluid could be further improved by dispersing different nanoparticles of different size, shape, and volume fraction.

Ma et al. [130] prepared monodisperse silver nanoparticles with a particle size of about 6–7 nm and low volatile multialkylated cyclopentanes (MACs) lubricant. The effect of silver nanoparticles as additive in MACs base oil on the friction and wear behavior of MACs was investigated. The friction and wear test of a steel disk sliding against the same steel counterpart ball was carried out on an optimal oscillating friction and wear tester. The morphology and elemental distribution of the worn surface of both the steel ball and steel disk and the chemical feature of typical element thereof were examined using a JEM-1200EX scanning electron microscope equipped with a Kevex energy dispersive X-ray analyzer attachment and X-ray photoelectron spectroscope, respectively. Friction and wear test indicates that the wear resistance and load-carrying capacity of MACs base oil were markedly raised and its friction coefficient changed little when 2% Ag nanoparticles were added in it. Results show that Ag nanoparticles were deposited on the friction pair surfaces to form low shearing stress metal Ag protective film in the rubbing process.

Zhang et al. [131] investigated the effect of surface-modified Ag nanoparticles as additives of multialkylated cyclopentanes (MACs) in air and vacuum by a vacuum four-ball tribometer. The results showed that both the MACs and MACs containing Ag nanoparticles exhibited steady and low friction coefficients and slight wear in air. However, under vacuum conditions, MACs showed the initial seizure-like high friction, while introducing Ag nanoparticles could effectively eliminate it. Confirmed by the surface analysis, the improved tribological performances caused by Ag nanoparticles could be ascribed to the metal Ag boundary film formed on the friction pair surfaces during tests. Sliding friction experiments were carried out under vacuum and in air lubricated by MACs and MACs containing Ag nanoparticles. Ag nanoparticles were effective in improving the tribological properties of MACs base oil, especially in eliminating the initial seizure-like high friction under vacuum condition. The tribological mechanism of Ag nanoparticles was the adsorption and deposition of Ag nanoparticles on the friction pair surfaces to form the boundary film, which decreased the shearing stress and effectively inhibited initial high friction under vacuum condition. Ag nanoparticles could be considered as a potential candidate for lubricating oil additives for space application.

7.6 SUMMARY

Nanoparticles are one of the most effective additives in several applications. They are used mostly as antimicrobial or biocidal agents to prevent the growth of bacteria, fungi, and yeast. Nanoparticles are also used in tribological systems and cutting fluids as biocides to prevent the growth of degradative microorganisms that decrease the efficiency of these fluids by degrading their different components. Furthermore, cutting fluids formulated by nanoparticle components are more efficient. Nanoparticles are a new trend in formulating cutting fluids and in improving the efficiencies of tribological systems.

REFERENCES

1. K.L.M. Buers, E.L. Prince, and C.J. Knowles, The ability of selected bacterial isolates to utilize components of synthetic metal-working fluids as sole sources of carbon and nitrogen for growth, *Biotechnol. Lett.*, *19*, 791–794 (1997).
2. J. Fox, H. Anderson, T. Moen, G. Gruetzmacher, L. Hanrahan, and J. Fink, Metal working fluid-associated hypersensitivity pneumonitis: An outbreak investigation and case-control study, *Am. J. Ind. Med.*, *35*, 58–67 (1999).
3. K. Kreiss and J. Cox-Ganser, Metal working fluid-associated hyper sensitivity pneumonitis: A workshop summary, *Am. J. Ind. Med.*, *32*, 423–432 (1997).
4. A.I. Awosika-Olumo, K.L. Trangle, and L.F. Fallon, Microorganism induced skin disease in workers exposed to metal working fluids, *Occup. Med.*, *53*, 35–40 (2003).
5. J.O. Falkinham, Mycobacterial aerosols and respiratory disease, *Emerg. Infect. Dis.*, *9*, 763–767 (2003).
6. G.B. Shelton, W.D. Flanders, and G.K. Morris, *Mycobacterium* sp. as a possible cause of hypersensitivity pneumonitis in machine workers, *Emerg. Infect. Dis.*, *5*, 270–273 (1999).
7. D. Trout, D.N. Weissman, D. Lewis, R.A. Brundage, A. Franzblau, and D. Remick, Evaluation of hypersensitivity pneumonitis among workers exposed to metal removal fluids, *Appl. Occup. Environ. Hyg.*, *18*, 953–960 (2003).
8. R.J. Wallace, Jr., Y. Zhang, R.W. Wilson, L. Mann, and H. Rossmoore, Presence of a single genotype of the newly described species *Mycobacterium immunogenum* in industrial metalworking fluids associated with hypersensitivity pneumonitis, *Appl. Environ. Microbiol.*, *68*, 5580–5584 (2002).
9. J.S. Yadav, I. Khan, F. Fakhari, and M.B. Soellner, DNA-based methodologies for rapid detection, quantification, and species- or strain-level identification of respiratory pathogens (mycobacteria and pseudomonads) in metalworking fluids, *Appl. Occup. Environ. Hyg.*, *18*, 966–975 (2003).
10. D.I. Bernstein Z.L. Lummus, G. Santilli, J. Siskosky, and I.L. Bernstein, Machine operator's lung. A hypersensitivity pneumonitis disorder associated with exposure to metalworking fluid aerosols. *Chest*, *108*, 636–641 (1995).
11. C. Buzea, I. Pacheco, and K. Robbie, Nanomaterials and nanoparticles: Sources and toxicity, *Biointerphases*, *2*, 17–71 (2007).
12. I.M. Yakutik, G.P. Shevchenko, and S.K. Rakhmanov, The formation of monodisperse spherical silver particles, *Colloids Surf. A Physicochem. Eng. Asp.*, *242*, 175–179 (2004).
13. K. Patel, S. Kapoor, D.P. Dave, and T. Mukherjee, Synthesis of nanosized silver colloids by microwave dielectric heating, *J. Chem. Sci.*, *117*, 53–60 (2005).
14. M.I. Baraton and L. Merhari, Advances in air quality monitoring via nanotechnology, *J. Nanoparticle. Res.*, *6*, 107–117 (2004).
15. F.E. Kruis, H. Fissan, and A. Peled, Synthesis of nanoparticles in the gas phase for electronic, optical and magnetic applications—A review, *J. Aerosol Sci.*, *29*, 511–535 (1998).
16. S.G. Penn, L. He, and M.J. Natan, Nanoparticles in bioanalysis, *Curr. Opin. Chem. Biol.*, *7*, 609–615 (2003).
17. P.V. Kamat and D. Meisel, Nanoscience opportunities in environmental remediation, *Comptes Rendus Chimie*, *6*, 999–1007 (2003).
18. B. Yin, H. Ma, S. Wang, and S. Chen, Electrochemical synthesis of silver nanoparticles under protection of poly(N-vinylpyrrolidone), *J. Phys. Chem. B*, *107*, 8898–8904 (2003).
19. R.M. Penner, Mesoscopic particles and wires by electrodeposition, *J. Phys. Chem. B*, *106*, 3339–3353 (2002).

20. P. Raveendran, J. Fu, and S.L. Wallen, Completely "green" synthesis and stabilization of metal nanoparticles, *J. Am. Chem. Soc.*, *125*, 13940–13941 (2003).
21. X.Z. Lin, X. Teng, and H. Yang, Direct synthesis of narrowly dispersed silver nanoparticles using a single-source precursor, *Langmuir*, *19*, 10081–10085 (2003).
22. Z. Jiang, W. Yuan, and H. Pan, Luminescence effect of silver nanoparticle in water phase, *Spectrochim. Acta A*, *61*, 2488–2494 (2005).
23. D.D. Evanoff and G. Chumanov, Size-controlled synthesis of nanoparticles. 1. 'Silver-only' aqueous suspensions via hydrogen reduction, *J. Phys. Chem. B*, *108*, 13948–13956 (2004).
24. R.K. Hailstone, Computer simulation studies of silver cluster formation on AgBr microcrystals, *J. Phys. Chem.*, *99*, 4414–4428 (1995).
25. Y. Shiraishi and N. Toshima, Colloidal silver catalysts for oxidation of ethylene, *J. Mol. Catal. A*, *141*, 187–192 (1999).
26. M. Kowshik, S. Ashtaputre, S. Kharrazi, W. Vogel, J. Urban, S.K. Kulkarni, and K.M. Paknikar, Extracellular synthesis of silver nanoparticles by a silver-tolerant yeast strain MKY3, *Nanotechnology*, *14*, 95–100 (2003).
27. F. Kruis, H. Fissan, and B. Rellinghaus, Sintering and evaporation characteristics of gas-phase synthesis of size-selected PbS nanoparticles, *Mater. Sci. Eng. B*, *69*, 329–334 (2000).
28. J. Jung, H. Oh, H. Noh, J. Ji, and S. Kim, Metal nanoparticle generation using a small ceramic heater with a local heating area, *J. Aerosol Sci.*, *37*, 1662–1670 (2006).
29. F. Mafune, J. Kohno, Y. Takeda, T. Kondow, and H. Sawabe, Structure and stability of silver nanoparticles in aqueous solution produced by laser ablation, *J. Phys. Chem. B*, *104*, 8333–8337 (2000).
30. S. Kim, B. Yoo, K. Chun, W. Kang, J. Choo, M. Gong, and S. Joo, Catalytic effect of laser ablated Ni nanoparticles in the oxidative addition reaction for a coupling reagent of benzylchloride and bromoacetonitrile, *J. Mol. Catal. A*, *226*, 231–234 (2005).
31. S. Link, C. Burda, B. Nikoobakht, and M. El-Sayed, Laser-induced shape changes of colloidal gold nanorods using femtosecond and nanosecond laser pulses, *J. Phys. Chem. B*, *104*, 6152–6163 (2000).
32. N. Tarasenko, A. Butsen, E. Nevar, and N. Savastenko, Synthesis of nanosized particles during laser ablation of gold in water, *Appl. Surf. Sci.*, *252*, 4439–4444 (2006).
33. M. Kawasaki and N. Nishimura, 1064-nm laser fragmentation of thin Au and Ag flakes in acetone for highly productive pathway to stable metal nanoparticles, *Appl. Surf. Sci.*, *253*, 2208–2216 (2006).
34. T. Tsuji, K. Iryo, N. Watanabe, and M. Tsuji, Preparation of silver nanoparticles by laser ablation in solution: Influence of laser wavelength on particle size, *Appl. Surf. Sci.*, *202*, 80–85 (2002).
35. T. Tsuji, T. Kakita, and M. Tsuji, Preparation of nano-Size particle of silver with femtosecond laser ablation in water, *Appl. Surf. Sci.*, *206*, 314–320 (2003).
36. B. Wiley, Y. Sun, B. Mayers, and Y. Xi, Shape-controlled synthesis of metal nanostructures: The case of silver, *Chem. Eur. J.*, *11*, 454–463 (2005).
37. D.D. Evanoff Jr. and G. Chumanov, Size-controlled synthesis of nanoparticles. 2. Measurement of extinction, scattering, and absorption cross sections, *J. Phys. Chem. B*, *108*, 13957–13962 (2004).
38. G. Merga, R. Wilson, G. Lynn, B. Milosavljevic, and D. Meisel, Redox catalysis on "naked" silver nanoparticles, *J. Phys. Chem. C*, *111*, 12220–12226 (2007).
39. M. Oliveira, D. Ugarte, D. Zanchet, and A. Zarbin, Influence of synthetic parameters on the size, structure, and stability of dodecanethiol-stabilized silver nanoparticles, *J. Colloid Interface Sci.*, *292*, 429–435 (2005).
40. A.A. Abd-Elaal, S.M. Tawfik, and S.M. Shaban, Simple one step synthesis of nonionic dithiol surfactants and theirself-assembling with silver nanoparticles: Characterization, surface properties, biological activity, *Appl. Surf. Sci.*, *342*, 144–153 (2015).

41. N.A. Negm, A.A. Abd-Elaal, D.E. Mohamed, A.F. El-Farargy, and S. Mohamed, Synthesis and evaluation of silver nanoparticles loaded with Gemini surfactants: Surface and antimicrobial activity, *J. Ind. Eng. Chem.*, *24*, 34–41 (2015).

42. M. Brust and C. Kiely, Some recent advances in nanostructure preparation from gold and silver particles: A short topical review, *Colloids Surf. A Physicochem. Eng. Asp.*, *202*, 175–186 (2002).

43. Y. Zhang, H. Peng, W. Huang, Y. Zhou, and D. Yan, Facile preparation and characterization of highly antimicrobial colloid Ag or Au nanoparticles, *J. Colloid Interface Sci.*, *325*, 371–376 (2008).

44. Y. Krutyakov, A. Olenin, A. Kudrinskii, P. Dzhurik, and G. Lisichkin, Aggregative stability and polydispersity of silver nanoparticles prepared using two-phase aqueous organic systems, *Nanotechnol. Russ.*, *3*, 303–310 (2008).

45. W. Zhang, X. Qiao, and J. Chen, Synthesis of nanosilver colloidal particles in water/oil microemulsion, *Colloids Surf. A Physicochem. Eng. Asp.*, *299*, 22–28 (2007).

46. P. Cozzoli, R. Comparelli, E. Fanizza, M. Curri, A. Agostiano, and D. Laub, Photocatalytic synthesis of silver nanoparticles stabilized by TiO_2 nanorods: A semiconductor/metal nanocomposite in homogeneous nonpolar solution, *J. Am. Chem. Soc.*, *126*, 3868–3879 (2004).

47. H. Huang and Y. Yang, Preparation of silver nanoparticles in inorganic clay suspensions, *Compos. Sci. Technol.*, *68*, 2948–2953 (2008).

48. Y. Zhou, S.H. Yu, C.Y. Wang, X.G. Li, Y.R. Zhu, and Z.Y. Chen, A novel ultraviolet irradiation photoreduction technique for the preparation of single-crystal Ag nanorods and Ag dendrites, *Adv. Mater.*, *11*, 850–852 (1999).

49. Y. Socol, O. Abramson, A. Gedanken, Y. Meshorer, L. Berenstein, and A. Zaban, Suspensive electrode formation in pulsed sonoelectrochemical synthesis of silver nanoparticles, *Langmuir*, *18*, 4736–4740 (2002).

50. D.G. Shchukin, I.L. Radtchenko, and G. Sukhorukov, Photo induced reduction of silver inside microscale polyelectrolyte capsules, *Chem. Phys. Chem.*, *4*, 1101–1103 (2003).

51. R. Jin, Y.C. Cao, E. Hao, G.S. Metraux, G.C. Schatz, and C. Mirkin, Controlling anisotropic nanoparticle growth through plasmon excitation, *Nature*, *425*, 487–490 (2003).

52. J.P. Malval, M. Jin, L. Balan, R. Schneider, D.L. Versace, H. Chaumeil, A. Defoin, and O. Soppera, Photoinduced size-controlled generation of silver nanoparticles coated with carboxylate derivatized thioxanthones, *J. Phys. Chem. C*, *114*, 10396–10402 (2010).

53. C. Johans, J. Clohessy, S. Fantini, K. Kontturi, and V.J. Cunnane, Electrosynthesis of polyphenylpyrrole coated silver particles at a liquid-liquid interface, *Electrochem. Commun.*, *4*, 227–230 (2002).

54. Y. Zhang, F. Chen, J. Zhuang, Y. Tang, D. Wang, Y. Wang, A. Dong, and N. Ren, Synthesis of silver nanoparticles via electrochemical reduction on compact zeolite film modified electrodes, *Chem. Commun.*, *24*, 2814–2815 (2002).

55. H. Ma, B. Yin, S. Wang, Y. Jiao, W. Pan, S. Huang, S. Chen, and F. Meng, Synthesis of silver and gold nanoparticles by a novel electrochemical method, *Chem. Phys. Chem.*, *24*, 68–75, 2004.

56. J.P. Abid, A.W. Wark, P.F. Brevet, and H.H. Girault, Preparation of silver nanoparticles in solution from a silver salt by laser irradiation, *Chem. Commun.*, *24*, 792–793 (2002).

57. S. Eutis, G. Krylova, A. Eremenko, N. Smirnova, A.W. Schill, and M. El-Sayed, Growth and fragmentation of silver nanoparticles in their synthesis with laser and CW light by photo-sensitization with benzophenone, *Photochem. Photobiol. Sci.*, *4*, 154–159 (2005).

58. P.K. Sudeep and P.V. Kamat, Photosensitized growth of silver nanoparticles under visible light irradiation: A mechanistic investigation, *Chem. Mater.*, *17*, 5404–5410 (2005).

59. J. Chen, K. Wang, J. Xin, and Y. Jin, Microwave-assisted green synthesis of silver nanoparticles by carboxymethyl cellulose sodium and silver nitrate, *Mater. Chem. Phys.*, *108*, 421–424 (2008).

60. S. Navaladian, B. Viswanathan, T.K. Varadarajan, and R.P. Viswanath, Microwave assisted rapid synthesis of anisotropic Ag nanoparticles by solid state transformation, *Nanotechnology*, *19*, 045603 (2008).

61. K.J. Sreeram, M. Nidhin, and B.U. Nair, Microwave assisted template synthesis of silver nanoparticles, *Bull. Mater. Sci.*, *31*, 937–942 (2008).

62. S. Komarneni, D. Li, B. Newalkar, H. Katsuki, and A.S. Bhalla, Microwave-polyol process for Pt and Ag nanoparticles, *Langmuir*, *18*, 5959–5962 (2002).

63. H. Yin, T. Yamamoto, Y. Wada, and S. Yanagida, Large-scale and size-controlled synthesis of silver nanoparticles under microwave irradiation, *Mater. Chem. Phys.*, *83*, 66–70 (2004).

64. M. Tsuji, K. Matsumoto, P. Jiang, R. Matsuo, S. Hikino, X.L. Tang, and K.S. Nor Kamarudin, The role of adsorption species in the formation of Ag nanostructures by a microwave polyol route, *Bull. Chem. Soc. Jpn.*, *81*, 393–400 (2008).

65. B. Hu, S.B. Wang, K. Wang, M. Zhang, and S.H. Yu, Microwave-assisted rapid facile "green" synthesis of uniform silver nanoparticles: Self-assembly into multilayered films and their optical properties, *J. Phys. Chem. C*, *112*, 11169–11174 (2008).

66. B. Soroushian, I. Lampre, J. Belloni, and M. Mostafavi, Radiolysis of silver ion solutions in ethylene glycol: Solvated electron and radical scavenging yields, *Radiat. Phys. Chem.*, *72*, 111–118 (2005).

67. S.P. Ramnami, J. Biswal, and S. Sabharwal, Synthesis of silver nanoparticles supported on silica aerogel using gamma radiolysis, *Radiat. Phys. Chem.*, *76*, 1290–1294 (2007).

68. D. Long, G. Wu, and S. Chen, Preparation of oligochitosan stabilized silver nanoparticles by gamma irradiation, *Radiat. Phys. Chem.*, *76*, 1126–1131 (2007).

69. P. Cheng, L. Song, Y. Liu, and Y.E. Fang, Synthesis of silver nanoparticles by γ-ray irradiation in acetic water solution containing chitosan, *Radiat. Phys. Chem.*, *76*, 1165–1168 (2007).

70. V. Hornebecq, M. Antonietti, T. Cardinal, and M. Treguer-Delapierre, Stable silver nanoparticles immobilized in mesoporous silica, *Chem. Mater.*, *15*, 1993–1999 (2003).

71. K.A. Bogle, S.D. Dhole, and V.N. Bhoraskar, Silver nanoparticles: Synthesis and size control by electron irradiation, *Nanotechnology*, *17*, 3204–3208 (2006).

72. Z.S. Pillai and P.V. Kamat, What factors control the size and shape of silver nanoparticles in the citrate ion reduction method, *J. Phys. Chem. B*, *108*, 945–951 (2004).

73. J.A. Jacob, H.S. Mahal, N. Biswas, T. Mukerjee, and S. Kapoor, Role of phenol derivatives in the formation of silver nanoparticles, *Langmuir*, *24*, 528–533 (2008).

74. H. Huang and X. Yang, Synthesis of polysaccharide-stabilized gold and silver nanoparticles: A green method, *Carbohydr. Res.*, *339*, 2627–2631 (2004).

75. N. Vigneshwaran, R.P. Nachane, R.H. Balasubramanya, and P.V. Varadarajan, A novel one-pot green synthesis of stable silver nanoparticles using soluble starch, *Carbohydr. Res.*, *341*, 2012–2018 (2006).

76. C. Tai, Y.H. Wang, and H.S. Liu, A green process for preparing silver nanoparticles using spinning disk reactor, *AIChE J.*, *54*, 445–452 (2008).

77. Y. Yin, Z.Y. Li, Z. Zhong, B. Gates, and S. Venkateswaran, Synthesis and characterization of stable aqueous dispersions of silver nanoparticles through the Tollens process, *J. Mater. Chem.*, *12*, 522–527 (2002).

78. L. Kvitek, R. Prucek, A. Panaček, R. Novotn, J. Hrbac, and R. Zbořil, The influence of complexing agent concentration on particle size in the process of SERS active silver colloid synthesis, *J. Mater. Chem.*, *15*, 1099–1105 (2005).

79. A. Panacek, L. Kvitek, R. Prucek, M. Kolar, R. Vecerova, and N. Pizurova, Silver colloid nanoparticles: Synthesis, characterization, and their antibacterial activity, *J. Phys. Chem. B*, *110*, 16248–16253 (2006).

80. L. Kvitek, A. Panacek, J. Soukupova, M. Kolar, R. Vecerova, R. Prucek, M. Holecova, and R. Zboril, Effect of surfactants and polymers on stability and antibacterial activity of silver nanoparticles (NPs), *J. Phys. Chem. C*, *112*, 5825–5834 (2008).

81. J. Soukupova, L. Kvitek, A. Panacek, T. Nevecna, and R. Zboril, Comprehensive study on surfactant role on silver nanoparticles (NPs) prepared via modified Tollens process, *Mater. Chem. Phys.*, *111*, 77–81 (2008).

82. A. Troupis, A. Hiskia, and E. Papaconstantinou, Synthesis of metal nanoparticles by using polyoxometalates as photocatalysts and stabilizers, *Angew. Chem. Int. Ed.*, *41*, 1911–1914 (2002).

83. P.R. Leroueil, S. Hong, A. Mecke, J.R. Baker, B.G. Orr, and M.M. Holl, Nanoparticle interaction with biological membranes: Does nanotechnology present a Janus face, *Acc. Chem. Res.*, *40*, 335–342 (2007).

84. A.M. Derfus, W.C.W. Chan, and S.N. Bhatia, probing the cytotoxicity of semiconductor quantum dots, *Nano Lett.*, *4*, 11–18 (2004).

85. A.M. Derfus, W.C.W. Chan, and S.N. Bhatia, Intracellular delivery of quantum dots for live cell labeling and organelle tracking, *Adv. Mater.*, *16*, 961–966 (2004).

86. K.H. Cho, J.E. Park, T. Osaka, and S.G. Park, Optimization of the sputter deposited platinum cathode for a direct methanol fuel cell, *J. Electrochim. Acta*, *51*, 956–960 (2005).

87. P. Mukherjee, A. Ahmad, D.S. Mandal, S. Senapati, R. Sainkar, M.I. Khan, R. Parishcha et al., Fungus-mediated synthesis of silver nanoparticles and their immobilization in the mycelial matrix: A novel biological approach to nanoparticle synthesis, *Nano Lett.*, *1*, 515–519 (2001).

88. N. Duran, P.D. Marcato, O.L. Alves, and G. Souza, Mechanistic aspects of biosynthesis of silver nanoparticles by several *Fusarium oxysporum* strains, *J. Nanobiotechnol.*, *3*, 1–8 (2005).

89. V. Alt, T. Bechert, P. Steinrücke, M. Wagener, P. Seidel, E. Dingeldein, E. Domann, and R. Schnettler, An in vitro assessment of the antibacterial properties and cytotoxicity of nanoparticulate silver bone cement, *Biomaterials*, *25*, 4383–4391 (2004).

90. P.L. Nadworny, J. Wang, E.E. Tredget, and R.E. Burrell, Anti-inflammatory activity of nanocrystalline silver in a porcine contact dermatitis model, *Nanomedicine*, *4*, 241–251 (2008).

91. J.V. Rogers, C.V. Parkinson, Y.W. Choi, J.L. Speshock, and S.M. Hussain, A preliminary assessment of silver nanoparticle inhibition of monkeypox virus plaque formation, *Nanoscale Res. Lett.*, *3*, 129–133 (2008).

92. G.A. Patil, M.L. Bari, B.A. Bhanvase, V. Ganvir, S. Mishra, and S.H. Sonawane, Continuous synthesis of functional silver nanoparticles using microreactor: Effect of surfactant and process parameters, *Chem. Eng. Process.*, *62*, 69–77 (2012).

93. S. He, H. Chen, Z. Guo, B. Wang, C. Tang, and Y. Feng, High-concentration silver colloid stabilized by a cationic Gemini surfactant, *Colloids Surf. A Physicochem. Eng. Asp.*, *429*, 98–105 (2013).

94. S. Sarkar, A.D. Jana, S.K. Samanta, and G. Mostafa, Facile synthesis of silver nanoparticles with highly efficient anti-microbial property, *Polyhedron*, *26*, 4419–4426 (2007).

95. R. Janardhanan, M. Karuppaiah, N. Hebalkar, and T.N. Rao, Synthesis and surface chemistry of nano silver particles, *Polyhedron*, *28*, 2522–2530 (2009).

96. N. Iqbal, M.R. Abdul Kadir, N.A. Malek, N.H.B. Mahmood, M.R. Murali, and T. Kamarul, Characterization and antibacterial properties of stable silver substituted hydroxyapatite nanoparticles synthesized through surfactant assisted microwave process, *Mater. Res. Bull.*, *48*, 3172–3177 (2013).

97. K.H. Cho, J.E. Park, T. Osaka, and S.G. Park, The study of antimicrobial activity and preservative effects of nanosilver ingredient, *Electrochim. Acta*, *51*, 956–960 (2005).

98. N. Duran, D.P. Marcato, H.I. De Souza, L.O. Alves, and E. Esposito, Antibacterial effect of silver nanoparticles produced by fungal process on textile fabrics and their effluent treatment, *J. Biomed. Nanotechnol.*, *3*, 203–208 (2007).

99. J.R. Morones, L.J. Elechiguerra, A. Camacho, K. Holt, B.J. Kouri, T.J. Ramirez, and J.M. Yocaman, The bactericidal effect of silver nanoparticles, *Nanotechnology*, *16*, 2346–2353 (2005).

100. K.Y. Yoon, J.H. Byeon, J.H. Park, and J. Hwang, Susceptibility constants of *Escherichia coli* and *Bacillus subtilis* to silver and copper nanoparticles, *Sci. Total Environ.*, *373*, 572–575 (2007).

101. P. Jain and T. Pradeep, Potential of silver nanoparticle-coated polyurethane foam as antibacterial water filter, *Biotechnol. Bioeng.*, *90*, 59–63, 2005.

102. Z. Khan, S.A. AL-Thabaiti, A.Y. Obaid, Z.A. Khan, and A.O. Al-Youbi, Effects of solvents on the stability and morphology of CTAB stabilized silver nanoparticles, *Colloids Surf. A Physicochem. Eng. Asp.*, *390*, 120–125 (2011).

103. J.W. Richard, B.A. Spencer, L.F. McCoy, E. Carina, J. Washington, and P. Edgar, Acticoat versus silver ion: The truth, *J. Burns Surg. Wound Care*, *1*, 11–20, (2002).

104. J.J. Castellano, S.M. Shafii, F. Ko, G. Donate, T.E. Wright, and R.J. Mannari, Comparative evaluation of silver-containing antimicrobial dressings and drugs, *Int. Wound J.*, *4*, 114–122 (2007).

105. H.J. Klasen, A historical review of the use of silver in the treatment of burns. Part I. Early uses, *Burns*, *30*, 1–9 (2000).

106. A.B.G. Landsdown, Silver I: Its antibacterial properties and mechanism of action, *J. Wound Care*, *11*, 125–138 (2002).

107. W.B. Hugo and A.D. Russell, Types of antimicrobial agents, Adam P. Fraise, Jean-Yves Maillard, Syed A. Sattar (eds.). In: *Principles and Practice of Disinfection, Preservation and Sterilization*, Blackwell Scientific Publications, Oxford, U.K. (1982).

108. R.H. Demling and L. DeSanti, Effects of silver on wound management, *Wounds*, *13*, 4–15 (2001).

109. I. Chopra, The increasing use of silver-based products as antimicrobial agents: A useful development or a cause for concern, *J. Antimicrob. Chemother.*, *59*, 587–590 (2007).

110. C.A. Moyer, L. Brentano, D.L. Gravens, H.W. Margraf, and W.W. Monafo, Treatment of large human burns with 0.5% silver nitrate solution, *Arch. Surg.*, *90*, 812–867 (1965).

111. C.G. Bellinger and H. Conway, Effects of silver nitrate and sulfamylon on epithelial regeneration, *Plast. Reconstr. Surg.*, *45*, 582–585 (1970).

112. C.L. Fox and S.M. Modak, Mechanism of silver sulfadiazine action on burn wound infections, *Antimicrob. Agents Chemother.*, *5*, 582–588 (1974).

113. C.G. Gemmell, D.I. Edwards, and A.P. Frainse, Guidelines for the prophylaxis and treatment of methicillin-resistant *Staphylococcus aureus* (MRSA) infections in the UK, *J. Antimicrob. Chemother.*, *57*, 589–608 (2006).

114. A.R. Shahverdi, A. Fakhimi, H.R. Shahverdi, and S. Minaian, Synthesis and effect of silver nanoparticles on the antibacterial activity of different antibiotics against *Staphylococcus aureus* and *Escherichia coli*, *Nanomed. Nanotechnol. Biol. Med.*, *3*, 168–171 (2007).

115. S. Shrivastava, T. Bera, A. Roy, G. Singh, P. Ramachandrarao, and D. Dash, Characterization of enhanced antibacterial effects of novel silver nanoparticles, *Nanotechnology*, *18*, 103–112 (2007).

116. S. Pal, Y.K. Tak, and J.M. Song, Does the antibacterial activity of silver nanoparticles depend on the shape of the nanoparticle? A study of the gram-negative bacterium *Escherichia coli*, *Appl. Environ. Microbiol.*, *27*, 1712–1720 (2007).

117. M. Rai, A. Yadav, and A. Gade, Silver nanoparticles as a new generation of antimicrobials, *Biotechnol. Adv.*, *27*, 76–83 (2009).

118. P. Gong, H. Li, X. He, K. Wang, J. Hu, and W. Tan, Preparation and antibacterial activity of Fe_3O_4/Ag nanoparticles, *Nanotechnology*, *18*, 604–611 (2007).

119. D.L. Leaper, Silver dressings: Their role in wound management, *Int. Wound J.*, *3*, 282–294 (2006).

120. M. Ip, S.L. Lui, V.K.M. Poon, I. Lung, and A. Burd, Antimicrobial activities of silver dressings: An in vitro comparison, *J. Med. Microbiol.*, *55*, 59–63 (2006).

121. J. Jun, D.Y. Yuan, W.S. Hai, Z.S. Feng, and W.Z. Yi, Preparation and characterization of antibacterial silver-containing nanofibers for wound dressing applications, *J. US-China Med. Sci.*, *4*, 52–54 (2007).

122. J. Tian, K.K.Y. Wong, C.M. Ho, C.N. Lok, W.Y. Yu, C.M. Che, J.F. Chiu, and P.K.H. Tam, Topical delivery of silver nanoparticles promotes wound healing, *Chem. Med. Chem.*, *1*, 171–180 (2006).

123. F. Furno, K.S. Morley, B. Wong, B.L. Sharp, P.L. Arnold, and S.M. Howdle, Silver nanoparticles and polymeric medical devices: A new approach to prevention of infection?, *J. Antimicrob. Chemother.*, *54*, 1019–1024 (2004).

124. A. Kumar, P.K. Vemula, P.M. Ajayan, and G. John, Silver-nanoparticle-embedded antimicrobial paints based on vegetable oil, *Nat. Mater.*, *7*, 236–241 (2008).

125. A. Gupta and S. Silver, Silver as a biocide: Will resistance become a problem?, *Nat. Biotechnol.*, *16*, 888–892 (1998).

126. Q.L. Feng, J. Wu, G.Q. Chen, F.Z. Cui, T.N. Kim, and J.O. Kim, A mechanistic study of the antibacterial effect of silver ions on *Escherichia coli* and *Staphylococcus aureus*, *J. Biomed. Mater. Res.*, *52*, 662–668 (2000).

127. T. Matsuura, Y. Abe, K. Sato, K. Okamoto, M. Ueshige, and Y. Akagawa, Prolonged antimicrobial effect of tissue conditioners containing silver zeolite, *J. Dent.*, *25*, 373–377 (1997).

128. H. Nikawa, T. Yamamoto Hamada, M.B. Rahardjo, and S. Murata Nakaando, Antifungal effect of zeolite-incorporated tissue conditioner against *Candida albicans* growth and/or acid production, *J. Oral Rehabil.*, *25*, 350–357 (1997).

129. N. Saravanakumar, L. Prabu, M. Karthik, and A. Rajamanickam, Experimental analysis on cutting fluid dispersed with silver nano particles, *J. Mech. Sci. Technol.*, *28*, 645–651 (2014).

130. J. Ma, Y. Mo, and M. Bai, Effect of Ag nanoparticles additive on the tribological behavior of multialkylated cyclopentanes, *Wear*, *266*, 627–631 (2009).

131. S. Zhang, L. Hu, H. Wang, and D. Feng, The anti-seizure effect of Ag nanoparticles additive in multialkylated cyclopentanes oil under vacuum condition, *Tribol. Int.*, *55*, 1–6 (2012).

Section III

Tribological Applications in Automotive, Petroleum Drilling, and Food

8 Correlating Engine Dynamometer Fuel Economy to Time-Dependent Tribological Data in Friction Modifier Studies

Frank J. DeBlase

CONTENTS

8.1 INTRODUCTION

8.1.1 Future Fuel Economy ASTM D7589 Sequence VIE Engine Dynamometer Testing

With recent effort to conserve natural resources by increasing fuel economy through the use of improved passenger car motor oils (PCMOs), the ASTM D7589 (Sequence VID) fuel economy engine test [1] was changed to now better evaluate lubricants intended for much longer drain intervals than in the past [2,3]. The latest Sequence VIE test was developed to insure that lubricants both meet increases in fuel economy targets and retain fuel economy with greater mileage accumulation (over a range of driving conditions). The final fuel economy targets will be set forth in the lubricant specification (GF-6) for the next generation of PCMO [4]. This improvement made in the earlier Sequence VID fuel economy test to generate the newer Sequence VIE test is presented in Figure 8.1 and was developed by the industry consortium charged with maintaining the lubricant specifications for fuel economy [5]. An ASTM D7589 monitoring task force has incorporated a number of changes in developing the latest version of the Sequence VIE engine test. The six stages to measure fuel economy differ in temperature, engine speed, and load, and are given in Table 8.1.

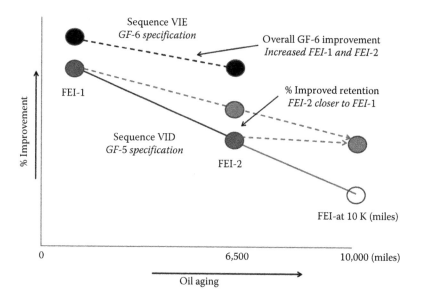

FIGURE 8.1 Concept of increased fuel economy retention illustrated by the extended durability measured by higher FEI-2 levels reaching levels for the extended mileage accumulation of 10,000 miles oil change from 6,500 miles. To evaluate PCMO with required durability, the ASTM D7589 Sequence VID test was modified to Sequence VIE, with more severe oil aging equivalent to 10,000 miles compared with 6,500 miles for previous VID version.

TABLE 8.1

Sequence VIE Test Parameters as Defined by ASTM D7589 Test Surveillance Panel

Parameter	Stage 1	Stage 2	Stage 3	Stage 4	Stage 5	Stage 6
Speed (r/min)	2000 ± 5	2000 ± 5	1500 ± 5	695 ± 5	695 ± 5	695 ± 5
Load cell (N-m)	105.0 ± 0.1	105.0 ± 0.1	105.0 ± 0.1	20.0 ± 0.1	20.0 ± 0.1	40.0 ± 0.1
Nominal (Power, kW)	22	22	16.5	1.5	1.5	2.9
Oil gallery (Temp., °C)	115 ± 2	65 ± 2	115 ± 2	115 ± 2	35 ± 2	115 ± 2
Coolant, (Temp., °C)	109 ± 2	65 ± 2	109 ± 2	109 ± 2	35 ± 2	109 ± 2
Stabilization, (Time, minutes)	60	60	60	60	60	60

Notes: Several stages (1 and 2) are run at high speeds and closer to elastohydrodynamic lubrication regime. Stage 3 is an intermediate stage with mixed lubrication, although a higher load also favors boundary lubrication. Stages 4–6 are run at lower speeds and closer to boundary lubrication, where friction modifiers play a significant role, especially in stages 4 and 6, which are also run at a higher temperature (115°C).

The key changes made to improve the durability of fuel economy can be summarized as follows:

1. A *1-hour engine running stabilization* before each of the six stages measuring fuel economy, including the initial Base Lubricant Blend testing (BLB1, BLB2), the candidate oil testing, and the final Base Lubricant After testing (BLA). These tests are used to calculate the fuel economy improvements (FEIs) FEI-1 and FEI-2 of the candidate relative to the baseline oil. The VIE is a 196-hour test, compared with 153 hours for the VID, and the high-temperature (120°C) oil aging is 27% longer in the VIE.

2. Aging is increased for FEI-2 from 84 hours (VID) to 109 hours (VIE) to match vehicle oxidation numbers. The total aging, 125 hours for the VIE compared with 100 hours for the VID, results in a rise in oil consumption compensated by greater oil volume in the oil pan.

3. The 2012 GM engine is different, with coatings on (main and rod) bearings and pistons, and a redesign of the head cam and oil lines to the timing chain, the use of 60 μm oil filters, and 100 kPa pressure of a 50:50 antifreeze/water coolant mixture.

4. Six baseline lubricant flushes are used after the final baseline lubricant (BLA) is tested at the end of the test to clean the engine of detergents such as overbased calcium sulfonates compared with only three flushes for VID test used earlier. These flushes are needed between starting the next test to insure there is no residual carryover from the previous test oil in the sample.

TABLE 8.2

Differences between Sequence VID and Next-Generation Test Sequence VIE

	Sequence VID and VIE Test Operating Conditions (ASTM D7589)					
	Initial Aging before FEI-1		Final Aging before FEI-2		Test Stage Stabilization	Total Test
Test	Time (hours)	Temp. (°C)	Time (hours)	Temp. (°C)	Time (hours)	Time (hours)
Seq. VID	16	120	84	120	1	106
Seq. VIE	16	120	109	120	1	131

Note: The major differences are in the time for final aging after FEI-1 and before FEI-2.

The specific engine test differences related to time and temperature are further compared in Table 8.2 for both VID and VIE generations of the fuel economy test. With the changes in the Sequence VIE test, the fuel economy durability of the lubricant can be better evaluated [6].

Lubricant additives that reduce friction, or commonly called friction modifiers (FMs), play a greater role in helping lubricant formulators reduce oil viscosity to improve fuel economy, while still maintaining boundary lubrication in these thinner oils [7]. To help develop better friction modifier additives *that help lubricants meet the GF-6 Sequence VIE fuel economy demands*, laboratory time-dependent tribological testing is used to characterize changes in the coefficient of friction (COF) with time, and therefore with extended duration of oil use. These tests include time-dependent isothermal tribological testing using both the Cameron Plint TE-77 tribometer and the Mini Traction Machine (MTM). Finally, in order to relate the laboratory time-dependent tribological testing reference frame or time scale to engine Sequence VID or VIE engine fuel economy, Fourier transform infrared (FTIR) spectroscopy is used as a common molecular analysis in both experiments. This spectroscopic technique is used to match test oil aging and oxidation during time-dependent tribological experiments to oxidation signatures of the same test oil after completing the FEI-1 or FEI-2 experiments of the ASTM D7589 VIE.

8.2 EXPERIMENTAL STUDIES OF FRICTION-REDUCING ADDITIVES' PERFORMANCE DURABILITY

8.2.1 MATERIALS

The studies presented apply to friction modifier performance in formulated PCMOs comprising Group III, Group III-plus, and Group IV base stocks with a range of viscosities. The PCMO of the formulations presented include the following:

1. For tribological studies (Figures 8.3, 8.5, and 8.6) (short-term and extended hold studies), a formulated Group III SAE 5W-30 oil with a kinematic viscosity (ν) of 66.12 centistokes (cSt) at 40°C, 9.43 cSt at 100°C, and a viscosity index (VI) = 122 was used.

2. For tribological study and Sequence VID fuel economy testing (Figures 8.7 and 8.11), a Group III SAE 5W-20 oil with a ν of 45.1 cSt at 40°C, 8.3 cSt at 100°C, and a VI = 161 was used.

3. For tribological studies and Sequence VIE fuel economy testing (Figures 8.4, 8.8, and 8.12), a Group III-plus SAE 0W-20 with a ν of 28.45 cSt at 40°C, 7.42 cSt at 100°C, and a VI = 246 was used.

The friction modifier additives presented in these studies include the following:

1. GMO, a glycerol monooleate (commercially supplied ester surfactant for lubricants) that is surface active and can arrange to form a tribofilm at metallic surfaces; however, it is susceptible eventually to hydrolysis and oxidation of the oleic acid tail group.

2. Experimental organic friction modifier, EXP-OFM1, a proprietary organic surfactant technology with a unique structure developed to target durable friction reduction with extended performance. Like GMO, it is an ashless technology that does not contain any sulfur, phosphorus, molybdenum, or any other metals. Essentially, this surfactant develops a sustainable friction-reducing tribofilm to provide retention of boundary lubrication.

3. Typical molybdenum-containing friction modifiers, specifically molybdenum dithiocarbamate (MoDTC), developed for research by Chemtura and also available commercially.

8.2.2 INSTRUMENTATION

The instruments used in these studies include those used internally for tribological and spectroscopic studies and those used externally in terms of engine fuel economy testing performed at the Southwest Research Institute (SWRI) and will be discussed separately.

8.2.2.1 Tribological Equipment

The friction testing equipment includes a Cameron Plint TE77 High-Frequency Machine equipped with a Kistler Model 9203 force transducer and a 5007 charge amplifier with a Gulton West 2050 Temperature controller, mechanical applied load, and both a dual trace oscilloscope and a three-pen Yt recorder for data acquisition (useful for extended tests). The specimens used were in friction mode of line contact pin-on-plate consisting of a 16 mm long nitride steel dowel pin (6 mm diameter, Rockwell C scale hardness 60), with the longer pin axis length, rubbed against a hardened ground steel plate (RC hardness 60/0.4 μm surface roughness). The dowel is translated over a chosen fixed amplitude (2.35 mm at a set reciprocating frequency of 5 Hz). Details of the type of friction studies using the TE77, including temperature, load, and time, are given in Tables 8.3 and 8.5.

Also used for friction studies is an MTM test using a 19 mm hardened stainless steel ball (SAE AISI 52100) rotated at set speeds and loads against a flat 46 mm diameter polished hardened steel disk (SAE AISI 52100) that is also rotating independently. The relative speeds of both the rotating ball and the disk can be changed to the desired percent slide-to-roll ratio (SRR). Typical Stribeck curves are obtained at a fixed load

TABLE 8.3

Cameron Plint (Model TE-77 Friction Instrument) Pin-on-Plate Friction Mode Test Parameters for Measurements of Coefficients of Friction in the Temperature Range 50°C–160°C

Test Conditions—Cameron Plint TE-77

Stage	Load (N, kg)	Temperature of Ramp (°C)	Ramp Time (minutes)	Hold Time (minutes)	Frequency (Hz)
1	0, 0	25 → 35	10	5	0
2	50, 5.1	35 → 50	10	5	5
3	100, 10.2	50 → 165	60	5	5

Note: The test involves three stages; the first warms the oil, without load or sliding motion; the second applies half the maximum load 50 N and warms the test sample further to 50°C while the dowel pin is rubbed against the plate with a frequency of 5 Hz. In the final stage, the load is increased to 100 N and the sample is linearly heated from 50°C to 160°C while maintaining rubbing contact.

TABLE 8.4

Mini Traction Machine Instrument (AISI 52100 Steel Ball on Disk) Parameter Set for Test Sample Held at a Temperature of 120°C While Operated in the Stribeck Curve Mode of Coefficient of Friction versus Mean Speed (Entrainment Speed)

Test Conditions—MTM

Load (N, kg)	Temperature (°C)	Speed Range (mm/s)	Slide-to-Roll Ratio (%)
30, 3.06	120	2000–5	50

Note: The measurements are made under a given constantly applied normal load (30 N) and percent slide-to-roll ratio (50%), over a range of speeds, from that of hydrodynamic to boundary lubrication conditions.

and SRR, and the speeds are reduced from a high initial speed (2000 mm/s) down to 6 mm/s, while recording the traction or COF. Details of the type of friction studies using the MTM, including load, speed, and SRR, are given in Tables 8.4 and 8.5.

8.2.2.2 Engine Dynamometer Testing (Sequence VID and VIE)

The experimental configurations used to measure fuel economy in the Sequence VID and in the latest version of the Sequence VIE are based on different engine designs and testing protocol differences, given in Table 8.2. In the Sequence VID, a General Motors 2009 3.0 L Cadillac SRX LY7 V6 six-cylinder engine is used. In contrast, the Sequence VIE uses a 3.6 L General Motors 2012 Malibu LY7 V6 six-cylinder engine with a controlled external heating system for the lubricant. The details regarding the speed (revolutions per minute [rpm]), power load (kilowatt [kW]), and lubricant and coolant temperatures are given for all testing stages in Table 8.1.

TABLE 8.5
Time-Dependent Laboratory Tribology Isothermal Hold Studies Carried Out for Both Cameron Plint TE-77 and the Mini Traction Machine Where Depending on Temperature the Duration of the Test Is Altered to Sufficiently Cover All the Changes in the COF with Time

Hold Test Method	Load (N)	Temperature Range (°C)	Minimum Isothermal Hold Time (hours)
Cameron Plint TE77	100	50–160	50
MTM Stribeck	30	150	48
Cameron Plint TE77	100	50–120	150

Note: These parameters are useful for both oil sump isothermal temperatures of 120°C–135°C and higher accelerated temperatures of 150°C and 160°C.

8.2.2.3 FTIR Spectroscopy and Scanning Electron Microscopy (SEM) Instruments

The infrared (IR) instrument used to obtain the mid-IR spectrum of the oil samples studied was a Perkin Elmer Spectrum 1000 spectrometer controlled by Spectrum v5.0.2 IR software running on Windows XP. Four wavenumber (cm^{-1}) resolutions from 600 to 4000 cm^{-1} were recorded by transmission spectroscopy using a 0.05 mm path length NaCl liquid cell. The growth of the carbonyl stretching peaks at the characteristic 1710, 1731, and 1777 cm^{-1} follows oxidation aging, developing a range of oxidized hydrocarbons from aldehydes to carboxylic acids and anhydrides. The scanning electron microscopy (SEM) system used to analyze changes in the molybdenum dithiocarbamate FM derived tribofilms during the extended Cameron Plint tribological study (Figures 8.16 and 8.17) is a Hitachi TM 3030 with Quantx70 Software for metals analysis.

8.2.3 LABORATORY TEST METHODS: ISOTHERMAL TRIBOLOGICAL STUDIES

Several new test procedures were developed for use with a Cameron Plint (CP) TE-77 tribological testing instrument (operating in friction mode) and a PCS Instruments' MTM (in Stribeck friction measurement configuration). These tests were developed in order to continuously measure the sustained reduction of the COF in a PCMO formulated with a range of friction modifier additives under varying normal load and temperature. This study involves continued monitoring of the COF during isothermal testing for a longer duration (up to several days in some cases) at a fixed elevated temperature. Samples of fully formulated 5W-30 oil with and without a friction modifier to be tested are placed on the test specimens and a two-step approach to the tribological testing is followed: First, a standard tribological COF test (ex. fixed or ramped temperature) is followed by a second extended COF test step under continued isothermal conditions. The standard test conditions of each method and its

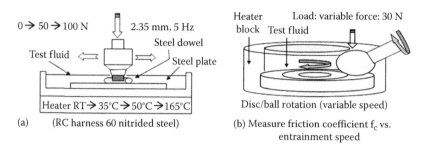

(a)   (RC harness 60 nitrided steel)   (b) Measure friction coefficient f_c vs.
                                                     entrainment speed

FIGURE 8.2 Experimental configuration of tribology geometry for Cameron Plint TE-77 (a) line contact kept under 100 N load as temperature is ramped following warm-up conditioning to 50°C and subsequently ramped to 165°C (see Table 8.3), and Mini Traction Machine (b) point contact kept under 30 N load at a fixed temperature up to 150°C, with a slide-to-roll ratio of 50% and programmed to measure Stribeck data from 2000 mm/s down to 5 mm/s (see Table 8.4).

isothermal extended studies are given in Tables 8.3 through 8.5 and Figures 8.2 through 8.4. For the extended tribological testing at the end of the standard test, with the final elevated temperature reaching 120°C, 135°C, and 160°C for the TE-77 and 150°C for the MTM, the load is maintained and the tribological measurements are continued *isothermally for an additional amount of time (typically several hours up*

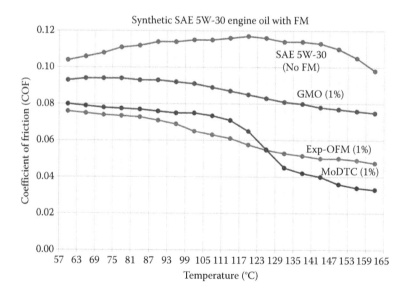

FIGURE 8.3 Cameron Plint TE77 friction measurement for line contact kept under 100 N load with a temperature ramp up to 165°C. A preconditioning step is run at 50°C in 15 minutes at half the full load (50 N). The specific heating rate and dwell times for the various stages are given in Table 8.3. A comparison is shown for SAE 5W-30 fully formulated without FM and also treated with 1% MoDTC and 1% Exp-OFM1. The data indicate that a temperature of ~120°C is required to induce trifilm formation and conversion of MoDTC to MoS2, which becomes very effective at reducing friction.

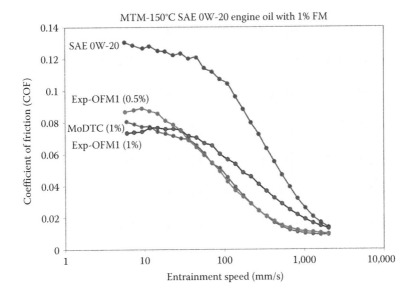

FIGURE 8.4 Mini Traction Machine (MTM) testing of 0W-20 PCMO with MoDTC-FM at 1% and Exp-OFM1 at two concentrations, 1% and 0.5% by weight. The conditions used to develop Stribeck curves were 30 N load force applied to the ball on the disc, test oil kept at 150 °C, with a 50% slide-to-roll ratio (ball and disc speed differences), at 50% slide-to-roll ratio, and entrainment speeds varied from 2000 mm/s (elastohydrodynamic), decreasing through (mixed lubrication), and down to 5 mm/s (boundary layer lubrication regime), where friction modifiers adsorb to boundary surfaces and maintain the separation of sliding surfaces, thereby lowering the COF.

to 48 hours or several days if desired). The change in the reduction of the COF is measured over the extended time to simulate mileage accumulation. If no further COF change occurred, the test was stopped.

8.2.3.1 Extended Cameron Plint TE-77 Isothermal Studies

The key specifications of the TE-77 used were discussed in Section 8.2.2.1; the specific types of experiments are discussed in the following text. Typical examples of the isothermal study of various friction modifiers held at 161°C–162°C (Figure 8.5) and also at 120°C (Figure 8.6) show changes in the COF with time, starting from 0 hours (at 50°C) heated to 160°C or 120°C, respectively, followed by continued isothermal Plint TE-77 testing at 160°C for 50 hours or 120°C for 150 hours (~2 or 6 days, respectively) of continued line contact measurement. The test fluids consist of 1% by weight of friction modifier added to a Group III, 5W-30 oil with all other additives at their typical specific concentrations. In addition to 5W-30, a lower-viscosity Group III, 5W-20 oil was studied for FM impact on the COF by time-dependent tribological measurements extended at 135°C. For comparison with real-world engine performance, this particular 5W-20 oil was also evaluated by ASTM D7589 Sequence VID fuel economy engine test. The data in Figure 8.7 show the COF changes in the isothermal tribological test for this oil at 135°C, extended up to 100 hours.

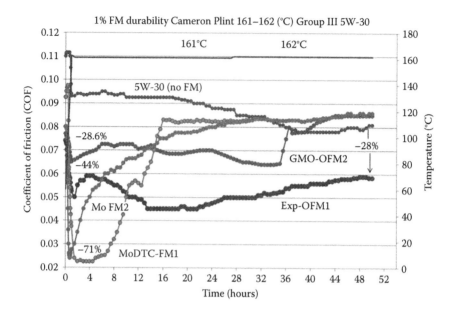

FIGURE 8.5 Results of a 160°C study of Group III 5W-30 with friction modifiers for 50 hours of COF monitoring under a 100 N load and a frequency of 5 Hz for a line contact tribological measurement. Glycerol monooleate (GMO), molybdenum dithiocarbamates (MoDTC-FM1 and Mo-FM2), and a new experimental organic friction modifier (Exp-OFM1) were compared in this study. Initially, MoDTC shows the greatest friction reduction; however, with extended time, this performance is lost, while organic FMs performance, and especially EXP-OFM1 performance, is retained, with only 28% reduction in the COF even after 52 hours. Differences in the performances between MoDTC-FM1 and MoDTC-FM2 may be due to the interaction with ZDDPs, which are known to impact the friction reduction performance of MoDTCs. (From Bec, S. et al., *Tribol. Lett.*, 17(4), 797, 2004.)

They also highlight the points in time that correlate with the duration to end of test (EOT) for the test oil run in the Sequence VID and if it were run in the VIE test, based upon correlated FTIR signatures. This temperature was chosen as an average oil sump temperature and the same oil was evaluated in the Sequence VID fuel economy engine test. In addition to these 5W-30 and 5W-20 PCMO time-dependent tribological isothermal studies, a 0W-20 fully formulated GF-5 PCMO was also evaluated by time-dependent tribological studies evaluated at 160°C (Figure 8.8). Again, there are noticeable improvements with the addition of an effective FM, whose durability in friction reduction is measured by its constant COF throughout the extended period of the test.

8.2.3.2 Mini Traction Machine (MTM) Test

In this test, a 19 mm hardened stainless steel ball (SAE AISI 52100) is rotated at set speeds and loads, against a flat 46 mm diameter polished hardened steel disk

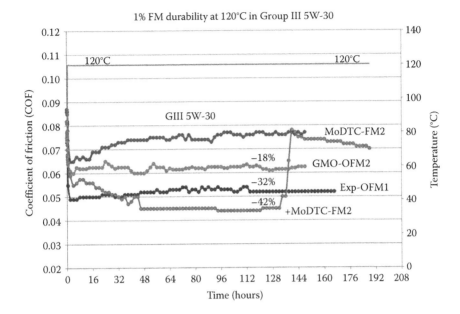

1% FM durability at 120°C in Group III 5W-30

FIGURE 8.6 Study of friction modifier–treated initial oil measured at 120°C and then subsequently held for at least 150 hours of continued COF monitoring under a 100 N load and a reciprocating frequency of 5 Hz for a line contact tribological measurement. Glycerol monooleate (GMO), molybdenum dithiocarbamates (MoDTC-FM2), and a new experimental organic friction modifier (Exp-OFM1) are compared in this study, which shows similar performance as the 160°C experiments; however, the time scale is longer, approaching over 150 hours, at which point the MoDTC performance breaks down again, while the EXP-OFM1 performance remains at 32% COF reduction.

(SAE AISI 52100) that is also rotating independently. The speeds of both the disk and the rotating ball change the percent SRR. While maintaining a set SRR, the instrument first operates with the disk running at a higher speed than the ball, and then reverses so that the ball is turning at a higher speed than the disk. The experimental parameters controlled include the sample temperature (°C), load Newtons (N), and SRR (%). The 150°C Stribeck measurement data are based on a 30 N applied load and on an SRR of 50%. Under these conditions, the mean speed (also called entrainment speed) is slowly reduced from 2000 mm/s to ~5 (mm/s), in order to move from the hydrodynamic dependent (bulk lubricant liquid–liquid internal friction or viscosity), through mixed, to the boundary layer lubrication regimes. An illustration of this MTM experiment in Stribeck configuration is given in Figure 8.4. This type of test measures the COF of a formulated lubricant in a rolling–sliding contact geometry which correlates with the region of the engine under similar contact. For example, it simulates the mechanism causing friction at the interface where the cam rubs against the cam-follower (or tappet). A corresponding MTM 150°C isothermal hold study is also shown in Figure 8.9. The data indicate a

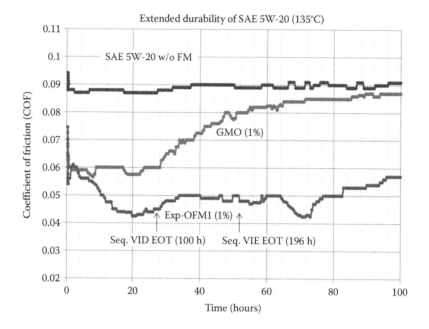

FIGURE 8.7 Study of friction modifier–treated initial oil measured at 135°C and then subsequently held for at least 100 hours of continued COF monitoring under a 100 N load and a reciprocating frequency of 5 Hz for a line contact tribological measurement. Based on further FTIR analysis, the time associated with the end of test (EOT) of both the Sequence VID and the Sequence VIE test is indicated by the arrows on the Exp-OFM1 versus COF curve, which show the corresponding VID and VIE EOT. A comparison is given between 1% glycerol monooleate (GMO) and an experimental organic friction modifier (Exp-OFM1), which demonstrates a continued reduction in the COF with Exp-OFM1, while GMO begins to lose its performance after 28 hours of testing.

breakdown in the friction reduction performance of molybdenum dithiocarbamate (MoDTC) with time, and this study is similar to the Cameron Plint TE-77 hold studies conducted at 160°C.

8.2.4 FTIR STUDIES OF LUBRICANT CHANGES IN TRIBOLOGICAL AND ENGINE TESTS

FTIR spectra were recorded on a Perkin Elmer 1000 spectrometer at four wavenumber (cm^{-1}) resolution from 4000 to 600 cm^{-1} using a 0.05 mm path length NaCl transmission cell. First, the spectrum of the baseline test oil (at time 0 hours) is recorded and used for spectral subtraction of each subsequent test sample spectrum. Spectral subtractions were performed to eliminate interference from the bulk oil C–H stretching and bending absorbances, so the analysis is focused on the changes in the fingerprint region of the spectrum. The growth of carbonyl stretching peaks at 1710, 1731, and 1777 cm^{-1} monitors oxidation,

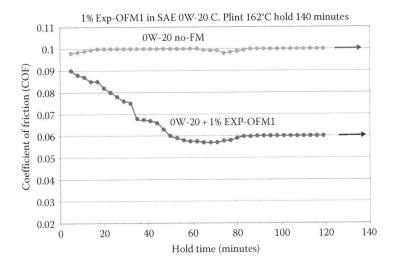

FIGURE 8.8 Study of friction modifier–treated 0W-20 oil measured at a ramp up to 162°C in 1 hour and then subsequently held isothermally at 162°C for up to 140 minutes of continued COF monitoring. The tribological hold conditions were under a 100 N load and a reciprocating frequency of 5 Hz for a line contact continued measurement. The sustained friction reduction durability of the 1% Exp-OFM1 concentration in SAE 0W-20 indicates that it should also provide high fuel economy improvement in the Sequence VIE engine test, which has a long oil aging stage, thus requiring high durability in the friction modifier performance. This same 0W-20 PCMO oil samples were then also evaluated in the ASTM D7589 Sequence VIE test for a follow-up to the fuel economy durability at the end of test and showed a high fuel economy improvement.

while nitration is typical of absorbances at 1633 cm^{-1}. These features can also be analyzed in oil nitro-oxidation bench test aging by various laboratory simulation experiments such as the ASTM ROBO test [8] and an in-house nitro-oxidation test reported earlier [9]. By comparing changes in the carbonyl oxidation region of the spectrum, FTIR was used to determine which point in time of tribological isothermal hold studies corresponds to the same level of oil aging oxidation in the engine dynamometer fuel economy Sequence VID or VIE test. Data presented (Figure 8.10) show that oil tested in the Cameron Plint isothermal tribological study after a 27 h holding time has an FTIR spectrum very close to the spectrum of the same oil at the end of the Sequence VID test—*both show good agreement in the extent of oxidation.*

8.2.5 FUEL ECONOMY ENGINE DYNAMOMETER TESTING, ASTM D7589

The engine testing protocol followed in ASTM D7589 of engine specification in Section 8.2.2.2 can be described as the following sequence of test steps for measuring FEIs at two stages, FEI-1 and FEI-2, where there is specific variation in temperature,

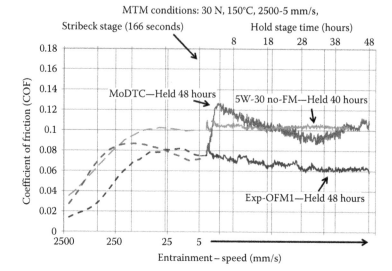

MTM conditions: 30 N, 150°C, 2500-5 mm/s,

FIGURE 8.9 Mini Traction Machine isothermal hold tribological studies comparing COF changes in PCMO with 1% (wt.) Exp-OFM1 and 1% (wt.) MoDTC. MoDTC shows that the friction reduction performance is lost when the oil is held at 150°C for 4 hours, where the COF even rises above that of the oil without any FM. After holding at 150°C, the MoDTC-treated oil never reaches the lower ~0.07 initial COF at the start of the hold measurements. It does regain some performance but not enough to maintain the COF level below that of the oil without FM even after 48 hours. This loss in the friction reduction performance with MoDTC is similar to the changes observed in the Cameron Plint hold studies at 160°C. In addition, the data show, as with the Cameron Plint hold studies, the 1% Exp-OFM1 maintains its lower COF of 0.06 for the entire 48 hours measured.

engine speed (rpm), and dynamometer load, as described in the following six engine test stages, which are detailed further in Table 8.2 [1]:

1. 20W-30 baseline fuel economy (FEI-BL) is measured twice before adding test oil (BLB-1, BLB-2).
2. Test oil is added and aged in the engine for 16 hours, at 2250 rpm engine speed and 120°C.
3. Test oil fuel economy is measured to determine FEI-1.
4. Test oil is aged further: 109 hours, 2250 rpm, 120°C in the VIE test and 84 hours in the VID test.
5. Test oil fuel economy is measured again to determine FEI-2.
6. Baseline 20W-30 oil FEI-BL is finally measured again after adding test oil (BLB-A).

The calculated FEIs over the higher-viscosity 20W-30 reference baseline oil is represented as a sum of the initial and extended aged oil measurements and is given by

$$FEI\text{-}Sum = FEI\text{-}1 + FEI\text{-}2 \qquad (8.1)$$

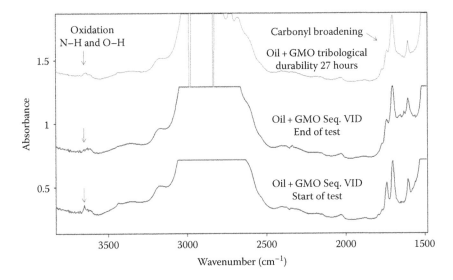

FIGURE 8.10 FTIR spectra showing changes in 1% GMO in samples measured at the start of the Sequence VID engine test and at the end of the test. The changes in the spectral carbonyl fingerprint region show a close match with the same test oil measured in the Cameron Plint tribology hold test for 27 hours. Interestingly, some higher-frequency N–H stretchings typical of alkylated diphenylamine antioxidants are decreased at 27 hours of tribological testing at 135°C.

Both FEI-Sum and FEI-2 are set values needed to be met in the GF-6 lubricant specification, and the minimum required values are set by the consortium specifically for each oil viscosity (i.e., 5W-30, 5W-20, 0W-20). Another parameter not part of the specification that can be evaluated relates to the FEI retention (FEI-R) between stages. Percent FEI-R is given by

$$FEI\text{-}R = \left(FEI\text{-}2/FEI\text{-}1\right) \times 100 \qquad (8.2)$$

The data in Figure 8.11 show an example of the FEI-Sum, FEI-1, and FEI-2 from a typical ASTM D7589 Sequence VID fuel economy test. The difference between FEI-1 and FEI-2, in addition, measures the performance durability after extended aging. This parameter is even more important in the Sequence VIE test, where the second aging is longer (109 hours) compared with the duration (84 hours) for the Sequence VID test. The results of Sequence VIE testing for the lower-viscosity 0W-20 oil (typically with less boundary layer lubrication, thus requiring higher friction protection) are given in Figure 8.12. The Sequence VIE test extends the second-phase aging duration and stresses, to a greater extent, the lubricant retention of the friction reduction performance.

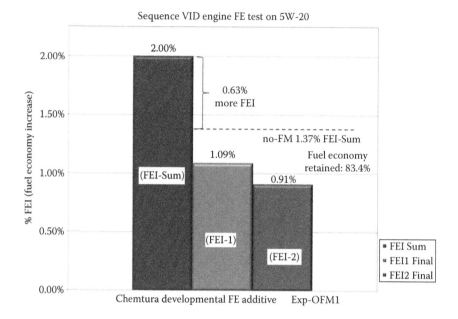

FIGURE 8.11 Results of 1% Exp-OFM1 in 5W-20 PCMO measured for fuel economy. The results of calculated FEI-Sum, FEI-1, and FEI-2 from this typical ASTM D7589 Sequence VID test show good durability (small drop in fuel economy performance). The difference between FEI-1 and FEI-2 shows durability after extended 84 hours of FEI-2 aging, and in fact, the fuel economy retention FEI-R is 83.4% of the value of the first stage of testing FEI-1. This parameter is even more important in the Sequence VIE test, where the second aging is longer (for 109 hours) compared with the duration (84 hours) for the Sequence VID test.

8.3 RESULTS AND DISCUSSION: ISOTHERMAL TRIBOLOGY, ENGINE FUEL ECONOMY, AND FTIR

8.3.1 CAMERON PLINT TE-77 AND MTM ISOTHERMAL HOLD EXTENDED TRIBOLOGY

The Cameron Plint TE-77 line contact studies (Figure 8.5) holding at 161°C–162°C show that although initially molybdenum-based friction modifiers (MoDTC-FM) chemically change to form MoS_2 friction-reducing tribofilms above 120°C, their performance is lost if held at 160°C and the testing continued. Differences in the performances between MoDTC-FM1 and MoDTC-FM2 may be due to the interaction with zinc dialkyldithiophosphates (ZDDPs), which are known to impact the friction reduction performance of molybdenum dithiocarbamates and improve MoDTC durability [10]. Very high concentrations of ZDDPs should be avoided since these compounds have been shown to increase friction at certain thicknesses of the tribolayer film they form. Industry standard GMO organic friction modifier shows less reduction in the COF, but it does so for a longer period of time initially until it again loses its performance with sufficient time. In contrast, an experimental ashless organic-based friction modifier, Exp-OFM1, appears

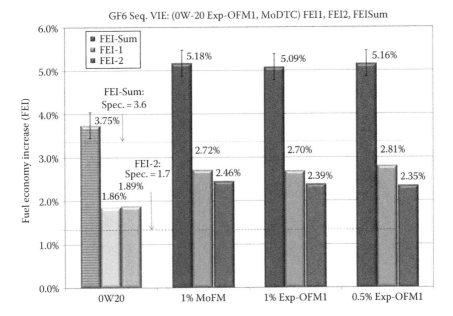

FIGURE 8.12 Results for 1% Exp-OFM1, 0.5% Exp-OFM1, and 1% MoDTC (MoFM) compared with results for a 0W-20 fully formulated passenger car motor oil (except without friction modifier), as measured for fuel economy. The FEI-Sum, FEI-1, and FEI-2 from this typical ASTM D7589 Sequence VIE test show good comparable fuel economy performance and durability of the organic Exp-OFM1 and molybdenum-containing FM, MoFM. In this formulation with FM added, in addition to sustained durable fuel economy, with small differences from FEI-1 and FEI-2 after extended 109 hours of FEI-2 aging, the FEI-2 values reached are considerably higher than the proposed minimum specification of FEI-2 for 0W-20 oil of 1.7%.

to maintain for a longer period a good level of the friction reduction performance. The results, when the temperature is reduced to 120°C (Figure 8.6), show similar differences between friction modifiers and their friction reduction performance. It takes longer for a similar durability breakdown to occur, with again the Exp-OFM1 showing the highest sustained level of sustained friction reduction.

In addition to the SAE 5W-30 weight oil, testing of lower-viscosity 5W-20 (Table 8.6; Figure 8.7) again shows the Exp-OFM1 friction reduction performance retention and changes with durability loss in the GMO at 135°C (typical average sump oil temperature). If we compare the COF increasing rapidly for GMO in the isothermal durability tribology (Figure 8.7) then during the time involved for the Sequence VID engine, one would expect some level of FEI-R for the new experimental friction modifier additive (EXP-OFM-1) and lacking or reduced significantly for GMO FM. The retention of fuel economy performance for this same 5W-20 oil was measured during the ASTM D7589 Sequence VID, as seen in Figures 8.11 and 8.13; Table 8.7.

Comparing the tribological hold studies on an even lower-viscosity oil 0W-20 (Figure 8.8) is of further interest in light of the drive to higher fuel economy through lower-viscosity oils. This SAE 0W-20 PCMO is fully formulated in a very stable base-stock oil which was found to be responsive to friction modifiers, and again,

TABLE 8.6
Table of PCMO Test Oil Compositional Information in Cameron Plint TE-77 and MTM Testing

Formulation	A	B	C	D
Base Lubricant	SAE 5W-30	SAE 5W-30	SAE 5W-30	SAE 5W-30
Friction modifier (FM)	None	GMO[a]	Exp-OFM1[b]	MoDTC[c]
% (wt.) FM	—	1%	1%	1%
Figures 8.3, 8.5, 8.6, and 8.9				
Base Lubricant	SAE 0W-20	SAE 0W-20	SAE 0W-20	SAE 0W-20
Friction modifier (FM)	None	GMO	Exp-OFM1	MoDTC
% (wt.) FM	—	1%	1%, 0.5%	1%
Figures 8.4 and 8.8				
Base Lubricant	SAE 5W-20	SAE 5W-20	SAE 5W-20	SAE 5W-20
Friction modifier (FM)	None	GMO	Exp-OFM1	MoDTC
% (wt.) FM	—	1%	1%	1%
Figure 8.7				

[a] GMO, glycerol monooleate organic friction modifier surfactant (C, H, O).
[b] Exp-OFM1, experimental organic friction modifier surfactant (*no P, S, Mo, or other metals*).
[c] MoDTC, molybdenum dithiocarbamate (MoDTC-FM1 monomer, MoDEC-FM2 dimer).

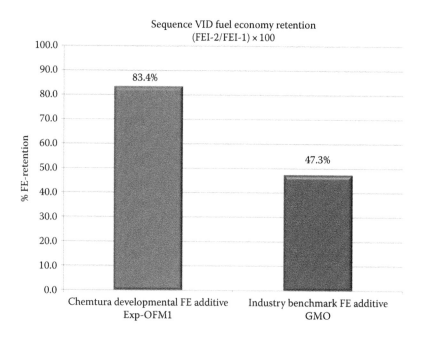

FIGURE 8.13 Percent retention in fuel economy improvement (FEI-R) for the Exp-OFM1 and a similar Sequence VID testing of glycerol monooleate (GMO) organic FM. FEI-R is calculated as FEI-2/FEI-1 × 100.

TABLE 8.7

Table of Proposed Next-Generation GF-6 Specification for Passenger Car Motor Oil for the Engine Dynamometer Fuel Economy Testing ASTM D7589

GF-6A Specification for Fuel Efficiency; Sequence VID or Equivalent (VIE) ASTM D7589		
Viscosity Grade	FEI-Sum	FEI-2
XW-20	3.6% minimum	1.7% minimum after 100-hour aging[a]
XW-30	2.9% minimum	1.4% minimum after 100-hour aging[a]
10W-30	2.5% minimum	1.1% minimum after 100-hour aging[a]

[a] FEI-2 is equivalent to 10,000 miles of oil aging.

we see the performance behavior of the friction modifier tested with time. The EXP-OFM1 again shows a significant reduction in the COF with a retention of performance and no increase at the elevated temperatures for over 2 hours of continuous testing. If we compare the more challenging engine dynamometer Sequence VIE measurement, for both the Exp-OFM1 and MoDTC additives, the results are very similar (Figure 8.12). In general, based on the comparisons of Cameron Plint TE-77 hold studies and engine FEI testing, it does appear that there is some agreement at points in time captured with the extended tribological hold studies that match the time of engine testing in the Sequence VIE test.

8.3.2 COMPLEMENTARY MTM ISOTHERMAL TRIBOLOGICAL STUDIES

When the standard Stribeck curves are measured in the MTM at 150°C, the trends are similar to that of the Cameron Plint TE-77 standard test. The MTM 150°C isothermal extended hold testing is also similar to the Plint TE77 extended tribological study at 161°C–162°C for the same 5W-30 oil. Again, a breakdown of MoDTC performance becomes evident with sufficient testing hold time for this same 5W-30 oil both in the Cameron Plint TE-77 and the MTM held at 150°C (Figures 8.5 and 8.9). This indicates that the geometry of the point contact with a sliding–rolling interaction, such as a Plint TE-77 line contact, is also affected by oil oxidation and changes resulting in an increase in the COF with time. The ability to run the MTM continuously at different speeds in the extended hold studies allows measurements of changes in the mixed, elastohydrodynamic lubrication regimes with time as well as boundary lubrication. The MTM isothermal hold studies at elastohydrodynamic conditions of speed and temperature relate to operating conditions during the Sequence VIE test when the engine speed is increased or the temperature of the lubricant is decreased and the elastohydrodynamic role influenced by the viscosity (internal friction) of the lubricant and fluid mechanics of flow creating hydraulic pressure to separate surfaces from contact.

8.3.3 FTIR TRACKING OF OIL OXIDATION AND AGING

The comparison of changes in the IR spectrum absorbances between oil aging during engine dynamometer testing and during tribological hold studies seems to support

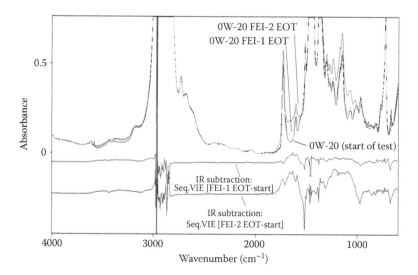

FIGURE 8.14   Top traces show the FTIR spectral overlay of the 0W-20 oil at the start of test, after FEI-1 and FEI-2 testing phases of the VIE test. The lower traces are generated by subtracting the starting 0W-20 oil spectrum from FEI-1 and FEI-2 phases to highlight changes in the carbonyl oxidation region. The data indicate oxidation changes that occur after Sequence VIE initial oil aging and fuel economy measurement in Phase 1 FEI-1 (end of test, EOT) at 84 hours (middle spectrum of the overlay). In addition, the spectrum indicates more oxidation after the longer FEI-2 oil aging at 109 hours. The 0W-20 oil at the start is subtracted from the oil after FEI-1 (red) and FEI-2 (blue). These difference spectra magnify the increasing absorbance of oxidation features as the test oil oxidizes in the engine both after FEI-1 testing and with further oxidation greatest after FEI-2 testing at the end of the test.

the concept that a point in time can be determined for a reasonably good correlation of oil aging effects. The spectra given in Figure 8.10 clearly show agreement after 27 hours of tribological testing with a good match to the Sequence VID EOT FTIR spectrum. The relative changes in the COF for the tribological test should then indicate that there is still good friction reduction performance and the VID fuel economy engine test data appear to support this (Figures 8.11 and 8.13). These studies indicate that there is a complementary approach to running the hold tribological test for evaluating potential friction modifiers and the engine FEI. In a similar manner, when FTIR spectra are compared from starting and ending test samples of the Sequence VIE fuel economy test of this 0W-20 PCMO, one can again measure the absorbance changes in the spectrum of the oil relating to different levels of aging oxidation brought about by engine testing (Figures 8.14 and 8.15).

8.3.4   ENGINE FEI ASTM D7589 SEQUENCE VID AND VIE

In both the Sequence VID testing (Figure 8.11) and the Sequence VIE testing (Figure 8.12) of the new Exp-OFM1, we can see considerable FEI compared with the reference oil used in the test (baseline 20W-30) as well as the same formulated oil without any friction modifier. There is considerable improvement exceeding the new proposed higher

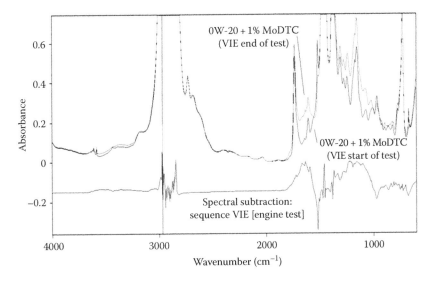

FIGURE 8.15 Top traces show the FTIR spectral overlay of start and end of test for Sequence VIE 1% MoDTC over 0W-20 at the start of test. The data indicate oxidation changes that occur after Sequence VIE oil aging and fuel economy measurements of FEI-2 are completed, as indicated by the differential IR at the bottom of figure, where 0W-20 at the start of test was subtracted from the end of test spectrum. The differential IR clearly indicates oxidation by virtue of the features in the oxidation carbonyl C=O stretching region from 1650 to 1850 wavenumbers (cm^{-1}) and C–O–C stretching absorbances around 1000 (cm^{-1}), typical of aldehydes, ketones, esters, carboxylic acids, and other oxygen-rich compounds resulting from oil oxidation and nitration.

fuel economy specification 3.6% FEI-Sum and 1.7% FEI-2 pointed out in the data for reference. A measurement of 5% FEI-Sum exceeds GF6 proposed oil specification by approximately 1.4 times the required value, and importantly, the retention in the FEI or FEI-R of 84%–90% is also desirable. This combination of adding a high-performing durable friction modifier to a responsive base oil 0W-20 formulation may offer substantial benefit for continuing to increase lubricant-derived fuel economy suited for long drain intervals. In order to extend mileage accumulation substantially between oil drain intervals, in addition to fuel economy which can be extended, lubricant detergency, dispersancy, and antioxidancy must also be addressed and improved. The friction modifier performance tested is further substantiated in that both 1.0 (wt.%) and lower concentrations of 0.5 (wt.%) show similar improvement in the tribological experiments and perform in the engine improving fuel economy equivalent to MoDTC but at half the concentration of MoDTC. In general, the 0W-20 with EXP-OFM1 overall performed well in this Sequence VIE test and benefits from the latter being an ashless totally organic friction modifier. No doubt, the key stages of the engine dynamometer test sensitive to friction modifiers are benefiting from durable COF reduction performance as well. Thus, overall, the engine testing data appear to correlate with isothermal hold studies assisted by the FTIR signature data analysis to pinpoint where in the extended tribological studies the test oils' COF will match the performance of the Sequence VIE engine.

8.4 SUMMARY

The coupled use of isothermal tribological hold studies at various temperatures and characterizing oil aging oxidation changes in the laboratory appears to be very useful in designing in general friction modifier chemistries that should also respond well in the engine. An analysis of some of the experiments given indicate that performance loss can occur in MoDTC-treated oils, which further supports the observation of a breakdown in the initial MoS_2 tribofilm formed. This type of tribological experimentation should serve to also test longer, more severe durability and extended drain cycles beyond those predicted in dynamometer testing. Since, as initially discussed, the goals of changing the Sequence VIE were to predict engine oil performance out to 10,000 mile oil drains (Figure 8.1), these tools should help predict the response of candidate molecular structures to meet these levels and learn how well they will prevent or slow the rate of loss in friction reduction performance. In an effort to fully understand the nature of performance loss as well, a close molecular-level examination of the test specimens used in the extended tribological test was done.

In addition to FTIR mapping of oxidation, an additional tool to directly measure tribofilm breakdown was investigated. To further understand the decrease in friction

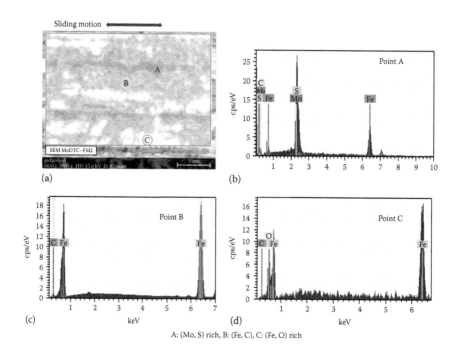

A: (Mo, S) rich, B: (Fe, C), C: (Fe, O) rich

FIGURE 8.16 SEM analysis of a Cameron Plint TE-77 specimen plate (a) following 162°C tribological isothermal hold studies of MoDTC-treated PCMO after performance is lost. One can see that within the sliding contact area, there is a limited area remaining where Mo and S can be identified (regions of point A analysis (b)). There are more areas where Fe, C, and O are present, without any Mo or S derived from molybdenum dithiocarbamate (areas of points B and C (b and d)).

reduction of MoDTC with time, SEM, coupled with X-ray fluorescence elemental analysis, was performed on the specimen plate following a typical isothermal hold test at 162°C (Figure 8.16). This SEM–X-ray analysis clearly mapped out how only limited areas existed where molybdenum and sulfur were still present, and within the contact area, there were large areas where only iron, carbon, and oxygen were present, further supporting the evidence of tribofilm degradation with extended contact rubbing in the tribological hold studies. In addition to the metals analysis, by analyzing the specimen with an FTIR microspectroscopy, focused down to a fraction of a 1 mm spot size, the surface was analyzed showing a significant spectrum of oxidized hydrocarbon, particularly at the ends of travel of the movement of the Cameron Plint dowel specimen in line contact with the plate. It is as if a debris field is generated and pushed to the outer boundary of the tribological testing path of rubbing in this experiment (Figure 8.17).

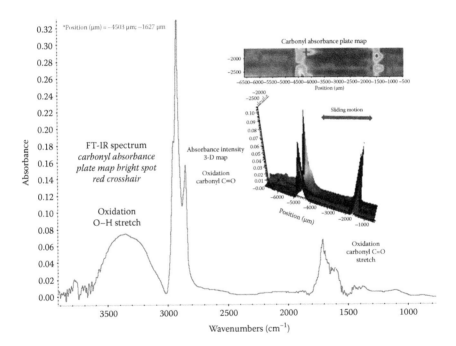

FIGURE 8.17 Micro FTIR analysis of a Cameron Plint TE-77 specimen plate following 162°C tribological isothermal hold studies of oil treated with 1% (wt.) MoDTC. The FTIR analysis correlates with the SEM analysis, showing areas of little Mo and S and more C and O associated with oxidation. Insert shows two-dimensional (2D) and three-dimensional (3D) maps of oxidation carbonyl absorbance (2D bright spots, 3D positive intensity peaks). Below the insert is a typical FTIR microspectroscopy of a small fraction of a 1 mm area in the spectral region showing high oxidation as evidenced by the absorbance at the hydroxyl O–H stretching vibration centered at 3350 cm⁻¹ and the carbonyl C=O stretching vibration centered at 1700 cm⁻¹ (of a point on the 2D, 3D map). The map shows a high oxidation area appearing at the ends of the tribological sliding motion developed during the line contact measurements of friction during extended testing at 160°C.

The concept of applying FTIR spectroscopy and SEM in situ analysis to help correlate extended tribological testing with real-world engine testing is useful for understanding frictional changes and friction modifier additives' performance in extended oil drain applications. The goal in future lubricant direction is likely to continue to assist in fuel conservation and in reducing harmful emissions while still protecting engine life. No doubt, future progress will be made through the use of combinations of friction modifiers [11,12], and their use in newly improved coatings applied to engine metallurgy as well [13,14].

ACKNOWLEDGMENTS

Special thanks to Cyril Migdal, PhD (Chemtura Corp.), for support and discussions regarding friction modifier performance evaluations, and to Mr. Brian Fox, Mr. Jon Goodell, and Mr. Mike Maselli (Chemtura Corp.) for tribological testing support. In addition, special thanks to Mr. Sergio deRooy (Shell Global Solutions US, Technical Center, Houston), Mr. Mike Baumbarger (Honda R&D Americas, Advanced Materials Research), Mr. Dan Worcester (Southwest Research Institute, SWRI), and Mr. James Linden (Linden Consulting, LLC) for their help and technical discussions of lubricant requirements and testing in both laboratory and operating engine environments, including ASTM D7589 Sequence VIE fuel economy testing.

REFERENCES

1. D7589-15 A. *Standard Test Method for Measurement of Effects of Automotive Engine Oils on Fuel Economy of Passenger Cars and Light-Duty Trucks in Sequence VID Spark Ignition Engine*, ASTM International, West Conshohocken, PA (2015).
2. S. Korcek, J. Sorab, M. D. Johnson, and R. K. Jensen. Automotive lubricants for the next millennium, *Industrial Lubrication and Tribology* 52, 209–220 (2000).
3. R. Thorn, K. Kollmann, W. Warnecke, and M. Frend. Extended oil drain intervals: Conservation of resources or reduction of engine life, SAE Technical Paper 951035 (1995), doi:10.4271/951035.
4. J. Van Rensselar. PC-11 and GF-6: New engines drive changes in oil specs, *Tribology & Lubrication Technology* 69, 30–34, 36–38 (January 2013).
5. Linden, J. and Leverett, C., *Consortium to Develop a New Sequence VID Fuel Efficiency Test for Engine Oils*, International Lubricants Standardization and Approval Committee, Washington, DC (2008).
6. G. Guinther and J. Styer. Correlation of the sequence VID laboratory fuel economy test to real world fuel economy improvements. No. 2013-01-0297. SAE Technical Paper, Warrendale, PA (2013).
7. N. Canter. Fuel economy: The role of friction modifiers and VI improvers, *Tribology & Lubrication Technology* 69, 14–27 (2013).
8. B. Kinker, R. Romaszewski, and P. Palmer. ROBO—A bench procedure to replace sequence IIINGA engine test, in *Automotive Lubricant Testing and Advanced Additive Development*, T. Simon, K. Bernard, and W. Mathias (Eds.), ASTM International, West Conshohocken, PA (2008).
9. F. DeBlase. Automotive lubricant friction modifiers: Additive durability studies, in *Surfactants in Tribology*, Vol. 4, G. Biresaw and K. L. Mittal (Eds.), CRC Press, Boca Raton, FL, pp. 283–312 (2014).

10. S. Bec, A. Tonck, J. Georges, and G. Roper. Synergistic effects of MoDTC and ZDTP on frictional behavior of tribofilms at the nanometer scale. *Tribology Letters* 17(4), 797–809 (2004).
11. K. Topolovec-Miklozic, T. Reg Forbus, and H. A. Spikes. Film thickness and roughness of ZDDP antiwear films. *Tribology Letters* 26(2), 161–171 (2007).
12. R. I. Taylor and R. C. Coy. Improved fuel efficiency by lubricant design: A review. *Proceedings of the Institution of Mechanical Engineers, Part J: Journal of Engineering Tribology* 214, 1–15 (2000).
13. M. K. Patel and V. J. Gatto. An optimized molybdenum dithiocarbamate technology in combination with organic friction modifiers for enhanced fuel efficiency possibilities, in *Proceedings of the 20th International Colloquium Tribology—Industrial and Automotive Lubrication*, TAE, Stuttgart, Germany, p. 25 (2016).
14. A. Neville, A. Morina, T. Haque, and M. Voong. Compatibility between tribological surfaces and lubricant additives—How friction and wear reduction can be controlled by surface/lube synergies. *Tribology International* 40, 10, 1680–1695 (2007).

9 Formation of Tribofilms from Surfactants with Different Degrees of Ethoxylation on Steel Surfaces in the Boundary Lubrication Regime

F. Quintero, J.M. González, J. de Vicente Álvarez, J.E. Arellano, and S. Rosales

CONTENTS

9.1   INTRODUCTION

Drilling fluids used in the oil industry are classified according to the nature of continuous base fluid phase as water based, oil based, and pneumatic or gas based.
Basically, drilling fluid formulations are composed of a base fluid (water or oil),
a weighting powder material (Ba_2SO_4, Fe_2O_3, or $CaCO_3$), and various additives to
control fluid properties such as rheology [1–4], fluid losses [5–7], shale inhibition
[8–11], and others. In essence, drilling fluids are complex mixtures of a base component and different additives. Each additive provides a specific property, which helps
to control the performance of the fluid and permits an efficient drilling operation.

The type of drilling fluid selected for a drilling operation depends on the geologic
formation being drilled, the depth, the mechanical resistance, and the wellbore's
pressure. The main functions of drilling fluid are to maintain hole stability, transport
the rock cuttings from the bottom hole to the surface, control formation pressure, and
cool and lubricate the drill bit [4].

The lubricity function of drilling fluid is extremely important due to the high frictional forces encountered at every stage of well construction (drilling, completion,
and maintenance). The sources of frictional forces include the pipe's resistance to
rotation (torque) and raising and lowering movements (drag) of the drill bit and string
inside the well due to contact with the wellbore (metal to rock) and the casing (metal
to metal). The friction generates a considerable amount of heat and drag forces [12].
Excessive torque and drag can cause an unacceptable loss of power, making oil well
operations inefficient.

Friction between the string and the hole is a critical factor for high-angle and extended-
reach wells. This can be minimized by increasing the lubricity of the circulating working

fluid and also by employing other friction reduction techniques. The improvement of the fluid lubricity can be achieved using lubricant additives, generally available as film-producing liquids, solid beads, powders, and fibers. Liquid additives employed for such applications include glycols, oils and esters [13], surfactants [14,15], and polymer-based lubricants [16]. Examples of solid additives for drilling fluids include graphite, calcium carbonate flakes, glass, and plastic beads [17,18].

Friction reduction tools (FRTs) are downhole drill string tools designed and applied to reduce rotating friction, casing friction, and pipe wear. These tools feature a nonrotating drill pipe protector that includes a sleeve on a lubricating bearing surface, which becomes the effective contact point for torque generation.

There are various types of FRTs that can be installed on the pipe or between the pipe connections. FRTs are used in various drilling and wellbore construction applications for a variety of reasons, including for rig limitations, complex well paths, differential sticking, buckling casing wear, torque reduction, axial drag, and, ultimately, operating costs. They can be classified as either fixed or roller type. FRTs are best used in the initial building section of the hole or in the deviated portion of the wellbore where the contact forces are excessive. FRTs improve drilling by increasing available weight and minimizing slip-stick. The placement and spacing of FRTs to achieve optimum performance depends on the well profile [12,15,16].

In oil well construction and maintenance processes, especially during drilling, all the equipment and fluid systems present different tribological phenomena and related problems. Deep hole drilling for petroleum applications presents a challenge in the choice of drilling materials because they are used in extremely harsh operating environment. Low wear is obviously desirable to increase shaft and casing life and to reduce maintenance, while low friction is desirable to reduce the energy needed for drilling. The dominant wear modes include impact wear, abrasion, and slurry erosion. If the wear mode cannot be controlled or predicted, it could cause catastrophic failures of the equipment and the wellbore and the ultimate loss of the hole [18–20]. Abrasion and erosion wears are caused mainly by the solid particles used as weighting material in the drilling fluid formulation.

As mentioned before, surfactants could be used as additives for water-based drilling fluids [21,22]. These additives improve the lubrication properties of water between rubbing surfaces under high loads by producing a physical and/or chemical effect on the surfaces of the friction pair. The result is a reduction in the friction coefficient and/or the wear rate. These surfactant additives are capable of forming layers and thus able to withstand the extreme pressure developed in the contact area.

This molecular layer is formed by surfactant adsorption on the worn surfaces. The lubricating effect or friction reduction is caused by the formation of a low-shear interface between the opposing surfaces, which produces a low-friction molecular layer on the surface. Strong adsorption ensures that almost every available surface site is occupied by the surfactant molecules to produce a dense and robust film. This film can bear a high load without being destroyed, thus resulting in minimum metal–metal contact between the two solid surfaces. However, this layer is so thin that the mechanics of asperity contact is identical to that of dry contact [23,24].

The adsorption of surfactant molecules on solid surfaces depends on several factors, including the surfactant structure, temperature, pH, electrolyte concentration, and polarity of solid surface [25–37]. As a result, the lubricating effect of surfactants will depend on the variables that control their adsorption property on the contacting surfaces.

In order to accurately and completely describe a lubricant, it is important to classify their behavior into three separate regimes as depicted in the Stribeck curve. The curve is a plot of friction (μ) or film thickness (h) between a lubricated tribopair as a function of a parameter Z derived from the viscosity of lubricant (η), speed (u), and load (L).

As shown in Figure 9.1, the Stribeck curve can be divided into three separate regimes: (1) The boundary regime observed at low sliding speeds and high loads (low Z in Figure 9.1), where friction is determined by both surface–surface asperities interaction and the ability of the lubricant to adsorb chemically onto the surface and form an interfacial film. Depending on the dynamic conditions, different sliding wear mechanisms may occur in this regime such as corrosive, fatigue, and adhesive wear. (2) The hydrodynamic lubrication (HL) regime occurs at high speeds (high Z), in which the surfaces are fully separated by a lubricant film, whose thickness depends on the viscosity (η) and entrainment speed (u). The friction is governed by the bulk rheological properties of the lubricant film in the contact zone. In the HL regime no direct physical contact occurs between the surfaces, so no sliding wear occurs but wear due to surface fatigue, cavitation, or fluid erosion occurs. (3) Between the boundary and hydrodynamic regimes is the mixed lubrication regime, where the thickness of the lubricant film carries part of the load, while the other part is supported by the surface asperities.

The objective of this investigation is to establish the structures of surfactants that can be employed to control friction and wear in drilling fluids. The effect of adsorption of surfactant mixtures with different degrees of ethoxylation on the tribological

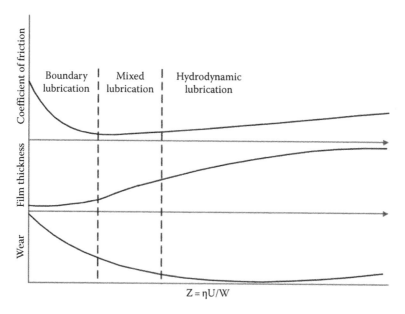

FIGURE 9.1 Schematic Stribeck curve.

properties of bilayers (tribofilms) formed on steel surfaces under boundary lubrication conditions was evaluated. The goal was to develop new surfactant-based additives to lower friction and wear for water-based drilling fluids. These surfactant additives must exhibit the following characteristics: highly effective in reducing friction and wear at low concentrations, compatible with other drilling fluid additives, able to support drilling conditions, low toxicity, and environmentally safe.

9.2   EXPERIMENTAL DETAILS

9.2.1   MATERIALS

9.2.1.1   Friction Reducer Additives

Commercial anionic and nonionic surfactant mixtures were used as received without further purification. They were obtained from the Petroleum and Petrochemical Service C.A. (PPS), Valencia, Venezuela. The anionic surfactants were of the ethoxylated lauryl phosphate ester family where the degrees of ethoxylation were 4, 7, and 9 ethylene oxides. The nonionic surfactants were ethoxylated lauryl alcohol where the degrees of ethoxylation were 4, 7, and 9 ethylene oxides. The surfactant mixtures A, B, and C of similar degree of ethoxylation contained 20% w/w ethoxylated lauryl phosphate ester and 80% w/w ethoxylated lauryl alcohol (Table 9.1).

9.2.1.2   pH Modifier

Sodium hydroxide (NaOH, 99% purity) was used to adjust pH and was obtained from Akzo Nobel, Germany. It was used to adjust the pH to 10 units.

9.2.1.3   Viscosifying Additives

Xanthan gum (Akzo Nobel, Germany) and glycerin (Sigma–Aldrich, United States) were used as viscosity modifier additives for tribological tests.

9.2.1.4   Tribopair

Stainless steel balls (AISI 316) and plates (AISI 304) were used for the tribological tests. The balls as well as the plates were supplied by Anton Paar (Germany). The specifications of the ball and plate are given in Table 9.2.

TABLE 9.1
Surfactant Mixtures Used in This Investigation

Surfactant Mixture	Description	Ratio Nonionic/ Anionic (% w/w)
A	Ethoxylated lauryl alcohol with four degrees of ethoxylation/ ethoxylated lauryl phosphate ester with four degrees of ethoxylation	80/20
B	Ethoxylated lauryl alcohol with seven degrees of ethoxylation/ ethoxylated lauryl phosphate ester with seven degrees of ethoxylation	
C	Ethoxylated lauryl alcohol with nine degrees of ethoxylation/ ethoxylated lauryl phosphate ester with nine degrees of ethoxylation	

TABLE 9.2
Physical Properties of Ball and Plate Used

Tribopair	Material	Dimensions	Elastic Modulus (GPa)	Poisson's Ratio	Roughness (μm)
Ball	AISI 316	D = 12.7 mm	193	0.3	0.160
Plate	AISI 304	6 × 15 × 3 mm	193	0.3	0.496

9.2.2 CLEANING PROCEDURE OF TRIBOPAIR FOR TRIBOLOGICAL TESTS

The cleaning procedure of tribopair (plate and ball) is an important step toward carrying out the different tribological tests because of any contaminant that could affect the measurements leading to erroneous interpretations or conclusions. The experimental protocol established for this investigation was divided into three general stages: (1) immersion cleaning, (2) ultrasonic cleaning, and (3) drying.

The immersion cleaning process consisted of placing the tribopair in contact with toluene for 5 min in order to remove water-insoluble solids such as fats, oils, and waxes. After that, the tribopair was placed in contact with isopropanol for 5 min to ensure the removal of any contaminant remaining. The second stage was the ultrasonic cleaning (10 min) for eliminating potential contaminants remaining from the previous step through vibration and the collapse of bubbles in immersion liquid in order to generate an effective scrubbing action for the removal of contaminants. The immersion liquid used in this stage was acetone. Finally, the third stage included drying the materials at 40°C using a convection oven. The cleaning protocol explained earlier was also applied to the self-positioning plate of tribo-rheometer.

9.2.3 METHODS

9.2.3.1 Friction Measurement Instrument

Friction measurements were carried out in a nonconforming ball-on-three-plates contact using a tribo-rheometer (model MCR 302, Anton Paar, Germany) under boundary lubrication conditions. A schematic of the instrument is given in Figure 9.2. The setup

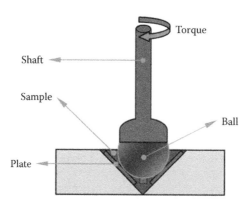

FIGURE 9.2 Basic components of tribo-rheometer used in this study.

is based on the ball-on-three-plates principle (or ball-on-pyramid) consisting of a geometry in which a ball is held, an inset where three small plates can be placed, and a bottom stage movable in all directions on which the inset can be fixed [38,39]. The flexibility of the bottom plate is required for the same normal load acting on all the three contact points of the upper ball. The rotating ball is adjusted automatically and the forces are evenly distributed on the three friction contacts. If required, the ball and the plates for the setup can be exchanged to adapt to several material combinations.

In this setup, a ball (radius R) is pressed at a given normal force F_N against three plates that are mounted on a movable stage. As a result, the same load is acting evenly on all three frictional pairs. After that, the ball is rotated at an increasing sliding speed V, while the plates are held at rest. The resulting torque can be correlated with the friction force by employing simple geometric calculations. The normal force of the tribo-rheometer is transferred into a normal load acting perpendicular to the bottom plates at the contact points. The tribology setup is temperature controlled by Peltier elements to ensure the same temperature at the bottom plates and at the upper ball [38,39].

To determine the actual load acting on the three plates in a stationary condition, it is sufficient to perform a decomposition of normal force acting on each plate (Figure 9.3). Equation 9.1 allows to relate the normal force and the load on a plate i (F_{Li}):

$$F_N = \sum_{i=1}^{3} F_{Li} \cos\alpha \tag{9.1}$$

$$F_N = \sum_{i=1}^{3} F_{Li} \sin\beta \tag{9.2}$$

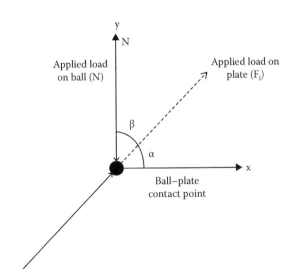

FIGURE 9.3 Loads acting on the ball-plate contact point.

Equations 9.1 and 9.2 can be rearranged as follows:

$$F_{Li} = \frac{F_N}{\cos \alpha} \tag{9.3}$$

$$F_{Li} = \frac{F_N}{\sin \beta} \tag{9.4}$$

where
$\alpha = \beta = 45°$ in correspondence to the inclination of the plates in this device
F_N is the normal force acting on each plate (N)
F_{Li} is the load on a plate i (N)

Since friction is the force that opposes the relative motion of two surfaces in contact, it is necessary to establish the equations governing the dynamic condition of the test method. Therefore, if the contact point between the ball and the plate is defined as the axis of rotation and the radius of the ball as a radius vector, this leads to an equation for determining the torque of the system and the frictional force. Figure 9.4 shows the forces involved when the system is subjected to a rotational movement [38].

According to the force diagram shown in Figure 9.4, the torque τ is

$$\tau = F_F \cdot b \tag{9.5}$$

where
τ is the torque (N·m)
F_F is the total friction force (N)
b is the moment arm (m)

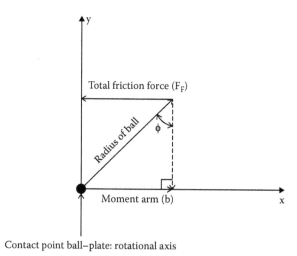

FIGURE 9.4   Schematic representation of the forces involved in a system of Cartesian coordinates when the system is subjected to a rotational movement.

The moment arm b is given by the following equation:

$$b = \sin \varphi \cdot R_{ball} \qquad (9.6)$$

where
$\phi = 45°$ in correspondence to the inclination of the plates in this device
R_{ball} is the radius of ball (m)

Rearranging Equation 9.3, the total friction force can be calculated as

$$F_F = \frac{\tau}{\sin \varphi \cdot R_{ball}} \qquad (9.7)$$

Finally, the friction coefficient μ can be obtained from Equations 9.4 and 9.7 as follows:

$$\mu = \frac{\tau}{R_{ball} \cdot F_N} \qquad (9.8)$$

where
μ is the friction coefficient (dimensionless)
τ is the torque (N·m)
R_{ball} is the radius of ball (m)
F_N is the normal force acting on each plate (N)

All tribological tests were performed under pure sliding conditions. Parameters related to contact radius a and maximum contact pressure p_{max} were calculated according to Hertz's contact theory [36,38,39]. These assumptions were established:

- Elastic bodies, small strains within elastic limit.
- There is smooth nonconforming contact.
- The contact half-width is much less than any other dimension.
- There is no friction.

According to Hertz's contact theory, the contact modulus (or reduced elastic modulus) E* can be calculated as follows:

$$E^* = \left(\frac{1-\upsilon_1^2}{E_1} + \frac{1-\upsilon_2^2}{E_2} \right)^{-1} \qquad (9.9)$$

where
E^* is the contact modulus (Pa)
υ_1 is Poisson's ratio for the ball (dimensionless)
E_1 is the elastic modulus for the ball (Pa)
υ_2 is Poisson's ratio for plate (dimensionless)
E_2 is the elastic modulus for plate (Pa)

Then, parameters related to contact radius a and maximum contact pressure p_{max} are given by the following equations:

$$a = \left(\frac{3}{4} \cdot \frac{R_{ball} \cdot F_{Li}}{E^*} \right)^{1/3} \qquad (9.10)$$

where
a is the contact radius (m)
R_{ball} is the radius of ball (m)
F_{Li} is the load on each surface (N)
E^* is the contact modulus (Pa)

$$p_{max} = \frac{3}{2} \cdot \left(\frac{F_{Li}}{\pi a^2} \right) \qquad (9.11)$$

where
p_{max} is the maximum contact pressure (Pa)
F_{Li} is the load on each surface (N)
a is the contact radius (m)

These values are a = 5.96×10^{-5} m and p_{max} = 6.34×10^8 Pa, respectively. The normal load was fixed at 10 N. The high-elastic modulus and high-contact pressure could indicate that the lubricating film is not present and the contact may operate in the boundary lubrication regime [36,38,39].

Friction measurements were carried out over a wide range of sliding speeds, which was progressively increased from 5×10^{-5} to 1 m/s in order to fully map the Stribeck friction curve. Data was collected at a rate of 300 points per second. The test was conducted for a total of 300 s. For aqueous surfactant solutions, the Stribeck curves were produced under the following test conditions: temperature = 25°C, 49°C, and 80°C and pH = 4, 7, and 10. For surfactant mixture A + xanthan gum and A + glycerin, the Stribeck friction curves were generated at pH 10 and 25°C.

The operational conditions of the tribo-rheometer for measuring the static friction coefficient were a load of 10 N and a torque sweep from 3 to 50 mN·m, performing one measurement per second for 300 s. The study of the stability of the tribofilm was performed by carrying out two tests: (1) load and constant speed over a specific time and (2) oscillatory strain at constant frequency and amplitude. In the first case, the stability of the tribofilm was evaluated at three speeds (2, 21, and 200 rpm) applying a constant load of 10 N and taking one measurement per second for 600 s. The second operational condition, the frequency was 1 Hz and 14.4% strain at a constant load of 10 N and taking one measurement per second over 600 s.

All data reported here for the different tribological measurement conditions correspond to the arithmetic average of three repeat tests.

9.2.3.2 Contact Angle Measurements by the Wilhelmy Plate Method

The Wilhelmy plate method was originally used to determine the surface tension of liquids. Subsequently, it has been modified to study wettability changes in three-phase systems (solid–liquid–gas) by determining changes in the forces involved at

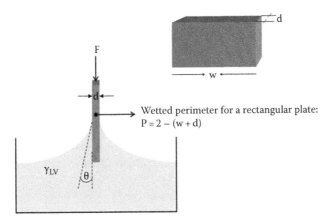

FIGURE 9.5   Schematic representation of the Wilhelmy's plate method to measure contact angle.

the triple contact point [40,41]. The forces acting on the plate are interfacial and buoyancy forces. The general equation for the Wilhelmy plate method is (Figure 9.5)

$$F = p \cdot \gamma_{lv} \cos \theta + F_F \qquad (9.12)$$

where
   F is the force acting on the plate (mN)
   p is the perimeter of the plate (m)
   γ_{lv} is the liquid–gas surface tension (mN/m)
   θ is the contact angle
   F_F is the buoyancy force (mN)

If the force exerted by the meniscus of the liquid on the plate and surface tension of the liquid are known and we assume the buoyancy force too negligible, the contact angle at the triple contact point of the gas–liquid–solid system can be calculated as follows:

$$\cos \theta = \frac{F}{p \cdot \gamma_{lv}} \qquad (9.13)$$

The Wilhelmy plate method to determine the contact angle provides a more detailed description regarding the dynamic behavior of a liquid on a solid surface. In contrast, using other methods where the contact angle is measured at a specific point of the solid, the Wilhelmy method allows for the dynamic contact angle to be measured over the entire plate surface. Thus, the test area is larger, which minimizes measurement error due to solid composition heterogeneity and roughness.

   In this test, the solid surface under study was held vertically on a microbalance. Subsequently, the liquid was kept in a cylindrical container and moved upwards at constant speed until the liquid reached the contact plate [42]. The microbalance was used to measure the force exerted by the contact line of the liquid on the solid surface in the forward and reverse directions (Figure 9.5).

9.2.3.3   Surface Analysis to Determine the Volume of Wear Scar

In order to determine the wear reducing efficiency of surfactant mixtures used in this work (Table 9.1), the morphology of the wear scar on the surface of the plates after tribological tests (static/dynamic) was examined by optical microscope analysis and analyzed to calculate the wear scar volume. Optical profilometry tests were performed on a noninvasive optical profiler (Zygo NewView, model 6000, United States). In tribological tests (ball–plate contact), there are three possible wear conditions: (1) only the ball suffers wear, (2) only the plate suffers wear, and (3) both the ball and the plate suffer wear (Figure 9.6) [43]. The volume of the worn scar was calculated by using the equation of spherical cap volume (Figure 9.7) [44] as follows:

$$V_c = \frac{1}{6}\pi \cdot h_c \left(3a^2 + h_c^2\right) \tag{9.14}$$

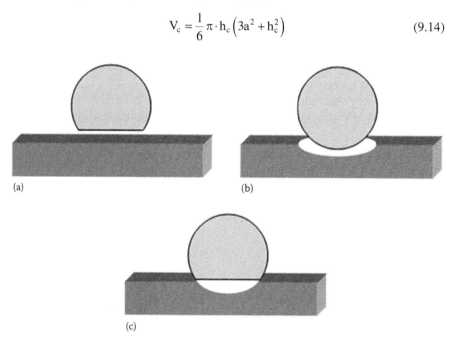

(a)   (b)

(c)

FIGURE 9.6   General wear conditions on ball-plate contact surfaces for friction tests: (a) only the ball suffers wear, (b) only the plate suffers wear, and (c) the ball as well as the plate suffer wear.

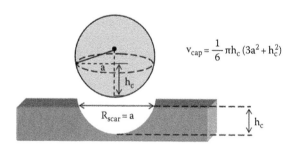

FIGURE 9.7   Spherical cap method of wear volume calculation.

where

V_c is the volume of the spherical cap (cm^3)

h_c is the height of the cap (cm)

a is the radius of the base of the cap (cm)

9.3 RESULTS AND DISCUSSION

9.3.1 SURFACTANT ADSORPTION KINETICS ON STEEL SURFACES

Adsorption is a process whereby a substance known as adsorbate accumulates on a surface of another substance called adsorbent. The result of this adsorption is the formation of a liquid or gaseous film of adsorbate on the surface of a solid or liquid known adsorbent. Adsorbate film formed on the adsorbent could be a monolayer or bilayer and produces numerous interfacial phenomena such as the dispersion of solid, emulsion formation, reduction of surface/interfacial tension, and wettability changes, among others. The speed at which the adsorption and desorption of adsorbate occurs affects the performance of the monolayer or bilayer as a promoter in the control agent or interface generation phenomena. The factors affecting the adsorption process include

- Solubility of adsorbate
- Environment of physicochemical conditions (pH and temperature)
- Chemical structure of the adsorbate
- Chemical composition of the adsorbent
- Nature of the adsorbent (hydrophilic/hydrophobic)
- Contact time between the adsorbent and adsorbate
- Molecular weight of the adsorbate

In order to determine the time for the surfactant mixtures (adsorbate) A, B, and C to reach the kinetic equilibrium of adsorption on the steel surfaces (adsorbent), the dynamic coefficient of friction (DCOF) was measured at 25°C for contact times of 5, 15, and 30 min and pH of 4, 7, and 10 units. Figure 9.8 and Table 9.3 show a summary of DCOF. The DCOF did not change significantly with contact time; that is, after 5 min, the aqueous surfactant solutions have reached the equilibrium adsorption on surfaces of the steel tribopairs.

9.3.2 STRIBECK CURVE FOR AQUEOUS SURFACTANT MIXTURES A, B, AND C

The effects of mixtures of surfactants with different degrees of ethoxylation (A, B, and C) on the lubricant behavior of aqueous surfactant solutions were investigated. Aqueous surfactants at concentration of 1% w/w; at pH 4, 7, and 10; and at temperatures of 25°C, 49°C, and 80°C were used in the test. Dynamic friction tests were conducted on steel surfaces under boundary conditions. The experimental data was analyzed and used to generate Stribeck curves for each surfactant mixture.

Figures 9.9 through 9.11 show the Stribeck curves for aqueous surfactant solutions A, B, and C at different pH and temperatures. As mentioned before, the dynamic

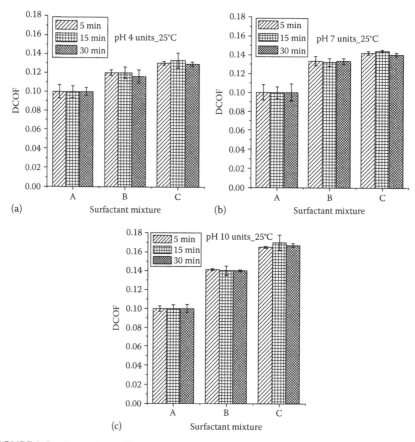

FIGURE 9.8 Dynamic coefficient of friction vs test time for aqueous surfactant mixtures: (a) pH 4, (b) pH 7, and (c) pH 10.

TABLE 9.3
Coefficient of Friction vs Test Time

Surfactant Mixture	pH	DCOF		
		Contact Time (min)		
		5	15	20
A	4	$0.100 \pm 7 \times 10^{-3}$	$0.100 \pm 6 \times 10^{-3}$	$0.100 \pm 4 \times 10^{-3}$
	7	$0.100 \pm 7 \times 10^{-3}$	$0.100 \pm 6 \times 10^{-3}$	$0.100 \pm 9 \times 10^{-3}$
	10	$0.100 \pm 2 \times 10^{-3}$	$0.100 \pm 8 \times 10^{-3}$	$0.100 \pm 2 \times 10^{-3}$
B	4	$0.100 \pm 8 \times 10^{-3}$	$0.100 \pm 6 \times 10^{-3}$	$0.100 \pm 9 \times 10^{-3}$
	7	$0.133 \pm 5 \times 10^{-3}$	$0.132 \pm 4 \times 10^{-3}$	$0.133 \pm 3 \times 10^{-3}$
	10	$0.142 \pm 2 \times 10^{-3}$	$0.144 \pm 1 \times 10^{-3}$	$0.140 \pm 2 \times 10^{-3}$
C	4	$0.100 \pm 3 \times 10^{-3}$	$0.100 \pm 4 \times 10^{-3}$	$0.100 \pm 4 \times 10^{-3}$
	7	$0.141 \pm 1 \times 10^{-3}$	$0.140 \pm 5 \times 10^{-3}$	$0.140 \pm 1 \times 10^{-3}$
	10	$0.165 \pm 1 \times 10^{-3}$	$0.170 \pm 8 \times 10^{-3}$	$0.167 \pm 2 \times 10^{-3}$

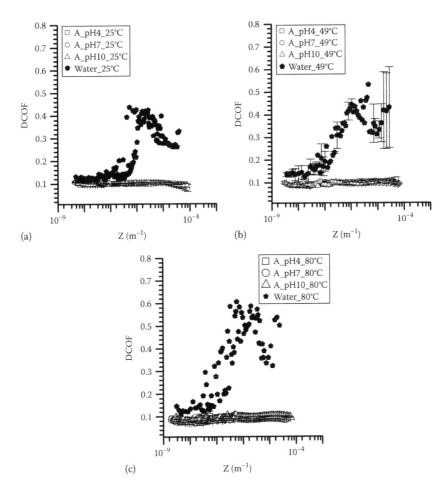

FIGURE 9.9   Z vs DCOF for surfactant mixture A: (a) 25 °C, (b) 49 °C, and (c) 80 °C.

friction tests were performed under pure sliding conditions where the sliding speed was increased from 5×10^{-5} to 1 m/s. The maximum sliding speed was determined from preliminary experiments in order to avoid a dry contact (absence of lubricant film). This precaution was taken because water is a poor lubricant and can evaporate as temperature increases in the contacting metal surfaces. The Stribeck curves for surfactant mixtures A, B, and C exhibited a boundary regime but did not show a transition to the mixed regime under the experimental conditions used in this study.

All solutions investigated exhibited an excellent performance of friction reduction of up to 67% relative to pure water (R). Surfactant mixture A showed the best performance as a friction reducing additive. The average DCOF was about 0.10, and this value is almost constant under the experimental conditions evaluated. Surfactant mixture A gave friction that was 67% lower than friction yielded by water and close to 25% lower than that from the other surfactant mixtures studied. This finding

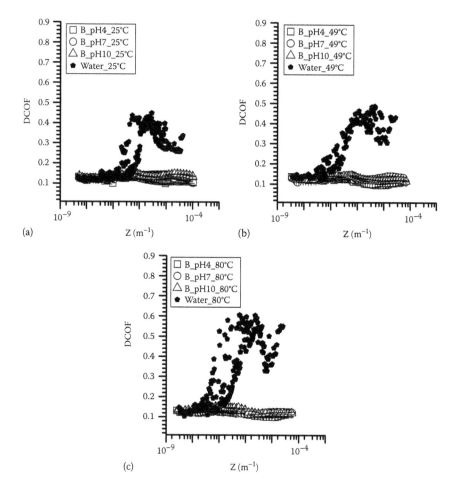

FIGURE 9.10 Z vs DCOF for surfactant mixture B: (a) 25 °C, (b) 49 °C, and (c) 80 °C.

reveals that the adsorption of surfactant mixture A generates a tribofilm, which was able to reduce friction in the entire temperature (25°C–80°C) and pH (4–10) ranges used in this investigation (Figure 9.12).

For surfactant mixture B, the average DCOF was about 0.13, and this value showed little variations under the experimental conditions evaluated. Surfactant mixture B reduced friction to 56% of pure water (R). The surfactant mixture B reduced friction at all temperature (25°C–80°C) and pH (4–10) ranges investigated (Figure 9.12).

In the case of surfactant mixture C, pH variations affected its effectiveness at reducing friction, as indicated Figure 9.12. The average DCOFs for mixture C were 0.13, 0.14, and 0.17 at pH 4, 7, and 10, respectively. This indicates that the adsorption of surfactant mixture C and the stability of the tribofilm depend on the pH of the aqueous media.

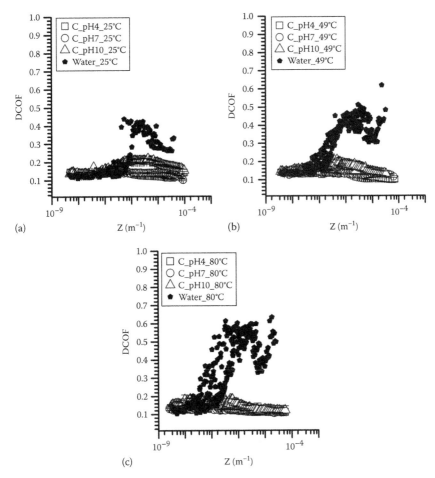

FIGURE 9.11   Z vs DCOF for surfactant mixture C: (a) 25 °C, (b) 49 °C, and (c) 80 °C.

9.3.3 Static Coefficient of Friction of Aqueous Surfactant Mixtures A, B, and C

If two solid surfaces are in contact with or without a lubricant film or friction reducing agent, they display two types of friction behavior: friction forces that oppose the initial motion of one surface under a specific load and friction force that continuously opposes the relative motion. These two forces are called static and dynamic friction forces, respectively. Several factors affect static and dynamic friction forces, such as surface roughness, lubrication regime, chemical property of the lubricant, stability of tribofilm, temperature, and load, among others.

The static coefficient of friction (SCOF) was determined for the surfactant systems (A, B, and C)/tribopair. Tests were conducted at pH 10 and 25°C, with a constant

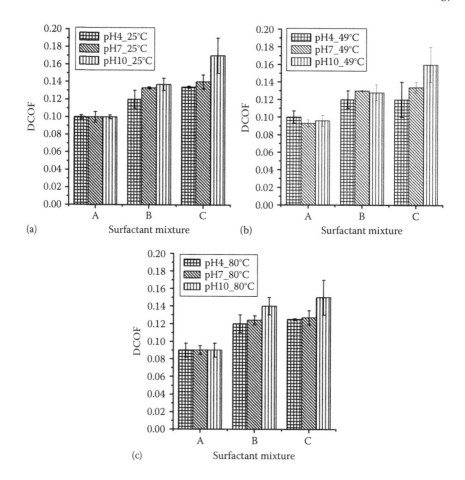

FIGURE 9.12 DCOF for surfactant mixtures A, B and C: (a) 25 °C, (b) 49 °C, and (c) 80 °C.

load of 10 N. The pH 10 was selected because the majority of water-based drilling fluids are applied at pH between 8 and 10 in order to avoid or minimize the corrosion of drill pipe and bacterial degradation of the polymer.

Figure 9.13 shows that 1% w/w of surfactant mixture A yielded the lowest SCOF, 0.13, whereas 1% w/w of surfactant mixtures B and C showed SCOF values of 0.14 and 0.15, respectively (Table 9.4). The difference in the performance of surfactants evaluated with respect to the reduction of static friction is related to the type of tribofilm adsorbed, the extent of hydrophilicity or hydrophobicity of the surface, and the wettability changes that can promote the surfactant [45,46]. Previous studies on the adsorption of phosphate ester surfactants on hematite surfaces have shown that at pH 10, a bilayer of surfactant covers the surface of the solid (hematite), providing excellent wear resistance properties to water-based drilling fluids for oil wells [21,22].

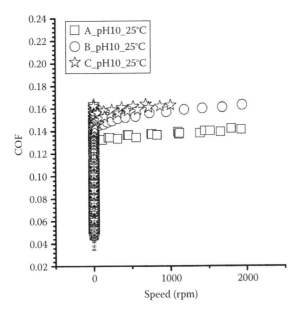

FIGURE 9.13   Static coefficient of friction for aqueous surfactant mixtures A, B and C as a function of pH at 25°C.

TABLE 9.4

Static and Dynamic Friction as a Function of the Surfactant Mixtures A, B, and C at pH 10 and 25°C

Surfactant Mixture	DCOF[a]	SCOF[b]
A	$0.010 \pm 2 \times 10^{-3}$	$0.130 \pm 4 \times 10^{-3}$
B	$0.137 \pm 7 \times 10^{-3}$	$0.150 \pm 7 \times 10^{-3}$
C	$0.17 \pm 2 \times 10^{-2}$	$0.150 \pm 7 \times 10^{-3}$

[a] Data from Figure 9.11.
[b] The test of static friction was carried out under the following experimental parameters: load = 10 N and range of speed from 0 to 2000 rpm.

9.3.4 Contact Angle of Surfactant Mixtures A, B, and C

Surfactants have the ability to adsorb onto a variety of surfaces. Surface active agent contains a hydrophobic portion, which may be a hydrocarbon chain or fluorocarbon, and a hydrophilic portion, which is polar. Hydrophilic groups can be cationic, anionic, nonionic, or amphoteric. The adsorption of surfactants on solid surfaces alters interfacial properties such as wettability.

The contact angles of surfactant mixtures A, B, and C at pH 10 and 25°C were 12°, 18°, and 22°, respectively. These results (Table 9.5) indicate that increasing the number of ethylene oxides in the surfactant molecule reduces its wetting or spreading

TABLE 9.5

Contact Angle of Aqueous Surfactant Mixtures A, B, and C at pH 10 and 25°C

Surfactant Mixture[a]	Contact Angle ± 2 (°)
A	12
B	18
C	22

[a] 80/20 nonionic ethoxylated lauryl alcohol/anionic ethoxylated lauryl phosphate ester surfactants.

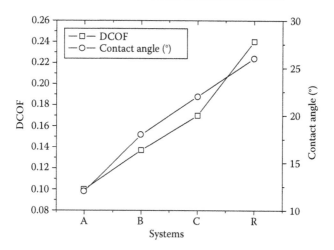

FIGURE 9.14 Effect of surfactant mixture structure on DCOF and contact angle of 1% w/w aqueous solution, pH 10 and 25°C.

property over the contact surface. Figure 9.14 compares DCOF and contact angle. It is observed that low contact angle correlated with low DCOF.

9.3.5 EVALUATION OF THE STABILITY OF TRIBOFILM OF THE MIXTURE SURFACTANTS

The surfactant mixtures studied (A, B, and C) are in a boundary lubrication regime. In this regime, the surfaces are separated by a film with a thickness less than the roughness of the surfaces in contact. When surfactants are employed, a physical adsorption occurs. The lubrication mechanism that predominates can be the friction modified by the adsorbed layer. The effectiveness of this layer or bilayer formed on the surface depends on its resistance to shear and reordering capability at the interface of the surfactant molecule [47–49].

9.3.5.1 Evaluation of the Stability of the Tribofilm with Shear Continuous

The stability of the film formed from aqueous surfactant mixtures was evaluated using the changes in the DCOF at different speeds at constant load, temperature, and pH.

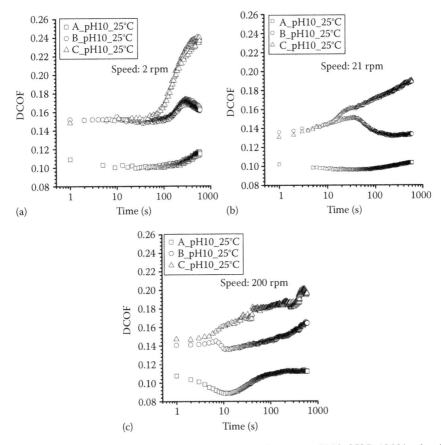

FIGURE 9.15 DCOF vs time for various surfactant mixtures at pH 10, 25°C, 10 N load and speed of: (a) 2 rpm; (b) 21 rpm and (c) 200 rpm.

Figure 9.15 shows the change in DCOF at speeds of 2, 21, and 200 rpm, with a load of 10 N for a period of 600 s. The DCOF of surfactant mixture A shows small changes when subjected to different shear velocities. This is an indicative that this surfactant mixture has excellent capability for its monomers to reform a tribofilm between the two surfaces. The initial and final coefficients of friction for this system at different speeds are given in Table 9.6.

For surfactant mixture B, significant changes of DCOF as a function of applied speed were observed. At 2 rpm, the DCOF was constant throughout the test duration. This means that the tribofilm was capable of constantly reforming and maintaining its monolayer or bilayer structure. The DCOF was between 0.15 and 0.16 (Figure 9.15a). Compared to surfactant mixture A, it resulted in 30% increased DCOF. At a shear rate of 21 rpm, surfactant mixture B showed that reordering of monomers to form a new film was not instantaneous. It took about 35 s for re-adsorption to occur. As a result, it displayed maximum and minimum DCOF values of 0.15 and 0.13, respectively (Figure 9.15b). Surfactant mixture B, at 200 rpm, displayed an increasing DCOF with time. This suggests that at this shear rate, this surfactant

TABLE 9.6

Initial (t = 0) and Final (t = 600 s) DCOF for Surfactant Mixtures A, B, and C at Different Speeds (2, 21, and 200 rpm) but Similar Load (10 N), Temperature (25°C), and pH (10)

Surfactant Mixture	2 rpm		21 rpm		200 rpm	
	$DCOF_i$	$DCOF_f$	$DCOF_i$	$DCOF_f$	$DCOF_i$	$DCOF_f$
A	$0.110 \pm 6 \times 10^{-3}$	$0.120 \pm 2 \times 10^{-3}$	$0.10 \pm 7 \times 10^{-2}$	$0.100 \pm 2 \times 10^{-3}$	$0.110 \pm 4 \times 10^{-3}$	$0.110 \pm 1 \times 10^{-3}$
B	$0.105 \pm 3 \times 10^{-3}$	$0.160 \pm 4 \times 10^{-3}$	$0.14 \pm 8 \times 10^{-2}$	$0.14 \pm 2 \times 10^{-2}$	$0.140 \pm 1 \times 10^{-3}$	$0.160 \pm 2 \times 10^{-3}$
C	$0.150 \pm 1 \times 10^{-3}$	$0.240 \pm 5 \times 10^{-3}$	$0.13 \pm 7 \times 10^{-2}$	$0.190 \pm 2 \times 10^{-3}$	$0.14 \pm 2 \times 10^{-2}$	$0.20 \pm 1 \times 10^{-2}$

mixture cannot regenerate the tribofilm between the two surfaces. It gave initial and final DCOF of 0.14 and 0.16, respectively (Figure 9.15c).

For surfactant mixture C, it can be concluded that the kinetics of re-adsorption was not effective at all shear rates evaluated. Thus, this surfactant system was unable to maintain a stable film between the moving surfaces and maintain a good lubrication. At 2 rpm, surfactant mixture C maintained a stable bilayer for approximately 58 s. It gave minimum and maximum coefficients of friction of 0.15 and 0.24, respectively. At speeds of 21 and 200 rpm, its behavior did not change, with DCOF increasing with time. Thus, the maximum DCOF at 21 rpm was 0.19 and at 200 rpm was 0.2.

9.3.5.2 Evaluation of the Stability of the Tribofilm through Oscillatory Strain

Oscillatory deformations of 1% and 14.4% at 1 Hz were used in order to evaluate the bilayer surfactant stability adsorbed on the tribopair based on its effect on DCOF.

Figure 9.16a and Table 9.7 show the initial and final average DCOF for the three surfactant mixture at 1% strain. The results for surfactant mixture A are 0.110 and 0.120, respectively. The result indicates that (1) the diffusion of the surfactant monomers from bulk to interface is fast enough to ensure bilayer formation on the tribopair and (2) the DCOF is constant during the test period, which indicates that the structure of the adsorbed surfactants on the surface is unchanged.

For surfactant mixture B at 1% strain, the DCOF data indicates that the adsorbed film on surface metallic presents a good stability. The DCOF does not show significant variation during the test period. The initial and final values of DCOF were 0.170 and 0.170, respectively. This value is larger by 35% from that for surfactant mixture A (Figure 9.16a; Table 9.7).

For surfactant mixture C, the DCOF at 1% strain increased with time, from DCOF of 0.170 initially to 0.190 at the end (Figure 9.16a; Table 9.7). This result might be due to the instability of the tribofilm, which it is not a desirable characteristic of lubricants that produce surface layers. The tribofilm of the surfactant mixture C was

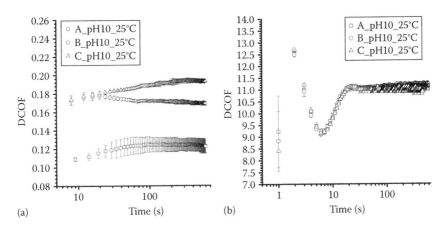

FIGURE 9.16 DCOF vs. time for surfactant mixtures at pH 10, 25°C, 10 N load under oscillatory strain deformation at 1 Hz of: (a) 1%, (b) 14.4%.

TABLE 9.7

Initial and Final DCOF for Surfactant Mixtures under Oscillatory Deformation

Surfactant Mixture[a]	Oscillatory Deformation	
	DCOF$_i$[b]	DCOF$_f$[b]
A	$0.110 \pm 3 \times 10^{-3}$	$0.120 \pm 8 \times 10^{-3}$
B	$0.170 \pm 7 \times 10^{-3}$	$0.170 \pm 2 \times 10^{-3}$
C	$0.170 \pm 6 \times 10^{-3}$	$0.190 \pm 1 \times 10^{-3}$

[a] Physicochemical conditions for all solutions evaluated: pH 10 and 25°C.
[b] The test of dynamic friction was carried out under the following experimental parameters: load 10 N, frequency 1 Hz, strain 1% and 21 rpm.

not able to quickly re-adsorb on the surface after it has been removed by the applied oscillatory deformation. This may be attributed to the large size of the surfactant molecules for this surfactant mixture. In addition, the area per molecule and the diffusion coefficient [25] for this surfactant mixture are higher and lower, respectively, than for surfactant mixtures A and B.

At 14.4% applied deformation and 1 Hz, all three surfactant mixtures showed similar DCOF values that were two orders of magnitude higher than for systems at 1% strain deformation and 1 Hz. This may be due to the high deformation preventing the recovery of the tribofilm, which will cause an increment of the DCOF (Figure 9.16b).

9.3.6 LUBRICATION REGIME FOR SURFACTANT MIXTURE A WITH XANTHAN GUM OR GLYCERIN VISCOSITY MODIFIERS

The main objective of this research was to characterize the friction and wear properties of surfactant mixtures under different conditions in water-based drilling fluids. Drilling fluids are complex mixtures of a base component and additives. Additives provide specific properties to help control the performance of the fluid and permit efficient drilling. Thus, it is important to determine the interaction between the major additives in the formulation to avoid possible interference with each other. To this end, the effect of the viscosity of the aqueous solution of surfactant mixture A in the presence of xanthan gum or glycerin on its lubrication properties was evaluated using dynamic friction tests on steel/steel surfaces.

In Section 9.3.2, it was demonstrated that surfactant mixtures A, B, and C exhibited boundary friction and did not show mixed film properties under the experimental conditions used. In order to produce the lower friction that corresponds to a mixed film regime, the viscosity of the aqueous surfactant mixtures was increased by adding viscosity modifiers. Increased viscosity will allow the formation of a lubricant film whose thickness could be slightly greater than the surface roughness. Such a film could exert significant pressure in the contact zone and partially separate the tribopair surfaces, producing a mixed film condition [50–52].

In order to evaluate the effect of viscosity, two aqueous lubricant solutions were prepared and tested as follows: (1) solution A + xanthan gum was formulated with

1% w/w of surfactant mixture A with 0.0079 g/mL of xanthan gum at pH 10 using sodium hydroxide and (2) solution A + glycerin with 1% w/w of surfactant mixture A at pH 10 using sodium hydroxide.

Figure 9.17 shows the plot of DCOF versus "Z" for a number of aqueous lubricant solutions at 25°C. The data lubricant A + xanthan gum exhibited a high DCOF. This means that the higher viscosity obtained by the addition of xanthan gum was insufficient to generate a lubricant film capable of partially separating the surfaces to produce a mixed film lubrication condition.

Table 9.8 summarizes its average DCOF values for the various aqueous lubricant systems. A DCOF of 0.10 was obtained for surfactant mixture A with added xanthan

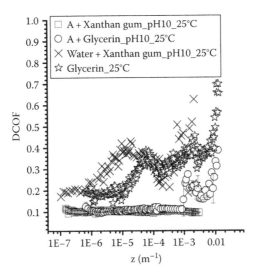

FIGURE 9.17 Dynamic COF vs Z for various aqueous lubricants with xanthan gum or glycerin viscosity modifiers.

TABLE 9.8
Tribological Properties of Aqueous Lubricant with Xanthan Gum and Glycerin Viscosity Modifier

System[a]	Viscosity[b] (Pa·s)	DCOF[c]	Lubrication Regime
A + Xanthan gum	$0.170 \pm 1 \times 10^{-3}$	$0.100 \pm 4 \times 10^{-3}$	Boundary
A + Glycerin	$0.13 \pm 2 \times 10^{-2}$	$0.14 \pm 6 \times 10^{-2}$	Boundary–dry contact
Water + Xanthan gum	$0.122 \pm 1 \times 10^{-3}$	$0.32 \pm 9 \times 10^{-2}$	Boundary–dry contact
Glycerin	$0.135 \pm 2 \times 10^{-3}$	$0.30 \pm 9 \times 10^{-2}$	Boundary–dry contact
A	$0.00100 \pm 3 \times 10^{-5}$	$0.100 \pm 2 \times 10^{-3}$	Boundary
Water	$0.00100 \pm 1 \times 10^{-5}$	$0.24 \pm 1 \times 10^{-2}$	Boundary–dry contact

[a] Physicochemical conditions for all solutions evaluated: pH 10 and 25°C.
[b] Measurement viscosity: 100 s^{-1} at 25°C.
[c] Conditions for lubrication testing: load 10 N and range of sliding speed from 5×10^{-5} to 1 m/s.

gum which is 68% lower than the reference system (water + xanthan gum). This means that surfactant mixture A maintained its adsorption and lubrication properties in the presence of xanthan gum.

A similar result was observed for the aqueous lubricant solution A + glycerin. The high viscosity was insufficient to generate a lubricant film to partially separate the tribopair surfaces and develop mixed lubrication conditions (Figure 9.17). Different to the formulation with xanthan gum, the aqueous lubricant solution with glycerin was capable to provide a liquid film only until a $Z = 1 \times 10^{-3}$ m^{-1}. At higher Z values, the tribopair develops a dry lubrication condition and the DCOF rises from 0.2 to 0.5. It is possible that glycerin affects not only the rate of surfactant adsorption but also the strength of the adsorbed film. Thus, at high Z values, the stability of the tribofilm is reduced due to the slow regeneration of the surfactant film adsorbed at the interface, causing a metal–metal dry contact.

Table 9.8 shows that aqueous lubricant A + glycerin gave a DCOF value of 0.14, which is 53% lower than the reference system (water + glycerin).

9.3.7 STATIC FRICTION OF AQUEOUS SURFACTANT MIXTURES A, B, AND C WITH XANTHAN GUM OR GLYCERIN VISCOSITY MODIFIERS

The effect of xanthan gum and glycerin viscosity modifiers on the static friction of surfactant mixtures was investigated. The force required to move the ball at rest with constant load of 10 N was determined.

Surfactant mixture A with the xanthan gum viscosity modifier displayed an SCOF of 0.118 (Figure 9.18; Table 9.9), which was similar to the values for system A without

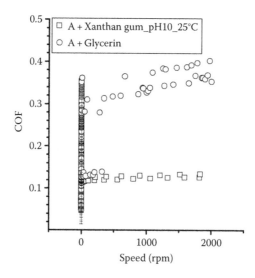

FIGURE 9.18 Static coefficient of friction vs speed for surfactant mixture A with added viscosity modifiers xanthan gum and glycerin at pH 10; 25°C; 10 N load.

TABLE 9.9

Static and Dynamic Coefficient of Friction for Surfactant Mixture with Xanthan Gum and Glycerin Viscosity Modifiers at pH 10 and 25°C

Surfactant Mixture A + Viscosity Modifiers	DCOF[a]	SCOF[b]
A + Xanthan gum	$0.100 \pm 4 \times 10^{-3}$	$0.118 \pm 9 \times 10^{-3}$
A + Glycerin	$0.14 \pm 6 \times 10^{-2}$	$0.33 \pm 7 \times 10^{-2}$

[a] Data from Figure 9.16.
[b] The test of static friction was carried out under the following experimental parameters: 10 N load and range of speed from 0 to 2000 rpm.

xanthan gum (Figure 9.13 and Table 9.4). This suggests that surfactant mixture A is able to form a layer or bilayer with the same tribological properties with or without added xanthan gum viscosity modifiers.

Surfactant mixture A in the presence of glycerin viscosity modifier showed an SCOF of 0.33, which was 60% higher than the value in the presence of xanthan gum viscosity modifier (Figure 9.18, Table 9.9).

9.3.8   STABILITY OF THE TRIBOFILM FOR SURFACTANT MIXTURE A WITH XANTHAN GUM OR GLYCERIN VISCOSITY MODIFIERS

The lubrication regimes for surfactant mixture A with xanthan and glycerin viscosity modifiers were found to be boundary and boundary dry, respectively (Table 9.8). The tribological performance of these mixtures depends on its ability to form tribofilm with suitable tribological properties between the contact surfaces. The effectiveness of this layer or bilayer on the tribopair depends on the shear resistance capacity (continuous or oscillatory) and the reordering ability of the surfactant molecules at the interface [47–49].

9.3.8.1   Evaluation of the Stability of the Tribofilm with Shear Continuous

Figure 9.19 shows the DCOF values for surfactant mixture A with xanthan gum and glycerin viscosity modifiers at a constant speed of 21 rpm and a load of 10 N for 600 s. It can be seen that the DCOF with added xanthan gum remains constant throughout the test. This shows an excellent ability of the surfactant monomers to rearrange and reform tribofilm, similar to the results without viscosity modifiers (Figure 9.15). The initial and final DCOF for this system are given in Table 9.10.

In the case where the system with the glycerin viscosity modifier was used, a significant change in DCOF occurred after 600 s of the test that resulted in the transition of the lubrication regime from the boundary to dry contact (Figure 9.19). It gave initial and final DCOF values of 0.117 and 0.21, respectively, which corresponds to a 44% change. The DCOF results indicate the instability of the tribofilm when it is subjected to continuous shear at 21 rpm and 10 N load. This instability can be attributed to an irreversible surfactant monomer desorption from the tribopair and by the low polarity of glycerin.

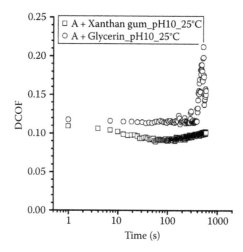

FIGURE 9.19 Dynamic coefficient of friction (DCOF) vs. time for surfactant mixture A with added xanthan gum or glycerin viscosity modifiers. Test conditions: pH 10, 25°C, 10 N load, constant speed of 21 rpm.

TABLE 9.10
Effect of Viscosity Modifiers on Initial and Final Values of the DCOF for Surfactant Mixture

Surfactant Mixture[a] + Viscosity Modifier	DCOF$_i$[b]	DCOF$_f$[b]
A + Xanthan gum	$0.109 \pm 4 \times 10^{-3}$	$0.100 \pm 5 \times 10^{-3}$
A + Glycerin	$0.117 \pm 5 \times 10^{-3}$	$0.21 \pm 2 \times 10^{-2}$

[a] Physicochemical conditions for all solutions evaluated: pH 10 and 25°C.
[b] The test of dynamic friction was carried out under the following experimental parameters: 10 N load and constant shear rate of 21 rpm.

9.3.8.2 Evaluation of Tribofilm Stability through Oscillatory Strain

The purpose of the application of oscillatory strains on the tribofilm formed is to evaluate its stability based on its effect on the DCOF. The results of these tests indicate that the effect on DCOF values were similar to the previous results, which showed the best performance at a frequency of 1 Hz and 1% strain (Figure 9.20 and Table 9.11).

9.3.9 WEAR EVALUATION

Wear due friction between two bodies in contact occurs due to the combined effects of various physical and chemical processes: single asperities, oxide films, and adsorbed species. Wear is classified into (1) adhesive, (2) abrasive, (3) corrosive, and (4) surface fatigue. Based on the physical characteristics of the tribopair evaluated, which are the lubrication regime (boundary), load, and sliding speed conditions, abrasive wear was expected in this study. On the other hand, the use of the surfactant mixtures A, B, and C was expected to result in reduced wear.

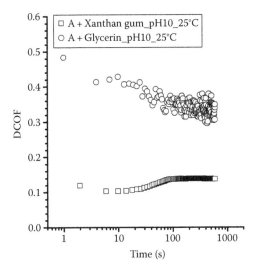

FIGURE 9.20 DCOF vs. time for surfactant mixture A with xanthan gum or glycerin viscosity modifiers under oscillatory deformation. Test conditions: pH 10; 25°C, 10 N load, frequency of 1 Hz and 1% strain.

TABLE 9.11
Initial and Final DCOF for Surfactant Mixture A with Viscosity Modifiers under Oscillatory Deformation

	Oscillatory Deformation	
Surfactant Mixture[a] + Viscosity Modifier	DCOF$_i$[b]	DCOF$_f$[b]
A + Xanthan gum	$0.119 \pm 3 \times 10^{-3}$	$0.138 \pm 8 \times 10^{-3}$
A + Glycerin	$0.414 \pm 7 \times 10^{-3}$	$0.374 \pm 2 \times 10^{-3}$

[a] Physicochemical conditions for all solutions evaluated: pH 10 and 25°C.
[b] The test of dynamic friction was carried out under the following experimental parameters: 10 N load, frequency 1 Hz, strain 1% and 21 rpm.

9.3.9.1 Wear Scar Volume

Figure 9.21 shows the effect of the surfactant mixtures evaluated in this work on the wear scar of the plate surface at pH 10 and at temperatures of 25°C, 49°C, and 80°C. All three surfactant mixtures exhibited excellent wear reduction up to 99% below to the reference fluid, which is surfactant-free water. At 25°C, a wear scar volume of 0.025 mm³ was observed for the reference fluid. Surfactant mixture A showed the lowest wear scar volume of 1×10^{-5} mm³ at 25°C (Figure 9.22), which corresponds to a 99.97% reduction from that of the reference fluid. At 49°C and 80°C, the reference fluid gave a similar wear scar volume of 0.03 mm³ (Figure 9.21). The best wear reduction at 49°C and 80°C was achieved by surfactant mixture A, where the wear scar volume was 1×10^{-4} and 2×10^{-4} mm³, respectively. These values correspond to wear reductions of about 99.62% (49°C) and 99.28% (80°C), respectively (Figure 9.22).

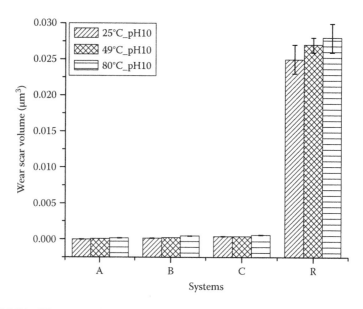

FIGURE 9.21   Wear scar volume of aqueous surfactant mixtures and water reference system.

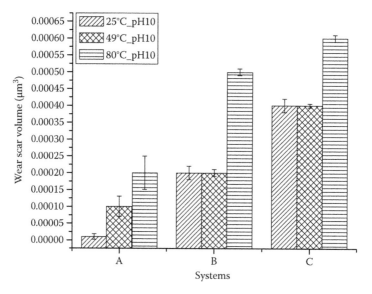

FIGURE 9.22   Wear scar volume of aqueous surfactant mixtures at different temperatures.

9.3.9.2   Wear Surface Optical Analysis

The optical surface profilometry images after the sliding wear test on the metal plate for all the fluids (R, A, B, and C) at 25°C, 49°C, and 80°C show a symmetrical scar with a characteristic pattern of abrasive wear (Figures 9.23 through 9.25). This pattern was created by metal–metal asperity contact due to the 10 N normal load applied on the ball and plate.

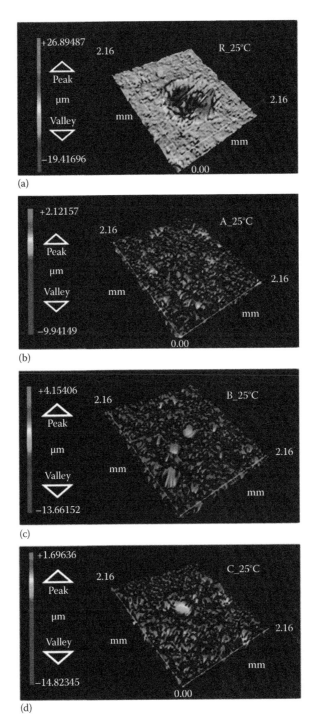

FIGURE 9.23 Optical surface profilometry images of plate after wear test using aqueous surfactant mixtures A, B and C at 25°C; pH 10, load 10 N and range of sliding speed from 5×10^{-5} to 1 m/s.

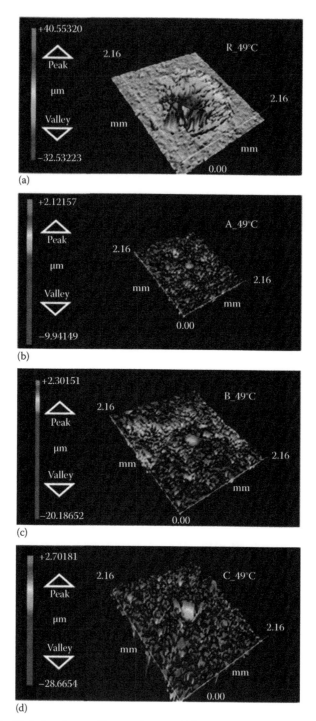

FIGURE 9.24 Optical surface profilometry images of plate after wear test using aqueous surfactant mixtures A, B and C at 49°C; pH 10, load 10 N and range of sliding speed from 5×10^{-5} to 1 m/s.

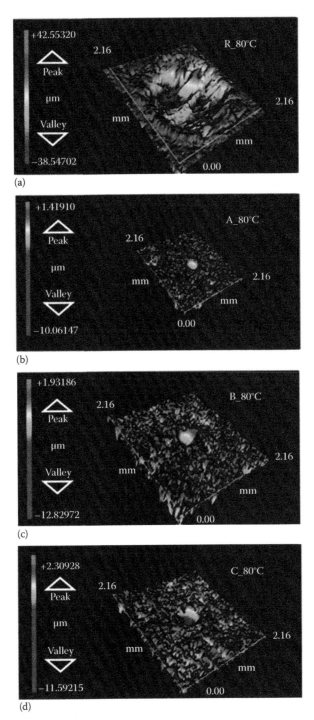

FIGURE 9.25   Optical surface profilometry images of plate after wear test using aqueous surfactant mixtures A, B and C at 80°C; pH 10, load 10 N and range of sliding speed from 5×10^{-5} to 1 m/s.

The smallest wear scar observed at all temperatures was from aqueous surfactant mixture A (Figures 9.23 through 9.25). It is known that surfactant adsorption on the surface depends on its surface affinity. Monomers and micelles fill up micro asperities generating a film strongly bound to the surface. Such film may enlarge the real area of contact distributing the load, which improves the load carrying capacity of the fluid.

Surfactant mixtures B and C produced the biggest wear scar, probably due to a change in the aggregation number and geometry of the adsorbed surfactants (Figures 9.23 through 9.25). This will produce a weak and unstable film with a poor load carrying property that allows contact of the sliding surfaces [14,53,54]. Possibly that surfactant mixture A adsorbed on the metal surfaces generated and a highly packed and dense monolayer or bilayer.

9.4 SUMMARY

The lubrication regime exhibited for aqueous surfactant mixtures A, B, and C is a boundary lubrication property that did not transition to the mixed film regime under the experimental conditions used in this work. The mechanism of lubrication of these surfactants involves the formation of a sacrificial layer on the tribo-surfaces. All surfactant solutions (A, B, and C) investigated exhibited an excellent friction reduction property of up to 67% relative to pure water. Surfactant mixture A showed the best friction reducing property.

The stability of the tribofilm formed by surfactant mixtures A and B on steel surfaces (ball–plate) was evaluated using dynamic friction tests at constant velocity and strain oscillation. The result showed minor changes in the coefficient of friction, which indicates an excellent capability of the surfactant monomers to re-adsorb and form the tribofilm on the two surfaces. On the other hand, the tribofilm of surfactant mixture C was not able to quickly reform on the surface after it has been removed by the applied shear strain (continuous and oscillatory). This may be attributed to the large molecular size of these surfactant mixture monomers, which are known to have a larger area per molecule than mixtures A and B.

The addition of xanthan gum viscosity modifier into surfactant mixture solution A did not generate a lubricant film capable of partially separating the friction surfaces and generate mixed film lubrication conditions with a lower DCOF. Similar results were also observed for the aqueous lubricant solution A with glycerin viscosity modifier.

The antiwear performance of surfactant mixtures A, B, and C was excellent, displaying a wear reduction of up to 99% relative to water under the experimental conditions used in this study. Likewise, optical surface profilometry showed that surfactant mixture A produced the smallest wear scar at 25°C, 49°C, and 80°C. Surfactant mixture B and C showed similar wear scars that were larger than system A. Based on the properties of the tribopair evaluated, the lubrication regime, load and sliding speed conditions, and optical profilometry results, it is concluded that the mechanism of wear for the system investigated here is abrasive wear.

ACKNOWLEDGMENT

This research was sponsored by PDVSA Intevep S.A., Venezuela, Strategic Research Management in Production.

REFERENCES

1. S. Baba Hamed and M. Belhadri, Rheological properties of biopolymers drilling fluids, *J. Petrol. Sci. Eng.*, 67, 84–90 (2009).

2. M. Dolz, J. Jiménez, M.J. Hernández, J. Delegido, and A. Casanovas, Flow and thixotropy of non-contaminating oil drilling fluids formulated with bentonite and sodium carboxymethyl cellulose, *J. Petrol. Sci. Eng.*, 57, 294–302 (2007).

3. V. Mahto and V.P. Sharma, Rheological study of a water based oil well drilling fluid, *J. Petrol. Sci. Eng.*, 45, 123–128 (2004).

4. R. Caenn and G.V. Chillingar, Drilling fluids: State of the art, *J. Petrol. Sci. Eng.*, 14, 221–230 (1996).

5. T. Hamida, E. Kuru, and M. Pickard, Filtration loss characteristics of aqueous waxy hull-less barley (WHB) solutions, *J. Petrol. Sci. Eng.*, 72, 33–41 (2010).

6. V.C. Kelessidis, C. Papanicolaou, and A. Foscolos, Application of Greek lignite as an additive for controlling rheological and filtration properties of water-bentonite suspensions at high temperatures: A review, *Int. J. Coal Geol.*, 77, 394–400 (2009).

7. H. Dehghanpour and E. Kuru, Effect of viscoelasticity on the filtration loss characteristics of aqueous polymer solutions, *J. Petrol. Sci. Eng.*, 76, 12–20 (2011).

8. Y. Qu, X. Lai, L. Zou, and Y.N. Su, Polyoxyalkyleneamine as shale inhibitor in water-based drilling fluids, *Appl. Clay Sci.*, 44, 265–268 (2009).

9. M. Khodja, J.P. Canselier, F. Bergaya, K. Fourar, M. Khodja, N. Cohaut, and A. Benmounah, Shale problems and water-based drilling fluid optimisation in the Hassi Messaoud Algerian oil field, *Appl. Clay Sci.*, 49, 383–393 (2010).

10. H. Zhong, Z. Qiu, W. Huang, and J. Cao, Shale inhibitive properties of polyether diamine in water-based drilling fluid, *J. Petrol. Sci. Eng.*, 78, 510–515 (2011).

11. J. Guo, J. Yan, W. Fan, and H. Zhang, Applications of strongly inhibitive silicate-based drilling fluids in troublesome shale formations in Sudan, *J. Petrol. Sci. Eng.*, 50, 195–203 (2006).

12. R. Samuel, Friction factors: What are they for torque, drag, vibration, bottom hole assembly and transient surge/swab analyses?, *J. Petrol. Sci. Eng.*, 73, 258–266 (2010).

13. J.D. Kercheville, A.A. Hinds, and W.R. Clements, Comparison of environmentally acceptable materials with diesel oil for drilling mud lubricity and spotting fluid formulations, Paper SPE 14797 in: *IADC/SPE Drilling Conference Proceedings*, Dallas, Texas (1986).

14. J.M. González, F. Quintero, J.E. Arellano, R.L. Márquez, C. Sánchez, and D. Pernía, Effects of interactions between solids and surfactants on the tribological properties of water-based drilling fluids, *Colloids Surf. A*, 391, 216–223 (2011).

15. W.E. Foxenberg, S.A. Ali, T.P. Long, and J. Vian, Field experience shows that new lubricant reduces friction and improves formation compatibility and environmental impact, Paper SPE 112483, in: *SPE International Symposium and Exhibition on Formation Damage Control Proceedings*, Lafayette, LA (2008).

16. M.S. Aston, P.J. Hearn, and G. McGhee, Techniques for solving torque and drag problems in today's drilling environment, Paper 48939, in: *SPE Annual Technical Conference and Exhibition Proceedings*, New Orleans, LA (1998).

17. P. Skalle, K.R. Backe, S.K. Lyomov, L. Kilaas, A.D. Dyrli, and J. Sveen, Microbeads as lubricant in drilling muds using a modified lubricity tester, Paper SPE 14797, in: *IADC/SPE Drilling Conference Proceedings* (1999).

18. J.J. Truhan, R. Menon, and P.J. Blau, The evaluation of various cladding materials for down-hole drilling applications using the pin-on-disk test, *Wear*, 259, 1308–1313 (2005).

19. S.N. Shah and S. Jain, Coiled tubing erosion during hydraulic fracturing slurry flow, *Wear*, 264, 279–290 (2008).

20. F. Quintero, J.M. González, J.E. Arellano, and M. Mas, Adsorption of surfactants with different degrees of ethoxylation on hematite weighting material and its effect on the tribological properties of water-based petroleum drilling fluids, in: *Surfactants in Tribology*, Vol. 4, G. Biresaw and K.L. Mittal (Eds.), pp. 349–384, CRC Press, Boca Raton, FL (2015).

21. F. Quintero and J.M. González, Adsorption of Surfactants on Hematite used as weighting material and the effects on the tribological properties of water-based drilling fluids, in: *Surfactants in Tribology*, Vol. 3, G. Biresaw and K. Mittal (Eds.), pp. 463–489, CRC Press, Boca Raton, FL (2013).

22. J.M. Gonzalez, F. Quintero, R. Márquez, and S.D. Rosales, Formulation effects on the lubricity of O/W emulsions used as oil well working fluids, in: *Surfactants in Tribology*, Vol. 2, G. Biresaw and K.L. Mittal (Eds.), pp. 241–265, CRC Press, Boca Raton, FL (2011).

23. U.C. Chen, Y.S. Liu, C.-C. Chang, and J.F. Lin, The effect of the additive concentration in emulsions to the tribological behavior of a cold rolling tube under sliding contact, *Tribol. Int.*, 35, 309–320 (2002).

24. M.W. Sulek and T. Wasilewski, Tribological properties of aqueous solutions of alkyl polyglucosides, *Wear*, 260, 193–204 (2006).

25. M.J. Rosen, *Surfactants and Interfacial Phenomena*, John Wiley & Sons, New York, 1978.

26. L. Zhang and P. Somasundaran, Adsorption of mixtures of nonionic sugar-based surfactants with other surfactants at solid/liquid interfaces: I. Adsorption of n-dodecyl-β-D-maltoside with anionic sodium dodecyl sulfate on alumina, *J. Colloid Interface Sci.*, 302, 20–24 (2006).

27. M.A. Muherei, Equilibrium adsorption isotherms of anionic, nonionic surfactants and their mixtures to shale and sandstone, *Mod. Appl. Sci.*, 3, 158–167 (2008).

28. V.M. Starov, Surfactant solutions and porous substrates: Spreading and imbibition, *Adv. Colloid Interface Sci.*, 111, 3–27 (2004).

29. M.R. Böhmer, L.K. Koopal, R.L. Janssen, E.M. Thomas, and A.R.R.K. Rennie, Adsorption of non-ionic surfactants on hydrophylic surfaces, *Langmuir* 8, 2228–2239 (1992).

30. S. Partyka, S. Zaini, M. Lindheimer, and B. Brun, The adsorption of non-ionic surfactants on a silica gel, *Colloids Surf.*, 12, 255–270 (1984).

31. L. Zhang, P. Somasundaran, J. Mielczarski, and E. Mielczarski, Adsorption mechanism of n-dodecyl-β-D-maltoside on alumina, *J. Colloid Interface Sci.*, 256, 16–22 (2002).

32. P. Somasundaran and S. Krishnakumar, Adsorption of surfactants and polymers at the solid–liquid interface, *Colloids Surf. A*, 123–124, 491–513 (1997).

33. R. Atkin, V.S.J. Craig, E.J. Wanless, and S. Biggs, Mechanism of cationic surfactant adsorption at the solid-aqueous interface, *Adv. Colloid Interface Sci.*, 103, 219–304 (2003).

34. A. Fan, P. Somasundaran, and N.J. Turro, Adsorption of alkyltrimethylammonium bromides on negatively charged alumina, *Langmuir*, 13, 506–510 (1997).

35. S. Paria and K.C. Khilar, A review on experimental studies of surfactant adsorption at the hydrophilic solid–water interface, *Adv. Colloid Interface Sci.*, 110, 75–95 (2004).

36. G. Stachowiak and A.W. Batchelor, *Engineering Tribology*, 3rd edn., Butterworth-Heinemann, Boston (2001).

37. S.M. Hsu, Molecular basis of lubrication, *Tribol. Int.*, 37, 553–559 (2004).

38. A.F. Bombard and J. de Vicente, Thin-film rheology and tribology of magnetorheological fluids in isoviscous-EHL contacts, *Tribol. Lett.*, 47, 149–162 (2012).

39. A.J.F. Bombard and J. de Vicente, Boundary lubrication of magnetorheological fluids in PTFE/steel point contacts, *Wear*, 296, 484–490 (2012).

40. P.R. Teasdale and G.G. Wallace, In situ characterization of conducting polymers by measuring dynamic contact angles with Wilhelmy's plate technique, *React. Polym.*, 24, 157–164 (1995).

41. A. Al-Shareef, P. Neogi, and B. Bai, Force based dynamic contact angles and wetting kinetics on a Wilhelmy plate, *Chem. Eng. Sci.*, 99, 113–117 (2013).

42. J. Shang, M. Flury, J.B. Harsh, and R.L. Zollars, Comparison of different methods to measure contact angles of soil colloids, *J. Colloid Interface Sci.*, 328, 299–307 (2008).

43. S. Sharma, S. Sangal, and K. Mondal, On the optical microscopic method for the determination of ball-on-flat surface linearly reciprocating sliding wear volume, *Wear*, 300, 82–89 (2013).

44. M. Kalin and J. Vizintin, Use of equations for wear volume determination in fretting experiments, *Wear*, 237, 39–48 (2000).

45. Z. Pawlak, W. Urbaniak, and A. Oloyede, The relationship between friction and wettability in aqueous environment, *Wear*, 271, 1745–1749 (2011).

46. A. Borruto, G. Crivellone, and F. Marani, Influence of surface wettability on friction and wear tests, *Wear*, 222, 57–65 (1998).

47. J. Schöfer, P. Rehbein, U. Stolz, D. Löhe, and K.H. Zum Gahr, Formation of tribochemical films and white layers on self-mated bearing steel surfaces in boundary lubricated sliding contact, *Wear*, 248, 7–15 (2001).

48. S.M. Hsu and R.S. Gates, Boundary lubricating films: Formation and lubrication mechanism, *Tribol. Int.*, 38, 305–312 (2005).

49. T. Kubo, S. Fujiwara, H. Nanao, I. Minami, and S. Mori, Boundary film formation from overbased calcium sulfonate additives during running-in process of steel-DLC contact, *Wear*, 265, 461–467 (2008).

50. M. Woydt and R. Wäsche, The history of the Stribeck curve and ball bearing steels: The role of Adolf Martens, *Wear*, 268, 1542–1546 (2010).

51. J. de Vicente, J.R. Stokes, and H.A. Spikes, Soft lubrication of model hydrocolloids, *Food Hydrocolloids*, 20, 483–491 (2006).

52. M. Kalin and I. Velkavrh, Non-conventional inverse-Stribeck-curve behaviour and other characteristics of DLC coatings in all lubrication regimes, *Wear*, 297, 911–918 (2013).

53. I. Samerski, J. Schöfer, and A. Fischer, The role of wear particles under multidirectional sliding wear, *Wear*, 267, 1319–1324 (2009).

54. R.I. Trezona, D.N. Allsopp, and I.M. Hutchings, Transitions between two-body and three-body abrasive wear: Influence of test conditions in the microscale abrasive wear test, *Wear*, 225, 205–214 (1999).

10 Load- and Velocity-Dependent Friction Behavior of Cow Milk Fat

Teresa Tomasi, Angela M. Tortora,
Cristina Bignardi, and Deepak H. Veeregowda

CONTENTS

10.1 INTRODUCTION

Oral cavity is subjected to severe chemical and mechanical perturbations throughout the day during the consumption of food and beverage products. The food components that constitute flavor and texture interact with the tongue, palate, and teeth to alter the oral sensory perception. Sensory perception is vital to the food transport process in the oral cavity. For example, the friction response of the taste buds, which are the sensory receptors on the tongue, helps in facilitating bolus formation for swallowing [1]. Also, friction of the taste buds is useful in differentiating food products based on creaminess [2], smoothness, and moistness [3]. Therefore, friction is an important factor in sensory perception of food products.

Milk is an important component of dairy food products, and its fat content can influence the oral sensory perception. For example, an increase in the fat content in milk has been related to an increase in the creaminess perception [2]. Furthermore, friction measurements on different milk fat samples under sliding–rolling testing condition showed that friction was lower for high-fat content (4%) compared with low-fat content (0.3%). In another friction study on milk fat, under pure sliding condition, it was observed that the friction of 3.5% and 1.5% fat milk was the same [4]. This study is in contrast with the understanding that high-fat milk gives lower friction compared with low-fat milk. Moreover, it seems that the friction response of milk fat can be influenced by the type of sliding motion at the rubbing surfaces.

In this study, a pin-on-disk instrument with sliding hydrophobic surfaces is used to determine the friction coefficient of 0.3% and 3.5% fat milk. The friction coefficient is measured at different loads and velocity decay rates (λ), where λ is a constant. Friction coefficient maps are developed and used to investigate the role of fat content in the friction mechanism of milk.

10.2 EXPERIMENTAL SECTION

10.2.1 MATERIALS

10.2.1.1 Milk Samples

Low-fat cow milk (0.3% fat) and high-fat milk (3.5% fat) were commercial products purchased from a local supermarket. These samples were from the same manufacturer or supplier. Note that the protein contents were the same for the low- and the high-fat milk. Demineralized water was used as control in all the experiments.

10.2.1.2 Pin-and-Disk Specimens

Polydimethylsiloxane (PDMS) pins and disks were used in friction measurements. They were prepared from a silicon elastomer kit (Sylgard 184, Dow corning, Midland, MI). The base and the curing agent were mixed in a ratio (w/w) of 10:1 and transferred into the molds of pins and disks. Overnight curing was conducted at 65°C in an oven. A mold with polystyrene hemispherical wells was used to prepare a hemispherical pin of 6 mm diameter and a polystyrene circular plate was used to prepare a disk of 60 mm diameter. The roughness of the PDMS disks and pins was measured by contact-mode atomic force microscopy. The average roughness was measured to be 3 and 5 nm for the disks and pins, respectively.

10.2.2 METHODS

10.2.2.1 Preparation of the Pin-and-Disk Specimens for Friction Test

PDMS specimens were cleaned before friction measurements. Refer to the illustrations in Figure 10.1 for a detailed description of the cleaning steps.

10.2.2.2 Friction Coefficient Measurements

Friction coefficient (μ) was determined using a pin-on-disk instrument (TR 20, Ducom Instruments Pvt. Ltd., Bengaluru, India). This instrument can operate under a normal load range of 1.5–40 N and a speed range of 1–1000 rpm. Winducom software (Ducom Instruments Pvt. Ltd., India) was used to determine the friction

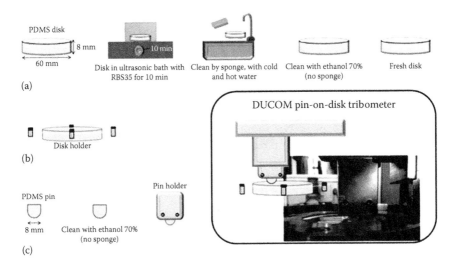

FIGURE 10.1   Cleaning and mounting of pin-and-disk specimens for friction measurements of cow milk fat: (a) disk cleaning (left to right), (b) disk mounting, and (c) pin mounting (left to right).

coefficient as the disk sliding velocity was continuously decreased with time from 80 to 2 mm/s at an exponential rate as follows:

$$V_f = V_i \cdot e^{-\lambda t}$$

where
   V_f and V_i are the initial and final velocity (mm/s), respectively
   the constant λ represents the velocity decay rate
   t represents the duration in seconds

The velocity exponential decay profile at a decay rate (λ) of 0.003, as acquired by the pin-on-disk instrument, is reproduced in Figure 10.2.

   The velocity decay rate was fixed at 0.003 for three loading conditions, that is, 1.5, 6.5, and 11.5 N. In another set of experiments, the load was fixed at 1.5 N for three different decay rate conditions, that is, 0.003, 0.005, and 0.009. Note that as λ is increased, the velocity sweep is quicker, which can influence the rate of recovery of perturbed surface layers. In all the experiments, the total volume of milk and demineralized water (control) was 4 mL. Note that demineralized water is deionized water without any minerals.

10.3   RESULTS

The friction coefficient at different loads was used to differentiate between 0.3% fat milk, 3.5% fat milk, and water (Figure 10.3). At 1.5 N load, during the start of the test (high sliding velocity), the friction coefficient of 0.3% fat milk was 0.01, that is, 100 times lower than that of water and five times lower than that

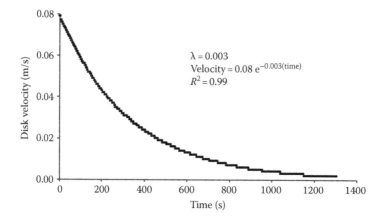

$\lambda = 0.003$
Velocity $= 0.08\ e^{-0.003(\text{time})}$
$R^2 = 0.99$

FIGURE 10.2 Exponential decay of disk sliding velocity as a function of time. The decay rate (λ) was 0.003, and initial and final disk velocities were kept at 80 and 2 mm/s, respectively.

of 3.5% fat milk (Figure 10.3a). As the velocity decreases, the low friction coefficient transits into its higher value, and this transition was delayed for 0.3% fat milk compared with 3.5% fat milk and water. At the lowest sliding velocity, the friction coefficient of 0.3% fat milk was still lower than those of 3.5% fat milk and water (Figure 10.4). At 6.5 N load and high sliding velocity, the friction coefficient was the same for 0.3% and 3.5% fat milk (Figure 10.3b). And at low sliding velocity, the friction coefficient of 0.3% fat milk was almost two times higher than that of 3.5% fat milk (Figure 10.4). At 11.5 N load, the friction coefficient trend for the 0.3% and 3.5% fat milk samples was similar and there was no difference in the friction coefficient at low sliding velocity (Figures 10.3c and 10.4). In general, the friction coefficient of 0.3% fat milk and 3.5% fat milk was lower than that of water at all loads applied in this study (Figure 10.4).

Friction coefficients at different velocity decay rates (λ) were used to differentiate 0.3% fat milk, 3.5% fat milk, and water (Figure 10.5). At the exponential decay rate of 0.003, two important trends were observed: (1) during high and low sliding velocities, the friction coefficient of 0.3% fat milk was lower than that of 3.5% fat milk, and (2) the transition point, that is, increase in the friction coefficient, was delayed for 0.3% fat milk compared with 3.5% fat milk (Figure 10.5a). As the decay rate is increased to 0.005, friction for 0.3% and 3.5% fat milk remains the same (Figures 10.5b and 10.6). Further increase in the decay rate to 0.009 will only partially reinstate the trends observed at 0.003 decay rate (Figure 10.5c). In general, the friction coefficient of water increases from 0.003 to 0.005 or 0.009 decay rate (Figure 10.6). At all decay rates, the friction coefficient of milk was lower than that of water.

10.4   DISCUSSION

The friction coefficient of PDMS surfaces decreases due to an increase in the normal load. Such a behavior is expected for low-modulus (or soft) hydrophobic surfaces due to the dominant adhesion forces arising from the interlocking

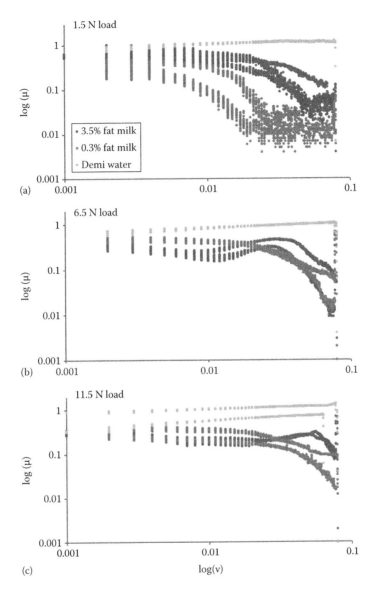

FIGURE 10.3   Change in the friction coefficient (μ) as a function of disk velocity for 3.5% fat milk, 0.3% fat milk, and control (water) at normal load of 1.5 N (a), 6.5 N (b), and 11.5 N (c). Data presented include two repetitions per load. The velocity decay rate (λ) was fixed at 0.003.

of asperities [5,6]. Furthermore, the friction coefficient is strongly influenced by the velocity decay rate (λ). An increase in λ decreased the rate of recovery of the sheared asperities, which will increase their deformation and increase the friction. The above mechanism is derived from friction studies on a water medium and it is not applicable to the milk medium. Therefore, we suspect the role of

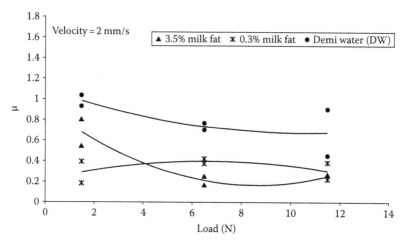

FIGURE 10.4 Relationship between the friction coefficient at low velocity (2 mm/s) and normal load for 3.5% fat milk, 0.3% fat milk, and control (water). Data presented include two repetitions per load.

milk components being dominant over the substrate effect. Milk is a mixture of proteins and fat molecules that can readily adsorb onto hydrophobic surfaces and reduce friction better than water.

Casein forms 80% of milk proteins and occurs in the form of micelles, which stabilize the fat molecules in milk [7]. This stability arises from its hydrophobic interaction with the fat molecules and electrostatic repulsion between the micelles. The stability is disrupted by process parameters such as heat, pressure, and shear [7,8]. These process parameters decrease the hydrophobic interaction between casein and fat molecules. This phenomenon is observed in our friction experiments. The friction force induces shear thinning of low- and high-fat milk. We use the term shear thinning because of the increase in the friction as the sliding velocity decreases, which is related to the decrease in film thickness [9]. Later, we hypothesize that shear thinning of milk will produce two different surface active layers on the PDMS, depending on the fat content (Figure 10.7). The friction force will separate casein (hydrophobic head and hydrophilic tail) from the fat molecules, and then the casein molecules adsorb on the PDMS surface via hydrophobic interaction. The surface layer in low-fat milk (0.3%) will have more casein than fat molecules (Figure 10.7a), whereas in high-fat milk (3.5%), there will be more fat molecules than casein (Figure 10.7b). Casein with its hydrophilic tail attracts water molecules; that is, more water molecules could be bound to the surface layer in low-fat milk. At low load (1.5 N), the shear plane runs through the hydrophilic tail with water molecules, and more water-bound hydrophilic tails are responsible for low friction, as determined for low-fat milk. However, at high load (6.5 N), the shear plane is pushed below the hydrophilic tail and runs through fat molecules. Here, more fat molecules on the surface are responsible for low friction, as shown for high-fat milk. An increase in the rate of decay (from 0.003 to 0.005) shows that more hydrophilic tails contribute to

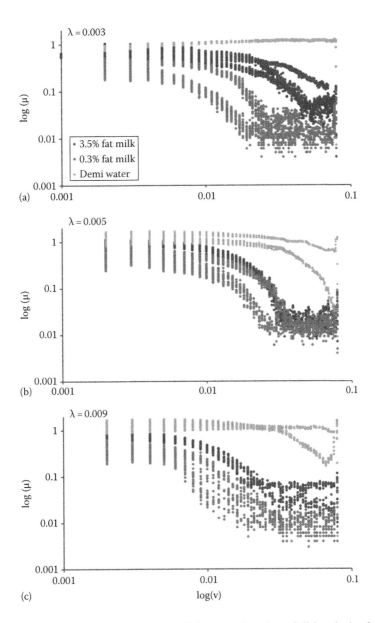

FIGURE 10.5 Change in the friction coefficient as a function of disk velocity for 3.5% fat milk, 0.3% fat milk, and control (water) at a velocity decay rate of 0.003 (a), 0.005 (b), and 0.009 (c). The normal load was fixed at 1.5 N. Data presented include two repetitions per decay rate.

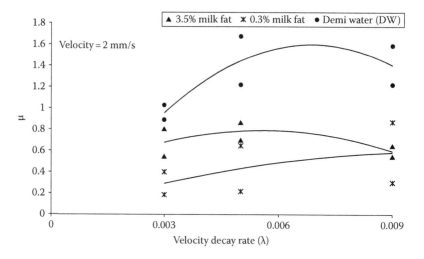

FIGURE 10.6 Relationship between the friction coefficient at low velocity (2 mm/s) and different velocity decay rates for 3.5% fat milk, 0.3% fat milk, and control (water). Data presented include two repetitions per decay rate.

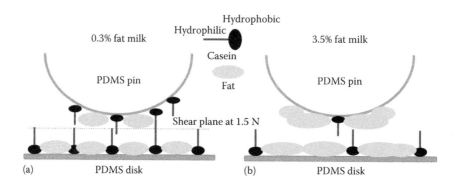

FIGURE 10.7 Schematic of the surface layer composed of casein and fat molecules, adsorbed on PDMS surfaces after shear thinning, for 0.3% fat milk (a) and 3.5% fat milk (b).

the quick recovery and replenishment of water at the shear plane, as low-fat milk maintains a low friction coefficient compared with high-fat milk. More experiments need to be conducted to verify this hypothesis.

10.5 CONCLUSION

Milk components reduce friction compared with water at all loads and velocities used in this study. Low-fat milk has more casein and high-fat milk has more fat molecules in the surface layer formed after shear thinning. Casein with its hydrophilic tail dominates the low friction behavior at low load, and fat molecules dominate the low friction behavior at high load. Moreover, hydrophilic tails are useful in quick replenishment of water molecules, which reduce friction compared with fat molecules on the surface. Overall, this study can be useful in investigating the role of friction in the sensory perception of dairy food products.

ACKNOWLEDGMENT

We thank Mr. Philipp T. Kuhn from the University of Groningen for preparing the PDMS pin-and-disk samples used in our experiments.

REFERENCES

1. K. D. Foster, J. M. V. Grigor, J. N. Cheong, M. J. Y. Yoo, J. E. Bronlund, and M. P. Morgenstern, The role of oral processing in dynamic sensory perception, *J. Food Sci.*, 76, 49–61 (2011).
2. A. Chojnicka-Paszun, H. H. J. De Jongh, and C. G. De Kruif, Sensory perception and lubrication properties of milk: Influence of fat content, *Int. Dairy J.*, 26, 15–22 (2012).
3. D. H. Veeregowda, H. C. Van der Mei, J. De Vries, M. W. Rutland, J. J. Valle-Delgado, P. K. Sharma, and H. J. Busscher, Boundary lubrication by brushed salivary conditioning films and their degree of glycosylation, *Clin. Oral Invest.*, 16, 1499–1506 (2012).
4. E. H. A. De Hoog, J. F. Prinz, L. Huntjens, D. M. Dresselhuis, and G. A. Van Aken, Lubrication of oral surfaces by food emulsions: The importance of surface characteristics, *J. Food Sci.*, 71, 337–341 (2006).
5. M. J. Adams, B. J. Briscoe, and S. A. Johnson, Friction and lubrication of human skin, *Tribol. Lett.*, 26(3), 239–253 (2007).
6. E. Van Der Heide, X. Zeng, and M. A. Masen, Skin tribology: Science friction?, *Friction*, 1(2), 130–142 (2013).
7. T. Huppertz, P. F. Fox, K. G. De Kruif, and A. L. Kelly, High pressure-induced changes in bovine milk proteins: A review, *Biochim. Biophys. Acta—Proteins Proteomics*, 1764, 593–598 (2006).
8. V. Raikos, Effect of heat treatment on milk protein functionality at emulsion interfaces. A review, *Food Hydrocolloids*, 24, 259–265 (2010).
9. J. R. Stokes, M. W. Boehm, and S. K. Baier, Oral processing, texture and mouthfeel: From rheology to tribology and beyond, *Curr. Opin. Colloid Interface Sci.*, 18, 349–359 (2013).

Section IV

Biobased Amphiphilic Materials in Tribology and Related Fields

11 Test Methods for Testing Biodegradability of Lubricants
Complete or Not Complete?

Ben Müller-Zermini and Gerhard Gaule

CONTENTS

11.1 INTRODUCTION

For the development of environmentally acceptable lubricants, collaboration among engineers, chemists, and biologists is necessary. Therefore, the development of biolubes is an interdisciplinary task. This fact leads to a high risk of misunderstandings. According to current norms and standards, the most important property of a biolube is rapid biodegradation. It can also be expected that a biolube is not toxic and that by using biolubes, valuable resources are saved and the principle of sustainability is realized. Particularly, the idea of sustainability actually is missing in most norms and standards of today.

Even measuring the biodegradation extent is not easy. As biodegradation laboratory tests are biological tests, the precision of most methods is very low compared with physical methods. In addition, some methods allow various possibilities to introduce the water-insoluble lubricant into the aqueous phase of the test flasks. However, the biodegradation extent of a water-insoluble sample strongly depends on the dispersion of the sample in the water, so different sample introductions lead to a high degree of variation in the test results, too [1]. In this chapter, we wish to compare different biodegradability test methods concerning their precision and quality of test results.

11.2   PRINCIPLES FOR TESTING BIODEGRADABILITY

There are many different test methods for testing the biodegradability of organic substances. All methods have one thing in common: they try to simulate the biodegradation process in the laboratory. For this, a mineral substrate, which is water with dissolved mineral salts, the oil sample, and bacteria are added into a bottle or flask. The bacteria are mostly obtained from a sewage treatment plant. Some test methods also allow the use of surface water from lakes or rivers. The prepared flasks are incubated at room temperature for a certain time period. Some tests prescribe 21 days, others 28 days, and sometimes even longer time periods.

The degradation process consists of a chain of many very complex biochemical reactions. The sum of all the reactions can be represented as follows:

$$\text{Organic substance} + \text{Oxygen} \rightarrow \text{Carbon dioxide} + \text{Water} \qquad (11.1)$$

The organic substance is degraded to carbon dioxide and water, and for this to happen, there is a need for oxygen.

For measuring the degradation extent, changes in the concentration of reactants and products are measured. For testing lubricants, it is very common to measure the concentration of the remaining sample and its nonvolatile degradation products (CEC-L-103-12 [2] or CEC-L-33-A-93 [3]). Other test methods measure the amount of oxygen consumed, for example, OECD 301 C [4] or DIN EN ISO 9408 [5]. Then, there are test methods that measure carbon dioxide evolution (OECD 301 B [4] or DIN EN ISO 9439 [6]).

Every test requires its own special test equipment: for measuring organic substance such as in CEC-L-103-12 [2], open flasks on a shaker are used. The laboratory analyzes the concentration of the organic substance before and after the 21-day incubation period. If gases are analyzed, the test flask must be airtight. Oxygen consumption can be measured by using a closed bottle or using a respirometer. In the case of the closed bottle, the concentration of oxygen is measured in the water. In the case of the respirometer, a sensor monitors the pressure drop due to the microorganism's oxygen consumption.

Carbon dioxide evolution is measured by trapping the produced carbon dioxide in a gas wash bottle filled with barium hydroxide solution, which acts like a trap for carbon dioxide. Subsequently, the amount of trapped carbon dioxide is measured by titration.

11.3 VIEWPOINTS OF THE DIFFERENT BIODEGRADATION TEST METHODS

The process of biodegradation is very complex. There are many reaction steps and, therefore, many intermediates. This makes measuring biodegradation of oils very complex. In the case of a simple mechanism, for example, radioactive decay, the reaction extent can be determined by measurement of the reactant concentration or the product concentration. In the case of biodegradation of a hydrocarbon, it is not just one reactant or one product; there are many intermediates. This fact makes measurement of the degradation extent very complex. Of course, the end product is carbon dioxide and the evolution of carbon dioxide can be measured. However, the formation of carbon dioxide occurs not only at the very end of the reaction chain, but also in the intermediate steps. This means that the formation of carbon dioxide is not an indicator of the complete degradation of the molecule. It is only an indicator of the complete degradation of one carbon atom.

The viewpoints of the various test methods are different, and this is illustrated in Figure 11.1.

In the case of a primary degradation test method, the observer would regard only the reactant molecules as not degraded. Therefore, the test method would need a specific analytical method that is able to detect only the sample molecules. This could be, for example, an enzymatic test such as a glucose test strip for measurement of glucose. In contrast to this, CEC-L-103-12 [2] and CEC-L-33-A-93 [3] use selective analytical methods that can detect all organic compounds that have C–H bonds. In the Coordinating European Council (CEC) test methods, not only the sample molecules are regarded as not degraded but also the nonvolatile, oil-soluble intermediates. Although the CEC test methods are often called "primary degradation tests," they do not measure primary degradation according to the original definition.

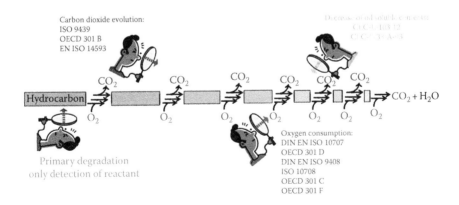

FIGURE 11.1 Viewpoints of different lubricant biodegradability test methods.

11.4 TERMS AND DEFINITIONS

11.4.1 MINERALIZATION

The International Union of Pure and Applied Chemistry (IUPAC) definition for the term mineralization is as follows:

> In the case of polymer biodegradation, this term is used to reflect conversion to CO_2 and H_2O and other inorganics. CH_4 can be considered as part of the mineralization process because it comes up in parallel to the minerals in anaerobic composting, also called methanization [7].

This means that an organic substance is converted into inorganic degradation products. Carbon dioxide is an inorganic reaction product. By measuring carbon dioxide evolution, the percentage of mineralization of an organic substance can be measured. Sometimes, biodegradation test methods that measure oxygen consumption are also called "mineralization methods." According to IUPAC definition, measurement of O_2 consumption is not mineralization; they do measure this process only indirectly. A chemical substance can absorb oxygen without releasing CO_2, H_2O, or CH_4, such as during rusting of iron or oxidation of alcohol to carboxylic acid.

11.4.2 PRIMARY DEGRADATION

There are different definitions of the term "primary degradation." Mostly, it is defined as the first step of a degradation process. In the detergent industry, detergents are primarily degraded when they have lost their surface activity. Another definition is the following: Measurement of primary degradation is measurement of the reactant concentration using a specific test method, for example, using enzymatic glucose strips for detection of glucose.

11.4.3 ULTIMATE BIODEGRADATION

According to the definition in the standard ISO 9439 [6], ultimate biodegradation is the "breakdown of a chemical compound or organic matter by microorganisms in the presence of oxygen to carbon dioxide, water and mineral salts of any other elements present (mineralization) and the production of new biomass" [6].

In the German language, this term is translated to "Vollständiger Abbau," and sometimes in the English literature, the term "complete degradation" is found. However, these terms can be misleading. Actually, there are two points of view:

1. Complete degradation regarding the amount of a substance: No test substance is left. The test substance has been completely degraded or consumed.
2. Complete degradation regarding degradation steps: One test sample molecule has been degraded totally to its end products through multiple steps. This is complete degradation in the meaning of ultimate biodegradation.

Figures 11.2 and 11.3 are illustrations of these two cases.

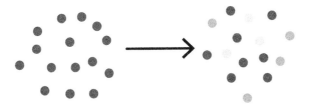

FIGURE 11.2 Complete degradation regarding the amount of a substance being degraded. All "red molecules" are degraded to "blue," "green," or "yellow molecules." There is no "red molecule" left. The "red molecules" have been completely degraded.

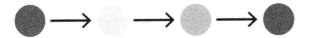

FIGURE 11.3 Complete degradation involving several degradation steps. Ultimate biodegradation. Only one molecule is regarded. This single "red molecule" is degraded to a "yellow molecule," which becomes a "green molecule" and finally becomes a "blue molecule." In this case, only the blue molecule is completely degraded.

11.5 DOES A TEST METHOD FOR MEASURING ULTIMATE BIODEGRADABILITY ALWAYS SHOW COMPLETE DEGRADATION?

To answer this question, we have to distinguish between two cases:

- Case 1: Test sample is a mixture.
- Case 2: Test sample is a single pure substance.

11.5.1 CASE 1: TEST SAMPLE IS A MIXTURE

The lubricant industry often uses ultimate biodegradation tests such as OECD 301 B [4] for testing formulated hydraulic fluids. These fluids are mixtures of structurally different substances: base oils and additives. In the case of a positive result of 60% or more biodegradation, the product is declared as "completely biodegradable." But is this correct?

For example, a mixture of 70% paraffinic and 30% aromatic hydrocarbons is tested according to OECD 301 B. Paraffinic and aromatic hydrocarbons are structurally different substances. Biodegradability of middle-chain and short-chain paraffinic hydrocarbons is generally very good. Within 28 days, it is possible that about 91% of these substances are degraded. However, biodegradability of aromatic hydrocarbons is really slow. It is possible that bacteria can only degrade 3% of these substances within 28 days. However, OECD 301 B measures carbon dioxide evolution regardless of the source. There is no information about the origin of the carbon dioxide.

The OECD 301 B test result of this mixture (70% paraffinic and 30% aromatic hydrocarbons) can achieve, for example, 65% biodegradation. This result would be interpreted as follows: The product is biodegradable, although there are components

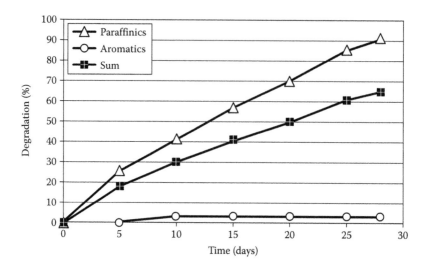

FIGURE 11.4 Degradation rates of the mixture and its components.

in the mixture that are not at all biodegradable. Figure 11.4 shows the degradation curves of the mixture and its components.

The example shows that in the case of measuring biodegradation of mixtures, the OECD 301B test does not show complete degradation. This phenomenon is already known and is called "sequential degradation." It is taken into consideration in the "Revised Introduction of the OECD Guidelines for Testing of Chemicals, Section 3," point 44: "Tests for ready biodegradability are not generally applicable for complex mixtures containing different types of chemicals" [8].

11.5.2 CASE 2: TEST SAMPLE IS A SINGLE PURE SUBSTANCE

Carbon dioxide evolution tests measure the carbon dioxide from the test substance produced by microorganisms. The C atoms in carbon dioxide molecules are all equal. Organic molecules may be very complex. There are different kinds of C atoms, which can react in different ways. It is possible that some of the C atoms of a single molecule are broken down more easily by microorganisms than other C atoms. This means that sequential biodegradation can occur even in a single molecule. Figure 11.5 shows an example molecule where sequential degradation can happen.

The molecule of this example consists of an aromatic and a polyglycol part. Degradation occurs step by step. The polyglycol part can be degraded easily and fast by microorganisms. The end product of polyglycol degradation is carbon dioxide. The aromatic part of the molecule (which is DDT [**d**ichloro**d**iphenyl-**t**richloroethane], which is a POP [persistent organic pollutant]) is very stable and most microorganisms cannot degrade it at all [9,10].

However, if the example molecule is tested using OECD 301 B method, it may be classified as biodegradable: The molecule consists of a total of 41 C atoms, of which 27 C atoms are part of the fast biodegradable polyglycol and only 14 C atoms are part of the aromatic DDT. If carbon dioxide evolution is measured, a biodegradation of

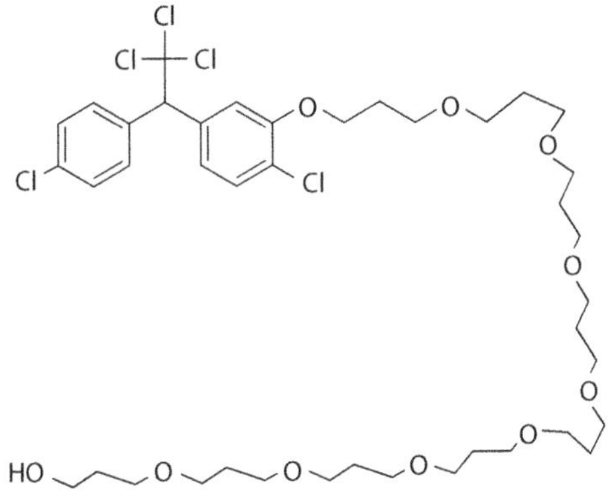

FIGURE 11.5 Example of a molecule that can be degraded sequentially. The molecule consists of an aromatic and a polyglycol part. The polyglycol part can be degraded easily. However, the aromatic part is a persistent organic pollutant.

27/41 = 66% could be found. Even in the case of a pure substance, the formation of persistent degradation products is possible, although the biodegradation, according to OECD 301 B, exceeds 60%.

The two examples mentioned earlier show that it is possible that biodegradation tests that measure ultimate biodegradability are not applicable to find persistent intermediates.

11.6 REPRODUCIBILITY, *R*: INDICATOR OF RELIABILITY OF TEST METHODS

How reliable are these test methods for measuring biodegradability? Compared with usual measurements in engineering or science, biological test methods are not very precise.

To determine the precision of a test method, there are round robins, which means that several laboratories measure the same sample and compare their results. The results of a round robin are two major statistical quantities:

1. Repeatability, *r*
2. Reproducibility, *R*

The repeatability refers to a single laboratory; the reproducibility compares the results of several laboratories. Reproducibility is defined as follows:

A reproducibility limit is the value below which the difference between two test results obtained under reproducibility conditions may be expected to occur with a probability of approximately 0.95 (95%) [11].

An example can clarify this: In a round robin for measuring the open-cup flash point, a reproducibility of 17°C had been obtained [12]. Two laboratories measure the flash point of the same sample. Laboratory A measures 181°C; laboratory B measures 169°C. The difference between these two results is 12°C. 12°C is less than 17°C, and this means that the measurements are in the normal range.

In another case, laboratory A measures 181°C and laboratory B measures 163°C. The difference between these two results is 18°C. 18°C is higher than 17°C. This means that the measurements are out of the normal range. Maybe this is one case out of the 5% which are not inside the range or there occurred an error in one of the laboratories.

11.6.1 Reproducibility of Lubrication Lab Methods

To compare the reproducibility of different lab methods, absolute reproducibility values can be related to typical results of the test method to obtain relative reproducibility. For example, the reproducibility of any density measurement according to ASTM D7042 [13] is 0.0015 g/mL and a typical value of the density of an oil is 0.85 g/mL, so the relative reproducibility is 0.0015/0.85 = 0.2%. Table 11.1 shows the relative reproducibilities of selected lubrication laboratory methods.

There are some lubrication test methods that are very precise, such as the measurement of density, refractive index, or viscosity. There are also others that are not very precise, such as friction on the four-ball test or the pour point.

11.6.2 Reproducibility of Biodegradation Test Methods

Most of today's biodegradation test methods do not even have any data about their reproducibility. The only data about reproducibility is found in the test methods CEC-L-103-12 [2], DIN EN ISO 14593 [22], and ASTM D 5864 [23]. Data about reproducibility of the OECD 301 [4] tests can be found in the publication of the round robin of 1988 [24]. Table 11.2 shows the relative reproducibilities of some biodegradation test methods.

With the exception of OECD 301 A [4] and CEC-L-103-12 [2] compared with the usual lubrication laboratory test methods, the reproducibilities of biodegradation test methods are very poor. It is notable that these two methods are not respirometric

TABLE 11.1
Relative Reproducibilities (*R*) of Selected Lubricant Test Methods

Method [Ref.]	Standard	*R* (Rel.) (%)[a]
Density [13]	ASTM D7042 [13]	0.2
Refractive index [14]	DIN 51423-1 [14]	0.4
Viscosity [13]	ASTM D7042 [13]	0.6
pH value [15]	DIN 51369 [15]	9
Saponification number [16]	DIN 51559-1 [16]	10
Flash point: open cup [12]	DIN EN ISO 2592 [12]	11
Simulated distillation [17]	DIN EN 15199-1 [17]	12
Acid number [18]	DIN 51558-1 [18]	15
X-ray fluorescence spectroscopy [19]	DIN 51396-2 [19]	30
Four-ball test [20]	DIN 51350-2 [20]	30
Pour point [21]	DIN ISO 3016 [21]	33

[a] The smaller values of *R* correspond to better reproducibility of the test method.

TABLE 11.2
Relative Reproducibilities of Some Biodegradation Test Methods

Standard [Ref.]	Test Substance	R (Rel.) (%)	Test Type
OECD 301 A [4]	Water soluble	23	Consumption
CEC-L-103-12 [2]	Poor water solubility	24	
DIN EN ISO 14593 [22]	Water soluble	48–67	Respirometric
OECD 301 B [4]	Water soluble	67	
ASTM D 5864-05 [23]	Poor water solubility	70	
OECD 301 F [4]	Water soluble	117	
OECD 301 D [4]	Water soluble	147	

methods. In OECD 301 A [4] and CEC-L-103-12 [2], the decrease in the test sample and its nonvolatile intermediates is measured.

It seems that the precision of respirometric test methods is much poorer. One explanation for this phenomenon could be the complexity of these methods. This can be shown by an example: Someone is having a meal. The task is to find out how much of the meal has been eaten already. There are two possibilities to complete the task:

A. Measurement of the carbon dioxide produced and calculation of the metabolized meal. For this, the man has to be placed in a closed system purged with carbon dioxide–free air. The exiting air has to be analyzed and the concentration of carbon dioxide has to be measured.
B. Measurement of the meal's weight using a balance. For this, only a suitable balance is necessary.

Figures 11.6 and 11.7 show the test equipment for cases A and B, respectively.

It is obvious that more mistakes can be made using method A than using method B.

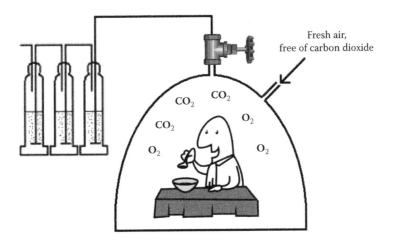

FIGURE 11.6 Measurement of carbon dioxide produced for calculation of the consumed meal.

FIGURE 11.7   Measurement of the meal's weight using a balance for calculation of the consumed meal.

11.7   CONCLUSION

Lubricants are mixtures of poorly water-soluble substances. Unfortunately, it is not possible to use OECD 301 A [4] to test poorly water-soluble substances. So far, the test method CEC-L-103-12 [2] is the only method for testing the biodegradability of lubricants that has a similar reproducibility, R, to most test methods used by the lubricant industry.

REFERENCES

1. A. de Morsier et al., Biodegradation tests for poorly soluble compounds, *Chemosphere* 16, 833 (1987).
2. CEC L-103-12, Biological degradability of lubricants in natural environment, April 16, 2013.
3. CEC L-33-A94, Biodegradability of two-stroke cycle outboard engine oils in water, January 1, 2002.
4. OECD Guideline for Testing of Chemicals, OECD 301 ready biodegradability, July 17, 1992.
5. ISO 9408: 1999, Water quality—Evaluation of ultimate aerobic biodegradability of organic compounds in aqueous medium by determination of oxygen demand in a closed respirometer, December 1999.
6. International Standard: ISO 9439, Water quality—Evaluation of ultimate aerobic biodegradability of organic compounds in aqueous medium—Carbon dioxide evolution test, 1999.
7. M. Vert et al., Terminology for biorelated polymers and applications (IUPAC Recommendations 2012), *Pure Appl. Chem.*, *84*(2), 377–410 (2012).
8. OECD, OECD Guidelines for the Testing of Chemicals, Section 3, Revised Introduction to the OECD Guidelines for Testing of Chemicals, Section 3, Paris, France (2006).
9. World Health Organization, DDT and its derivatives, Environmental Health Criteria monograph No. 009, Geneva, Switzerland (1979).

10. L. Ritter et al., Persistent organic pollutants an assessment report (pdf), United Nations Environment Programme, Retrieved May 29, 2017.
11. A.D. McNaught et al., *IUPAC, Compendium of Chemical Terminology*, 2nd edn., Blackwell Scientific Publications, Oxford, U.K. (1997).
12. International Standard: DIN EN ISO 2592, Petroleum products—Determination of flash and fire points—Cleveland open cup method (ISO 2592:2000), 2002.
13. ASTM D 7042-04, Standard test method for dynamic viscosity and density of liquids by stabinger viscometer, West Conshohocken, PA (2004).
14. DIN 51423-1, Measurement of the relative refractive index with the precision refractometer, 2010.
15. DIN 51369, Entwurf Kühlschmierstoffe—Bestimmung des pH-Wertes von wassergemischten Kühlschmierstoffen, 2010.
16. DIN 51559-1, Entwurf-Mineralöle—Bestimmung der Verseifungszahl—Verseifungszahl über 2 Farbindikator Titration, 2008.
17. DIN EN 15199-1, Mineralölerzeugnisse—Gaschromatographische Bestimmung des Siedeverlaufs—Teil 1 Mitteldestillate und Grundöle, 2007.
18. DIN 51558-1, Mineralöle—Bestimmung der Neutralisationszahl Farbindikator Titration, 1979.
19. DIN 51396-2, Schmieröle—Bestimmung von Abriebelementen—Wellenlängendispersive Röntgenfluoreszenz Analyse (RFA), 2008.
20. DIN 51350-2, Entwurf—Schmierstoffe—Prüfung im Shell Vierkugel Apparat-Bestimmung der Schweißkraft von flüssigen Schmierstoffen, 2008.
21. DIN ISO 3016, Entwurf—Mineralölerzeugnisse-Bestimmung des Pourpoint, 2012.
22. DIN EN ISO 14593, CO_2 Headspace Test-Biologische Abbaubarkeit mittels Bestimmung des anorg. Kohlenstoffs in geschlossenen Flaschen, 2005.
23. ASTM D5864-05, Standard test method for determining aerobic aquatic biodegradation of lubricants or their components.
24. OECD, OECD Ring-test of methods for determining ready biodegradability: Chairman's Report (M. Hashimoto; MITI) and final report (M. Kitano and M. Takatsuki; CITI), Paris, France (1988).

12 Biosynthesis and Derivatization of Microbial Glycolipids and Their Potential Application in Tribology*

Daniel K.Y. Solaiman, Richard D. Ashby, and Girma Biresaw

CONTENTS

12.1 INTRODUCTION

Microbial glycolipids (MGLs) are amphiphilic molecules biologically synthesized as secondary metabolites by many microorganisms. Since MGLs have good surface-lowering activity, they have been extensively researched and developed as a bio-based alternative to the synthetic surfactants that are used extensively in household and industrial cleaning products, cosmetics, oral hygiene consumer products, foods, and environmental remediation [1–4]. The value-added properties of MGLs such as strong antimicrobial, *in vitro* skin-cell rejuvenating, anticancer and antiviral, and

---

* Mention of trade names or commercial products in this article is solely for the purpose of providing specific information and does not imply recommendation or endorsement by the U.S. Department of Agriculture. USDA is an equal opportunity provider and employer.

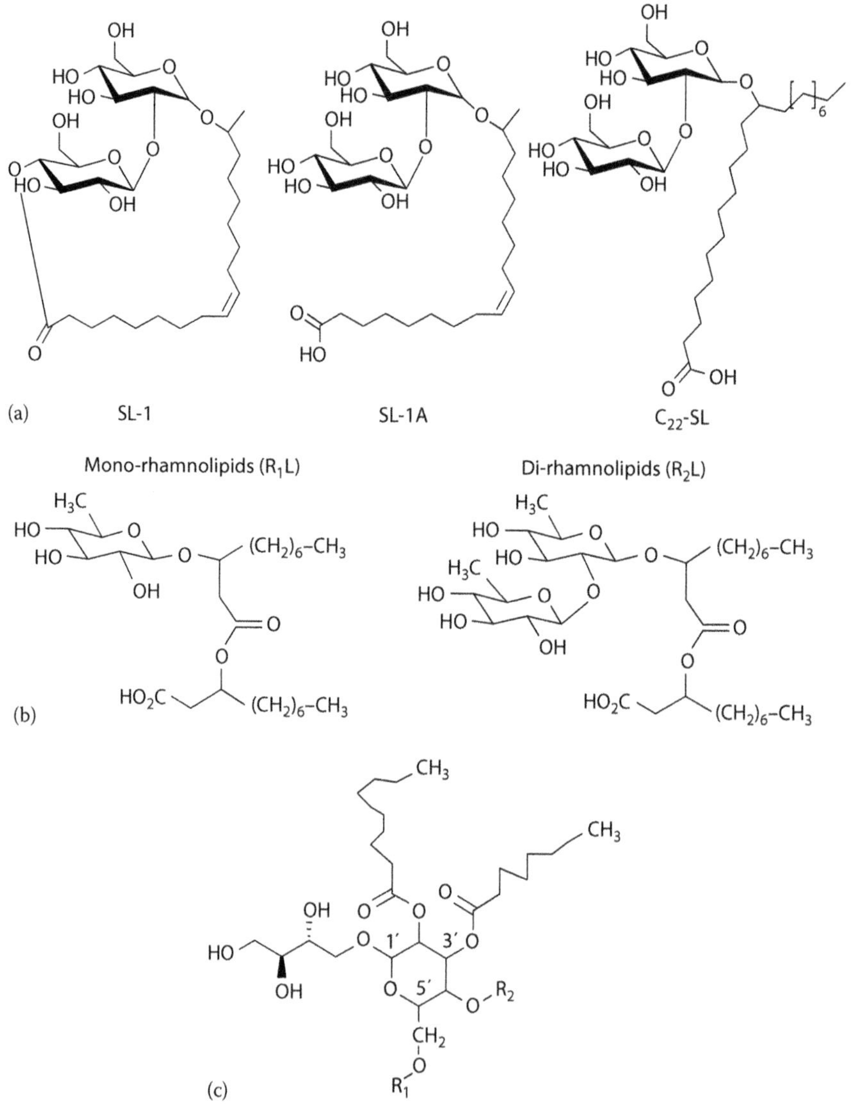

FIGURE 12.1 (a) Structures of sophorolipids from *Starmerella bombicola* (i.e., 17-L-[(2'-O-β-glucopyranosyl-β-D-glucopyranosyl)-oxy]-9-octadecenoic acid 6',6"-diacetate sophorolipids in the 1',4"-lactone form (SL-1) and the free-acid form (SL-1A)) and from *Rhodotorula bogoriensis* (i.e., 13-L-[(2'-O-β-glucopyranosyl-β-D-glucopyranosyl)-oxy]-docosanoic acid (C$_{22}$-SL)). (b) Structures of mono-rhamnolipid (i.e., α-L-rhamnopyranosyl-3-hydroxydecanoyl-3-hydroxydecanoate, R$_1$L) and di-rhamnolipid (i.e., α-L-rhamnopyranosyl-(1→2)-α-L-rhamnopyranosyl-3-hydroxydecanoyl-3-hydroxydecanoate, R$_2$L). (c) Structure of MEL (i.e., 4-O-β-(2',3'-di-O-alka(e)noyl-6'-O-D-mannopyranosyl)-erythritol).

wound-healing promoting activities all add to the attractiveness of these biosurfactants as commercially feasible biobased chemicals [5–9]. Various techno-economic assessments placing some of these MGLs at estimated production costs at $2–5 lb^{-1} further stimulate interest in these biobased products [10–13]. The three most widely studied MGLs are sophorolipids (SLs), rhamnolipids (RLs), and mannosylerythritol lipids (MELs) [14–16]. SL and RL are of most immediate interest because of their emerging commercial potential [12]. SLs contain a sophorose (dimer of glucose) linked through a glycosidic bond to the hydroxyl group of a hydroxy fatty acid and are synthesized by various species of yeasts [17]. Figure 12.1a shows the structures of SL in lactonic (SL-1) and free-acid (SL-1A) forms as produced by *Candida bombicola* yeast when grown on oleic acid (a $C_{18:1}$ fatty acid) and glucose [18,19]. The extent of acetylation at the C-6' and C-6" positions and the structure of the fatty acid of SLs can vary depending on the yeast species, fermentation conditions, and substrates used. For example, the use of *Rhodotorula bogoriensis* as a producer yeast results in SLs in which the fatty acid moiety is a 13-hydroxydocosanoic acid [20,21]. RL is the other MGL most studied by researchers. In RL, the sugar moiety is composed of one (mono-rhamnolipid or R_1L) or two (di-rhamnolipid or R_2L) rhamnose molecules, and the lipid moiety is usually a dimer of 3-hydroxy fatty acids with 8–14 carbon chain lengths (see Figure 12.1b for the representative structures of RLs). As with SL, the structures of RL produced in fermentation are dictated by the producer strain, fermentation conditions, and the substrate used [22]. Finally, MELs produced by certain yeasts and fungi such as *Pseudozyma* and *Ustilago* species are composed of a mannosylerythritol disaccharide esterified at C-2' and C-3' of the mannose moiety with fatty acids of varying carbon chain lengths [3,23,24]. In this review paper, the structural versatility of MGLs from biosynthesis and downstream modification point-of-view is highlighted, and a perspective for their potential applications in tribology is presented.

12.2   BIOSYNTHETIC PRODUCTION OF SELECT MICROBIAL GLYCOLIPIDS

A good surfactant must possess a chemical structure that will induce measurable reductions in surface and interfacial tensions, lower critical micelle concentrations (CMCs), lower Krafft-point temperatures, possess good solubility in hot and cold water, demonstrate fast kinetics for their self-assembly, exhibit high biodegradability and biocompatibility, provide a good environmental profile, and have a low cost-to-performance ratio [25]. One class of surfactants that is garnering attention is the MGLs, which are a class of amphiphilic molecules that are produced in high yields from renewable resources via fermentation. These molecules are typically produced from microbial sources, are composed of both a carbohydrate and lipid moiety, and possess unique characteristics and functionality that make them attractive as potential substitutes for petroleum-based surfactants. Currently, SLs, RLs, and MELs are three MGLs that are of major industrial interest owing to their favorable surfactant properties, renewable nature, widespread application potential, and low environmental impact.

12.2.1 Sophorolipids

SLs are extracellular MGLs that typically consist of a hydrophobic fatty acid tail generally from 16 to 18 carbon atoms in length (saturated or unsaturated) and a hydrophilic disaccharide headgroup (sophorose; 2-O-β-D-glucopyranosyl-β-D-glucopyranose), which may be acetylated at the C-6′ and/or C-6″ hydroxy group of each glucose moiety (Figure 12.1a). The hydroxy fatty acid tail is linked to the sophorose sugar through a β-glycosidic linkage between the 1′ hydroxy group of the sophorose and the ω or ω-1 hydroxy group of the fatty acid. In addition, the carboxylic acid group of the fatty acid may be lactonized to the disaccharide ring at C-4″ or remain in the open-chain form.

The historical chronology of SL production can be seen in Table 12.1. SLs were first reported in the early 1960s [26,27] from the yeast *Torulopsis magnoliae* (reclassified in 1968 as *Torulopsis apicola* [28] and currently recognized as *Candida apicola*). The SL product from those early trials was structurally identified as a partially acetylated 2-O-β-D-glucopyranosyl-β-D-glucopyranose unit attached to 17-L-hydroxystearic acid or 17-L-hydroxyoleic acid through a β-glycosidic linkage [27,28]. At the same time, a new SL derivative was found produced from *Candida bogoriensis* (since reclassified as *Rhodotorula bogoriensis*) whose fatty acid component was 13-hydroxybehenic acid [20]. *Torulopsis bombicola*, a third SL-producing yeast strain, was identified in 1970 (since reclassified as *Candida bombicola* and more recently as *Starmerella bombicola*) whose SL products and production schemes closely approximated those of *Candida apicola* [31]. More recently, SL biosynthesis has been demonstrated in a strain of *Wickerhamiella domericqiae* whose structural characteristics showed a variety of chemical variations but whose predominant chemical species was 17-L-hydroxystearic acid 1′,4″-lactone 6′,6″-diacetate, which is also commonly found in SLs produced by *Starmerella bombicola* when grown under appropriate conditions [33]. Then, in 2008 and 2010 additional *Candida* strains were documented to produce SLs. *Candida batistae* was shown to produce predominantly

TABLE 12.1
Historical Chronology of SL Biosynthesis

Year	Producing Strain (Reclassification)[a]	References[b]
1961	*Torulopsis magnoliae*	[26]
1962	*Torulopsis magnoliae* (*Torulopsis apicola*, 1968; *Candida apicola*, 1978)	[27] ([28,29])
	Candida bogoriensis (*Rhodotorula bogoriensis*, 1979)	[20] ([30])
1970	*Torulopsis bombicola* (*Candida bombicola*, 1978; *Starmerella bombicola*, 1998)	[31] ([29,32])
2006	*Wickerhamiella domericqiae*	[33]
2008	*Candida batistae*	[34]
2010	*Candida riodocensis*, *Candida stellata*, *Candida* sp. Y-27208	[17]

[a] Parenthetical information denotes the reclassification of the strain and the year in which the reclassification occurred.

[b] Parenthetical references are those describing the reclassification.

acidic SLs with the majority of the fatty acid side chains representing 18-L-([2′-*O*-β-D-glucopyranosyl-β-D-glucopyranosyl]-oxy)-oleic acid, 6′, 6″-diacetylate [34]. Since then, *Candida riodocensis*, *Candida stellata*, and *Candida* sp. Y-27208 have also been documented to produce predominantly di-acetylated acidic SLs [17].

Currently, *S. bombicola* has garnered the most interest for SL biosynthesis owing to its high production yields and its ability to grow in the presence of a range of substrates. In fact, *S. bombicola* is the strain most utilized in industry to produce large-scale quantities of SLs. As such, the information included in the balance of this section will refer to results obtained from *S. bombicola*. Figure 12.2 shows a proposed overview of the metabolic processes involved in SL biosynthesis. *S. bombicola* can utilize many different hydrophobic substrates including *n*-alkanes, alcohols, aldehydes, and triacylglycerols in the production of SLs. Generally, *S. bombicola*

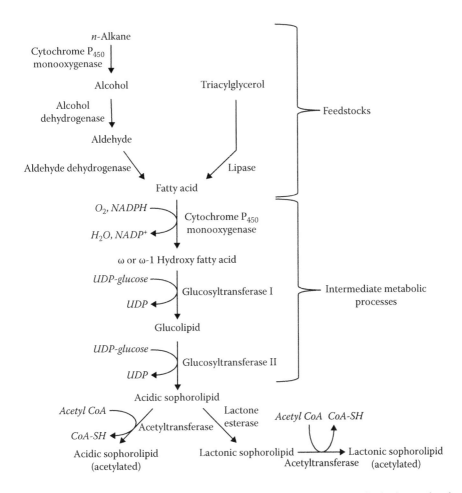

FIGURE 12.2 Proposed overview of the metabolic reactions necessary for both acetylated and non-acetylated SLs in the free-acid and lactone forms. (Adapted from van Bogaert, I.N.A. et al., *Appl. Microbiol. Biotechnol.*, 76, 23, 2007 [126].)

production yields are maximized when 16-carbon and 18-carbon fatty acids are utilized (primarily palmitic acid, stearic acid, oleic acid, or linoleic acid; linolenic acid has not been shown to be a quality hydrophobe for SL production). However, when hydrophobic substrates such as *n*-alkanes are utilized in the production media, *S. bombicola* has the genetic capability to oxidize the alkane to produce fatty acids that can then undergo β-oxidation to form the preferred fatty acid substrates for SL synthesis. These reactions also apply to alcohols and aldehydes that may be used as feedstock. If no hydrophobic carbon source is present, fatty acids can be synthesized *de novo* from the acetyl-CoA intermediates derived from glycolysis.

Once formed, the fatty acids are oxidized at the ω or ω-1 position by an reduced nicotinamide adenine dinucleotide phosphate (NADPH)-dependent cytochrome P450 monooxygenase enzyme, which results in a hydroxy fatty acid. This enzyme is one of a family of cytochrome P450 monooxygenases that have been found in different strains of *Candida* and have been classified into the CYP52 family of cytochrome P450 enzymes capable of hydroxylating alkanes and fatty acids [35]. Once the hydroxy fatty acids are created, the fatty acids are glycosidically attached to a glucose moiety at C-1′ through the action of glucosyltransferase I using uridine 5′-diphosphate (UDP)-glucose as the glucosyl donor [36]. Subsequently, a second glucose molecule is introduced into the molecule through another glycosidic linkage between C-2′ of the forming SL molecule and C-1′ of the newly attached glucose molecule. This second glucosyl-transfer may be catalyzed by a second glucosyltransferase enzyme (glucosyltransferase II), which has been purified but whose activities seem to be analogous [35,37]. This enzymatic action results in the non-acetylated free-acid form of the SL. At this stage, the free-acid form can be acetylated through the action of acetyltransferase enzymes utilizing acetyl-CoA as the acetyl donor or enzymatically converted to the 1′,4″ lactone through the action of a lactone esterase that can then be acetylated. Acetylation of the final SL products may involve acetyl-transfer to both, one, or neither of the 6′ and 6″ hydroxy groups resulting in di-, mono-, or non-acetylated structural variants.

Production yields are a very important parameter to understand when assessing biological processes for large-scale use. These yields are closely associated with the starting feedstock materials and the specific fermentation protocols used for synthesis. SLs are generally regarded as having production capacities far greater than other glycolipid molecules. Table 12.2 shows some of the hydrophobic feedstocks utilized in SL biosynthesis. Glucose is typically utilized as the hydrophilic substrate as it can be easily utilized for SL biosynthesis without further enzymatic conversions. Some researchers have used combinations of glucose and other polar molecules to increase production yields to more than 400 g/L [38,39], but reported yields typically fall between 50 and 200 g/L.

In addition, crude glycerol and soy solubles have been used as glucose substitutes to produce SLs [40,41]. Crude glycerol is the coproduct of the chemical transesterification reactions involved in biodiesel production. Depending on the efficiency of the reaction and the recovery process, crude glycerol streams can be compositionally distinct with respect to glycerol, free fatty acids, and mono-, di-, and triacylglycerols. Since glycerol can be easily transported into the cell and assimilated into the central metabolic system, the crude glycerol stream did result in SL biosynthesis but only at a fraction of the yield provided by glucose. On a molar basis, four molecules of glycerol are required

TABLE 12.2

Reported Sophorolipid Yields by *S. bombicola* from Various Hydrophobic Substrates

Yield (g/L)	Hydrophobic Substrate	References
>400	Rapeseed oil[a]	[38]
	Corn oil[b]	[39]
300–400	Rapeseed oil ethyl ester	[127]
200–300	Rapeseed oil	[127]
	Sunflower oil methyl ester	[127]
	Palm oil methyl ester	[127]
100–200	Sunflower oil	[127]
	Safflower oil	[128]
	Canola oil[c]	[129]
	Animal fat	[130]
	Oleic acid	[131]
	Linseed oil methyl ester	[127]
50–100	Palm oil	[127]
	Fish oil	[127]
	Waste frying oil	[132]
	Stearic acid	[133]
	Hexadecane	[134]
<50	Soybean oil	[18,134]
	Palmitic acid	[133]
	Linoleic acid	[133]
	Stearic acid methyl ester	[18]
	2-Dodecanol	[135]

[a] Hydrophilic carbon source was deproteinized whey.
[b] Hydrophilic carbon source was a mixture of glucose and honey.
[c] Hydrophilic carbon source was lactose.

to produce a single sophorose sugar molecule, while only two glucose molecules are required. Because of this difference, SL volumetric yields with crude glycerol were only 60% of what was obtained using glucose and oleic acid as substrates.

Soy molasses is a coproduct derived from soybean processing. It is derived from the whole bean after flaking and removal of the crude soy oil through solvent extraction. Then, after a grinding and sizing step to produce soy flour and soy protein/concentrate, the remaining material is soy molasses. Soy molasses is high in potentially fermentable carbohydrates and is primarily composed of sucrose, raffinose, and stachyose. When using soy molasses as a glucose substitute in SL biosynthesis, it was found that soy molasses could indeed function as a substitute for glucose [41] but only the sucrose fraction of the soluble carbohydrates was utilized by the organism resulting in volumetric yields that were only 75% compared to the glucose/oleic acid system [42].

12.2.2 Rhamnolipids

RLs are surface-active amphiphilic glycolipids produced primarily by *Pseudomonas aeruginosa* [43] but have also been documented to be produced by other bacterial strains such as *Pseudomonas putida* [44], *Pseudomonas chlororaphis* [45], *Burkholderia pseudomallei* [46], and *Burkholderia plantarii* [47]. They are generally produced as mixtures of structural congeners (more than 60 congeners and homologues have been described [48]) with a conserved mono- or di-rhamnose headgroup linked to a variable tailgroup comprised most often of a monomer or dimer of β-D-(hydroxyalkanoyloxy)alkanoic acid [49] (Figure 12.1b).

The biological function of RLs remains unclear, but many theories have arisen linking them to roles as virulence factors in *Pseudomonas* infections [50], in antimicrobials [51], in biofilm development [52], and in swarming motility [53] and may improve the aqueous solubility of hydrophilic compounds for metabolic processes [54]. However, their chemical structures provide natural surfactancy to these molecules that can benefit numerous industrial applications with advantages over chemo-surfactants including biodegradability, lower toxicity, stability in extreme environments, and structural diversity.

The RL biosynthetic pathway in *Pseudomonas aeruginosa* is shown in Figure 12.3. This pathway is transcriptionally regulated through quorum sensing, which is dependent upon cell density and the formation of autoinducer molecules (e.g., butanoyl-homoserine lactone and 3-oxo-dodecanoyl-homoserine lactone) that bind to the regulatory proteins RhlR and LasR and activate the expression of *rhlAB* and *rhlC* genes. RLs are synthesized via a quadriphasic system involving (1) the formation of the fatty acid dimer moiety (HAA) through the *de novo* fatty acid biosynthetic process and the action of the gene products RhlG (β-ketoacyl reductase) [55] and RhlA (rhamnosyltransferase A) [56]; (2) the synthesis of dTDP-L-rhamnose, the carbohydrate donor, from the central metabolic pathway by the action of AlgC and the *rmlBCAD* gene products [57]; (3) the formation of mono-rhamnolipids from dTDP-L-rhamnose and HAA by the action of RhlB (rhamnosyltransferase B) [58]; and (4) the production of di-rhamnolipids via RhlC (rhamnosyltransferase C) [59].

12.2.3 Mannosylerythritol Lipids

First described in 1956 [60], MELs are amphiphilic molecules that are produced primarily by the basidiomycetous yeasts of the genus *Pseudozyma* (as a major metabolite) [61,62] and by the smut fungus *Ustilago* sp. (as a relatively low-level metabolite) [63,64]. These molecules are composed of either 4-*O*-β-D-mannopyranosyl-erythritol or 1-*O*-β-D-mannopyranosyl-erythritol as a hydrophilic headgroup and fatty acid derivatives as the hydrophobic tailgroups. As with the other producer organisms, a range of different MEL molecules can be produced differing in alkyl chain length (generally ranging in length between 6 and 18 carbons) [23] and degree of acetylation. These molecules may be di-acetylated (at C-4' and C-6' of the mannose sugar moiety, MEL-A; Figure 12.1c), mono-acetylated at C-6' (MEL-B), mono-acetylated at C-4' (MEL-C), or non-acetylated (MEL-D) [65].

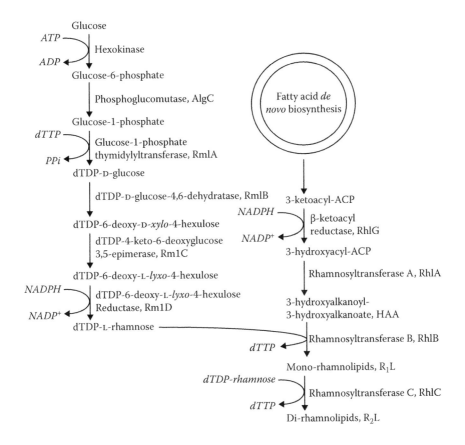

FIGURE 12.3 Metabolic pathways involved in RL biosynthesis.

The biochemical pathway for MEL biosynthesis has been proposed for *U. maydis* and can be seen in Figure 12.4. This enzymatic sequence is the result of a gene cluster that encodes for a mannosyltransferase (Emt1), two acyltransferases (Mac1 and Mac2), a transporter protein (Mmf1), and an acetyltransferase enzyme (Mat1) [66]. A homologous gene cluster has been discovered in *Pseudozyma antarctica* T-34 [67] and *Pseudozyma hubeiensis* [68] indicating the presence of a conserved biosurfactant metabolism among the *Pseudozyma* and *Ustilago* strains. According to the sequence of reactions, MELs are synthesized through a four-step scheme. The first step is the mannosylation of erythritol catalyzed by mannosyltransferase (Emt1) that transfers GDP-mannose directly onto C-4′ of the erythritol, a stereospecific reaction where only mannosyl-D-erythritol is produced. Steps 2 and 3 involve the sequential addition of two fatty acid acyl groups through the action of two regioselective acyltransferases (Mac1 and Mac2). These enzymes react with fatty acids of different chain lengths to acylate C-2′ of the mannose moiety with a fatty acid 6 or 8 carbons in length and C-3′ of the mannose moiety with fatty acids 10 or more carbons in length [64]. The sequence of

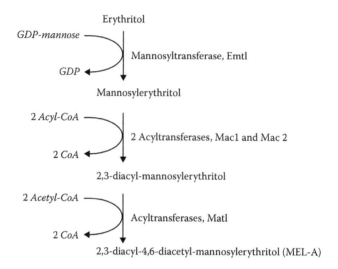

FIGURE 12.4 Proposed biosynthetic route to MELs.

activity of these enzymes is still not clear, but the lack of one or both of these gene products abolishes MEL production completely. Finally, an acyltransferase (Mat1) catalyzes the acetylation of MEL at C-6′ and C-4′. Interestingly, the fatty acids provided in the biosynthetic pathway are the product of a partial peroxisomal β-oxidation or the product of the complete mitochondrial β-oxidation of fatty acids of a much longer chain [69–71].

Like SLs, MEL can be produced in yields of more than 100 g/L, which is advantageous for potential industrial applications. Interestingly, strains that synthesize MEL in large yields produce MEL-A as the predominant component. Table 12.3 shows the maximum yield and fatty acid profiles of those MEL producers whose yields exceed 100 g/L.

TABLE 12.3
Maximum Yields and Structural Compositions of MELs Derived from High-Producing (>100 g/L) *Pseudozyma* Strains

Organism	MEL Yield (g/L)	Substrate	Predominant MEL Type	Predominant FA Profile	References
P. aphidis	165	Soybean oil	MEL-A	C8–C10	[65,136]
P. rugulosa	142	Soybean oil	MEL-A	C8–C10	[137]
P. antarctica	140	*n*-alkanes (even#)	MEL-A	C8–C10	[138]
		n-alkanes (odd#)	MEL-A	C9–C11	[138]
P. parantarctica	107	Soybean oil	MEL-A (di-acetyl.)	C8–C12	[139]
			MEL-A (tri-acetyl.)	C8–C12, C16–C18	[139]

12.3 CHEMICAL AND ENZYMATIC DERIVATIZATION AND MODIFICATION

It is expected that the physicochemical properties (e.g., foaming, surface tension lowering, and solubility) and bioactivities (e.g., antimicrobial and antitumor activity) of MGLs can be altered by structural derivatization or modification in order to be better suited in certain intended applications. The chemical structures of MGLs in fact render these molecules highly amenable to derivatization and modification by chemical and enzymatic means [72]. The saccharide groups of these molecules contain both the primary and the secondary hydroxyl functional groups. Similarly, the lipid moiety of the glycolipids could present reactive centers in the form of double bond and also with its carboxylic acid functional group. Together these functional groups make possible either homologous intermolecular reaction or the attachment of heterologous pendant groups. Furthermore, the lipid and saccharide moieties could also be dissociated to yield separate entities that are themselves novel value-added biomolecules. The following subsections provide examples of the wide-ranging possibilities of derivatization and modification of MGLs.

12.3.1 SACCHARIDE GROUP

The free hydroxyl functional groups of the saccharide of MGLs are the prime targets for acylation or esterification either chemically or enzymatically. Some MGLs as biosynthesized by the producing microorganisms might have already contained variously acetylated hydroxyl group(s). In this case, these acyl groups can also be removed chemically or enzymatically to expose the free hydroxyl for subsequent reaction if desired.

Among the three major MGLs, the derivatization and modification efforts are found most with SL because of the availability of this MGL class in large quantities and in relatively clean form. Zerkowski et al. [73,74] reported the chemical derivatization of SL to yield new glycolipids in which an amino acid of choice is attached to the saccharide moiety; the derivatized SLs are more water soluble than the parent molecule without any significant changes in their surfactant activities and critical-micelle-concentration values. In their work, the stearic sophoroside (i.e., SL in which a stearic acid molecule constitutes the lipid component of the molecule) produced by fermentation was used as the starting material to avoid the added complexity of the double bond in the widely studied oleic sophoroside (i.e., SL with an oleic acid component). Existing functional groups on the amino acid units were first appropriately protected, and a reactive para-aminobenzoic acid linker was then attached to the amino acid. Using carbodiimide-mediated coupling methods, Zerkowski et al. demonstrated the successful attachment of these amino acids onto the stearic sophoroside to obtain a list of new SLs wherein the saccharide moiety is chemically linked to an amino acid having a positively or negatively charged, or a zwitterionic group (Figure 12.5).

Peng et al. [75] employed a cross-metathesis approach to attach uncharged substitute groups on the sugar moiety of SL. For this purpose, the fatty acid moiety of the SL must contain a double-bond such as in the oleic SL. A ruthenium-based

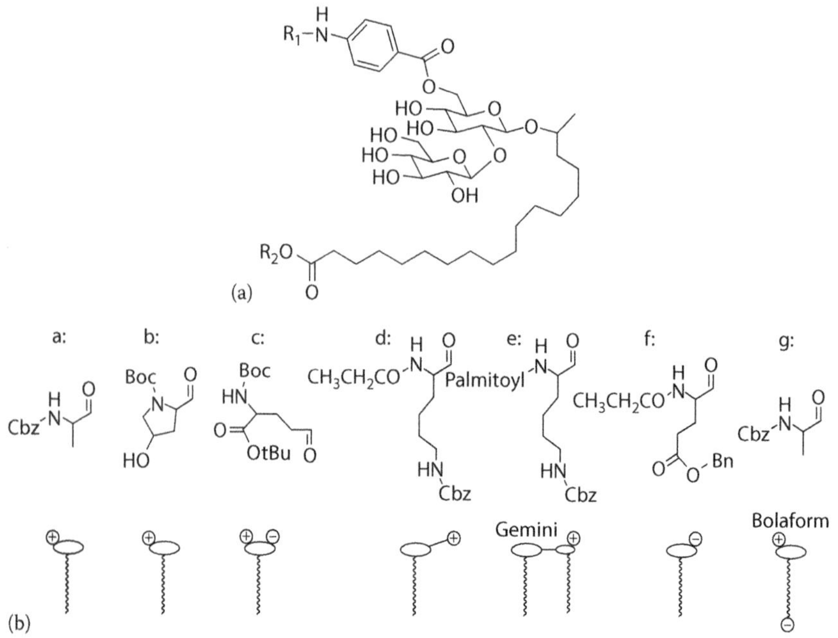

FIGURE 12.5 (a) Basic structure of modified sophorolipids with different amino acid linked through coupling agent. (b) Schematic of amino acids with various charge configurations coupled to the modified sophorolipids. (Adapted from Zerkowski, J.A. et al., Charged sophorolipids and sophorolipid containing compounds, U.S. Patent Number 7,718,782, 2010.)

metathesis catalyst $RuCl_2[C_{21}H_{24}N_2][C_{15}H_{10}][P(C_6H_{11})_3]$ was used to catalyze the reaction of SL with alkene and n-alkyl acrylates. For ethylene, however, it was found that the first-generation Grubbs catalyst, that is, bis(tricyclohexylphosphine)benzylidine ruthenium(IV) dichloride, was preferred, which afforded a complete conversion of the starting SL. The Grubbs reaction products are SL in which the C-1′ is glycosidically linked to an alkenyl group and the C-4″ is acylated also with another alkenyl chain. On further alcoholysis followed by hydrogenation reaction, SL containing a single hydrophobic alkyl chain of 10, 12, or 14 carbons glycosidically bonded to C-1′ was obtained. This approach demonstrated an exquisite method to obtain SLs with a medium-chain-length (C10–C14) hydrophobic tail.

Enzymatic approach was also widely explored to bring about the modification of the sugar moiety of SL. Bisht et al. [76] enzymatically synthesized derivatives of SL in which the sophorose saccharide was esterified to hydrophobic acyl group. After screening seven commercially available lipases, the authors selected *Candida antartica* lipase enzyme (Novozym 435) as the best catalyst to carry out the derivatization under the reaction conditions in tetrahydrofuran (THF). The starting materials used in the study were SLs esterified at the carboxylate end of the molecule with a methyl, ethyl, or butyl group. Upon Novozym 435–catalyzed condensation with the selected acylating agents, the 1°-alcohol groups on the sophorose (i.e., at the C-6′ and C-6″ positions) of the SL esters were found esterified to the corresponding acyl group such as acetate, succinate, or acrylate. It is noteworthy that a slight variation of the reaction protocol led to the formation of a sophorolactone in which the carboxyl carbon of the fatty acid chain is esterified to the C-6″ hydroxyl group [76].

Following the similar lipase-catalysis approach as Bisht et al. [76], Singh et al. [77] further demonstrated the selective modification of the C-6′, C-6″, or both the C-6′ and C-6″ of SLs to contain an ester or amide group(s). The starting material for the enzymatic reactions was the ethyl ester of SL molecules obtained via a reaction of microbial SLs with sodium ethoxide. Two lipases, that is, Novozym 435 and Lipase PS-C, were studied as the catalysts where THF was the reaction medium. The researchers found that Lipase PS-C functioned regioselectively to affect the acylation of the C-6″, while Novozym 435 was less discriminate in carrying out the amidation or acylation of the C-6′ and/or C-6″ of the SL esters.

Instead of using an alkyl ester of microbial SLs as the starting material, Recke et al. [78] performed lipase-catalyzed acylation of the saccharide moiety of alkyl sophorosides as obtained *in situ* from the fermentation of a medium-chain alcohol, that is, 2-dodecanol, as a co-substrate by *Candida bombicola*. Using Novozym 435 lipase as a catalyst in toluene medium, it was shown that 3-hydroxydecanoic acid can be attached to both the C-6′ and C-6″ of the sophorose moiety of the alkyl sophoroside to form a di-acylated derivative. Similarly, 17-hydroxyoctadecanoic acid was esterified to the C-6′ or both the C-6′ and C-6″ of the alkyl sophoroside using Lipozyme IM 20 lipase in toluene solvent. Furthermore, the investigators demonstrated that a glucose unit of the alkyl sophoroside could be removed using a snail glucuronidase to yield the corresponding alkyl glycoside, whose C-6 position could then be esterified to a molecule of sebacic acid using Novozym 435 lipase in toluene solvent.

Unlike with SL, the modification study on RL is very limited perhaps because of the lower availability and consequently a higher cost of the materials. While production yield of SL has been reported at >400 g/L of fermentation culture [39], reported yields of RL were often <100 g/L [79]. In one study, the researchers described using naringinase enzyme of *Penicillium decumbens* to convert di-rhamnolipid into mono-rhamnolipid, presumably through the cleavage of the glycosidic bond between the two rhamnoses even though the supposedly released rhamnose was not identified. The same study further reported that with mono-rhamnolipid as a substrate, naringinase was capable of catalyzing the hydrolysis of the glycosidic bond between the rhamnose and the lipid moiety to release the dimeric 3-(3-hydroxydecyloxy)decanoic acid).

Similarly, only limited studies could be found on the modification of MELs and mostly limited to the degree of acetylation of the C-4′ and C-6′ of the mannose sugar moiety (Figure 12.1c). Depending on the pattern of acetylation, MEL exhibits physicochemical properties unique to the structure [80]. Fukuoka et al. [81] demonstrated that using a lipase enzyme, Novozym 435, it was possible to deacetylate MEL-B to MEL-D at >99% conversion rate. With MEL-A as the starting material, however, the lipase enzyme could only remove the acetyl group at C-6′ to yield MEL-C, indicating that the protocol using Novozym 435 is effective in deacetylating the C-6′ of MELs. Recke et al. [82] carried out lipase-mediated modification of MELs with a different goal in mind. In their study, they attempted the acylation of MEL-A with (*R*)-3-hydroxydecanoic acid (obtained from RL) or (*S*)-17-hydroxyoctadecanoic (from SL) using Novozym 435 lipase and obtained acylated MEL-A derivatives at the C1 of the erythritol moiety of the MEL. With MEL-B as a starting material and Lipozyme as biocatalyst, however, acylation at the C1 of erythritol moiety was successful only when (R)-3-hydroxydecanoic acid (but not (S)-17-hydroxyoctadecanoic) was used as an acyl donor.

12.3.2 LIPID MOIETY

The lipid moiety of MGLs is basically an acyl chain of various lengths (10–22 carbons). When the carboxylate function of the acyl chain is free, it constitutes an active reaction center for modification. Unsaturated carbon bonds that could be present in the acyl chain also present reactive sites for potential derivatization. Azim et al. [83] exploited the reactivity of the carboxylate functional group in the lipid moiety of SL to synthesize a series of amino acid conjugates. Carbodiimide coupling reaction was used to form the amide linkage between an incoming amino acid and the carboxyl group of the recipient SL. A total of 13 amino acid–SL conjugates containing glycine, serine, leucine, phenylalanine, aspartate, and glutamate were synthesized using this method. Among these 13 conjugates, some became esterified with an alkyl group on the carboxyl group of its amino acid moiety, and others had lost the unsaturated carbon–carbon bond in the alkyl chain. Delbeke et al. [84] took advantage of the reactive unsaturated carbon–carbon double bond in the alkyl chain of SL to generate through an ozonolysis reductive workup the aldehyde intermediate for further derivatization. By the nature of the scission ozonolysis reaction, the starting alkyl chain of 18 carbons is consequently shortened to 9 carbons at the site of the original double bond. Using the aldehyde intermediate thus obtained, these researchers then carried out reductive amidation with secondary amines followed by alkylation with alkyl iodide to finally obtain the quaternary ammonium salts of SLs. These quaternary ammonium compounds were evaluated for antimicrobial activity against Gram-positive bacteria with great results, but other physicochemical properties such as surface tension-lowering activity and critical micelle concentration were not determined in comparison to the starting materials.

12.3.3 HOMOLOGOUS INTERMOLECULAR REACTION

The modification of the individual hydrophilic sugar and hydrophobic fatty acid moieties of MGLs has been highlighted in preceding sections. In this unit, reactions in the entire MGL molecule with itself are discussed. In one instance, SLs were strung together to form a poly(SL) polymeric biomaterial consisting of repeating units of 18-carbon oleic acid and sophorose alternately linked [85]. The polymeric biomaterial has a number-average molecular weight (M_n) of 37,000 g/mol with a polydispersity index (M_w/M_n) of 2.1, and the repeating units assume an interesting asymmetric bola-amphiphilic structure. This unique structural feature conferred unusual crystallinity properties to the novel biomaterial. The authors' laboratory subsequently demonstrated the applicability of this approach to synthesize poly(SL) having various substituent groups at the C-6′ and C-6″ of the sophorose moieties [86]. In this latter work, SL-based monomers with variously modified C-6′ and C-6″

of the sophorose moiety were first enzymatically synthesized. These include 6'- and/ or 6"-alkylated SLs synthesized enzymatically using lipase or cutanase, and 6'- and/ or 6"-dehydroxylated and -iodinated/-azidated SLs obtained by chemical synthesis routes. The polymerization of these monomers in various combinations was then carried out using a second-generation Grubbs catalyst. The resultant homo- or hetero-copolymers have number-average molecular weights (M_n) of 36,000–84000 g/mol and polydispersity indices of 2.2–2.9.

12.4 POTENTIAL APPLICATIONS IN TRIBOLOGY

Lubricants play many critical roles in modern economy, including that of allowing for efficient manufacturing and transportation of goods and services. The widespread use of lubricants results in a significant reduction in energy demand, thereby reducing the greenhouse gases (GHG) emitted from energy production and manufacturing. Lubricants can also contribute to the GHG production due to accidental or unavoidable release (e.g., lubricant processing, blending, transportation) or when used in processes (e.g., metalworking) that are known to release volatile organic carbon. One way of reducing the negative environmental impact of lubricants is to replace petroleum-based lubricants, which are net positive GHG emitters, with biobased formulations, which are net negative GHG emitters [87]. Currently, more than 35 million metric tons per year of lubricants is consumed worldwide [88], of which less than 1% is biobased. Thus, there is a need and large market incentive for replacing petroleum-based lubricants with environmentally friendly biobased lubricants.

Current biobased lubricant development work is aimed at replacing both petroleum-based base oils and additives used in lubricant formulations for various applications such as engine oils and hydraulics. Thus, a number of farm products are actively being pursued for the development of biobased base oils [89–93] as well as biobased additives [94–108].

To be used as a base oil, the biobased candidate must be liquid at room temperature and also must satisfy a number of critical requirements. The ideal candidates for such applications are vegetable oils and their derivatives (e.g., estolides) [109,110]. Thus, a number of approaches are being pursued to develop vegetable oils and their derivatives to match the performance and cost of petroleum base oils. These include sesquiterpenes, microbially synthesized from sugars, which have been successfully developed into very competitive biobased base oils [111,112].

To be used as additives, the biobased candidates must have reasonable solubility in base oils and also must demonstrate one of many additive properties [113] (e.g., antioxidant, pour point depressant [PPD]). To date, reported biobased additives under development include viscosity index improvers and/or PPD [94–98], biobased dispersants [99,100], biobased overbased detergents [101,102], biobased friction modifiers [103], biobased antiwear additives [104–106], biobased extreme pressure (EP) additives [104–106], and biobased multifunctional additives [107,108].

Table 12.4 presents the selected physical, chemical, and biological properties of representative MGLs. As can be seen in Table 12.4, most MGLs are solid at room temperature, which excludes their development as biobased base oils. However, as discussed in Section 12.3, MGLs can be chemically and/or enzymatically modified

TABLE 12.4
Selected Properties of MGLs[a]

Property	Sophorolipid (*S. bombicola*)	Rhamnolipid	Mannosylerythritol Lipid
Molecular weight, g/mol	~690	~540 (R_1L)–704 (R_2L)	~634
Density, g/mL	(Solid)	(Solid)	(Solid)
Melting point, °C	49	n/a	n/a
Boiling point, °C	n/a	n/a	n/a
RT solubility, % w/w, in			
Water	<0.3	<0.1	<0.1
Soybean oil	n/a	n/a	n/a
Hexadecane	n/a	n/a	n/a
50%–70% ethanol	≥45	n/a	n/a
Min. water surface tension, dyn/cm	34–40	26–29	23.7–33.8
CMC, mol/L	0.5–4×10^{-4}	7–285×10^{-6}	1.2–360×10^{-6}
Min. oil–water interfacial tension, dyn/cm			
Water–hexadecane (concentration)	5 (10 mg/L)	≤1 (0.1 wt%)	
Water–tetradecane (concentration)			2.1–2.4 (n/a)
Water–soybean oil (concentration)	1–2 (n/a)		
Kerosene (concentration)		0.2–3.2 (0.05%–0.1%)	0.1
Microbial activity	(+)	(+)	(+)
Min. inhibitory concentration for			
Staphylococcus aureus	>512 µg/mL	128 µg/mL	12.5–25 µg/mL
Escherichia coli	>512 µg/mL	32 µg/mL	>400 µg/mL
Bacillus subtilis	4 ppm	64 µg/mL	2.5–6.2 µg/mL
Streptococcus mutans	1 ppm	n/a	n/a
Propionibacterium acnes	0.5 ppm	n/a	n/a
Micrococcus luteus	0.48 µg/mL	32 µg/mL	3.1–12.5 µg/mL
Mycobacterium rhodochrous	n/a	n/a	25 µg/mL

[a] Data were from References 3,119,120,123,124,140–145.

to produce new products. Such modification can be undertaken on the glycolipid molecule as a whole or on its saccharide or lipid moiety, separately. Even though, at this time, we don't have an example of such modification that produced a viable base oil, the possibility of producing one through a directed development effort cannot be discounted.

Examination of the data in Table 12.4 indicates that MGLs have the potential to be used in lubricant formulations as additives. Based on the properties indicated in Table 12.4, there are three additive functions available for glycolipids without subjecting them to chemical and/or enzymatic modification. These are friction modifiers, emulsifiers, and biocides.

To be used as friction modifier additive, the material must have functional groups to render it polar. It also must have good solubility in base oils, both petroleum based (e.g., hexadecane) and biobased (e.g., soybean oil). As shown in Figure 12.1, glycolipids comprise a wide range of functional groups in their structures including hydroxy and carboxylic acids and esters. As shown in Table 12.4, glycolipids have excellent solubility in base oils. Thus, because of their polarity and good solubility in base oils, glycolipids are suitable for use as friction modifiers. The function of friction modifiers is to adsorb on friction surfaces and reduce friction and wear. Previous work [103,114–116] has shown excellent friction modifier property for a number of biobased materials, such as vegetable oils, fatty acids, FAMEs, and estolides, that have excellent solubility in base oils and also contain functional groups (esters, hydroxyl, epoxy groups, etc.). Thus, glycolipids will be part of these biobased materials with excellent friction modifier additive properties.

MGLs are surface active as demonstrated by their ability to lower the surface tension of water, form micelles in aqueous solutions, display a critical micellar concentration, and lower the interfacial tension between water and oils (Table 12.4) [117–120]. This property allows MGLs to be used as emulsifiers, which is an important group of additives for use in lubricant formulations. Emulsifiers allow oil-based lubricant formulations to be dispersed in water and produce water-based lubrication fluids [121,122]. Water-based lubricants are used in a variety of lubrication processes including as metalworking and hydraulic fluids. They are particularly important in lubrication processes that generate a large amount of heat (e.g., hot rolling of steel, aluminum). Such processes cannot be safely carried out with oil-based lubricants due to a potential fire hazard. The water-based lubricant, besides providing the lubrication needs (lower friction and lower wear), is also used as a coolant to remove the heat generated during the process and allow the lubrication process to be carried out safely. Depending on the requirement of the lubrication process, water-based fluid formulations can be categorized as dispersions of oil in water, w/o emulsions, w/o microemulsions, or micellar solutions [121,122]. MGLs of varying surface-active properties, by themselves or in combination with other emulsifiers (surfactants and co-surfactants), can be used to formulate any one of the listed categories of water-based lubricant formulations.

The third potential application area for MGLs is as biocides. As shown in Table 12.4, MGLs have excellent antimicrobial properties for a variety of microorganisms [123,124]. Lubrication fluids, particularly water-based metalworking fluids, operate under conditions of humidity and temperature that allow for the formation and growth of bacteria and other organisms [125]. The use of biobased oils in the lubricant formulations, which are easier to metabolize by many microorganisms, makes the problem worse. Microbial attack causes the lubricant to be degraded, thereby altering its composition, properties (e.g., viscosity), pH, and many other characteristics. As a result, the lubricant will not perform as expected and must be discarded and replaced with a new batch, which is an expensive undertaking. In addition, the presence of bacteria and other microorganisms in the fluid can produce offensive odors and be a potential health hazard to metalworking machine operators. Thus, MGL biocides can be used to prevent and/or control microorganism growth in lubricants and thereby allow for the effective use of water-based lubricant fluids.

In addition to the three potential additive functions described earlier, MGLs can be tailored for use for many other additive functions (e.g., antioxidant, antiwear, extreme pressure, PPD). This can be accomplished through directed chemical/enzymatic modification of MGLs and/or their saccharide/lipid moieties. Currently, there are no reports where such modifications have been undertaken to produce new biobased additives.

12.5 SUMMARY

MGLs are ecologically friendly biobased materials with a promising tribology application. They are amphiphilic molecules with surface-active property useful as surfactant and emulsifier. Their antimicrobial activity can be exploited to prevent microbial contamination and degradation of biobased lubricants. Fermentation process technologies have been developed to biosynthesize the MGLs in various configurations having different components and substituent groups using renewable fermentative feedstocks. Chemical and enzymatic modification methods have also been devised to further impart variation to the molecules. Process model analyses showed that the estimated production costs for these materials are not prohibitive. We are thus optimistic that continued R&D will result in the use of these ecologically friendly and sustainable materials in biobased lubricant formulations.

REFERENCES

1. S. Shekhar, A. Sundaramanickam, and T. Balasubramanian, Biosurfactant producing microbes and their potential applications: A review, *Crit. Rev. Environ. Sci. Technol.*, 45(14), 1522–1554 (2015).
2. E.J. Gudiña, V. Rangarajan, R. Sen, and L.R. Rodrigues, Potential therapeutic applications of biosurfactants, *Trends Pharmacol. Sci.*, 34(12), 667–675 (2013).
3. T. Morita, T. Fukuoka, T. Imura, and D. Kitamoto, Production of mannosylerythritol lipids and their application in cosmetics, *Appl. Microbiol. Biotechnol.*, 97 (11), 4691–4700 (2013).
4. R. Marchant and I.M. Banat, Microbial biosurfactants: Challenges and opportunities for future exploitation, *Trends Biotechnol.*, 30(11), 558–565 (2012).
5. G. Dey, R. Bharti, R. Sen, and M. Mandal, Microbial amphiphiles: A class of promising new-generation anticancer agents, *Drug Discovery Today*, 20(1), 136–146 (2015).
6. X. Zhang, X. Fan, D.K.Y. Solaiman, R.D. Ashby, Z. Liu, S. Mukhopadhyay, and R. Yan, Inactivation of *Escherichia coli* O157:H7 *in vitro* and on the surface of spinach leaves by biobased antimicrobial surfactants, *Food Control*, 60, 158–165 (2016).
7. D.K.Y. Solaiman, R.D. Ashby, J.A. Zerkowski, A. Krishnama, and N. Vasanthan, Control-release of antimicrobial sophorolipid employing different biopolymer matrices, *Biocat. Agric. Biotechnol.*, 4, 342–348 (2015).
8. T. Stipcevic, A. Piljac, and G. Piljac, Enhanced healing of full-thickness burn wounds using di-rhamnolipid, *Burns*, 32, 24–34 (2006).
9. T. Piljac and G. Piljac, Use of rhamnolipids in wound healing, treating burn shock, atherosclerosis, organ transplants, depression, schizophrenia and cosmetics, U.S. Patent 7,262,171 (2007).
10. R.D. Ashby, A.J. McAloon, D. Solaiman, W.C. Yee, and M.L. Reed, A process model for approximating the production costs of the fermentative synthesis of sophorolipids, *J. Surf. Deterg.*, 16, 683–691 (2013).
11. M. Henkel, M.M. Müller, J.H. Kügler, R.B. Lovaglio, J. Contiero, C. Syldatk, and R. Hausmann, Rhamnolipids as biosurfactants from renewable resources: Concepts for next-generation rhamnolipid production, *Proc. Biochem.*, 47, 1207–1219 (2012).

12. Transparency Market Research, Microbial biosurfactants market (rhamnolipids, sopho-rolipids, mannosylerythritol lipids (MEL) and other) for household detergents, indus-trial & institutional cleaners, personal care, oilfield chemicals, agricultural chemicals, food processing, textile and other applications—Global industry analysis, size, share, growth, trends and forecast, 2014–2020 (2014).

13. L. Dobler, L.F. Vilela, R.V. Almeida, and B.C. Neves, Rhamnolipids in Perspective: Gene regulatory pathways, metabolic engineering, production and technological fore-casting, *New Biotechnol.*, 33(1), 123–135 (2016).

14. D. Hayes, D. Kitamoto, D. Solaiman, and R. Ashby (Eds.), *Biobased Surfactants and Detergents: Synthesis, Properties, and Applications*, AOCS Publishing, Urbana, IL (2010).

15. M.M. Müller, J.H. Kügler, M. Henkel, M. Gerlitzki, B. Hörmann, M. Pöhnlein, C. Syldatk, and R. Hausmann, Rhamnolipids—Next generation surfactants?, *J. Biotechnol.*, 162, 366–380 (2012).

16. I.N.A. Van Bogaert, J. Zhang, and W. Soetaert, Microbial synthesis of sophorolipids, *Proc. Biochem.*, 46, 821–833 (2011).

17. C.P. Kurtzman, N.P.J. Price, K.J. Ray, and T.M. Kuo, Production of sophorolipid bio-surfactants by multiple species of *Starmerella (Candida) bombicola* yeast clade, *FEMS Microbiol. Lett.*, 311, 140–146 (2010).

18. H.-J. Asmer, S. Lang, F. Wagner, and V. Wray, Microbial production, structure eluci-dation and bioconversion of sophorose lipids, *J. Am. Oil Chem. Soc.*, 65, 1460–1466 (1988).

19. A. Nuñez, R. Ashby, T.A. Foglia, and D.K.Y. Solairnan, Analysis and characterization of sophorolipids by liquid chromatography with atmospheric pressure chemical ionization, *Chromatographia*, 53, 673–677 (2001).

20. A.P. Tulloch, J.F.T. Spencer, and M.H. Deinema, A new hydroxy fatty acid sophoroside from *Candida bogoriensis*, *Can. J. Chem.*, 46, 345–348 (1968).

21. A. Nuñez, R. Ashby, T.A. Foglia, and D.K.Y. Solairnan, LC/MS analysis and lipase modification of the sophorolipids produced by *Rhodotorula bogoriensis*, *Biotechnol. Lett.*, 26(12), 1087–1093 (2004).

22. R.R. Saikia, H. Deka, D. Goswami, J. Lahkar, S.N. Borah, K. Patowary, P. Baruah, and S. Deka, Achieving the best yield in glycolipid biosurfactant preparation by selecting the proper carbon/nitrogen ratio, *J. Surfact. Deterg.*, 17, 563–571 (2013).

23. J.I. Arutchelvi, S. Bhaduri, P.V. Uppara, and M. Doble, Mannosylerythritol lipids: A review, *J. Ind. Microbiol. Biotechnol.*, 35, 1559–1570 (2008).

24. M. Yu, Z. Liu, G. Zeng, H. Zhong, Y. Liu, Y. Jiang, M. Li, X. He, and Y. He, Characteristics of mannosylerythritol lipids and their environmental potential, *Carbohydr. Res.*, 407, 63–72 (2015).

25. J. Scheibel, The impact of feedstocks on future innovation in surfactant technology for the detergent market, in: *98th American Oil Chemists' Society Meeting*, Quebec City, Quebec, Canada (2007).

26. P.A.J. Gorin, J.F.T. Spencer, and A.P. Tulloch, Hydroxy fatty acid glycosides of sopho-rose from *Torulopsis magnoliae*, *Can. J. Chem.*, 39, 846–855 (1961).

27. A.P. Tulloch, J.F.T. Spencer, and P.A.J. Gorin, The fermentation of long-chain com-pounds by *Torulopsis magnoliae*, *Can. J. Chem.*, 40, 1326–1338 (1962).

28. A.P. Tulloch and J.F.T. Spencer, Fermentation of long-chain compounds by *Torulopsis apicola*. IV. Products from esters and hydrocarbons with 14 and 15 carbon atoms and from methyl palmitoleate, *Can. J. Chem.*, 46, 1523–1528 (1968).

29. D. Yarrow and S.A. Meyer, Proposal for amendment of the diagnosis of the genus *Candida* Berkhout nom. cons., *Int. J. Syst. Bacteriol.*, 28, 611–615 (1978).

30. J.A. Von Arx and A.C.M. Weijman, Conidiation and carbohydrate composition in some *Candida* and *Torulopsis* species, *Antonie Van Leeuwenhoek*, 45, 547–555 (1979).

31. J.F.T. Spencer, P.A.J. Gorin, and A.P. Tulloch, *Torulopsis bombicola* sp. n., *Antonie Van Leeuwenhoek*, 36, 129–133 (1970).

32. C.A. Rosa and M.-A. Lachance, The yeast genus *Starmerella* gen. nov. and *Starmerella bombicola* sp. nov., the teleomorph of *Candida bombicola* (Spencer, Gorin & Tullock) Meyer & Yarrow, *Int. J. Syst. Bacteriol.*, 48, 1413–1417 (1998).

33. J. Chen, X. Song, H. Zhang, Y.B. Qu, and J.Y. Miao, Production, structure elucidation and anticancer properties of sophorolipid from *Wickerhamiella domercqiae*, *Enzyme Microb. Technol.*, 39, 501–506 (2006).

34. M. Konishi, T. Fukuoka, T. Morita, T. Imura, and D. Kitamoto, Production of new types of sophorolipids by *Candida batistae*, *J. Oleo Sci.*, 57, 359–369 (2008).

35. D.R. Nelson, Cytochrome P450 nomenclature, *Meth. Mol. Biol.*, 107, 15–24 (1998).

36. T.B. Breithaupt and R.J. Light, Affinity-chromatography and further characterization of the glucosyltransferases involved in hydroxydocosanoic acid sophoroside production in *Candida bogoriensis*, *J. Biol. Chem.*, 257, 9622–9628 (1982).

37. T.W. Esders and R.J. Light, Glucosyl- and acetyltransferases involved in the biosynthesis of glycolipids from *Candida bogoriensis*, *J. Biol. Chem.*, 247, 1375–1386 (1972).

38. H.-J. Daniel, M. Reuss, and C. Syldatk, Production of sophorolipids in high concentration from deproteinized whey and rapeseed oil in a two stage fed batch process using *Candida bombicola* ATCC 22214 and *Cryptococcus curvatus* ATCC 20509, *Biotechnol. Lett.*, 20, 1153–1156 (1998).

39. G. Pekin, F. Vardar-Sukan, and N. Kosaric, Production of sophorolipids from *Candida bombicola* ATCC 22214 using Turkish corn oil and honey, *Eng. Life Sci.*, 5, 357–362 (2005).

40. R.D. Ashby, A. Nuñez, D.K.Y. Solaiman, and T.A. Foglia, Sophorolipid biosynthesis from a biodiesel co-product stream, *J. Am. Oil Chem. Soc.*, 82, 625–630 (2005).

41. D.K.Y. Solaiman, R.D. Ashby, A. Nuñez, and T.A. Foglia, Production of sophorolipids by *Candida bombicola* grown on soy molasses as substrate, *Biotechnol. Lett.*, 26, 1241–1245 (2004).

42. D.K.Y. Solaiman, R.D. Ashby, J.A. Zerkowski, and T.A. Foglia, Simplified soy molasses-based medium for reduced-cost production of sophorolipids by *Candida bombicola*, *Biotechnol. Lett.*, 29, 1341–1347 (2007).

43. F.G. Jarvis and M.J. Johnson, A glyco-lipide produced by *Pseudomonas aeruginosa*, *J. Am. Chem. Soc.*, 71, 4124–4126 (1949).

44. B.K. Tuleva, G.R. Ivanov, and N.E. Christova, Biosurfactant production by a new *Pseudomonas putida* strain, *Z. Naturforsch.*, 57, 356–360 (2002).

45. N.W. Gunther, A. Nunez, W. Fett, and D.K.Y. Solaiman, Production of rhamnolipids by *Pseudomonas chlororaphis*, a nonpathogenic bacterium, *Appl. Environ. Microbiol.*, 71, 2288–2293 (2005).

46. S. Haussler, M. Rohde, N. von Neuhoff, M. Nimitz, and I. Steinmetz, Structural and functional cellular changes induced by *Burkholderia pseudomallei* rhamnolipid, *Infect. Immun.*, 71, 2970–2975 (2003).

47. B. Hormann, M.M. Muller, C. Syldatk, and R. Hausmann, Rhamnolipid production by *Burkholderia plantarii* DSM 9509, *Eur. J. Lipid Sci. Technol.*, 112, 674–680 (2010).

48. A.M. Abdel-Mawgoud, F. Lepine, and E. Deziel, Rhamnolipids: Diversity of structures, microbial origins and roles, *Appl. Microbiol. Biotechnol.*, 86, 1323–1336 (2010).

49. K. Zhu and C.O. Rock, RhlA converts β-hydroxyacyl-acyl carrier protein intermediates in fatty acid synthesis to the β-hydroxydecanoyl-β-hydroxydecanoate component of rhamnolipids in *Pseudomonas aeruginosa*, *J. Bacteriol.*, 190, 3147–3154 (2008).

50. L. Zulianello, C. Canard, T. Kohler, D. Caille, J.-S. Lacroix, and P. Meda, Rhamnolipids are virulence factors that promote early infiltration of primary human airway epithelial by *Pseudomonas aeruginosa*, *Infect. Immun.*, 74, 3134–3147 (2006).

51. E. Haba, A. Pinazo, O. Jauregui, M.J. Espuny, M.R. Infante, and A. Manresa, Physicochemical characterization and antimicrobial properties of rhamnolipids produced by *Pseudomonas aeruginosa* 47T2 NCBIM 40044, *Biotechnol. Bioeng.*, 81, 316–322 (2003).

52. M.E. Davey, N.C. Caiazza, and G.A. O'Toole, Rhamnolipid surfactant production affects biofilm architecture in *Pseudomonas aeruginosa* PAO1, *J. Bacteriol.*, 185, 1027–1036 (2003).

53. N.C. Caiazza, R.M.Q. Shanks, and G.A. O'Toole, Rhamnolipids modulate swarming motility patterns of *Pseudomonas aeruginosa*, *J. Bacteriol.*, 187, 7351–7361 (2005).

54. R.M. Maier and G. Soberon-Chavez, *Pseudomonas aeruginosa* rhamnolipids: Biosynthesis and potential applications, *Appl. Microbiol. Biotechnol.*, 54, 625–633 (2000).

55. J. Campos-Garcia, A.D. Caro, R. Najera, R.M. Miller-Maier, R.A. Al-Tahhan, and G. Soberon-Chavez, The *Pseudomonas aeruginosa rhlG* gene encodes an NADPH-dependent β-ketoacyl reductase which is specifically involved in rhamnolipid synthesis, *J. Bacteriol.*, 180, 4442–4451 (1998).

56. E. Deziel, F. Lepine, S. Milot, and R. Villemur, *rhlA* is required for the production of a novel biosurfactant promoting swarming motility in *Pseudomonas aeruginosa*: 3-(3-hydroxyalkanoyloxy)alkanoic acids (HAAs), the precursors of rhamnolipids, *Microbiology*, 149, 2005–2013 (2003).

57. R. Rahim, L.L. Burrows, M.A. Monteiro, M.B. Perry, and J.S. Lam, Involvement of the *rml* locus in core oligosaccharide and O polysaccharide assembly in *Pseudomonas aeruginosa*, *Microbiology*, 146, 2803–2814 (2000).

58. U.A. Ochsner, A. Fiechter, and J. Reiser, Isolation, characterization, and expression in *Escherichia coli* of the *Pseudomonas aeruginosa rhlAB* genes encoding a rhamnosyltransferase involved in rhamnolipid biosurfactant synthesis, *J. Biol. Chem.*, 269, 19787–19795 (1994).

59. R. Rahim, U.A. Ochsner, C. Olvera, M. Graninger, P. Messner, J.S. Lam, and G. Soberon-Chavez, Cloning and functional characterization of the *Pseudomonas aeruginosa rhlC* gene that encodes rhamnosyltransferase 2, an enzyme responsible for di-rhamnolipid biosynthesis, *Mol. Microbiol.*, 40, 708–718 (2001).

60. B. Boothroyd, J.A. Thorn, and R.H. Haskins, Biochemistry of the ustilaginales: XII. characterization of extracellular glycolipids produced by *Ustilago* sp., *Can. J. Biochem. Physiol.*, 34, 10–14 (1956).

61. T. Morita, M. Konishi, T. Fukuoka, T. Imura, H.K. Kitamoto, and D. Kitamoto, Characterization of the genus *Pseudozyma* by the formation of glycolipid biosurfactants, mannosylerythritol lipids, *FEMS Yeast Res.*, 7, 286–292 (2007).

62. T. Morita, T. Fukuoka, T. Imura, and D. Kitamoto, Production of glycolipid biosurfactants by basidiomycetous yeasts, *Biotechnol. Appl. Biochem.*, 53, 39–49 (2009).

63. T. Morita, Y. Ishibashi, T. Fukuoka, T. Imura, H. Sakai, M. Abe, and D. Kitamoto, Production of glycolipid biosurfactants, mannosylerythritol lipids, using sucrose by fungal and yeast strains, and their interfacial properties, *Biosci. Biotechnol. Biochem.*, 73, 2352–2355 (2009).

64. S. Spoeckner, V. Wray, M. Nimitz, and S. Lang, Glycolipids of the smut fungus *Ustilago maydis* from cultivation on renewable resources, *Appl. Microbiol. Biotechnol.*, 51, 33–39 (1999).

65. U. Rau, L.A. Nguyen, S. Schulz, V. Wray, M. Nimitz, H. Roeper, H. Koch, and S. Lang, Formation and analysis of mannosylerythritol lipids secreted by *Pseudozyma aphidis*, *Appl. Microbiol. Biotechnol.*, 66, 551–559 (2005).

66. S. Hewald, U. Linne, M. Scherer, M.A. Marahiel, J. Kamper, and M. Bolker, Identification of a gene cluster for biosynthesis of mannosylerythritol lipids in the basidiomycetous fungus *Ustilago maydis*, *Appl. Environ. Microbiol.*, 72, 5469–5477 (2006).

67. T. Morita, H. Koike, Y. Koyama, H. Hagiwara, E. Ito, T. Fukuoka, T. Imura, M. Machida, and D. Kitamoto, Genome sequence of the basidiomycetous yeast *Pseudozyma antarctica* T-34, a producer of the glycolipid biosurfactants mannosylerythritol lipids, *Genome Announc.*, 1, e00064-13 (2013).

68. M. Konishi, Y. Hatada, and J. Horiuchi, Draft genome sequence of the basidiomycetous yeast-like fungus *Pseudozyma hubeiensis* SY62, which produces an abundant amount of the biosurfactant mannosylerythritol lipids, *Genome Announc.*, 1, e00409-13 (2013).

69. D. Kitamoto, H. Isoda, and T. Nakahara, Functions and potential applications of glycolipid biosurfactants from energy-saving materials to gene delivery carriers, *J. Biosci. Bioengin.*, 94, 187–201 (2002).

70. D. Kitamoto, H. Yanagishita, K. Haraya, and K. Kitamoto, Contribution of a chain-shortening pathway to the biosynthesis of the fatty acids of mannosylerythritol lipid (biosurfactant) in the yeast *Candida antarctica*: Effect of β-oxidation inhibitors on biosurfactant synthesis, *Biotechnol. Lett.*, 20, 813–818 (1998).

71. R.J.A. Wander, P. Vreken, S. Ferdiandusse, G.A. Jansen, H.R. Waterham, C.W.T. van Roermund, and E.G. van Grunsven, Peroxisomal fatty acid α- and β-oxidation in humans: Enzymology, peroxisomal metabolite transporters and peroxisomal diseases, *Biochem. Soc. Trans.*, 29, 250–267 (2001).

72. M. Pöhnlein, R. Hausmann, S. Lang, and C. Syldatk, Enzymatic synthesis and modification of surface-active glycolipids, *Eur. J. Lipid Sci. Technol.*, 117(2), 145–155 (2015).

73. J.A. Zerkowski, D.K.Y. Solaiman, R.D. Ashby, and T.A. Foglia, Head group-modified sophorolipids: Synthesis of new cationic, zwitterionic, and anionic surfactants, *J. Surfact. Deterg.*, 9(1), 57–62 (2006).

74. J.A. Zerkowski, D.K.Y. Solaiman, R.D. Ashby, and T.A. Foglia, Charged sophorolipids and sophorolipid containing compounds, U.S. Patent Number 7,718,782 (2010).

75. Y. Peng, F. Totsingan, M.A.R. Meier, M. Steinmann, F. Wurm, A. Koh, and R.A. Gross, Sophorolipids: Expanding structural diversity by ring-opening cross-metathesis, *Eur. J. Lipid Sci. Technol.*, 117(2), 217–228 (2015).

76. K.S. Bisht, R.A. Gross, and D.L. Kaplan, Enzyme-mediated regioselective acylations of sophorolipids, *J. Org. Chem.*, 64(3), 780–789 (1999).

77. S.K. Singh, A.P. Felse, A. Nunez, T.A. Foglia, and R.A. Gross, Regioselective enzyme-catalyzed synthesis of sophorolipid esters, amides, and multifunctional monomers, *J. Org. Chem.*, 68(14), 5466–5477 (2003).

78. V.K. Recke, M. Gerlitzki, R. Hausmann, C. Syldatk, V. Wray, H. Tokuda, N. Suzuki, and S. Lang, Enzymatic production of modified 2-dodecyl-sophorosides (biosurfactants) and their characterization, *Eur. J. Lipid Sci. Technol.*, 115(4), 452–463 (2013).

79. M.M. Müller, B. Hörmann, C. Syldatk, and R. Hausmann, *Pseudomonas aeruginosa* PAO1 as a model for rhamnolipid production in bioreactor systems, *Appl. Microbiol. Biotechnol.*, 87(1), 167–174 (2010).

80. T. Imura, N. Ohta, K. Inoue, and N. Yagi, Naturally engineered glycolipid biosurfactants leading to distinctive self-assembled structures, *Chem. Eur. J.*, 12, 2434–2440 (2006).

81. T. Fukuoka, T. Yanagihara, T. Imura, and T. Morita, Enzymatic synthesis of a novel glycolipid biosurfactant, mannosylerythritol lipid-D and its aqueous phase behavior, *Carbohydr. Res.*, 346, 266–271 (2011).

82. V.K. Recke, C. Beyrle, M. Gerlitzki, R. Hausmann, C. Syldatk, V. Wray, H. Tokuda, N. Suzuki, and S. Lang, Lipase-catalyzed acylation of microbial mannosylerythritol lipids (biosurfactants) and their characterization, *Carbohydr. Res.*, 373, 82–88 (2013).

83. A. Azim, V. Shah, G.F. Doncel, N. Peterson, W. Gao, and R. Gross, Amino acid conjugated sophorolipids: A new family of biologically active functionalized glycolipids, *Bioconjug. Chem.*, 17, 1523–1529 (2006).

84. E.I.P. Delbeke, B.I. Roman, G.B. Marin, K.M. Van Geem, and C.V. Stevens, A new class of antimicrobial biosurfactants: Quaternary ammonium sophorolipids, *Green Chem.*, 17(6), 3373–3377 (2015).

85. E. Zini, M. Gazzano, M. Scandola, S.R. Wallner, and R.A. Gross, Glycolipid biomaterials: Solid-state properties of a poly(sophorolipid), *Macromolecules*, 41(20), 7463–7468 (2008).

86. Y. Peng, D.J. Munoz-Pinto, M. Chen, J. Decatur, M. Hahn, and R.A. Gross, Poly(sophorolipid) structural variation: Effects on biomaterial physical and biological properties, *Biomacromolecules*, 15(11), 4214–4227 (2014).

87. M.C. McManus, G.P. Hammond, and C.R. Burrows, Life-cycle assessment of mineral and rapeseed oil in mobile hydraulic systems, *J. Ind. Ecol.*, 7, 163–177 (2004).

88. Anon., Lubricants industry factbook, *Lubes N Greases*, 21(Suppl. 8), 1–52 (2015).

89. M.P. Schneider, Plant-oil-based lubricants and hydraulic fluids, *J. Sci. Food Agric.*, 86, 1769–1780 (2006).

90. M.A. Schmidt, C.R. Dietrich, and E.B. Cahoon, Biotechnological enhancement of soybean oil for lubricant applications, in: *Synthetics, Mineral Oils, and Bio-Based Lubricants. Chemistry and Technology*, L.R. Rudnick (Ed.), Chapter 23, pp. 389–397, CRC Press, Taylor & Francis Group, Boca Raton, FL (2006).

91. S.Z. Erhan, A. Adhvaryu, and B.K. Sharma, Chemically functionalized vegetable oils, in: *Synthetics, Mineral Oils, and Bio-Based Lubricants. Chemistry and Technology*, L.R. Rudnick (Ed.), Chapter 22, pp. 361–387, CRC Press, Taylor & Francis Group, Boca Raton, FL (2006).

92. B.J. Bremmer and L. Plonsker, *Bio-Based Lubricants: A Market Opportunity Study Update*, Omni-Tech International, Midland, MI (2008).

93. J.C.J. Bart, E. Gucciardi, and S. Cavallaro, Renewable feedstocks for lubricant production, in: *Woodhead Publishing Series in Energy, Biolubricants*, Chapter 5, pp. 121–248, Woodhead Publishing, Swaston, Cambridge, U.K. (2013).

94. P. Ghosh, T. Das, D. Nandi, G. Karmakar, and A. Mandal, Synthesis and characterization of biodegradable polymer—Used as a pour point depressant for lubricating oil, *Int. J. Polym. Mater. Polym. Biomater.*, 59(12), 1008–1017 (2010).

95. P. Ghosh, T. Das, G. Karmakar, and M. Das, Evaluation of acrylate-sunflower oil copolymer as viscosity index improvers for lube oils, *J. Chem. Pharm. Res.*, 3(3), 547–556 (2011).

96. P. Ghosh, M. Das, M. Upadhyay, T. Das, and A. Mandal, Synthesis and evaluation of acrylate polymers in lubricating oil, *J. Chem. Eng. Data*, 56, 3752–3758 (2011).

97. S. Khalkar, D.N. Bhowmick, and A. Pratap, Synthesis of polymers from fatty alcohol and acrylic acid and its impact on tribological properties, *J. Oleo Sci.*, 62(3), 167–173 (2013).

98. P. Ghosh and G. Karmakar, Evaluation of sunflower oil as a multifunctional lubricating oil additive, *Int. J. Ind. Chem.*, 5, 7–16 (2014).

99. A. Beck, M. Bubálik, and J. Hancsók, Development of multifunctional detergent-dispersant additives based on fatty acid methyl ester for diesel and biodiesel fuel, in: *Biodiesel—Quality, Emissions and By-Products*, G. Montero (Ed.), pp. 153–170, InTech, Rijeka, Croatia (2011).

100. J. Hancsók, M. Bubálik, and Á. Beck, Development of multifunctional additives based on vegetable oils for high quality diesel and biodiesel, in: *Proceedings of European Congress of Chemical Engineering (ECCE-6)*, Copenhagen, Denmark, pp. 1–11 (2007).

101. Y. Wang and W. Eli, Synthesis of environmentally friendly over-based magnesium oleate detergent and high alkaline dispersant/magnesium oleate mixed substrate detergent, *Ind. Eng. Chem. Res.*, 49, 8902–8907 (2010).

102. K.C. Dohhen, H.C. Bhatia, K.K. Swami, R. Sarin, D.K. Tuli, M.M. Rai, and A.K. Bhatnagar, Process for the preparation of calcium phenate detergents from cashew nut shell liquid, U.S. Patent 5,910,468 A (1999).

103. G. Biresaw, A. Adhvaryu, S.Z. Erhan, and C.J. Carriere, Friction and adsorption properties of normal and high oleic soybean oils, *J. Am. Oil Chem. Soc.*, 79(1), 53–58 (2002).

104. G.B. Bantchev, J.A. Kenar, G. Biresaw, and M. Han, Free radical addition of butanethiol to vegetable oil double bonds, *J. Agric. Food Chem.*, 57(4), 1282–1290 (2009).

105. G. Biresaw, G.B. Bantchev, and S.C. Cermak, Tribological properties of vegetable oils modified by reaction with butanethiol, *Tribol. Lett.*, 43(1), 17–32 (2011).

106. G. Biresaw and G.B. Bantchev, Tribological properties of biobased ester phosphonates, *J. Am. Oil Chem. Soc.*, 90(6), 891–902 (2013).

107. G. Biresaw, J.A. Laszlo, K.O. Evans, D.L. Compton, and G.B. Bantchev, Tribological investigation of lipoyl glycerides, *J. Agric. Food Chem.*, 62(10), 2233–2243 (2014).

108. G. Biresaw, D. Compton, K. Evans, and G. Bantchev, Lipoate ester multi-functional lubricant additives, *Ind. Eng. Chem. Res.*, 55, 373–383 (2016).

109. S. Cermak and T. Isbell, Estolides: The next biobased functional fluid, *Inform*, 15(8), 515–517 (2004).

110. T. Isbell and S. Cermak, Synthesis of triglyceride estolides from lesquerella and castor oils, *J. Am. Oil Chem. Soc.*, 79(12), 1227–1233 (2002).

111. M. Bomgardner, Smooth running with soybeans, *Chemical Engineering News*, pp. 19–21, October 28 (2013).

112. C. Challener, Green lubricants continue progress, *ICIS Chemical Business*, pp. 21–24, February 17–23 (2014).

113. L.R. Rudnick (Ed.), *Lubricant Additives Chemistry and Applications*, 2nd edn., CRC Press, Boca Raton, FL (2009).

114. A. Adhvaryu, G. Biresaw, B.K. Sharma, and S.Z. Erhan, Friction behavior of some seed oils: Bio-based lubricant applications, *Ind. Eng. Chem. Res.*, 45(10), 3735–3740 (2006).

115. T.L. Kurth, J.A. Byars, S.C. Cermak, B.K. Sharma, and G. Biresaw, Non-linear adsorption modeling of fatty esters and oleic estolides via boundary lubrication coefficient of friction measurements, *Wear*, 262(5–6), 536–544 (2007).

116. K.O. Evans and G. Biresaw, Quartz crystal microbalance investigation of the structure of adsorbed soybean oil and methyl oleate onto steel surface, *Thin Solid Films*, 519(2), 900–905 (2010).

117. D.I. Comas, J.R. Wagner, and M.C. Tomas, Creaming stability of oil in water (O/W) emulsion: Influence of pH on soybean protein–lecithin interaction, *Food Hydrocoll.*, 20, 990–996 (2006).

118. C.-L. Xue, D.K.Y. Solaiman, R.D. Ashby, J. Zerkowski, J.-H. Lee, S.-T. Hong, D. Yang, J.-A. Shin, C.-M. Ji, and K.-T. Lee, Study of structured lipid-based oil-in-water emulsion prepared with sophorolipid and its oxidative stability, *J. Am. Oil Chem. Soc.*, 90, 123–132 (2013).

119. S.L.K.W. Roelants, K. Ciesielska, S.L. De Maeseneire, H. Moens, B. Everaert, S. Verweire, Q. Denon et al., Towards the industrialization of new biosurfactants: Biotechnological opportunities for the lactone esterase gene from *Starmerella bombicola*, *Biotechnol. Bioeng.*, 113(3), 550–559 (2016).

120. D.W.G. Develter and L.M.L. Lauryssen, Properties and industrial applications of sophorolipids, *Eur. J. Lipid Sci. Technol.*, 112(6), 628–638 (2010).

121. J.P. Byers, *Metalworking Fluids*, 487pp., Marcel Decker, New York (1994).

122. J.A. Schey, *Tribology in Metalworking Friction, Lubrication and Wear*, American Society of Metals, Metals Park, OH (1983)

123. J.N. Sleiman, S.A. Kohlhoff, P.M. Roblin, S. Wallner, R. Gross, M.R. Hammerschlag, M.E. Zenilman, and M.H. Bluth, Sophorolipids as antibacterial agents, *Ann. Clin. Lab. Sci.*, 39(1), 60–63 (2009).

124. K. Kim, D. Yoo, Y. Kim, B. Lee, D. Shin, and E.-K. Kim, Characteristics of sophorolipid as an antimicrobial agent, *J. Microbiol. Biotechnol.*, 12(2), 235–241 (2002).

125. W.R. Schwingel and A.C. Eachus, Antimicrobial additives for metalworking fluids, in: *Lubricant Additives Chemistry and Applications*, 2nd edn., L.R. Rudnick (Ed.), pp. 383–397, CRC Press, Boca Raton, FL (2009).

126. I.N.A. van Bogaert, K. Saerens, C. de Muynck, D. Develter, W. Soetaert, and E.J. Vandamme, Microbial production and application of sophorolipids, *Appl. Microbiol. Biotechnol.*, 76, 23–34 (2007).

127. A.-M. Davila, R. Marchal, and J.-P. Vandercasteele, Sophorose lipid production from lipidic precursors: Predictive evaluation of industrial substrates, *J. Ind. Microbiol.*, 13, 249–257 (1994).

128. Q. Zhou, V. Kleckner, and N. Kosaric, Production of sophorose lipids by *Torulopsis bombicola* from safflower oil and glucose, *J. Am. Oil Chem. Soc.*, 69, 89–91 (1992).

129. Q. Zhou and N. Kosaric, Utilization of canola oil and lactose to produce biosurfactant with *Candida bombicola*, *J. Am. Oil Chem. Soc.*, 72, 67–71 (1995).

130. M. Deshpande and L. Daniels, Evaluation of sophorolipid biosurfactant production by *Candida bombicola* using animal fat, *Bioresour. Technol.*, 54, 143–150 (1995).

131. U. Rau, C. Manzke, and F. Wagner, Influence of substrate supply on the production of sophorose lipids by *Candida bombicola* ATCC 22214, *Biotechnol. Lett.*, 18, 149–154 (1996).

132. S. Fleurackers, On the use of waste frying oil in the synthesis of sophorolipids, *Eur. J. Lipid Sci. Technol.*, 108, 5–12 (2006).

133. R.D. Ashby, D.K.Y. Solaiman, and T.A. Foglia, Property control of sophorolipids: Influence of fatty acid substrate and blending, *Biotechnol. Lett.*, 30, 1093–1100 (2008).

134. Y. Hu and L.-K. Ju, Sophorolipid production from different lipid precursors observed with LC-MS, *Enz. Microb. Technol.*, 29, 593–601 (2001).

135. A. Brakemeier, D. Wullbrandt, and S. Lang, *Candida bombicola*: Production, of novel alkyl glucosides based on glucose/2-dodecanol, *Appl. Microbiol. Biotechnol.*, 50, 161–166 (1998).

136. U. Rau, L.A. Nguyen, H. Roeper, H. Koch, and S. Lang, Fed-batch bioreactor production of mannosylerythritol lipids secreted by *Pseudozyma aphidis*, *Appl. Microbiol. Biotechnol.*, 68, 607–613 (2005).

137. T. Morita, M. Konishi, T. Fukuoka, T. Imura, and D. Kitamoto, Discovery of *Pseudozyma rugulosa* NBRC 10877 as a novel producer of the glycolipid biosurfactants, mannosylerythritol lipids, based on rDNA sequence, *Appl. Microbiol. Biotechnol.*, 73, 305–313 (2006).

138. D. Kitamoto, T. Ikegami, G.T. Suzuki, A. Sasaki, Y. Takeyama, Y. Idemoto, N. Koura, and H. Yanagishita, Microbial conversion of *n*-alkanes into glycolipid biosurfactants, mannosylerythritol lipids, by *Pseudozyma* (*Candida*) *antarctica*, *Biotechnol. Lett.*, 23, 1709–1714 (2001).

139. T. Morita, M. Konishi, T. Fukuoka, T. Imura, H. Sakai, and D. Kitamoto, Efficient production of Di- and Tri-acetylated mannosylerythritol lipids as glycolipid biosurfactants by *Pseudozyma parantarctica* JCM 11752, *J. Oleo Sci.*, 57, 557–565 (2008).

140. R.T. Otto, H.-J. Daniel, G. Pekin, K. Müller-Decker, G. Fürstenberger, M. Reuss, and C. Syldatk, Production of sophorolipids from whey: II. Product composition, surface active properties, cytotoxicity and stability against hydrolases by enzymatic treatment, *Appl. Microbiol. Biotechnol.*, 52(4), 495–501 (1999).

141. H.-S. Kim, J.-W. Jeon, S.-B. Kim, H.-M. Oh, T.-J. Kwon, and B.-D. Yoon, Surface and physico-chemical properties of a glycolipid biosurfactant, mannosylerythritol lipid, from *Candida antarctica*, *Biotechnol. Lett.*, 24, 1637–1641 (2002).

142. D. Kitamoto, H. Yanagishita, T. Shinbo, T. Nakane, C. Kamisawa, and T. Nakahara, Surface active properties and antimicrobial activities of mannosylerythritol lipids as biosurfactants produced by *Candida antarctica*, *J. Biotechnol.*, 29, 91–96 (1993).

143. D. Kitamoto, T. Morita, T. Fukuoka, and T. Imura, Self-assembling properties of glycolipid biosurfactants and their functional developments, in: *Biobased Surfactants and Detergents: Synthesis, Properties, and Applications*, D. Hayes, D. Kitamoto, D. Solaiman, and R. Ashby (Eds.), Chapter 9, pp. 231–272, AOCS Publishing, Urbana, IL (2009).
144. T.T. Nguyen and D.A. Sabatini, Microemulsions of rhamnolipid and sophorolipid biosurfactants, in: *Biobased Surfactants and Detergents: Synthesis, Properties, and Applications*, D. Hayes, D. Kitamoto, D. Solaiman, and R. Ashby (Eds.), Chapter 5, pp. 107–127, AOCS Publishing, Urbana, IL (2009).
145. N.M. Pinzon, Q. Zhang, S. Koganti, and L.-K. Ju, Advances in bioprocess development of rhamnolipid and sophorolipid production, in: *Biobased Surfactants and Detergents: Synthesis, Properties, and Applications*, D. Hayes, D. Kitamoto, D. Solaiman, and R. Ashby (Eds.), Chapter 4, pp. 77–105, AOCS Publishing, Urbana, IL (2009).

13 Biofuels from Vegetable Oils as Alternative Fuels
Advantages and Disadvantages

*Nabel A. Negm, Maram T.H. Abou Kana,
Mona A. Youssif, and Mona Y. Mohamed*

CONTENTS

13.1   TRIBOLOGY AND SURFACE PROPERTIES OF BIODIESEL

Biodiesel, a mixture of fatty acid monoalkyl esters, is obtained by transesterification of vegetable oils, animal fats, or used frying oils, with short-chain alcohols, such as methanol or ethanol, in the presence of a catalyst [1]. Because biodiesel properties strongly depend upon the fatty acid profiles of the feedstock, they can be tuned using raw materials containing components that will provide more favorable properties to biodiesels. Thus, prediction of the physical property for biodiesels makes it possible to optimize biodiesel production and blending processes, with the final aim of improving the fuel performance in the engine, particularly during atomization.

One of the most important properties of biodiesel in the application field is atomization. Atomization is the breakup of bulk liquid jets into small droplets using an atomizer or spray [2]. Adequate atomization enhances mixing and complete combustion in a direct injection engine, and therefore it is an important factor in engine emission and efficiency. This applies to microturbines and gas turbines as witnessed in the need for an atomizer in gas turbines when diesel is being used. A correct atomization allows for proper mixing and complete combustion in an injection engine, reducing emissions and increasing the engine efficiency [3].

Surface tension property has a major impact on fuel atomization, that is, the first stage of combustion [4]. Higher surface tensions make the drop formation of biodiesel difficult, leading to an inefficient fuel atomization. Furthermore, just like most biodiesel properties, surface tension increases with long fatty acid hydrocarbon (HC) chains and a level of unsaturated bonds [5], more unsaturated biodiesel fuels will present a higher surface tension.

The atomization of fuel is crucial in the combustion and emission in an engine, but the atomization processes in an engine and in microturbine are completely different [6]. Both microturbine and diesel engine have the same fundamentals where both operate through combustion, but the principle of the atomization process in both cases varies because the fuel injector for microturbine and diesel engine are not similar. For microturbine, the combustion is continuous, so the fuel atomization in microturbine is continuous without any cycles or strokes. Atomization plays a major role in combustion and emission in a microturbine. By modifying the atomization process, the gas turbine can produce a lower emission of nitrogen oxides (NO_x) and carbon monoxide (CO) [7].

Adequate atomization enhances mixing and complete combustion in a direct injection gas turbine, and therefore it is an important factor in gas turbine emission and efficiency [8]. Otherwise, the properties of a liquid fuel that affect atomization in a gas turbine are viscosity, density, and surface tension. For a gas turbine biodiesel injector at a fixed operating condition, the use of fuel with higher viscosity delays atomization by suppressing the instabilities required for the fuel jet to break up. An increase in fuel density adversely affects atomization, whereby higher fuel surface tension opposes the formation of droplets from the liquid fuel.

The viscosity of the fuel, on the other hand, is of great importance in controlling both the formation of the continuous film immediately after exit from the nozzle and of the subsequent ligament disruption into individual droplets. The viscous forces decrease the rate of breakup and distortions in the liquid and decrease the rate of disruption of the droplets formed initially and increase the final droplet size. The temperature relationships for kinematic viscosity show that vegetable oils have viscosities higher than that of conventional gas oil (diesel), thus tending to produce larger droplets. Viscosity has by far the greatest effect on jet atomization with high-viscosity fuels provoking deterioration in the quality of atomization. Of the relevant fuel properties, density is generally found to have relatively little influence on spray formation.

Biodiesel has more massive fragments and less fine droplets than those of diesel fuel due to its high liquid viscosity, resulting in a high mean droplet size. Consequently, it can be postulated that the breakup characteristic is strongly dominated by not only the surface tension but also the friction flow inside a droplet [9]. To increase the poor atomization of biodiesel fuel compared to diesel due to the larger Sauter mean diameter (SMD), ethanol can be blended together with biodiesel to produce a smaller SMD. This is because ethanol has a lower kinematic viscosity with active interaction with ambient gas. In other words, blending ethanol with biodiesel will enhance atomization characteristics [10].

13.2   DIESEL AND BIODIESEL

Petroleum diesel, also called petrodiesel or fossil diesel, is the most common type of diesel fuel. It is produced by fractional distillation of crude oil between 200°C and 350°C at atmospheric pressure, resulting in a mixture of carbon chains that typically contain between 8 and 21 carbon atoms per molecule.

13.2.1   CHEMICAL COMPOSITION

Petroleum-derived diesel is composed of about 75% saturated HCs (primarily paraffins including *n*-, *iso*-, and cycloparaffins) and 25% aromatic HCs (including naphthalenes and alkyl benzenes). The average chemical formula for common diesel fuel is $C_{12}H_{23}$, ranging approximately from $C_{10}H_{20}$ to $C_{15}H_{28}$. Petroleum is a useful chemical substance for many important purposes, but it is also a nonrenewable resource with a highly toxic composition. It poses significant problems when used in huge volumes throughout the industrialized world.

Biodiesel is an alternative fuel similar to conventional or "fossil" diesel. Biodiesel can be produced from straight vegetable oil, animal oil/fats, tallow, and waste cooking oil. The process used to convert these oils to biodiesel is called transesterification.

The chemical composition of biodiesel is monoalkyl esters of long-chain fatty acids derived from vegetable oils or animal fats.

13.3 TYPES OF BIOFUELS

13.3.1 BIODIESEL

Biodiesel is the most common biofuel in Europe. It is produced from oils or fats using transesterification and is a liquid similar in properties to fossil/mineral diesel [11].

13.3.2 GREEN DIESEL

Green diesel is produced through hydrocracking [12] biological oil feedstocks, such as vegetable oils and animal fats. Hydrocracking is a refinery method that uses elevated temperatures and pressure in the presence of a catalyst to break down larger molecules, such as those found in vegetable oils, into shorter HC chains used in diesel engines. It may also be called renewable diesel, hydrotreated vegetable oil, or hydrogen-derived renewable diesel. Green diesel has the same chemical properties as petroleum-based diesel.

13.3.3 BIOETHERS

Bioethers are produced by the reaction of reactive iso-olefins, such as iso-butylene, with bioethanol. Bioethers are created from wheat or sugar beet [13]. They also enhance engine performance while significantly reducing engine wear and toxic exhaust emissions.

13.3.4 BIOGAS

Biogas is methane produced by the process of anaerobic digestion of organic material by anaerobes [14]. It can be produced either from biodegradable waste materials or by the use of energy crops fed into anaerobic digesters to supplement gas yields.

13.3.5 SYNGAS

Syngas, a mixture of CO, hydrogen, and other HCs, is produced by partial combustion of biomass, that is, combustion with an amount of oxygen (O_2) that is not sufficient to convert the biomass completely to carbon dioxide (CO_2) and water [15].

13.3.6 SOLID BIOFUELS

Examples include wood, sawdust, grass trimmings, domestic refuse, charcoal, agricultural waste, nonfood energy crops, and dried manure.

13.4 RESOURCES FOR BIODIESEL PRODUCTION

There are different possible feedstocks for biodiesel production. Edible and nonedible vegetable oils are the most common sources for biodiesel. Currently, biodiesel is mainly prepared from *rapeseed* in Canada, *soybean* in the United States, *sunflower* in Europe, and *palm* in Southeast Asia.

13.4.1 Edible Oils

Edible vegetable oils as raw materials for first-generation biodiesel are a concern. This is because it raises the food-versus-fuel debate that may cause high food prices, particularly in developing countries. It can also cause other environmental problems due to the use of a wide area of arable land available. This problem can create serious ecological imbalances as countries worldwide convert forests to farmland (deforestation). Therefore, the nonedible vegetable oil or second-generation raw materials have become more attractive for the production of biodiesel [16].

13.4.2 Nonedible Plant Oils

Technologies are being developed to exploit cellulosic materials for the production of biodiesel (biodiesel, second generation) such as leaves and stems of plants, biomass derived from waste, and also oilseeds from nonedible plants [17]. Nonedible biodiesel crops are expected to use lands that are largely unproductive and those that are located in poverty-stricken areas and in degraded forests. Moreover, nonedible oil plants are well adapted to arid and semiarid conditions and require low fertilizer and moisture demand to grow [18]. Added to this, nonedible oils are not suitable for human consumption due to the presence of toxic components in the oils [19]. For all these reasons, the use of nonedible oils as raw material is a promising way in biodiesel production.

There are a large number of plants that produce nonedible oils. From a list of 75 plant species [20] containing oil in their seeds or kernels, 26 species are potential sources for biodiesel production. Some examples of nonedible oilseed crops are *Jatropha curcas*, *Calophyllum inophyllum*, *Sterculia*, *viotida Indica*, *Madhuca longifolia* (mahua), *Pongamia*, licorice, Kourosh seeds, linseed, *Pongamia* feather (karanja), the Brazilian *Hevea* (rubber seeds), neem, *Camelina sativa*, *vindleri liskoerila*, *Nicotiana tabacum* (tobacco), Deccan hemp, *Ricinus communis*, *Simmondsia chinensis*, babassu (jojoba), *Arca sativa* L., *Cerbera odollam* (sea mango), coriander (*Coriandrum sativum* L.), *Croton megalocarpus*, pilu, *Crambe*, *Syringa*, *Schleichera triguga* (kusum), *Stillingia*, *Shorea robusta*, *Terminalia bellirica* Roxb., *Cuphea*, *Camellia*, champaca, *Simarouba glauca*, *Garcinia indica*, rice bran, hingan (*Balanites*), desert date, cardoon, *Asclepias syriaca* (milkweed), *Guizotia abyssinica*, radish, Ethiopian mustard, *Syagrus*, tung, *Idesia polycarpa* var. vestita, algae, *Argemone mexicana* L., *Putranjiva roxburghii*, *Sapindus mukorossi* (soap nut), *Copaiba*, milk bush, almond, piqui, *Brassica napus*, tomato seed, and *Zanthoxylum bungeanum* [21]. Among these, *Jatropha*, *Moringa*, and castor oils are the most used in biodiesel production.

13.4.3 Used Edible Oils

The production of biodiesel from waste cooking oil to partially substitute petroleum diesel is one of the measures for solving the twin problems of environmental pollution and energy shortage [22]. Also, in order to reduce the cost of biodiesel production, waste cooking oil would be a good choice as raw material since it is cheaper [23] than virgin vegetable oils and other feedstocks.

13.4.4 MICROALGAE

Microalgae are photosynthetic microorganisms that convert sunlight, water, and CO_2 to algal biomass [24]. Microalgae are classified as diatoms (Bacillariophyceae), green algae (Chlorophyceae), golden brown (Chrysophyceae), and blue green algae (Cyanophyceae) [25]. Microalgae have long been recognized as potentially good sources for biofuel production because of their high oil content (more than 20%) and rapid biomass production [26]. Algae biomass can play an important role in solving the problem between the production of food and that of biofuels in the near future. The cultivation of microalgae does not need much land as compared to plants.

13.4.5 ANIMAL FATS

Animal fats used to produce biodiesel include tallow, choice white grease or lard, fish fat (in Japan), and chicken fat [27]. Compared to plant crops, these fats frequently offer an economic advantage because they are often priced favorably for conversion into biodiesel. Animal fat methyl ester has some advantages such as high cetane number (CN) and noncorrosive, clean, and renewable properties. Animal fats tend to be low in free fatty acids and water, but there is a limited amount of these oils available, meaning these would never be able to meet the fuel needs of the world.

13.5 BIODIESEL PRODUCTION

Bolonio and coworkers [28] tested diesel fuel (D), diesel (80%)–microalgae biodiesel (20%) (by volume) (D80-B20), diesel (70%)–microalgae biodiesel (20%)–butanol (10%) (D70-B20-But10), and diesel (60%)–microalgae biodiesel (20%)–butanol (20%) (D60-B20-But20) fuels to evaluate the effects of the fuel blends on the performance and exhaust emissions of a diesel engine. Engine performance parameters and exhaust gas emissions such as NO_x, CO, and smoke opacity were evaluated. These showed that, although butanol addition caused a slight reduction in torque and brake power values, the emission values of the engine were improved. Therefore, butanol can be used as a very promising additive to diesel–microalgae biodiesel blends.

The conversion reaction of nonedible *Jatropha curcas* de-oiled seed cake (the residue left after extraction of oil from the seed) into bio-oil and bio-char through pyrolysis process was conducted using particle size range, sweep gas flow rate, and operating temperature of 0.5–0.99 mm, 8 L/min, and 450°C, respectively. It resulted in a maximum yield of bio-oil (48%) and bio-char (35.1%) [29]. The crude bio-oil showed lower viscosity (1.98 cSt), higher moisture content (31%), and higher density (1040 kg/m³) than commercial petroleum diesel. The crude bio-oil was upgraded to make it suitable for use in engine application.

Crude *Calophyllum inophyllum* oil has been evaluated as a potential feedstock for biodiesel production [30]. *C. inophyllum* oil has a high acid value, which is 59.30 mg KOH/g. *C. inophyllum* biodiesel can become an alternative fuel in the future.

Sakthivel and others [31] investigated the feasibility of using biodiesel prepared from fish oil. They evaluated various properties such as viscosity, density, calorific value, flash point, and cetane value of biodiesel and biodiesel–diesel blends of different proportions. They showed that fish oil can indeed become an appropriate good source for biodiesel, with environmental benefits.

The transesterification process for the production of rice bran oil methyl ester was investigated [32]. The biodiesel obtained under optimum conditions has comparable properties to substitute mineral diesel. Rice bran oil methyl ester biodiesel could be recommended as a mineral diesel fuel substitute for compression ignition engines in transportation as well as in the agriculture sector.

Hazelnut soap stock/waste sunflower oil methyl ester was partially substituted for diesel fuel at most operating conditions and displayed acceptable performance parameters and emissions without any engine modification and preheating of the blends [33].

Microalgae methyl esters, which contain some long-chain, polyunsaturated fatty acid methyl esters (C22:5 and C22:6) not commonly found in terrestrial crop–derived biodiesels, were tested as biodiesel by Ardebili and his coworkers [34]. All fuel properties were satisfied or were very close to the ASTM D6751-12 and EN 14214 standards. Therefore, microalgae-derived biodiesel/petroleum blends of up to 50% are projected to meet all fuel property standards, engine performance, and emission. Results from this study clearly show microalgae methyl esters are suitable for regular use in diesel engines.

Gautam et al. [35] tested the biodiesel produced from edible and nonedible oils *Pongamia pinnata* (karanja), *Madhuca indica* (mahua), and *Sesamum indicum* (til) oilseeds. The tested properties were density, viscosity, calorific value, acid value, cloud point (CP), pour point (PP), fire point, flash point, cold filter plugging point (CFPP), thermogravimetric analysis, and differential scanning calorimetry. It was found that *til* oil produced a maximum amount of biodiesel followed by *karanja* and *mahua* oils. Interestingly, the calorific value of *til* biodiesel was the highest, followed by *mahua* and *karanja* biodiesel. The blending of biodiesel with conventional diesel fuel could improve the calorific value and increase the fire and flash points rendering it safer for handling and transportation. In comparing the properties of the biodiesels, it was concluded that all three oils can produce good quality biodiesel. However, they exhibited variable properties for engine application.

In the same trend, Kaul et al. [36] studied the reactive extraction of biodiesel from two potential nonedible oilseeds (*karanja* and *Simarouba glauca*). The smallest seed size resulted in high biodiesel yield, whereas the optimum methanol-to-oil ratio and catalyst concentration depended on the type of oil feed. The reaction temperature of 65°C was found to be optimum. The ^{1}H-NMR and gas chromatographic analyses results were comparable. It was found from ^{1}H-NMR analysis that biodiesel produced from reactive extraction of *Simarouba* had better oxidation stability than that from *karanja*.

Calophyllum inophyllum was investigated as a promising feedstock for biodiesel production [37]. Several aspects were studied including physical and chemical properties of crude *Calophyllum inophyllum* oil and methyl ester, fatty acid composition, blending, and engine performance and emissions of *Calophyllum inophyllum* methyl ester. *Calophyllum inophyllum* appears to be an acceptable feedstock for future biodiesel production.

Naik et al. [38] discussed the mechanism of a dual process adopted for the production of biodiesel from *karanja* oil containing free fatty acids up to 20%. The first step is acid-catalyzed esterification using 0.5% H_2SO_4, alcohol 6:1 molar ratio with respect to the high free fatty acid *karanja* oil to produce methyl ester by lowering the acid value. The next step is alkali-catalyzed transesterification. The yield of biodiesel from high free fatty acid *karanja* oil by the dual step process was found to be 97%. Ethanol-based biodiesel produced from a low-cost sludge *palm* oil using locally produced *Candida cylindracea* lipase from fermentation of *palm* oil mill effluent–based medium was investigated [39]. The sludge *palm* oil and ethanol have a promising potential for the production of renewable biodiesel using enzymatic-catalyzed esterification and transesterification.

A biodiesel production using two-step transesterification process on a laboratory scale was studied [40]. This study reveals that biodiesel production from *soybean* oil methyl ester, as one of nonedible feedstock, is able to be an alternative for petrodiesel.

The optimization of transesterification process parameters for the production of *Manilkara zapota* methyl ester using Taguchi experimental design was evaluated [41]. The biodiesel *Manilkara zapota* methyl ester produced with the optimized process parameters meets the global standards for biodiesel (EN 14214) and hence could be considered as a suitable substitute for fossil diesel in unmodified diesel engine applications.

Indian milkweed oil was evaluated as a potential feedstock for biodiesel production. The biodiesel can be only used in tropical countries due to the poor cold-flow properties [42]. The physical properties of the produced *Jatropha curcas* methyl esters such as viscosity, density, flash point, CP, PP, calorific value, acid value, iodine value, carbon residue, and sulfate ash were determined [43]. Overall, the properties of *Jatropha curcas* biodiesel were in the range that could be accepted and have met ASTM D6751 and EN 14214 standards.

13.6 TECHNOLOGICAL OPTIONS TO UTILIZE VEGETABLE OILS AS ALTERNATIVE FUEL FOR DIESEL ENGINES

13.6.1 TRANSESTERIFICATION (ALCOHOLYSIS)

Transesterification or alcoholysis is defined as the chemical reaction of alcohol with vegetable oils. In this reaction, methanol and ethanol are the most commonly used alcohols because of their low cost and availability. This reaction has been widely used to reduce the viscosity of vegetable oil and the conversion of the triglycerides into ester.

The production of biodiesel from the fat extracted from beef tallow waste using alkali-catalyzed transesterification employs two methods of transesterification, single step and two step. In both methods, KOH or NaOH with methanol was investigated. The reactions were performed at two temperatures (32°C and 60°C) for a fixed duration of 1 h. The fuel properties of the produced biodiesel were investigated. The results indicated that both methods of transesterification were successful to enhance the fuel properties of the tallow as compared to the direct use of tallow as a fuel [44].

The transesterification of *soybean* oil into biodiesel (methyl esters) was utilized using waste coral as a source of calcium oxide (CaO). The study showed that the optimum conditions from the experiment were a calcination temperature of 900°C, catalyst concentration of 6 wt%, and methanol-to-oil ratio of 12:1. Under these conditions, methyl ester content reached to 100 wt%. The catalyst was capable of being reused up to four times without much loss in activity [45].

Maneerung et al. [46] developed environmentally and economically benign heterogeneous catalysts for biodiesel production via transesterification of *palm* oil using CaO catalyst. The CaO catalysts exhibited high biodiesel production activity; over 90% yield of methyl ester can be achieved at the optimized reaction condition. Experimental kinetic data fit well the pseudo-first-order kinetic model. The activation energy of the transesterification reaction was calculated to be 83.9 kJ/mol. Moreover, the CaO catalysts derived from woody biomass gasification bottom ash can be reutilized up to four times, offering the efficient and low-cost catalysts, which could make biodiesel production process more economic and environmentally friendly.

Sarantopoulos et al. [47] studied two full factorial designs of experiments. Six variables, namely, esterification time (60–120 min), H_2SO_4 concentration (20–40 wt%), MeOH-free fatty acid (15:1–23:1), transesterification time (30–60 min), KOH concentration (1–2 wt%), and MeOH–triglyceride (6:1–9:1), that typically affect the production process, were used to investigate a two-step homogeneous base-catalyzed waste lard transesterification reaction for low-cost biodiesel production. The results showed that the esterification step is significantly affected by the reaction time and the MeOH-free fatty acid ratio value. Specifically, an increase of reaction time and MeOH-free fatty acid brings a reduction of the free fatty acid content. Likewise, the transesterification step is positively affected primarily by three independent variables, namely, reaction time, KOH concentration, and MeOH–triglyceride ratio. Two empirical models describing the evolution of the two-step transesterification reaction were developed. The models can become useful tools for further scaling up the process by predicting its reaction yield within a 95% of confidence level.

The methanolysis of vegetable oil to produce a fatty acid methyl ester (biodiesel fuel) was catalyzed by commercial ionic liquid and its chloride modification [48]. The imidazolium chloride ionic liquid was frequently chosen for the synthesis of biodiesel. The dual-functionalized ionic liquid is prepared by a direct combination reaction between imidazolium cation and various metal chlorides such as $CoCl_2$, $CuCl_2$, $NiCl_2$, $FeCl_3$, and $AlCl_3$. Imidazolium tetrachloroferrate was proved to be a selective catalyst for the methanolysis reaction at a yield of 97% when used at 1:10, catalyst-to-oil ratio for 8 h at 55°C. Operational simplicity, reusability of the used catalyst for at least eight times, high yields, and no saponification are the key features of this methodology. The dynamic viscosity and density of the upgraded vegetable oil decreased from 32.1 cP and 0.9227 g/cm³ to 10.2 cP and 0.9044 g/cm³, respectively, compared to those of the base vegetable oil. The objective of this study was the synthesis and characterization of biodiesel using commercial ionic liquid and its chloride modification. The ionic liquid catalysts were characterized using Fourier transformer infrared spectroscopy (FTIR), Raman spectroscopy, Differential scanning calorimetry (DSC), thermogravimetry, and UV spectroscopy.

Transesterification of *Sesamum indicum* L. oil was carried out with methanol in the presence of sodium methoxide, and the parameters affecting the reaction, vegetable oil/methanol molar ratio, catalyst concentration, reaction temperature, and time, were fully optimized by employing central composite design method [49]. A quadratic polynomial was developed to predict the response as a function of independent variables and their interactions, and only the significant factors affecting the yield were fitted to a second-order response surface reduced 2FI model. At the optimum condition of 1:6 oil-to-methanol molar ratio, catalyst concentration of 0.75%, and reaction time of 30 min, a biodiesel yield of 87.8% was achieved. Selected fuel properties were within the range set by ASTM and EN bodies.

Zhang and coworkers [50] studied a direct transesterification process using 75% ethanol and cosolvent to reduce the energy consumption of lipid extraction process and to improve the conversion yield of the microalgae biodiesel, using *n*-hexane as cosolvent. They showed that at the optimal reaction condition of *n*-hexane to 75% ethanol volume ratio 1:2, mixed solvent dosage 6.0 mL, reaction temperature 90°C, reaction time 2.0 h, and catalyst volume 0.6 mL, the direct transesterification process of microalgal biomass resulted in a high conversion yield up to 90.02 ± 0.55 wt%.

The production of *Jatropha* oil methyl esters via alkali-catalyzed transesterification route was investigated by Ahmed and coworkers [51]. They showed that a high percentage conversion (96.1%) of fatty acids into esters was achieved under optimized transesterification conditions of 6:1 oil-to-methanol ratio and 0.9 wt% NaOH for 50 min at 60°C.

A comparative study was conducted on the activity of La_2O_3 and two kinds of nano La_2O_3 catalysts prepared using sonochemical (nano La_2O_3-S) and hydrothermal methods in the transesterification to produce biodiesel [52]. Nano La_2O_3-S was selected for further optimization due to its simple preparation procedure and short preparation time. The fatty acid methyl ester content and yield obtained were successively 97.6% and 90.3% under optimal conditions. Moreover, nano La_2O_3-S showed a remarkable tolerance to free fatty acid.

The production of biodiesel through non-catalytic transesterification of microalgae oil in supercritical methanol and ethanol was studied [53]. The response surface methodology combined with a five-parameter-five-level central composite design was applied to optimize the temperature (270°C–350°C), pressure (80–200 bar), alcohol-to-oil molar ratio (10:1–42:1), residence time (10–50 min), and water content (0–10 wt%). They showed that optimal biodiesel yields are obtained at 90.8% and 87.8% with methanol and ethanol, respectively.

Bilgin and others [54] determined the transesterification reaction parameters to produce the lowest kinematic viscosity waste cooking oil biodiesel by using sodium hydroxide as a catalyst and ethanol. They investigated the effect of catalyst concentration (0.50%–1.75%), reaction temperature (60°C–90°C), reaction time (60–150 min), and alcohol/oil molar ratio (6:1–15:1) on the kinematic viscosity of produced biodiesels. They showed that reaction parameters giving the lowest kinematic viscosity of 4.4 cSt were obtained at 1.25% catalyst concentration, 70°C reaction temperature, 120 min reaction time, and 12:1 alcohol/oil molar ratio.

Nonedible oil source milk thistle (*Silybum marianum* L. Gaert.) plant was investigated [55]. The protocol for experiment was adjusted as follows: temperature (60°C), time of reaction (2 h), stirring (600 rpm), and the oil-to-alcohol molar ratio was fixed 1:6. They showed that the highest conversion percentage to biodiesel was 89.5% and 87.4% using solid base catalyst sodium hydroxide (0.75%) and potassium hydroxide (1.0%), respectively.

The enzymatic alcoholysis of *soybean* oil with methanol and ethanol using a commercial, immobilized lipase was investigated [56]. The investigation showed that the best conditions were obtained in a solvent-free system with an ethanol/oil molar ratio of 3.0, a temperature of 50°C, and an enzyme concentration of 7.0% (w/w). A three-step batch ethanolysis was most effective for the production of biodiesel with esters of about 60% after 4 h of reaction.

Chakraborty and Banerjee [57] investigated and developed an efficient method for maximizing biodiesel yield from used frying *soybean* oil containing high free fatty acid. They conducted separately a sequential biodiesel synthesis process comprising H_2SO_4-catalyzed esterification of physically pretreated used frying *soybean* oil followed by NaOH-catalyzed transesterification of unreacted triglyceride to maximize the overall yield of fuel-grade biodiesel. They showed that the yield of used frying *soybean* oil methyl ester was 94.7% and 97% at the best combination of operating parameters governing the esterification and transesterification steps, that is, 1, 4 h reaction time; 0.30, 0.20 v/v alcohol-to-oil ratio; 0.75, 1.0 wt% catalyst concentration; 50°C, 60°C reaction temperature; and 600, 400 rpm stirrer speed, respectively.

Vicente et al. [58] made a comparison of different base catalysts (sodium methoxide, potassium methoxide, sodium hydroxide, and potassium hydroxide) for methanolysis of sunflower oil. They showed that the biodiesel purity was near 100 wt% for all catalysts. However, near 100 wt% biodiesel yields were obtained only with the methoxide catalyst. According to the material balance of the process, yield losses were due to triglyceride saponification and methyl ester dissolution in glycerol. Although all the transesterification reactions were quite rapid and the biodiesel layers achieved nearly 100% methyl ester concentrations, the reactions using sodium hydroxide were the fastest.

Transesterification of *soybean* oil to biodiesel using CaO as a solid base catalyst was studied [59]. The results showed that a 12:1 molar ratio of methanol to oil, addition of 8% CaO catalyst, 65°C reaction temperature, and 2.03% water content in methanol gave the best results and the biodiesel yield exceeded 95% after 3 h. The catalyst lifetime was longer than that of calcined K_2CO_3/γ-Al_2O_3 and KF/γ-Al_2O_3 catalysts. CaO maintained sustained activity even after being repeatedly used for 20 cycles, and the biodiesel yield at 1.5 h was not affected much in the repeated experiments.

On the other hand, the transesterification of *soybean* oil to biodiesel using SrO as a solid base catalyst showed that the yield of biodiesel produced with SrO as a catalyst was in excess of 95% at temperatures below 70°C within 30 min [60]. SrO had a long catalyst lifetime and could maintain sustained activity even after being repeatedly used for 10 cycles. Transesterification of *soybean* oil to biodiesel using SrO as a catalyst is a commercially viable way to decrease the costs of biodiesel production.

Lewis acids ($AlCl_3$ or $ZnCl_2$) are also used to catalyze the transesterification of *canola* oil with methanol in the presence of tetrahydrofuran as cosolvent [61]. $AlCl_3$ catalyzes both the esterification of long-chain fatty acid and the transesterification of vegetable oil with methanol. This suggests that the catalyst is suitable for the preparation of biodiesel from vegetable oil containing high amounts of free fatty acids. The optimum conditions with $AlCl_3$ achieved 98% conversion at 24:1 molar ratio, 110°C, and 18 h reaction time with tetrahydrofuran as cosolvent. $ZnCl_2$ was far less effective as a catalyst compared to $AlCl_3$, which was attributed to its lesser acidity. Nevertheless, the presence of tetrahydrofuran minimized the mass transfer problem normally encountered in heterogeneous systems. Statistical analysis showed that the conversion with the use of $ZnCl_2$ differs only with reaction time but not with molar ratio.

Deep eutectic solvent consisting of choline chloride and glycerol (1:2 M ratio) was prepared and used as the cosolvent for transesterification of *rapeseed* oil to biodiesel catalyzed by sodium hydroxide [62]. The results suggested that up to 98% fatty acid methyl ester yields could be obtained at optimum conditions of 6.95 methanol/oil molar ratio, 1.34 wt% catalyst concentration, and 9.27 wt% deep eutectic solvent concentrations. The addition of deep eutectic solvent in the transesterification reaction substantially improved the fatty acid methyl ester yield, reduced the side reactions (such as saponification), and enabled a straightforward biodiesel separation and purification.

Another highly efficient solid base catalyst was prepared from waste carbide slag and tested for biodiesel production from *soybean* oil with methanol [63]. The waste carbide slag was calcinated at 650°C (CS-650) and exhibited 91.3% yield of fatty acid methyl esters at a reaction temperature of 65°C, a methanol/oil molar ratio of 9, and a catalyst/oil mass ratio of 1.0 wt% within a short reaction time of 30 min. The yield was much higher than that with commercial CaO. The merits of high catalytic activity, low cost, and abundant storage make the waste carbide slag a promising catalyst in the production of biodiesel.

Al-Hamamre and Yamin [64] studied waste frying oil conversion to biodiesel by alkali-catalyzed transesterification. They showed that the optimum conditions for biodiesel manufacturing were MeOH/oil ratio 0.4 v/v (corresponds to 9.5 M ratio), with 1.0% (% w/v) KOH (corresponds to 0.83% w/w), temperature of 50°C, and reaction time between 20 and 40 min. Under these conditions, the obtained biodiesel yield was approximately 98%.

Similarly, Karabas [65] used the corn *kernel* oil with high free fatty acid content as feedstock to produce biodiesel via process transesterification. Two stages were used to produce biodiesel after obtaining the corn *kernel* oil. The optimal process was found to be a catalyst concentration of 0.7 wt%, 8:1 alcohol-to-oil molar ratio, 50°C reaction temperature, and 40 min of reaction time using KOH catalyst. The corn *kernel* oil methyl ester yield was 90% under the optimal process parameters obtained by the Taguchi method.

Transesterification to produce biodiesel was performed using clay-based catalysts [66]. The preparation of the catalysts was performed by impregnation of CaO between aqueous solution and Suratthanee black clay and Ranong kaolin with a controlling catalyst. The CaO impregnated on Ranong kaolin showed better catalyst performance over that of Suratthanee black clay.

The effect of various process parameters such as the amount of catalyst, temperature, and time on biodiesel production of waste tallow as low-cost sustainable potential feedstock for biodiesel production was monitored [67]. The authors found that the optimal conditions for processing 5 g of tallow were: temperature 50°C and 60°C, oil/methanol molar ratio 1:30 and 1:30, amount of H_2SO_4 1.25 and 2.5 g for chicken and mutton tallow, respectively. Under optimal conditions, chicken and mutton fat methyl esters formation of 99.01% ± 0.71% and 93.21% ± 5.07%, respectively, was obtained after 24 h in the presence of acid.

The transesterification reactions of *castor* oil with ethanol and methanol as transesterification agents in the presence of several classical catalytic systems were studied [68]. The study indicates that biodiesel can be obtained by transesterification of *castor* oil using either ethanol or methanol as the transesterification agent. Similar yields of fatty acid esters may be obtained following ethanolysis or methanolysis. However, the reaction times required to attain similar yield are very different, with methanolysis being much more rapid.

Maleki and coworkers [69] compared the yield of solvent-free enzymatic methanolysis of *castor* oil, which is highly soluble in methanol, with *soybean* and *palm* oils that have low solubility in methanol. The reaction was studied under different operating conditions such as enzyme dosage, solvent, and acyl acceptor. They performed all reactions at 45°C and agitation rate of 200 rpm for 24 h using Lipozyme as a catalyst. They found that the yield of methanolysis of *castor* oil was remarkably high compared to *soybean* and *palm* oils, especially at lower dosages of enzyme. *Castor* oil was the most effective oil with the highest methyl ester yield of 67.6% at 15% of enzyme dosage. The high yield of methanolysis of *castor* oil has been thought to be due to its excellent solubility in methanol.

Ciudad and coauthors [70] investigated the application of waste frying oil mixed with *rapeseed* oil as a feedstock for the effective production of fatty acid methyl esters in a lipase-catalyzed process. They showed that the optimal conditions that would reach 100% fatty acid methyl ester were a methanol-to-oil molar ratio of 3.8:1, 100% (wt) waste frying oil, and incubation at 44.5°C for 12 h with agitation at 200 rpm.

The production of methyl esters from *castor* oil and methanol after neutralization of *castor* oil with glycerol was evaluated [71]. The reaction was carried out under atmospheric pressure and ambient temperature in a batch reactor, employing potassium hydroxide as a catalyst. The results showed a high yield of *castor* oil conversion into methyl esters after neutralization of *castor* oil with glycerol, and the highest yield was 92.5% after 15 min of reaction. The best operating condition was obtained applying an alcohol-to-oil molar ratio of 6.0% and 0.5% w/w of catalyst.

Biodiesel was produced from raw *castor* oil by homogeneous alkaline transesterification [72]. The experiment was conducted to evaluate the influence of temperature and reaction time on product yield and quality. The best reaction temperature and reaction time to produce biodiesel from raw *castor* oil were 65°C and 8 h, respectively. Models predict a product yield of 73.6% (w/w) and a purity of 83.4%.

The influence of catalyst concentration, methanol-to-oil molar ratio, reaction temperature, and reaction time on the methyl ester content reached by *castor* oil transesterification were evaluated [73]. The result showed that the most effective variables were catalyst concentration and methanol-to-oil molar ratio. The optimum conditions were 0.064 mol/L of CH_3OK, 18.8:1 of methanol-to-oil molar ratio, 45°C, and 10 min of reaction time. In these conditions, 97 wt% methyl ester content biodiesel was obtained.

Bateni and Karimi [74] applied *castor* plant to an integrated biodiesel and ethanol production and transesterified the extracted oil with ethanol produced through fermentation of the *castor* plant residue using 8% w/v sodium hydroxide at 100°C for 60 min to improve the ethanol production yield from 27.2% to 71.0%. They showed that the optimum biodiesel yield was 85.0% ±1.0%, obtained at 62.5°C, using an ethanol-to-oil mass ratio of 0.29:1, for 3.46 h, which was in agreement with the predicted yield (84.4%). Accordingly, 1 kg of *castor* plant resulted in the production of 149.6 g biodiesel and at least 30.1 g of ethanol as the final product with no extra alcohol feedstock requirement.

Rodríguez-Guerrero and coauthors [75] studied the biodiesel obtained by the reaction of *castor* oil with ethanol under sub- and supercritical conditions (200°C–350°C at endogenous pressure) using small amounts of sodium hydroxide as a catalyst. They showed that the maximum ethyl ester yield obtained was 98.9% for the catalytic process and 56.2% for the non-catalytic process.

The optimization of biodiesel production from *castor* oil using full factorial design in a batch laboratory-scale reactor by alkaline-catalyzed transesterification process was studied by Kılıç et al. [76]. They showed that this process gave an average yield of more than 90% biodiesel.

Extracted *castor* oil for biodiesel production by transesterification was investigated, including the effects of operating conditions on biodiesel production yield such as methanol-to-oil ratio, temperature, and reaction time [77]. The optimum biodiesel yield was 88.2%, obtained at 0.4:1 methanol-to-oil mass ratio at 40°C for 90 min. This yield corresponded to 155 g biodiesel per kg *castor* plant.

The process of producing biodiesel from *castor* oil, methanol and *castor* oil as reactants with 10:1 molar ratio and 3% sulfuric acid as a catalyst by transesterification reaction under mild conditions, was studied [78]. The authors showed the purity of more than 94% esters for all conducted experiments, which is a success for oil with a very complicated structure than other raw vegetable oils.

The effect of the reaction variables on methyl ester content, viscosity, acidity, and water content of biodiesel produced from *castor* oil was investigated [79]. The methanolysis of *castor* oil at 60°C in a batch reactor and the effect of three alkaline catalysts (CH_3ONa, NaOH, and KOH) and a cosolvent (hexane) were established. The authors concluded that sodium methoxide leads to a considerably higher methyl ester content than the other catalysts. Besides, when utilizing a cosolvent, the methyl ester content increases close to the value (95.5%) established by the European standard (EN 14214) (>96.5%). This has been ascribed to a significant improvement on oil–methanol contact.

Dairo and coauthors [80] carried out optimization of in situ biodiesel production from raw *castor* oil from raw *castor* bean oilseed (37.9% oil content) by alkaline-catalyzed in situ transesterification with sodium hydroxide as a catalyst and ethanol as a solvent in a laboratory batch processor. Reaction variables studied include reaction time (30–120 min), alcohol/seed weight ratio (0.5–2.0), catalyst concentration (0.3%–1.5%), and reaction temperature (40°C–70°C). They predicted that the highest yield of *castor* ethyl ester was 99.5% of expressible oil at the following optimized reaction conditions: alcohol/seed weight ratio of 0.5, catalyst/seed weight ratio of 1.31, reaction temperature of 60°C, and reaction time of 81.7 min.

Sousa et al. [71] evaluated the production of methyl esters from *castor* oil and methanol after neutralization of *castor* oil with glycerol under atmospheric pressure and ambient temperature in a batch reactor, employing potassium hydroxide as a catalyst. They showed a high yield of *castor* oil conversion into methyl esters after neutralization of *castor* oil with glycerol. The highest yield observed was of 92.5% after 15 min. The best operating condition was obtained at an alcohol-to-oil molar ratio of 6.0% and 0.5% w/w of catalyst.

The results of the transesterification reactions of *castor* oil with ethanol and methanol as transesterification agents indicated that biodiesel can be obtained by the transesterification of *castor* oil using either ethanol or methanol as the transesterification agent. Similar yields of fatty acid esters may be obtained following ethanolysis or methanolysis. The reaction times required to attain them are very different, with methanolysis being much more rapid [81].

The effects of ethanol, methanol, and their blends at different percentage mixtures on the properties and yields of biodiesel at varied transesterification times and temperatures using sodium hydroxide as a base catalyst were investigated by Efeovbokhan and coworkers [82]. They showed that at 70°C, the optimum yields were 88.4%, 94.2%, 94.8%, and 95.2% for ethanol and 90.6%, 95.6%, 96.0%, and 96.4% for methanol at 1, 2, 3, and 4 h, respectively. The biodiesel yields increased as the time of reaction progressed for both solvents, but the yields obtained from methanol were generally higher than those from ethanol. A mixture of the two solvents at 50% each produced the highest biodiesel yield of 98.6% at 70°C and after 4 h, compared to each solvent alone at the same time and temperature.

A microwave-assisted transesterification study of *castor* oil in the presence of ethanol and of potassium hydroxide as a catalyst was conducted [83]. The authors analyzed the effects of various reaction parameters such as reaction time, catalyst concentration, reaction temperature, and ethanol–oil molar ratio. They showed that ethyl esters were successfully produced by microwave-assisted transesterification. The maximum yield was 80.1% at 60°C, 10:1 alcohol-to-oil molar ratio, 1.5% potassium hydroxide, and 10 min, a reduction in reaction time for microwave-assisted transesterification as compared to conventional heating. Yields were slightly affected by temperature from 40°C to 70°C; this indicates a significant effect of microwaves even at low temperatures.

Da Silva and his coworkers [84] studied biodiesel production from *castor* oil with bioethanol from sugar cane in the presence of sodium ethoxide and sodium hydroxide as catalysts. The studied variables were reaction temperature, catalyst concentration, and ethanol-to-*castor* oil molar ratio. They showed the catalyst concentration as the most important variable, and the model obtained predicts the ethyl ester concentration as function of the reaction temperature, the ethanol-to-*castor* oil molar ratio, and the catalyst concentration. A conversion of 99% of the weight of ethyl ester was obtained at 30°C, with a mechanical stirrer, 1 wt% of sodium ethoxide, ethanol-to-*castor* oil molar ratio of 16:1, and at 30 min of reaction.

Jeong and Park [85] optimized the reaction factors for biodiesel synthesis from *castor* oil. They showed the maximum quantity of *castor* biodiesel (92 wt%)

after 40 min at 35.5°C, with an oil-to-methanol molar ratio of 1:8.24, and a catalyst concentration of 1.45% of KOH by weight of *castor* oil.

The effect of solid base catalyst KOH/NaY on the transesterification of *castor* oil to biodiesel was studied [86]. The effects of molar ratio of methanol to *castor* oil, catalyst dosage (catalyst mass/oil mass), reaction temperature, and reaction time on the conversion rate of transesterification were studied. They showed that under the optimal conditions of reaction temperature of 65°C, molar ratio of methanol to oil 9:1, mass fraction of catalyst 10%, and reaction time 3 h, the conversion rate of transesterification was 86.9%. KOH/NaY catalyst could be reused at least for three cycles with a significant conversion.

The production of biodiesel from waste oils by enzymatic conversion using immobilized lipase based on *Rhizopus oryzae* was studied [87]. The authors indicated that methanol/oil ratio of 1:4, immobilized lipase/oils of 30 wt%, and 40°C are suitable for waste oils under 1 atm., under the optimum conditions; the yield of methyl esters was around 88%–90%.

Shah and his colleague studied the conversion of waste cooking oil to biodiesel via simultaneous esterification and transesterification reaction over silica sulfuric acid as a solid acid catalyst [88]. The study showed highest fatty acids methyl esters obtained under the optimized condition was 98.7%. The reaction was found to follow the pseudo-first-order kinetics, and the rate constant of the reaction under optimum condition was 0.00852 min^{-1}.

Miao and Wu [89] introduced an integrated method for the production of biodiesel from microalgal oil. The obtained biodiesel was comparable to conventional diesel. They suggested that the new process, which combined bioengineering and transesterification, was a feasible and effective method for the production of high quality biodiesel from microalgal oil.

Vivek and Gupta [90] investigated the potential of *karanja* oil as a biodiesel source. They showed that optimum conditions were found to be pressure 1 atmosphere, temperature 68°C–70°C, reactant ratio 8–10 (MeOH-to-oil), reaction time 30–40 min, and catalyst (KOH) 1.5%.

The effects of mass transfer during the transesterification reaction of *canola* oil with methanol to form fatty acid methyl esters using a sulfuric acid catalyst at a MeOH/oil molar ratio of 6:1 were investigated [91]. The reactions rates were slow enough to permit the effects of mass transfer on the transesterification reaction to become more evident than at higher temperatures. For the two-phase experiments, it was postulated that the reaction occurred at the interface between the phases where the triglycerides, MeOH, and H_2SO_4 were in contact with one another.

13.6.2 Dilution of Vegetable Oils with Petroleum Diesel: Performance and Emission

Nonedible oil can be diluted with petroleum diesel to reduce its viscosity and improve the performance of the engine. This method does not require any chemical process. Therefore, blending of 20%–25% vegetable oil to petroleum diesel has been found to give good results for diesel engine.

De and Panua [92] analyzed the performance and emission characteristics of pure petroleum diesel, *Jatropha* oil, and *Jatropha* oil–diesel blended fuels with various blend ratios in engine laboratory single-cylinder, four-stroke, direct injection diesel engine. They showed that various parameters such as thermal efficiency and CO and NO_x emissions are very close to petroleum diesel for lower *Jatropha* concentrations. However, for higher *Jatropha* concentrations, the performance and emissions were much inferior to petroleum diesel.

Nagaraja and colleagues [93] investigated the performance and emission characteristics of a direct injection variable compression ratio engine when fueled with preheated *palm* oil and its 5%, 10%, 15%, and 20% blends with petroleum diesel (on a volume basis). They investigated the effects of compression ratio on brake power, mechanical efficiency, indicated mean effective pressure, and emission characteristics. They found that the blend with 20% *palm* oil (O20) is found to give maximum mechanical efficiency at higher compression ratio, which was 14.6% higher than diesel. Also, the brake power of blend O20 was found to be 6% higher than standard petroleum diesel at a higher compression ratio. The mean effective pressure of blend O20 was found to be lower than diesel at a higher compression ratio. Exhaust gas temperature is low for all the blends compared to diesel. The emission of CO and HCs dropped with an increase in blending ratio and compression ratio of maximum load. CO_2 emission was found to be higher than diesel.

Diesel and *soap nut* oil (10%, 20%, 30%, and 40%) fuel blends were used to conduct short-term engine performance and emission tests at varying loads (25% load increments from no load to full loads). Among the blends investigated, *soap nut* oil showed a better performance with respect to brake thermal efficiency (BTE) and brake specific energy consumption. All blends have showed higher HC emissions after about 75% load. *Soap nut* oil–diesel (10%–90%) and *soap nut* oil–diesel (20%–80%) showed lower CO emissions at full load. NO_x emission for all blends was lower, and *soap nut* oil–diesel (40%–60%) blend achieved a 35% reduction in NO_x emission. *Soap nut* oil–diesel (10%–90%) has an overall better performance with regard to both engine performance and emission characteristics [94].

Mbarawa [95] investigated the effects of the *clove stem* oil–diesel blended fuels on the engine BTE, brake specific fuel consumption (BSFC), specific energy consumption, exhaust gas temperatures, and exhaust emissions. The author revealed that the engine BTE and BSFC of the *clove stem* oil–diesel blended fuels were higher than the pure diesel fuel, while at the same time they exhibited lower specific energy consumption than the latter over the entire engine load range. The variations in exhaust gas temperatures between the tested fuels were significant only at medium-speed operating conditions. Furthermore, the HC emissions were lower for the *clove stem* oil–diesel blended fuels than for the pure diesel fuel, whereas the NO_x emissions increased remarkably when the engine was fueled with the 50% *clove stem* oil–diesel blended fuel.

Pine oil was used as blends with petroleum diesel (50% diesel and 50% *pine* oil) in a diesel engine [96]. A significant decrease was observed in CO, HC, and smoke emission by 45.9%, 32.4%, and 41.5%, respectively. However, NO_x emission was noted to have increased.

Ozaktas [97] investigated a used *sunflower* oil–diesel fuel blend (20:80 v/v%) in Pancar Motor diesel engine to determine engine characteristics and exhaust emission and the effect of the compression ratio on ignition delay (ID) characteristics and smoke emissions of blend fuel compared to a reference grade No. 2-D diesel fuel.

Pramanik [98] studied blends of varying proportions of *Jatropha curcas* oil and diesel and compared these with pure petroleum diesel fuel. The results established that 40%–50% of *Jatropha* oil can be substituted for diesel without any engine modification and preheating of the blends.

Tests were presented on a single-cylinder direct injection engine operating on diesel fuel, *Jatropha* oil, and blends of diesel and *Jatropha* oil in proportions of 97.4%–2.6%, 80%–20%, and 50%–50% by volume [99]. The results showed that 97.4% diesel 1%–2.6% *Jatropha* fuel blend produced maximum values of the brake power and BTE as well as minimum values of the BSFC. The 97.4%–2.6% fuel blend yielded the highest CN and even better engine performance than the diesel fuel suggesting that *Jatropha* oil can be used as an ignition-accelerator additive for diesel fuel.

Nwafor and Rice presented the results of an engine test on three fuels blends [100]. Test runs were also made on neat *rapeseed* oil and diesel fuel as bases for comparison. The tests showed an increase in BTE and reduction of power output by increasing the amount of *rapeseed* oil in the blends.

Labeckas and Slavinskas [101] examined the effect of *rapeseed* oil inclusion into diesel fuel on engine performance parameters. They showed that the BSFC has increased by 0.104%, 0.134%, and 0.156% for every 1% increase in *rapeseed* oil inclusion into diesel fuel. The maximum thermal efficiency values remain within 0.37% and 0.39% intervals. The maximum NO_x emission increases with the mass percent of O_2 in the fuel blend and for *rapeseed* oil and its blends *rapeseed* oil–diesel fuel (25%–75%) and *rapeseed* oil–diesel fuel (50%–50%) are higher by 9.2%, 20.7%, and 5.1%, respectively. Emissions of nitrogen dioxide (NO_2) increase with an increasing content of RO_2 (Peroxides) into diesel fuel. When operating on pure *rapeseed* oil and its blends, *rapeseed* oil–diesel fuel (25%–75%) and *rapeseed* oil–diesel fuel (50%–50%), the maximum CO emission reduces by 40.5%–52.9% and 7.2%–15.0%, respectively. The smoke opacity generated from *rapeseed* oil and its blends was also lower by 27.1%–34.6% and 41.7%–51.0%, respectively.

13.7 THERMAL CRACKING (PYROLYSIS) AND CATALYST CRACKING

Pyrolysis is the thermal conversion of the organic matters in the absence of O_2 and in the presence of a catalyst. The pyrolyzed material can be vegetable oils, animal fats, natural fatty acids, or methyl esters of fatty acids. The thermal decomposition of triglycerides produces alkanes, alkenes, alkadienes, aromatics, and carboxylic acids. The liquid fractions of the thermally decomposed vegetable oils are likely to have properties that approach diesel fuels.

The effects of pyrolysis parameters including temperature and catalyst on the product yields were investigated [102]. The pyrolysis of *Alcea pallida* stems was performed in a fixed-bed tubular reactor with and without catalyst at three different temperatures. The study revealed that catalysts had different effects on product yields and composition of bio-oils. Liquid yields were increased in the presence of zinc chloride and alumina but decreased with calcium hydroxide. The highest bio-oil yield (39.35 by weight) including aqueous phase was produced with alumina catalyst at 500°C.

Xu and coauthors [103] studied the catalytic cracking reactions of several kinds of woody oils. The products analyzed by gas chromatography-mass spectroscopy (GC-MS) and FTIR showed the formation of olefins, paraffins, and carboxylic acids. Several kinds of catalysts were compared. They found that the fractional distribution of product was modified by using base catalysts such as CaO. The catalytic cracking of woody oils generates fuels with physical and chemical properties comparable to those specified for petroleum-based fuels. The catalytic pyrolysis of *soybean* oil over mesoporous catalytic material (Me-Al-MCM-41, Me = La, Ni, or Fe) was accomplished [104]. Pyrolysis experiments were performed in a tubular reactor at the constant conditions of temperature 450°C, weight hourly space velocity 6 h^{-1}, and reaction time 4 h, in the absence and presence of the above catalysts. The catalytic activity of the catalysts on the yields and composition of the pyrolysis products was investigated. The biofuel obtained from the pyrolysis experiment under Ni-Al-MCM-41 has a good prospect to serve as an alternative for traditional fossil fuel.

In another study, Sang cracked *palm* oil at atmospheric pressure, at a reaction temperature of 450°C, and a weight hourly space velocity of 2.5 h^{-1} to produce biofuel in a fixed-bed microreactor [105]. The reaction was carried out over microporous HZSM-5 zeolite, mesoporous MCM-41, and composite micromesoporous zeolite as catalysts in order to study the influence of catalyst pore size and acidity over biofuel production. The products obtained were gas, organic liquid, water, and coke. The organic liquid product was composed of HCs that have gasoline, kerosene, and diesel boiling point range. The author showed that the maximum conversion of *palm* oil was 99 wt% and the gasoline yield was 48 wt%.

The pyrolytic conversion of *Chlorella* algae to liquid fuel precursor in the presence of a catalyst (Na_2CO_3) was studied using thermogravimetric analyzer coupled with mass spectrometer [106]. Liquid oil samples were collected from pyrolysis experiments in a fixed-bed reactor and characterized for water content and heating value. In the presence of Na_2CO_3, gas yield increased and liquid yield decreased compared to non-catalytic pyrolysis at the same temperatures.

Li and coauthors [107] compared the catalytic cracking process with thermal cracking process on the production of gaseous HC and gasoline conversion from *cottonseed* oil. They discussed the difference in composition of products from catalytic cracking versus thermal cracking. They showed that the compositions of the reaction products are heavily dependent on the catalyst type (catalyst activation) and reaction conditions. The products ranged from dry gas to light distillate, such as dry gas, liquefied petroleum gas, and gasoline. When the temperature of the catalytic cracking was above 460°C, the process of thermal cracking must be considerable.

On the other hand, fast pyrolysis of corncob with and without catalyst in a fluidized bed was investigated to determine the effects of pyrolysis parameters (temperature, gas flow rate, static bed height, and particle size) and a HZSM-5 zeolite catalyst on the product yields and the qualities of the liquid products [108]. The optimal conditions for liquid yield (56.8%) were a pyrolysis temperature of 550°C, a gas flow rate of 3.4 L/min, a static bed height of 10 cm, and a particle size of 1.0–2.0 mm. The presence of the catalyst increased the yields of noncondensable gases, water, and coke, while it decreased the liquid and char yields. Elemental analysis showed a more than 25% decrease in O_2 content of the collected liquid in the second condenser in the presence of HZSM-5 catalyst compared with that obtained without catalyst.

The production of light diesel fractions by thermal catalytic cracking of crude *palm* oil (*Elaeis guineensis*) in pilot scale was investigated [109]. The investigation was carried out under the following conditions: cracking reactor 143 L, operating in batch mode at 450°C, and atmospheric pressure, using 20% (w/w) sodium carbonate (Na_2CO_3) as a catalyst. The process yield of organic liquid products was 65.86% (w/w) with an acid value of 1.02 mg KOH/g, kinematic viscosity of 1.48 mm²/s, 30.24% (w/w) noncondensable gases, 2.5% (w/w) water, and 1.4% (w/w) coke. The yield on green diesel obtained by distillation was averaged at 24.9% (w/w), presenting an acid value of 1.68 mg KOH/g. GC-MS analysis indicated that the green diesel was composed of 91.38% (w/w) of HCs (31.27% normal paraffins, 54.44% olefins, and 5.67% of naphthenes) and 8.62% (w/w) of oxygenated compounds.

Biswas and Sharma [110] studied the conversion of a nonedible plant seed oil, *Jatropha* oil, by cracking and the utilization of cracked liquid product obtained as transportation fuel. They used three catalysts, namely, ZSM-5, ZSM-5 + SiAl, and NiMo/SiAl. They showed the formation of 36% gasoline range HCs (C_7–C_{11}) and 58% diesel range HCs (C_{12}–C_{22}) in the cracked liquid when the catalyst ZSM-5 + SiAl was used.

Yigezu and Muthukumar [111] investigated the utilization of metal oxides for the biofuel production from vegetable oil and determined the physical and chemical properties of the diesel-like products obtained and the influence of reaction variables on the product distribution. They employed six different metal oxides (Co_3O_4, KOH, MoO_3, NiO, V_2O_5, and ZnO) as catalysts. They indicated that the metal oxides are suitable for catalyzing the conversion of oil into organic liquid products.

The catalytic cracking of *rubber seed* oil to produce liquid HC fuels using ultra-stable Y-zeolite as a heterogeneous catalyst was studied [112]. Under the optimum cracking conditions of *rubber seed* oil 10 g, m(ultra-stable-Y-zeolite)/m(*rubber seed* oil) = 1:50, and 420°C for 90 min, the yield of liquid product reached 75.6%. The chemical composition and properties of the liquid fuel were similar to those of gasoline-based fuels (C_8–C_9 content >70%, low acid value, good cold-flow properties, and high calorific value).

The production of biofuel by catalytic cracking of *soybean* oil over a basic catalyst in a continuous pyrolysis reactor at atmospheric pressure was studied [113]. The authors designed experiments to study the effect of different types of catalysts on the yield and acid value of the diesel and gasoline fractions from the pyrolytic oil. The study indicated that the base catalyst gave a product with a relatively low acid number. These pyrolytic oils were further reacted with alcohol in order to decrease their acid value.

Tang and Wei [114] studied the fluid catalytic cracking process of waste edible oil. An environmentally friendly process was developed to produce clean fuel (such as gasoline and diesel) and high-value chemicals (such as propylene) in a small hot model of downer reactor at a temperature of 500°C, pressure of 1.1×10^5 Pa, resident time of 1 s, ratio of catalyst to oil of 1:2, and MA-83 catalyst. The results showed that the waste edible oil and *cottonseed* oil fluid catalytic cracking process have a product distribution of liquefied petroleum gas 10.3 wt%, gasoline 36.4 wt%, and diesel oil wt 23%. The desired products of gasoline and diesel showed trace quantities of sulfur, nitrogen, and heavy metals and have similar molecular weight and chemical structure as gasoline and diesel from petroleum-based fuel. Thus, waste edible oil fluid catalytic cracking is a promising process, with both economic and environmental benefits.

Doronin and others [115] investigated the routes of transformation of vegetable oils under catalytic cracking conditions. They revealed the influence of the component composition of the catalyst on distribution of the desired cracking products and their chemical group composition. They showed that the introduction of HZSM-5 zeolite into the composition of the oil cracking catalyst promotes the formation of C_2–C_4 light olefins; the yield of propylene and butylenes mainly increases. The high yield of both gasoline fractions and light olefins was obtained during cracking on zeolite catalysts. The study clarified the relationship between the unsaturation index of the vegetable oils and the product distribution during their catalytic cracking. Oils with a high index of unsaturation are cracked on the zeolite catalyst yielding primarily mono- and polyaromatic HCs.

The research group of Tian [116] carried out catalytic transformation of oils and fats in a laboratory-scale two-stage riser fluid catalytic cracking unit. They showed that most of the properties of produced gasoline and diesel oil fuel meet the requirements of national standards, containing low sulfur.

Kraiem and his coauthors [117] carried out measurements of high heating value, viscosity, density, flash point, acid index, moisture content, ash content, FTIR, and GC-MS to characterize the pyrolytic oil obtained during waste fish pyrolysis at 500°C. They revealed that waste fish fats can be considered an important feedstock for biofuel production. The bio-oil properties indicated good calorific value (9391 kcal/kg) relative to European biodiesel specifications. On the other hand, higher acidity (103 mg KOH/g) and viscosity (7 cSt) were obtained relative to conventional fuels. This limits the direct use of bio-oils as alternative fuel in diesel engines. An efficient mixture with fossil fuel is a promising solution to improve the fuel properties. Hence, pyrolysis seems to be an eco-friendly process to recover the fish fatty wastes.

The pyrolysis of waste sludges was investigated using mass spectrometry/thermogravimetry and a fixed-bed reactor [118]. Two types of sludge were used, namely, mixed sludge and oil sludge. The authors showed that degradation of organic structures of sludge took place in the first step, while inorganic materials of sludge were mainly decomposed in the second step (above 500°C). The pyrolysis of oil sludge produced a larger amount of oil with aliphatic compounds and high calorific value. On the other hand, the pyrolysis of mixed sludge gave a smaller amount of oil that was rich in polar compounds. The gaseous products from pyrolysis consisted of a high amount of combustible gases. Landfilling was found to be the best alternative to dispose the pyrolytic char obtained from pyrolysis.

The effects of pyrolysis parameters, including temperature and catalyst, on the product yields of *Alcea pallida* stems were investigated in a fixed-bed tubular reactor with and without catalyst at three different temperatures [102]. The authors found that a higher temperature resulted in lower liquid (bio-oil) and solid (bio-char) yields but higher gas yields.

Prado and Filho [119] evaluated the thermal cracking and the thermal catalytic cracking of *soybean* oil using bauxite, a high-acidity and low-cost catalyst. The products generated by the thermal catalytic cracking process showed better results than the thermal cracking products because of the low quantity of acids present. The catalyst used acted in the secondary cracking process, in which the fatty acids decompose and generate.

Apricot pulps were pyrolyzed at a heating rate of 5°C/min in a fixed-bed reactor under different pyrolysis conditions, to determine the role of final temperature, sweeping gas flow rate, and steam velocity on the product yields and liquid product composition [120]. Final temperature range studied varied between 300°C and 700°C, and the optimum conditions of pyrolysis of peach pulp were studied. Liquid products obtained under the most suitable conditions were characterized by FTIR and ^{1}H-NMR spectroscopy. The highest liquid product yield was obtained at 550°C. Liquid product yield increased significantly under nitrogen and steam atmospheres. Characterization showed that the bio-oil could be a potential source for synthetic fuels and chemical feedstock.

The characteristics of intermediate pyrolysis oils derived from sewage sludge and de-inking sludge (a paper industry residue), for use as fuels in a diesel engine, were studied [121]. The organic fraction of the oils were separated from the aqueous phase and characterized. This included elemental and compositional analysis, heating value, cetane index, density, viscosity, surface tension, flash point, total acid number, lubricity, copper corrosion, water content, carbon residue, and ash content. The study concluded that both intermediate pyrolysis oils are able to provide sufficient heat when used in diesel engine. However, poor combustion and carbon deposition may be encountered. Blending of these pyrolysis oils with petroleum diesel or biodiesel could overcome these problems.

Correia and coauthors [122] analyzed the thermal stability and rheological property of *sunflower* oil obtained from the pyrolysis process. Oil samples were subjected to thermal degradation, and the reaction product was evaluated by the thermogravimetric technique, at temperatures between 300°C and 900°C. The results from the rheological analyses confirmed Newtonian rheological behavior.

Sunflower (Helianthus annuus L.)-extracted bagasse pyrolysis experiments were performed in a fixed-bed reactor [123]. The effects of heating rate, final pyrolysis temperature, particle size, and pyrolysis atmosphere on the pyrolysis product yields and chemical compositions were investigated. The maximum oil yield of 23% was obtained under nitrogen atmosphere at a pyrolysis temperature of 550°C and a heating rate of 7°C/min. Chemical characterization showed that the oil obtained from sunflower-extracted bagasse may be potentially valuable as fuel and chemical feedstock.

Fast pyrolysis was performed for waste fish oil in a continuous pyrolysis pilot plant [124]. The experiment was carried out under steady-state conditions in which 10 kg of biomass was added at a feed rate of 3.2 kg/h. The authors showed that the bio-oil yield of 72%–73% was obtained with a controlled reaction temperature of 525°C. Fast pyrolysis process represents an alternative technique for the production of biofuels from waste fish oil with characteristics similar to petroleum fuels.

Fast pyrolysis of *soybean* oil in a pilot plant was investigated. The experimental runs were conducted using an experimental design with variable temperature (from 450°C to 600°C) and concentration of water (from 0% to 10%). The liquid products were analyzed by gas chromatography and by true boiling point distillation. The obtained biofuels were similar to fossil fuels. Thermal analysis showed that it is possible to use these products as an energy source for the process [125].

In a fixed-bed reactor, corn husk and waste oil were co-pyrolyzed at temperatures of 500°C, 550°C, and 600°C, respectively, under nitrogen atmosphere [126]. Co-pyrolysis products with focus on the physical and chemical properties of oil products were characterized by GC-MS and elemental analysis. The pyrolysis temperature of 550°C seems to be the optimum for maximum bio-oil yield and properties. Co-pyrolysis of corn husk and waste oil produced more liquid and less solid residue than the pyrolysis of only *corn* oil. When the weight ratio of waste oil-to-corn husk increased from 0 to 0.87, bio-oil yield increased dramatically from 44.7 to 70.62 wt%, with increasing acid content and decreasing phenols, alcohols, and ketones. The upgraded bio-oil has the potential to be an alternative fuel for engine after upgrading with zeolite.

Thermal cracking experiments were carried out to promote the conversion of waste cooking oil into a biofuel with similar property as gasoline (light naphtha) [127]. At higher temperatures and with longer residence times, the benzene concentration increases, but the values obtained were close to the European and Brazilian specification limits.

Fast pyrolysis of waste cooking oil in a continuous pilot plant, operating under isothermal conditions, was performed by varying the temperature (475°C, 500°C, and 525°C) and the residence time (5–70 s) [128]. The data were needed to construct a kinetic model based on chemical lumps. The proposed kinetic model could be used for future scale-up studies and process development for the industrial application of the thermal cracking of waste cooking oil for the production of biofuels and renewable chemicals.

Saponified *palm, olive, rapeseed,* and *castor* oils were pyrolyzed at 750°C for 20 s by pyrolysis gas chromatography with mass selective and flame ionization detection to clarify their thermochemical behaviors [129]. The liquefiable compounds recovered from *palm, olive,* and *rapeseed* oils mainly contained linear alkenes (up to C_{19}) and alkanes (up to C_{17}), both similar to those found in gasoline (C_4–C_{10}) and diesel fuel (C_{11}–C_{22}) boiling range fractions of petroleum. In the case of *castor* oil, a significant amount of undesired O_2-containing products (e.g., ketones and phenols) was formed. The obtained data on reaction mechanisms can also be utilized in applications where various biofuels are produced.

Pyrolysis of *neem* seeds (*Azadirachta indica*) was investigated to study the physical and chemical characteristics of the biofuel produced and to determine its feasibility as a commercial fuel [130]. Thermal pyrolysis of neem seeds was conducted in a semi-batch reactor at a temperature range of 400°C–500°C and a heating rate of 20°C/min. The physical properties of the biofuel obtained were close to those of petroleum fractions.

Silva and coworkers [131] used slow pyrolysis to obtain *castor* seed cakes pyrolysis oil and determine its chemical composition. Elemental analyses showed

HC (13.2%), nitrogen (9.5%), O_2 (8.4%), and sulfur (0.2%). The calorific value of the pyrolysis oils (37.5 MJ/kg) was similar to petroleum fuels used commercially (43–46 MJ/kg) and greater than typical wood-derived crude bio-oils (17 MJ/kg). Thus, this oil may be potentially useful for the production of chemicals and energy, after upgrading, proving to be a great choice destination of this waste.

13.8 MICROEMULSIFICATION OF VEGETABLE OILS

The effect of hot surface temperatures on IDs of microemulsion of *coconut* oil at various ambient air pressures and temperatures, which would have reached under off diesel engine conditions, was investigated [132]. The measurement of ID characteristics of conical fuel sprays impinging on hot surface in cylindrical combustion chamber showed that at a higher injection pressure, the ID of microemulsion of *coconut* oil and pure diesel attains a lower value at the same ambient air pressure inside the combustion chamber. At fixed injection pressure and higher ambient temperature conditions, the ID variations of microemulsion of *coconut* oil with chamber pressure show approximately the same trend as that of conventional diesel fuel.

The effect of combustion, performance, and emission characteristics of a direct injection diesel engine with biodiesel and bio-oil emulsions was studied [133]. Emulsions made from wood pyrolysis oil and methyl ester of *Jatropha* in a single-cylinder, four-stroke, and air-cooled diesel engine developed a power of 4.4 kW at 1500 rpm. Three emulsions with 5%, 10%, and 15% of wood pyrolysis oil on a volume basis were emulsified with 95%, 90%, and 85% of methyl ester of *Jatropha*. The methyl ester of *Jatropha*–wood pyrolysis oil emulsions showed an increased thermal efficiency and reduced HC and CO emissions. The NO_x emissions for the methyl ester of *Jatropha*–wood pyrolysis oil emulsions were found to be higher than those of diesel operation, but reduced with the addition of wood pyrolysis oil in methyl ester of *Jatropha*. The smoke emissions showed a declining trend compared to that of diesel operation in the same engine. The methyl ester of *Jatropha*–wood pyrolysis oil emulsions exhibited shorter ID than that of diesel operation at full load. The values of peak pressure were higher by 9.8%, 7.9%, and 8.8% than those of diesel operation at full load for the three emulsions, respectively.

Xu and coauthors [134] investigated the friction and wear behaviors of emulsified bio-oil, which is a very promising alternative fuel for engines. Emulsified bio-oil had better tribological properties than diesel oil and crude bio-oil. Contact load and oscillation frequency significantly influenced the friction coefficient, wear volume, and wear damage pattern. The friction coefficient decreased with an increase in load and increased with an increase in oscillation frequency. Furthermore, the wear volume slightly increased with an increase in load or oscillation frequency. The damage mechanism was attributed to adhesive wear under low load, and to abrasive wear under high load. The transition in the wear mechanism was related to the adsorption of the emulsified bio-oil molecules to the microstructural contact surface and to the mechanical actions.

A new microemulsions system comprising diesel and *palm* oil methyl ester with a potential to be used as alternative fuels for diesel engines was developed [135]. The microemulsions were formed and stabilized with a mixture of nonionic surfactants at a weight ratio of 80:20 at 20% (w/w), and with mixed cosurfactants at a weight ratio of 25:75, 20:80, and 10:90 for B0, B10, B20, and B30. The particle size, kinematic viscosity at 40°C, refractive index, density, heating value, CP, PP, and flash point of the selected water-in-diesel microemulsion were 19.40 nm (polydispersity of 0.012), 12.86 mm²/s, 1.435, 0.8913 g/mL, 31.87 MJ/kg, 7.15°C, 10.5°C, and 46.5°C. The corresponding values of the water-in-diesel–*palm* oil methyl ester were 20.72–23.74 nm, 13.02–13.29 mm²/s, 1.442, 0.8939–0.8990 g/mL, 31.45–27.34 MJ/kg, 7.2°C–6.8°C, 8.5°C–1.5°C, and 47.5°C–52.0°C.

The environmental life cycle assessment of microemulsion-based biofuel (ME50) from *palm* oil–diesel blends with ethanol was compared with alternative biofuels, neat biodiesel (B100) and biodiesel–diesel blends (B50) [136]. The surfactant used was a key element in the microemulsion formation due to its significant impact on land use and environmental toxicity. In comparing the impacts among ME50, B100, and B50, it was found that microemulsion fuel production has lower environmental impact than B100 and B50, except in terms of land use and fossil depletion.

Chen and others [137] studied the performance and emissions of two typical diesel engines using glucose solution emulsified diesel fuel. Emulsified diesel with a 15% glucose solution by mass fraction was used in diesel engines and compared with pure diesel. The brake thermal efficiencies were improved using emulsified diesel fuel. Emulsified fuel decreased NO_x and dust emissions except under a few specific operating conditions. HC and CO emissions were increased. For the automotive diesel engine, performance and emissions were measured using the 13-mode European Stationary Cycle. It was found that brake thermal efficiencies of emulsified diesel and pure diesel were comparable at 75% and 100% load. Dust emissions decreased significantly, while NO_x emissions decreased slightly. HC emissions increased, while CO emissions decreased under some operating conditions.

Raheman and Kumari [138] prepared an emulsified fuel containing 10% and 15% water by volume from a diesel blend with 10% *Jatropha* biodiesel to evaluate the combustion characteristics of a 10.3 kW, single-cylinder, four-stroke, water-cooled, direct injection diesel engine. The 10% *Jatropha* biodiesel and its emulsified fuel exhibited similar combustion stages as that of diesel, with no undesirable combustion features such as an increase in cylinder gas pressure. ID was longer at higher engine loads by increasing water percentage. CO, CO_2, HC, and NO_x emissions were reduced in emulsified fuel compared to 10% *Jatropha* biodiesel. Emulsified biodiesel can be recommended for use in place of biodiesel.

Lin and Lin [139] investigated emulsified *castor* biodiesel spray characteristics on direct injection engine emission and deposit formation. The biodiesel generator operated on emulsified *castor* biodiesel can improve the fossil diesel emissions. NO_x emission of emulsified *castor* biodiesel was solved by water–biodiesel emulsion technology. The biofuels deposit simulator provided potential deposit control additives for emulsified *castor* biodiesel. Without changing the engine structure, when the injection pressure was increased by 5%–10%, the optimum combination was 82.8% of castor biodiesel, 15% of water, 2% of bioethanol, and 0.2% of composite surfactant Span–Tween.

Cheng et al. [140] compared the effect of applying a biodiesel with either 10% blended methanol or 10% fumigation methanol, using waste cooking oil on a four-cylinder naturally aspirated direct injection diesel engine operating at a constant speed of 1800 rpm with five different engine loads. They observed a reduction of CO_2, NO_x, and particulate mass emissions and a reduction in mean particle diameter, in both cases, compared to petroleum diesel fuel.

Liu et al. [141] discussed the influence of additive concentration, methanol content, rotational speed, liquid flow rate of diesel on the rheological characteristics, surface tension, and stabilization. Emulsified fuel appeared to display Newtonian behavior. The rheological characteristics of the emulsions were significantly dependent on the percentages of surfactants and methanol added. The surface tension decreased obviously with the increase of the surfactants concentration. With higher additive amount, lower methanol content, higher rotational speed, and higher liquid flow rate, the methanol–diesel emulsified fuel showed better stabilization.

13.9 PHYSICAL AND CHEMICAL PROPERTIES OF BIODIESEL

Several properties are used to determine the quality of the biodiesel, petroleum diesel, and their blends. These properties are either chemical or physical. The physical properties include CP, PP, CFPP, kinematic viscosity, and density. Chemical properties include flash point; CN; acid number; iodine value; carbon residue; sulfur content; free and total glycerin; phosphorous, calcium, and magnesium content; and moisture, impurities, and unsaponifiable (MIU) content are given in the following text.

13.9.1 CLOUD POINT AND POUR POINT

The CP and PP are important for low-temperature applications for fuel. The CP is defined as the temperature at which a cloud of wax crystals first appear when the fuel is cooled under controlled conditions during a standard test. The PP is the temperature at which the amount of wax in the solution is sufficient to gel the fuel. Thus, it is the lowest temperature at which the fuel can flow. In general, biodiesel has higher CP and PP than petroleum diesel fuel, which is a disadvantage.

13.9.2 COLD FILTER PLUGGING POINT

CFPP is used as an indicator of low-temperature operability of fuels. The CFPP refers to the temperature at which the test filter starts to plug due to fuel components that have started to gel or crystallize. CFPP defines the fuels limit of filterability, having a better characteristic than CP for biodiesel as well as petroleum diesel. Usually, the CFPP of a fuel is lower than its CP. The CFFP tested is measured using ASTM D6371 method.

13.9.3 FLASH POINT

Flash point is another important property for biodiesel fuel. The flash point of a fuel is the temperature at which the fuel will ignite when exposed to a flame or a spark. Flash point varies inversely with the fuel's volatility. The flash point is the lowest temperature at which fuel emits enough vapors to ignite. Biodiesel has a high flash point, which is usually above 150°C, while generally conventional diesel fuel has a flash point of 55°C–66°C. Flash point is measured according to ASTM D93 and EN ISO 3679 specifications.

13.9.4 KINEMATIC VISCOSITY

Viscosity is defined as the resistance of liquid to flow. It refers to the thickness of the oil and is determined by measuring the amount of time it takes for a given volume of oil to pass through an orifice of a specified size. Kinematic viscosity is the most important property of biodiesel since it affects the operation of fuel injection equipment, particularly at low temperatures when an increase in viscosity affects the fluidity of the fuel. Moreover, a high viscosity may lead to the formation of soot and engine deposits due to insufficient fuel atomization. The kinematic viscosity of biodiesel is determined using ASTM D445 and EN ISO 3104 specifications.

13.9.5 CETANE NUMBER

The CN is a measure of the ignition quality of diesel fuel during combustion ignition. It provides information about the ID time of a diesel fuel upon injection into the combustion chamber. A high CN implies short ID. Fuels with low CN tend to cause knocking and show increased gaseous and particulate exhaust emissions due to incomplete combustion. CN is calculated relative to two compounds with assigned CN. These are hexadecane with a CN of 100 and heptamethyl nonane with a CN of 15.

13.9.6 DENSITY

Density is the relationship between the mass and volume of a liquid or a solid and can be expressed in units of grams per milliliter (g/mL). The density of diesel oil is important because it gives an indication of the delay between the injection and combustion of the fuel in a diesel engine (ignition quality) and the energy per unit mass (specific energy). This influences the efficiency of the fuel atomization for airless combustion systems. ASTM D1298 and EN ISO 3675/12185 standard test methods are used to measure the density of the biodiesel. According to these standards, density should be tested at the reference temperature of 15°C.

13.9.7 ACID NUMBER

The acid number is a measure of the amount of carboxylic acid groups such as a fatty acid, in a chemical compound, or in a mixture of compounds. Acid number can provide an indication of the level of lubricant degradation while the fuel is in service. Acid value or neutralization number is expressed in mg KOH required to neutralize 1 g of fatty acid methyl esters and is set to a maximum value of 0.5 mg KOH/g of oil in the European standard (EN 14104) and ASTM D664. A higher acid number can cause severe corrosion to a fuel supply system and internal combustion engine.

13.9.8   CARBON RESIDUE

Carbon residue test is used to indicate the extent of carbon deposits resulting from the combustion of a fuel. Carbon residue that is formed by decomposition and subsequent pyrolysis of the fuel components can clog the fuel injectors. Biodiesel made from vegetable oil feedstock has a carbon residue limit of 0.050% (by weight) according to ASTM D4530 and 0.3% (m/m) according to EN ISO10370 specification.

13.9.9   IODINE NUMBER

The iodine number is an index of the number of double bonds in a biodiesel chemical structure, which determines the degree of unsaturation of the biodiesel (number of unsaturated bonds in the alkyl chains of fatty acids). This property can greatly influence the oxidation stability and polymerization of glycerides, which can lead to the formation of deposits in diesel engine injectors. Iodine value is directly correlated to biodiesel viscosity, CN, and cold-flow characteristics (CFPP). The iodine number is set to a maximum value of 120 mg I_2/g according to EN 14111 specification.

13.9.10   CALORIFIC VALUE

Calorific value is an important parameter in the selection of a fuel. The calorific value of biodiesel is lower than that of biodiesel because of its higher O_2 content. Calorific value is measured according to ASTM D7544 specification in MJ/Kg of biofuel.

13.9.11   SULFUR CONTENT

The combustion of fuel containing sulfur causes emissions of sulfur oxides. Most of vegetable oil–and animal fat–based biodiesels have very low levels of sulfur content. However, specifying this parameter is important for engine operability. Sulfur content is measured according to ASTM D7039 specification in ppm.

13.9.12   SULFATE ASH CONTENT

The ash content describes the amount of inorganic contaminants such as abrasive solids and catalyst residues and the concentration of soluble metal soaps contained in a fuel sample. Sulfate ash content is set to a maximum value of 0.020% by the mass of biodiesel according to ASTM D874 specification.

13.9.13   WATER AND SEDIMENT CONTENT

The presence of water and sediment has two forms, which are either dissolved water or suspended water droplets. While biodiesel is generally considered to be insoluble in water, it actually takes up considerably more water than petroleum diesel fuel. The water content of biodiesel reduces the heat of combustion and will cause corrosion of vital fuel system components (fuel pumps, injector pumps, fuel tubes, etc.). Moreover, the sediment may consist of suspended rust and dirt particles, or it may originate from the fuel as insoluble compounds formed during fuel oxidation. The standard test methods for water content and sediment in biodiesel are ASTM D2709 and EN ISO 12937, respectively. Both methods limit the amount of water to 0.05 (vol.%).

13.9.14 FREE AND TOTAL GLYCERIN

Free and total glycerin is a measurement of how much triglyceride remains unconverted into methyl esters. Total glycerin is calculated from the amount of free glycerin, monoglycerides, diglycerides, and triglycerides. Structurally, triglyceride is a reaction product of glycerol molecule with fatty acid molecules, yielding three molecules of water and one molecule of triglyceride. Biodiesel made from vegetable oil feedstock has a total glycerin limit of 0.02 wt% according to ASTM D6584 specification.

13.9.15 PHOSPHOROUS, CALCIUM, AND MAGNESIUM CONTENT

Phosphorous, calcium, and magnesium are minor components that are typically associated with phospholipids and gums that may act as emulsifiers or cause sediment, lowering yields during transesterification process. The specifications from ASTM D6751 state that phosphorous content in biodiesel must be less than 10 ppm and calcium and magnesium combined must be less than 5 ppm. Phosphorous is determined using ASTM D4951 and EN 14107; calcium and magnesium are determined using EN 14538.

13.9.16 MOISTURE, IMPURITIES, AND UNSAPONIFIABLE CONTENT

MIU contents are the amount of water, filterable solids (such as bone fragments, food particles, or other solids), and other non-triglycerides in an oil that cannot be converted to monoalkyl fatty esters by esterification or transesterification. Hence, MIU contents must be removed before biodiesel production or during ester purification. Moisture is a minor component that can be found in all feedstocks. Moisture can react with the catalyst during transesterification, which can lead to soap formation and emulsions. Moisture in the biodiesels is measured in accordance to ASTM E203 and has a specification of 0.050 wt% maximum.

13.10 EFFECT OF CHEMICAL STRUCTURE ON BIODIESEL PERFORMANCE

Biodiesel performance depends on two main parameters included in the chemical structure of the triglyceride fatty acids. These are the unsaturation degree and content and the extent of hydroxyl groups in the HC chains of the fatty acids.

Benjumea et al. [142] indicated that the degree of unsaturation of biodiesel fuels did not significantly affect engine performance and the start of injection, but it had a noticeable influence on combustion characteristics and emissions, via its effect on the CN. A higher degree of unsaturation of biodiesel fuels led to a longer ID and, consequently, a more retarded start of combustion. Regardless of the engine operating mode, an almost constant start of injection was attained, while the premixed portion of combustion, peak heat release rate (HRR), maximum pressure gradient, peak in-cylinder bulk-gas-averaged temperature, total HC emissions, smoke opacity, and NO_x emissions increased with increasing degree of unsaturation.

Martins and coworkers [143] determined values of specific mass, kinematic viscosity, water content, acidity level, flash point, oxidative stability, and calorific value of *fish* oil biodiesel and assessed mandatory parameters regulated by the Brazilian National Agency of Petroleum, Natural Gas and Biofuels. They reported that biodiesel calorific value analysis levels were similar to those of diesel. Other parameters that characterize *fish* oil biodiesel physically and chemically were evaluated and indicated that *fish* oil is a promising alternative for biodiesel manufacturing.

Abbott [144] studied the effect of the degree of unsaturation of a biodiesel fuel (in term of iodine number) on pollutant emissions and combustion timing. Four biodiesel fuels with iodine numbers ranging from 90 to 125, pure and blended (30% and 70% biodiesel content) with a petroleum diesel reference fuel, in a four-cylinder, 2.2 L, turbocharged, direct injection diesel engine showed reductions in particle mass and opacity (60%–70%) and a slight increase in both fuel consumption (around 15% in mass) and NO_x emissions (9%). Additionally, the degree of unsaturation of biodiesel fuels was found to have significant effects on these emissions. As the biodiesel fuel became more unsaturated, NO_x emissions increased by 10% and particle mass emissions decreased by 20%. Regarding particle size distributions, biodiesel fuels showed a smaller mean diameter and delayed start of ignition but a higher HRR as biodiesel iodine number and unsaturation was increased.

Altun [145] tested three biodiesel fuels with iodine numbers ranging from 59 to 185 in a direct injection diesel engine–powered generator set at constant speed of 1500 rpm under variable load conditions to investigate the effect of the degree of unsaturation of biodiesel fuels. The tested fuel blends were quantified by the iodine number, on the performance and exhaust emissions of a diesel engine. Biodiesel fuels with increasing iodine value resulted in lower emissions of NO_x, CO, and smoke opacity, with some increase in emissions of unburned HCs. With their low energy content, neat biodiesel fuels resulted in an increase in fuel consumption compared to the conventional ultralow-sulfur diesel fuel.

The effect of the degree of unsaturation of biodiesel fuels on engine emissions, via its effect on the CN and adiabatic flame temperature, while engine performance was not significantly affected by the type of biodiesel fuel or its degree of unsaturation, was reported [146]. The biodiesel having lowest iodine number had highest CN, and lowest density and adiabatic flame temperature, which was good to reduce NO_x emissions. Additionally, more unsaturated biodiesel fuels showed higher NO_x emissions, smoke opacity, and lower HC emissions. It was observed that CN and adiabatic flame temperature are responsible for such results.

The composition and degree of unsaturation of the methyl ester present in biodiesel plays an important role in the chemical composition of emitted particulate matter. It was observed that linseed biodiesel produces more particulate matter and HCs than *palm* oil biodiesel as a consequence of more unsaturated compounds in its composition, which favor the dust precursor's formation in the combustion zone [147]. Thermogravimetric analysis showed that the amount of volatile material in the dust from biodiesel fuels was slightly lower than that of petroleum diesel fuel, but no significant differences were observed among biodiesels. Similarly, the chemical characteristics of the HCs of volatile material present in the particulate matter showed an increase in the aliphatic component as the degree of unsaturation of the fatty acid methyl ester increased.

The effect of biodiesel chemical structure (saturation degree and chain length) on diesel engine combustion was investigated [148]. Selected fatty acid methyl ester fuels included neat methyl esters and blends with *rapeseed* oil methyl esters to obtain the most significant properties in terms of particulate matter and NO_x emissions in a single-cylinder mechanical direct injection diesel engine for combustion and exhaust emissions. A correlation was obtained between exhaust emissions and the most significant fuel properties ranging from 88% to 97%. NO_x emissions were found to increase as chain length (with exception of C18:0) and the degree of unsaturation of fatty acid methyl esters increased. Total HC, CO, volatile organic fraction, and dust emissions increase as the chain length of HCs increases. Dust produced by short-chain length fatty acid methyl esters is easier to oxidize than dust from long-chain fatty acid methyl esters. The recommendation is that shorter-chain saturated fatty acid methyl ester fuels would be preferable for a combined emissions context.

Redel-Macías and colleagues [149] studied the effects of the chemical properties of biodiesel on exhaust and noise emissions at several engine operating conditions, using raw materials including high, medium, and low saturated fatty acids, that is, *palm, sunflower, olive pomace, coconut*, and *linseed* oil. The methyl esters were blended with petroleum diesel fuel and tested in a direct injection diesel engine. The use of biodiesel/petroleum diesel fuel blends reduced the emissions of CO and noise, thus improving the sound quality, while NO_x emission increased. Particularly, it was found that the higher the degree of unsaturation, the higher the emission of NO_x and the lower the CO emissions. The use of unsaturated biodiesel as partial diesel fuel substitute to improve CO emissions, besides engine sound quality to make the engine sound more pleasant, is recommended.

Islam et al. [150] scrutinized the influence of fatty acid composition and individual fatty acids on fuel properties using standard procedures (ASTM D6751 and EN 14214). The results confirmed the influences of the fatty acid profile. Furthermore, the results showed that polyunsaturated fatty acids increased the iodine value and had a negative influence on CN. Kinematic viscosity was negatively influenced by some long-chain polyunsaturated fatty acids such as C20:5 and C22:6 and some of the more common saturated fatty acids, such as C14:0 and C18:0.

The effect of fatty acid carbon chain length of biodiesel on the CNs of 29 samples of straight-chain, branched esters of C_1–C_4 and 2-ethylhexyl esters of various common fatty acids in an ignition quality tester was examined [151]. The CNs of these esters are not significantly affected by branching in the alcohol moiety. Therefore, branched esters, which improve the cold-flow properties of biodiesel, can be employed without greatly influencing ignition properties compared to the more common methyl esters. Unsaturation in the fatty acid chain was again the most significant factor causing lower CNs.

Knothe [152] determined the CN of neat trans fatty acid methyl esters of methyl-9-octadecenoate, methyl-9,12-octadecenoate, and C18:1 isomers methyl-6-octadecenoate and methyl-11-octadecenoate. The CNs of the positional and geometric isomers of methyl oleate were close to the CN of methyl oleate.

Tong et al. [153] used the measured CNs of pure fatty acid methyl esters and fatty acid methyl ester compositions and the reported CN for 59 different biodiesels collected from literature. They showed that the dependence of the CN on the carbon number varies with the degree of unsaturation of the fatty acid chain.

It was reported that CN, viscosity, and heating value of biodiesel increase with the increase of molecular weight, and these physical properties decrease as the number of double bonds increases [154]. Unlike the above properties, density decreases as molecular weight increases but increases as the degree of unsaturation increases.

A similar study reported that the viscosity of biodiesel fuels reduced considerably with an increase in unsaturation [155]. Contamination with small amounts of glycerides significantly affects the viscosity of biodiesel fuels.

Kinematic viscosity of biodiesel was reported to increase with the increasing chain length of either the fatty acid or alcohol moiety in a fatty ester or in an aliphatic HC [156]. The increase in kinematic viscosity with the increasing number of carbons was smaller in aliphatic HCs than in fatty compounds. The kinematic viscosity of unsaturated fatty compounds strongly depends on the nature and number of double bonds, with double bond position affecting viscosity less. Terminal double bonds in aliphatic HCs have a comparatively small viscosity-reducing effect. Branching in the alcohol moiety does not significantly affect viscosity compared to straight-chain analogues. Free fatty acids or compounds with hydroxyl groups possess a significantly higher effect on viscosity. The viscosity range of fatty compounds is greater than that of the various HCs found in conventional diesel. The effect of sulfur-containing compounds (dibenzothiophene) in conventional diesel fuel on the viscosity of toluene is less than that of fatty esters or long-chain aliphatic HCs.

Wang et al. [157] showed that fuel properties such as CN, iodine number, and oxidation stability are mainly determined by the degree of unsaturation of fatty acid in raw oils. When the degree of unsaturation (ratio between number of single and double bonds in the alkyl chain) of raw oils fatty acid is lower than 133.13, CN and iodine number meet the standards (GB/T 20828-2007). With the increase of the long-chain saturated fatty acids, CFPP of the biodiesel product tends to be higher. When the chain length saturated factors (the ratio between number of double and single bonds in the alkyl chain of fatty acid) is less than 8.41 and 2.72, it could meet the standards of 0°C and −10°C of the CFPP. The raw materials for high-quality biodiesel have a high content of monounsaturated fatty acids.

Hoekman et al. [158] reported that several fuel properties—including viscosity, specific gravity, CN, iodine value, and low temperature performance metrics—are highly correlated with the average unsaturation of the fatty acid methyl ester profiles. Due to the opposing effects of certain fatty acid methyl ester features, it is not possible to define a single composition that is optimum with respect to all important fuel properties. However, to ensure satisfactory in-use performance with respect to low-temperature operability and oxidative stability, biodiesel should contain relatively low concentrations of long-chain saturated and polyunsaturated fatty acid methyl esters.

Sokoto et al. [159] reported that some critical fuel parameters like oxidation stability, CN, iodine value, and viscosity correlated with the methyl ester composition and structural configuration. The CN and oxidation stability of the produced biodiesel is a function of the degree of unsaturation and long-chain saturated factor.

Levine et al. [160] sought to relate the effect of chain length, degree of unsaturation, and branching to the critical fuel property of the gross heat of combustion. It was found that heat of combustion (kJ/g) increased with chain length. A linear relationship was found between wt% carbon and hydrogen and heat of combustion.

Pinz et al. [161] studied the ideal biodiesel composition and concluded that the high presence of monounsaturated fatty acids (as oleic and palmitoleic acids), reduced presence of polyunsaturated acids, and controlled saturated acids content are recommended. In this sense, C18:1 and C16:1 are the best-fitting acids in terms of oxidative stability and cold weather behavior, among many other properties.

In order to obtain more oxidative stable biodiesel, Dantas et al. [162] measured the biodiesel oxidative stability by mixing *soybean* and *castor* ethyl esters. They showed that the introduction of *castor* oil biodiesel increased the blend's oxidative stability. The use of biodiesel blends was a good alternative in the correction of the oxidative stability of the final product without the need of antioxidant addition.

Biodiesels produced from a variety of real-world feedstocks as well as pure (technical grade) fatty acid methyl and ethyl esters were investigated for emissions performance in a heavy-duty truck engine [163]. The objective was to understand the impact of the biodiesel chemical structure, specifically fatty acid chain length and number of double bonds, on emissions of NO_x and particulate matter (PM). Seven biodiesels produced from real-world feedstocks and 14 produced from pure fatty acids were tested in a heavy-duty truck engine using the U.S. heavy-duty federal test procedure (transient test). It was found that the molecular structure of biodiesel can have a substantial impact on emissions. The properties of density, CN, and iodine number were found to be highly correlated with one another. For neat biodiesels, PM emissions were essentially constant at about 0.07 g/bhp-h for all biodiesels as long as density was less than 0.89 g/cm^3 or CN was greater than about 45.

In the suitability of biodiesel as engine fuel, Arumugam and Sriram [164] reported that the *castor* oil methyl ester with diesel blend up to 20% (B20) does not affect the engine oil performance with respect to viscosity, total acid number, dust, and wear. The performance of *castor* oil biodiesel is found to be comparable to that of petroleum diesel fuel, and the additional lubricity of biodiesel fuel due to higher viscosity as compared to diesel fuel resulted in lower wear of moving parts and thus improved the engine durability.

Castor biodiesel has interesting properties (low cloud and pour points) that show that this fuel is very suitable for use in extreme winter temperatures [165]. The increase in kinematic viscosity over a certain number of carbons is smaller in aliphatic HCs than in fatty compounds. The kinematic viscosity of unsaturated fatty compounds strongly depends on the nature and number of double bonds, with double bond position affecting viscosity less. Terminal double bonds in aliphatic HCs have a comparatively small viscosity-reducing effect. Branching in the alcohol moiety does not significantly affect viscosity compared to straight-chain analogues. Free fatty acids or compounds with hydroxyl groups possess a significantly higher viscosity.

The viscosity range of fatty compounds is greater than that of various HCs comprising petroleum diesel. The effect of dibenzothiophene, a sulfur-containing compound found in petroleum diesel fuel, on viscosity of toluene is less than that of fatty esters or long-chain aliphatic HCs. To further assess the influence of the nature of oxygenated moieties on kinematic viscosity, compounds with 10 carbons and varying oxygenated moieties were investigated. Overall, the sequence of influence on kinematic viscosity of oxygenated moieties is COOH $\approx$ C–OH > COOCH$_3$ $\approx$ C=O>C–O–C > no oxygen [166].

The effects of density and viscosity of biodiesels and their blends on various components of the engine fuel supply system such as fuel pump, fuel filters, and fuel injector using *rapeseed* oil biodiesel, *corn* oil biodiesel, and waste oil biodiesel blends were quantified by Tesfa and coauthors [167]. The effects of density and viscosity on the performance of the fuel supply system were modeled. The higher density and viscosity of biodiesel have a significant impact on the performance of fuel pumps and fuel filters as well as on air–fuel mixing behavior of compression ignition engine.

Concerning the effect of temperature, Farias and his colleagues [168] evaluated biodiesel blends of *passion fruit* and *castor* oils in different proportions and their thermal stability. Biodiesel blends of passion fruit and *castor* oils presented parameters in the standards of the Brazilian National Agency of Petroleum, Natural Gas and Biofuels. They indicated that *castor* oil biodiesel was more stable. Passion fruit biodiesel has a high content of oleic and linoleic acids, which are more susceptible to oxidation. The biodiesel blend of passion fruit and *castor* oils 1:1 increased the thermal stability in relation to passion fruit biodiesel. The biodiesel blend of passion fruit and *castor* oils 1:2 presented a higher thermal stability, because *castor* oil has a high content of ricinoleic acid.

Dantas et al. [169] measured the biodiesel oxidative stability using mixed *soybean* and *castor* ethyl esters. The introduction of *castor* oil biodiesel increased the blend stability. The use of biodiesel blends was a good alternative to improve the oxidative stability of the final product without the need of antioxidant addition.

13.11   STORAGE STABILITY OF BIODIESEL

The stability of biodiesel upon storage is an important property that affects the quality and efficiency of the biodiesel. Several factors affect the stability of biodiesel including the degree of unsaturation, O$_2$ content, impurities, and alkyl chains of fatty acids of triglycerides. Long storage time creates several problems to the biofuel such as oxidation into aldehydes and ketones, increase of water content, emulsification, and biodegradation by microorganisms.

The biodegradation extent of both aromatic and aliphatic HC fractions in saturated sandy microcosm spiked with diesel/biodiesel blends (D, B10, B20, B30, B40, B50, B60, B70, B80, B90, and B100, where D is a commercial petroleum diesel fuel and B is a commercial biodiesel blend) augmented with a bacterial consortium of petroleum degraders was evaluated [170]. The biodegradation kinetics of blends was based on measuring the amount of emitted CO$_2$ after 578 days. The biodegradation extents of both aliphatic and aromatic HCs were uninfluenced by the addition of biodiesel, regardless of the concentration used. Blending with biodiesel does not impact the long-term biodegradation of specific diesel oil fractions.

The biodiesel degradation characteristics under different storage conditions were investigated depending on the data obtained from monitoring the qualities of 12 bio-diesel samples, which were divided into three groups and stored at different temperatures and environments at a regular interval over a period of 52 weeks. The biodiesel under test degraded less than 10% within 52 weeks for those samples stored at 4°C and 20°C, while nearly 40% degradation was found for those samples stored at a higher temperature, that is, 40°C. High temperature and air exposure greatly increase the biodiesel degradation rate. The temperature or air exposure alone, however, had little effect on biodiesel degradation [171].

Yang and coworkers [172] investigated the storage stability of two commercially available biodiesels and their blends with diesel spiked with different impurities, which were stored at two different temperatures (15°C and 40°C) with air-tight and light screen. These samples were periodically monitored during the whole storage period by measuring a number of properties, such as acid value, induction time, and composition of fatty acid methyl esters. They reported the following observations:

1. Acid values increased but induction time decreased with the extension of storage for all samples without copper; however, both induction time and acid values kept nearly constant for all samples with copper.
2. The presence of water did not contribute significantly to the degradation of all tested samples over time.
3. A higher temperature (40°C) was favorable to the degradation of unsaturated fatty acid methyl esters, accompanied with the altered acid values and induction time in comparison with the same samples stored at 15°C.

Faster degradation of fatty acid methyl esters in blended samples than those in pure biodiesels may be partially due to the diluting effects of antioxidants in bio-diesel. However, the presence of any impurities did not affect the degradation rates of fatty acid methyl esters, which was not in agreement with the abovementioned acid values and induction time series. The addition of copper affected the acid values and induction time. Therefore, for samples with copper, fatty acid methyl ester profiles can represent their quality more appropriately than acid values and induction time.

The influences of oxidative variables on the burning characteristics of *palm* oil biodiesel were investigated by Lin and Chiu [173]. *Palm* oil biodiesel was stored at a constant-temperature water bath at either 60°C or 20°C continuously for 3000 h to observe the variation in its oxidative degradation and burning characteristics. Carbon residue and the amount of heat released, with storage time, and the effect of antioxidant addition on the burning characteristics were measured. The *palm* oil biodiesel suffered greater oxidative degradation at higher storage temperatures and in the absence of an antioxidant, which resulted in a faster decrease in the amount of heat released as the storage time elapsed. The oxidative stability of the *palm* oil biodiesel was worst at higher storage temperature, a longer storage time, and in the absence of an antioxidant, which caused a more extensive formation of sediments of oxidative products and hence a larger carbon residue after burning. The greater oxidative degradation also caused a more extensive decomposition of unsaturated fatty acids into oxidative products, which decreased the flash point and the cetane index of the fuel.

The changes in chemical structure upon storage, the mass and chemical composition changes in neat biodiesel from *soybean* and *canola* oil, diesel, and blends with diesel (5% and 20%, volume/volume), at ambient temperature, with air exposure lasting for 190 days, were monitored [174]. The losses in both the pure diesel and the blends of biodiesel and diesel were measured, but negligible changes in mass were observed for pure biodiesel. The chemical composition of the diesel petroleum HCs followed the expected evaporative-loss profile. However, fatty acid methyl esters did not show significant loss even after 190 days of exposure. The rate of evaporation of fatty acid methyl esters for pure biodiesel or blends was significantly lower than conventional diesel because of the higher boiling point of the biodiesel component compared with some petroleum HCs. Additionally, losses due to microbial activity and chemical degradation were negligible as the chemical composition of fatty acid methyl esters did not have significant variation from the starting materials even after 190 days exposure to air.

The influence of storage time of various blending biodiesel ratio under different storage temperature on fuel properties and exhaust emissions was investigated [175]. The biodiesel samples were stored in a clinical compartment, at different temperatures, and were monitored at regular intervals over a period of 60 days. Blends of biodiesel were varied from 5 vol% of biodiesel to 45 vol% of biodiesel and storage temperature from 5°C to 30°C. The effects of storage conditions on properties of biodiesel such as density, kinematic viscosity, acid value, water content, and flash point were discussed. The biodiesel density and viscosity are affected by storage duration and storage temperature. High blending ratio exhibits relative variant in emissions.

The influence of metal content on the storage of different biodiesels was studied by Kivevele et al. [176]. They evaluated the effect of metal contents and antioxidant additives on the storage stability of *croton* oil methyl ester and *Moringa* oil methyl ester biodiesel, doped with antioxidants hydroxy benzene (pyrogallol) and 3,4,5-trihydroxy benzoic acid (propyl gallate). They also mixed different transition metals (Fe, Ni, Mn, Co, and Cu). The samples were stored indoors for 6 months in completely closed and open translucent plastic bottles. The oxidation stability was measured every month. The fuel properties were measured according to EN 14214 and ASTM D6751 biodiesel standards. The freshly produced *croton* oil methyl ester and *Moringa* oil methyl ester without any additives displayed oxidation stability of 2.5 and 5.3 h, respectively. The oxidation stability of *croton* oil methyl ester did not meet the minimum requirement prescribed in ASTM D6751 and EN 14214 standards of 3 and 6 h, respectively. This was because it was rich in linoleic methyl esters (70.5%), which are prone to oxidation. The antioxidant propyl gallate was more effective than pyrogallol. Fe displayed a least detrimental effect on the oxidation stability of *croton* oil methyl ester and *Moringa* oil methyl ester, while Cu had great detrimental impact. The samples in the completely closed (air-tight) bottles recorded higher oxidation stability compared to the ones kept in the open bottles, because exposure to air enhanced the oxidation of the samples.

13.12   FUEL PROPERTIES

The fuel properties of biodiesel depend on the chemical structure, unsaturation, hydroxyl group content, HC chain length, and others. Several research groups have investigated the physical parameters of biodiesel and their blends.

Tesfa et al. [177] investigated the effects of types, fractions, and physical properties of waste oil, *rapeseed* oil and *corn* oil, and petroleum diesel on the combustion and performance characteristics of a compression ignition engine and on a four-cylinder, four-stroke, direct injection, and turbocharged diesel engine. The biodiesel types do not result in any significant differences in peak cylinder pressure and brake specific fuel combustion. The study also indicated that the engine running with biodiesel have a slightly higher in-cylinder pressure and HRR than the engine running with petroleum diesel. The brake specific fuel combustion for the engine running with neat biodiesel was higher than the engine running with normal diesel by up to 15%. It is also noticed that the physical properties of biodiesel affect the performance of the engine.

The effect of biodiesel molecular weight, structure, and the number of double bonds on the diesel engine operation characteristics was studied by Puhan and others [178]. The biodiesels were prepared and analyzed for fuel properties according to the standards. A constant speed diesel engine, which develops 4.4 kW of power, was run with biodiesels, and its performance was compared with petroleum diesel fuel. The *linseed* oil methyl ester with high linolenic acid content (unsaturated fatty acid ester) did not suit best for diesel engine due to high NO_x emission and low thermal efficiency.

Shahabuddin et al. [179] reported that biodiesel has an early start of combustion and shorter ID between 1°C–5°C and 0.25°C–1.0°C, respectively. A higher CN, lower compressibility, and fatty acid composition of biodiesel have been identified as the main elements for the early start of combustion and shorter ID. In addition, it was also found that the HRR of biodiesel was slightly lower than petroleum diesel owing to the lower calorific value, lower volatility, shorter ID, and higher viscosity. HRR is defined as the rate at which heat is generated by fire in diesel engines.

Qzturk [180] investigated performance, emissions, combustion, and injection characteristics of a diesel engine fueled with blends of diesel fuel No. 2 and a mixture of *canola* oil–hazelnut soap stock biodiesels. The hazelnut soap stock biodiesel was mixed with the *canola* oil biodiesel (diesel/biodiesel blends containing 5% [B5] and 10% [B10] biodiesel fuels) to improve some properties of the *canola* oil biodiesel and to reduce the cost of the fuel. The injection and ignition delays and the maximum HRRs decreased with the biodiesel addition, while the injection and combustion durations increased. In addition, it was determined that the O_2 content of B5 enhanced the combustion resulting in increased NO_x emission and decreased the total HC, CO, and smoke emissions. However, B10 fuel deteriorated the combustion due to higher density, viscosity, and surface tension. Therefore, total HC, CO, and smoke emissions increased, while NO_x emission decreased. CO_2 emissions for both blends were very similar to those of No. 2 diesel.

Allen et al. [181] investigated the effect of fatty ester composition on the combustion behavior of biodiesel fuel sprays of *soybean*-based methyl esters, canola-based methyl esters, and canola-based butyl esters. The appearance of first heat release from the fuels was less sensitive to O_2 concentration as the reaction zone temperature is increased. Growth in the ester alkoxy chain length from one to four carbon atoms had minimal effect on the ID. For tests at 12% O_2, the increased polyunsaturation of *soybean* methyl esters relative to canola methyl esters led to longer IDs. This was not observed at 18% O_2, where physical transport processes were more important. The canola butyl esters exhibited the largest peak values for HRRs, with the distinction becoming clear at reaction zone temperatures above 750 K. Normalization on ID and input energy bases indicates that variation of the maximum apparent HRRs among the fuels is primarily due to unique mixing times and fuel heating values.

Muralidharan et al. [182] estimated the performance, emission, and combustion characteristics of a single-cylinder, four-stroke variable compression ratio multi-fuel engine fueled with waste cooking oil methyl ester and its blends with standard diesel. Tests have been conducted using fuel blends of 20%, 40%, 60%, and 80% biodiesel with petroleum diesel, with an engine speed of 1500 rpm and fixed compression ratio 21, and at different loading conditions. They found that waste cooking oil methyl ester showed considerable improvement in the performance parameters as well as exhaust emissions. The blends, when used as fuel, resulted in the reduction of CO, HC, and CO_2 at the expense of NO_x emissions. It was found that the combustion characteristics of waste cooking oil methyl ester and its diesel blends closely followed those of petroleum diesel.

Ong et al. [30] tested the *C. inophyllum* biodiesel and petroleum diesel blends on the engine performance and emission characteristic. The performance and emission of 10% *C. inophyllum* biodiesel blends gave a satisfactory result in diesel engines as the brake thermal increase 2.30% and fuel consumption decrease 3.06% compared to petroleum diesel. Besides, 10% *C. inophyllum* biodiesel blends reduced CO and smoke opacity compared to petroleum diesel.

Kumar and Chauhan [183] showed that different chemical compositions of biodiesel depending on their origin lead to variation in their properties and performance and emission characteristics. Biodiesel produced from saturated feedstock reduced NO_x emission and resistance to oxidation but exhibited poor atomization.

Altun [145] tested three biodiesel fuels with iodine numbers ranging from 59 to 185 in a direct injection diesel engine–powered generator set at a constant speed of 1500 rpm under variable load conditions. The author investigated the effect of the degree of unsaturation of biodiesel fuels. The biodiesel fuels resulted in lower emissions of NO_x, CO, and smoke opacity, with some increase in the emissions of unburned HCs. With their low energy content, neat biodiesel fuels resulted in an increase in fuel consumption compared to the conventional petroleum diesel fuel (ultralow-sulfur diesel). The degree of unsaturation of biodiesel fuels had effects on engine emissions via its effect on the CN and adiabatic flame temperature. The engine performance was not significantly affected by the type of biodiesel fuel or its degree of unsaturation. The biodiesel having lowest iodine number had highest CN, and lowest density and adiabatic flame temperature, which was good to reduce NO_x emissions. Additionally, more unsaturated biodiesel fuels showed higher NO_x emissions, smoke opacity, and lower HC emissions. CN and adiabatic flame temperature are responsible for such results.

Canakci [184] compared the combustion characteristics and emissions of two different petroleum diesel fuels (No. 1 and No. 2) and biodiesel from *soybean* oil. The fuels were tested at steady-state conditions in a four-cylinder turbocharged direct injection diesel engine at full load and 1400 rpm engine speed. The biodiesel provided significant reductions in particulate matter, CO, and unburned HCs, while NO_x increased by 11.2%. Biodiesel had a 13.8% increase in BSFC due to its lower heating value. The biodiesel may be blended with No. 1 diesel fuel to be used without any modification on the engine.

Zhang et al. [185] investigated the combustion and emission characteristics of a turbocharged, common rail diesel engine fueled with diesel–biodiesel–diethyl ether blends. The BSFC of diesel–biodiesel–diethyl ether blends increased with the increase of oxygenated fuel fractions in the blends. Brake thermal efficiency shows little variation when operating on different diesel–biodiesel–diethyl ether blends. At a lower load, the NO_x emission of the diesel–biodiesel–diethyl ether blends exhibited little variation compared to the biodiesel fraction. The NO_x emission increased slightly with an increase in the biodiesel fraction in diesel–biodiesel–diethyl ether blends at medium load. NO_x emission increased remarkably with the increase in the biodiesel fraction at a high load. Particle mass concentration decreased significantly with the increase in the oxygenated fuel fraction at all engine speeds and loads; particle number concentration decreased remarkably with the increase in the oxygenated fuel fraction. HC and CO emissions decreased with increasing oxygenated fuel fraction in these blends.

Arslan [186] studied the use of waste cooking oil methyl ester as an alternative fuel in a four-stroke turbo diesel engine with four cylinders, direct injection. An engine fueled with diesel and three different blends of diesel/biodiesel (B25, B50, and B75) made from waste cooking oil showed that the biodiesel fuels produced slightly less smoke than the conventional diesel fuel, which could be attributed to better combustion efficiency. The use of biodiesel resulted in lower emissions of total HC and CO and increased emissions of NO_x. The exhaust emissions of diesel/biodiesel blends were lower than those of the diesel fuels, which indicate that biodiesel has more favorable effects on air quality.

Qi et al. [187] showed that the power output of biodiesel (*soybean* oil) was almost identical with that of petroleum diesel. The BSFC was higher for biodiesel due to its lower heating value. Biodiesel provided significant reduction in CO, HC, NO_x, and smoke under speed characteristic at full engine load. Based on the study, biodiesel can be used as a substitute for petroleum diesel in diesel engine.

Ejim et al. [3] showed that coconut biodiesel (B100) had similar atomization characteristics to petroleum diesel fuels No. 2 (D2) because of its similar properties, that is, density, surface tension, and viscosity. No significant differences in drop size were observed for all B5 blends (5% biodiesel–95% petroleum diesel), B20 blends (20% biodiesel–80% petroleum diesel), and B100 biodiesels of *palm, soybean, cottonseed, peanut,* and *canola.* These stocks of biodiesels and their blends can be used in direct injection engine with similar atomization characteristics. Ternary biodiesel blends, with 10 wt% petroleum diesel, can yield equal drop sizes as some binary blends with large quantities of two different petroleum diesel fuels No. 1 and No. 2 (D1 and D2). The ternary biodiesel blends reduce pollution from exhaust emissions better than the biodiesel blends with petroleum diesel fuels No. 1 and No. 2.

Rahman et al. [188] investigated the effect of the physical properties and chemical composition of biodiesels on engine exhaust particle emissions. Alongside with neat petroleum diesel, four biodiesels with variations in carbon chain length and degree of unsaturation at three blending ratios of B100 (pure biodiesel), B50 (50% biodiesel–50% petroleum diesel), and B20 (20% biodiesel–80% petroleum diesel) were investigated. The particle emission increased with increase in carbon chain length. However, for similar carbon chain length, particle emissions from biodiesel having a relatively high average unsaturation were found to be slightly less than that of low average unsaturation. Particle size was also found to be dependent on fuel type. Fuel or fuel mixture was responsible for higher particle mass, particle number emissions, and larger median particle size. Particle emissions reduced consistently with fuel O_2 content regardless of the proportion of biodiesel in the blends, whereas it increased with fuel viscosity and surface tension only for higher diesel–biodiesel blend percentages (B100, B50). Overall, it was evident from the results presented that the chemical composition of biodiesel was more important than its physical properties in controlling particle emissions.

Nguyen and Pham [189] studied the link between the O_2 content and engine performance as well as emission concentrations when using petroleum diesel and biodiesel blends B10 and B20 (0%, 10%, and 20% of biodiesel by volume in biodiesel–petroleum diesel mixtures, respectively). The biodiesel was derived from residues of the manufacturing process of *palm* cooking oil. The biodiesel was used on a modern common-rail, single-cylinder engine operating under a fixed injection condition (injection timing, pressure, and duration) but with a wide range of engine speeds. The O_2 enrichment fuel blends lower the engine power, which is attributable to lower heating values of the biodiesel blends compared to that of petroleum diesel. However, the O_2 content in the blends suppresses particle formation and decreases HC and CO concentrations.

Masjuki and Abdulmunin [190] showed that polyunsaturated fatty acids increased the iodine value and had a negative influence on CN. Kinematic viscosity was negatively influenced by some long-chain polyunsaturated fatty acids such as C20:5 and C22:6 and some of the more common saturated fatty acids such as C14:0 and C18:0.

Pattamaprom et al. [191] compared the storage degradation characteristics of biodiesels derived from two palm products, which are palm olein and palm stearin. The evaluation was performed in terms of chemical properties, engine performance, and exhaust emission by keeping biodiesels in dark closed-lid containers at room temperature for up to 6 months. The O_2 present in the container led to slow degradation of biodiesels through oxidative reaction with the double bonds in biodiesel. Within 6 months, the majority of oxidative products were composed of shorter hydroperoxide compounds and other short secondary products. These changes resulted in lower heating value and higher density of biodiesels, which in turn caused reductions in fuel combustion efficiency and fuel economy. In terms of emission, the degraded biodiesel produced more complete combustion as indicated by lower emissions of black smoke and CO but with higher emission of NO_x. In terms of *palm* oil type, even though palm olein biodiesel possessed a higher degree of unsaturation and produced higher peroxide value and acid values from the degradation, its combustion efficiency and fuel economy were superior to the biodiesel produced from palm stearin possibly due to its higher chain lengths.

Karavalakis et al. [192] investigated the regulated and unregulated emissions profile and fuel consumption of an automotive diesel and biodiesel blends, prepared from *rapeseed* methyl ester and a *palm* methyl ester. They evaluated the impact of biodiesel chemical structure on the emissions, as well as the influence of the applied driving cycle on the formation of exhaust emissions and fuel consumption. The study showed that NO_x emissions were influenced by certain biodiesel properties such as cetane and iodine numbers. NO_x emissions followed a decreasing trend over both cycles, where the most beneficial reduction was obtained with the application of the more saturated biodiesel. Particulate matter emissions decreased with the *palm*-based biodiesel blends over both cycles, with the exception of the 20% blend, which was higher compared to diesel fuel. Petroleum diesel–biodiesel blends led to increases in PM emissions over the nonlegislated Athens Driving Cycle.

Ra et al. [193] showed the significant effects of the fuel physical properties on ID and burning rates under various engine operating conditions. There is no single physical property that dominates the differences of ID between diesel and biodiesel fuels. However, among the different properties of the fuel, the simulation results were found to be most sensitive to liquid fuel density, vapor pressure, and surface tension through their effects on the mixture preparation processes.

Wang et al. [194] studied experimentally and analytically the spray characteristics of biodiesels (from *palm* and cooked oil) and petroleum diesel under ultrahigh injection pressures of up to 300 MPa. The estimation on spray droplet size showed that biodiesels generated larger SMD due to higher viscosity and surface tension.

Gao et al. [195] studied the spray characteristics of inedible oil using experimental and simulation methods. The spray was more concentrated, due to higher viscosity and surface tension of the biodiesel, compared to conventional diesel fuel. The macroscopic and microscopic spray properties of blended fuels containing 5%, 10%, and 20% biodiesel were similar to petroleum diesel.

Yoon et al. [196] performed an experimental investigation to analyze the effects of undiluted biodiesel fuel (100% methyl ester of *soybean* oil) on spray, combustion, and exhaust emission characteristics in a direct injection common-rail diesel engine with a cooled exhaust gas recirculation (EGR) system. EGR system is an engine system that reuses a portion of an engine's exhaust gas in the engine cylinders to decrease NO_x emission. The injection characteristics of test fuels such as injection rate and mass results were very similar for diesel and biodiesel fuels derived from *soybean* oil. However, when comparing the start of injection, biodiesel resulted in a short injection delay. The SMD of biodiesel fuel was higher than that of diesel fuel, and biodiesel showed a longer spray tip penetration as compared to that of diesel fuel. The combustion of diesel and biodiesel fuels indicated similar patterns of combustion pressure and rate of heat release. However, the combustion of biodiesel showed lower peak combustion pressures and peak HRRs than those of diesel fuel because of its lower heating value. The comparison of emission characteristics of biodiesel and diesel fuels showed that the biodiesel fuel emitted higher indicated specific NO_x and a remarkably low level of soot.

Suh and Lee [197] investigated the effect of injection parameters on the characteristics of dimethyl ether as an alternative fuel in a diesel engine with experimental and analytical models based on empirical equations. The investigation focused on the atomization characteristics of dimethyl ether and compared experimental and predicted results for spray development obtained by empirical models for diesel and dimethyl ether fuels. The atomization characteristics indicate that dimethyl ether showed better spray characteristics than conventional diesel fuel. Also, the fuel injection delay and maximum injection rate of dimethyl ether fuel were shorter and lower than those of diesel fuel under the same injection conditions.

Suh et al. [198] studied the factors influencing the spray and combustion characteristics of biodiesel fuel in experiments involving exhaust emissions and engine performance at various biodiesel blending ratios and injection conditions for engine operating conditions. The macroscopic and microscopic spray characteristics of biodiesel fuel such as injection rate, split injection effect, spray tip penetration, droplet diameter, and axial velocity distribution were compared with the results from conventional diesel fuel. The results showed that the spray tip penetration of biodiesel fuel was similar to that of diesel. The atomization characteristics of biodiesel showed that it has higher SMD and lower spray velocity than conventional diesel fuel, due to high viscosity and surface tension. The peak combustion pressures of diesel and blending fuel increased with advanced injection timing, and the combustion pressure of biodiesel fuel was higher than that of diesel fuel. As the pilot injection timing was retarded to 15° that was closed by the top dead center. The dissimilarities of diesel and blending fuels combustion pressure are reduced. The pilot injection enhanced the deteriorated spray and combustion characteristics of biodiesel fuel caused by different physical properties of the fuel.

Wang et al. [199] investigated spray, ignition, and combustion characteristics of biodiesel fuels under a simulated diesel engine condition (885 K, 4 MPa) in a constant volume combustion vessel. Two biodiesel fuels were used, which originated from *palm* oil and used cooking oil. The study showed that biodiesel fuels give appreciably longer liquid lengths and shorter IDs. At low injection pressure (100 MPa), biodiesel fuels gave shorter lift-off lengths than those of diesel. While at high injection pressure (200 MPa), the lift-off length of biodiesel fuel from *palm* oil gave the shortest value and that of biodiesel from used cooking oil gave the longest one. Air entrainment upstream of lift-off length of three fuels was estimated and compared to soot formation distance. They revealed that the viscosity and ignition quality of biodiesel fuel have great influences on jet flame structure and dust formation tendency.

Kim et al. [200] investigated the spray behaviors of biodiesel and dimethyl ether fuels using image processing and atomization performance analysis of the two fuel sprays injected through a common-rail injection system under various ambient pressure conditions in a high-pressure chamber. They showed that the ambient pressure had a significant effect on the spray characteristics of the fuels under various experimental conditions. The spray tip penetration and spray area decreased as the ambient pressure increased. The contour plot of biodiesel and dimethyl ether sprays showed a high light intensity level in the central regions of the sprays. The study revealed that the atomization performance of the biodiesel spray was inferior to that of dimethyl ether spray under the same injection and ambient conditions.

Gumus et al. [201] discussed the effects of fuel injection pressure on the exhaust emissions and BSFC of a direct injection diesel engine. BSFC is a measure of the fuel efficiency of any prime mover that burns fuel and produces rotational, or shaft, power. It is typically used for comparing the efficiency of internal combustion engines with a shaft output. It is the rate of fuel consumption divided by the power produced. The engine with biodiesel–petroleum diesel blends was running at four different fuel injection pressures (18, 20, 22, and 24 MPa) and four different engine loads (12.5, 25, 37.5, and 50 kPa). The BSFC, CO_2, NO_x, and O_2 emissions increased, whereas smoke opacity, unburned HC, and CO emissions decreased and thus were attributed to the fuel properties and combustion characteristics of biodiesel. The increased injection pressure caused a decrease of BSFC for the high percentage biodiesel–diesel blends (such as B20, B50, and B100), smoke opacity, and emissions of CO and unburned HC. It also increased the emissions of CO_2, O_2, and NO_x. The increased or decreased injection pressure caused the increase in BSFC values compared to the original injection pressure of diesel fuel and low percentage biodiesel–diesel blends (B5).

Park et al. [202] investigated the spray atomization characteristics of an undiluted biodiesel fuel of *soybean* oil methyl ester in a diesel engine when compared with that of diesel fuel ultralow-sulfur diesel. The experimental results were compared with numerical results predicted by the KIVA-3V code. The SMD was analyzed using a droplet analyzer system to investigate the atomization characteristics. SMD is defined as the average diameter of solid particles produced from the ignition of the fuel. They found that the peak injection rate increased and advances when the injection pressure increases due to the increase of the initial injection momentum. The injection rate of the *soybean* methyl ester was higher than that of diesel fuel despite its low injection velocity. The high ambient pressure induced the shortening of the spray tip penetration of the *soybean* methyl ester. The SMD of the *soybean* methyl ester decreased along the axial distance. The predicted local and overall SMD distribution patterns of diesel and *soybean* methyl ester fuels illustrated similar tendencies when compared to the experimental droplet size distribution patterns. KIVA is a program used for determining the fuel parameters of engines with vertical or canted valves, and KIVA-3V is the most mature version of KIVA with an improved version of the earlier Federal Laboratory Consortium Excellence in Technology Transfer and extended to model vertical or canted valves in the cylinder head of a gasoline or diesel engine.

Chen et al. [203] investigated spray and atomization characteristics for commercial No. 2 diesel fuel, biodiesel derived from waste cooking oil (B100), 20% biodiesel blended diesel fuel (B20), renewable diesel fuel produced in house, and civil aircraft jet fuel (Jet-A). All experiments were conducted by employing a high-pressure common-rail fuel injection system with a single-hole nozzle under room temperature and pressure. The biodiesel and jet fuel had different features compared with diesel. Longer spray tip penetration and larger droplet diameters were observed for B100. The smaller droplet size of the Jet-A were believed to be caused by its low viscosity and surface tension. B20 showed similar characteristics to diesel but with slightly larger droplet sizes and shorter tip penetration.

Kegl and Pehan [204] discussed the influence of biodiesel on the injection, spray, and engine characteristics with the aim to reduce harmful emissions on bus diesel engine with injection system. The injection, fuel spray, and engine characteristics obtained with biodiesel were compared to those obtained with petroleum diesel under various operating regimes using neat biodiesel from *rapeseed* oil. The obtained results were used to analyze the most important injection, fuel spray, and engine characteristics. The injection characteristics were determined numerically under the operating regimes. The study indicated that biodiesel reduces harmful emissions (NO_x, CO, smoke, and hydrocarbons) to some extent by adjusting the injection pump timing properly.

Shojaeefard et al. [205] indicated that lower blends of biodiesel provide acceptable engine performance and even improve it. Meanwhile, exhaust emissions are much decreased. Finally, a 15% blend of *castor* oil biodiesel was picked as the optimized blend of biodiesel–diesel. It was found that lower blends of *castor* biodiesel are an acceptable fuel alternative for the engine.

13.13 ADVANTAGES AND DISADVANTAGES OF BIODIESEL FUEL

13.13.1 ADVANTAGES OF BIODIESEL

13.13.1.1 Produced from Renewable Resources

Biodiesel is a renewable energy source unlike petroleum products that will deplete in the years to come. Since it is made from animal and vegetable fat, it can be produced on demand and also causes less pollution than petroleum diesel [206].

13.13.1.2 Can Be Used in Existing Diesel Engines

One of the main advantages of using biodiesel is that it can be used in existing diesel engines with little or no modifications at all. It can replace fossil fuels to become the most preferred primary transport energy source. Biodiesel can be used in 100% (B100) or in blends with petroleum diesel. For example, B20 is 20% blend of biodiesel with 80% petroleum diesel fuel. It improves engine lubrication and increases engine life since it is virtually sulfur-free [207].

13.13.1.3 Less Greenhouse Gas Emissions

Fossil fuels when burnt release greenhouse gases like CO_2 in the atmosphere that raises the temperature and causes global warming. To protect the environment from further heating up, many people have adopted the use of biofuels. Experts believe that using biodiesel instead of petroleum diesel can reduce greenhouse gases up to 78% [208].

13.13.1.4 Grown, Produced, and Distributed Locally

Fossil fuels are limited and may not be able to fulfill our demand for coal, oil, and natural gas beyond a certain period. Biodiesel can work as an alternative form of fuel and can reduce our dependence on foreign suppliers of oil as it is produced from domestic energy crops. It is produced in local refineries, which reduces the need to import expensive finished product from other countries [209].

13.13.1.5 Cleaner Biofuel Refineries

When oil is extracted from underground, it has to be refined to run diesel engines. You cannot use it straight away in the crude form. When it is refined, it releases many chemical compounds including benzene and butadiene into the environment, which are harmful to animals, plants, and human life [210].

13.13.1.6 Biodegradable and Nontoxic

When biofuels are burnt, they produce significantly less carbon output and few pollutants. As compared to petroleum diesel, biodiesel produces less soot (particulate matter), CO, unburned HCs, and sulfur dioxide (SO_2). The flash point for biodiesel is higher than 150°C, whereas it is about 52°C for petroleum diesel, which makes it less combustible. It is therefore safe to handle, store, and transport [211].

13.13.1.7 Better Fuel Economy

Vehicles that run on biodiesel achieve 30% more fuel economy than petroleum-based diesel engines, which means they make fewer trips to gas stations and run more miles per gallon [212].

13.13.1.8 Positive Economic Impact

Biofuels are produced locally, and thousands of people are employed in a biofuel production plant. Since biodiesel is produced from crops, an increase in demand for biodiesel leads to increase in demand for biofuel crops. Moreover, it creates less emission by reducing the amount of suspended particles in the air, which consequently reduces the cost of healthcare [213].

13.13.1.9 Reduced Foreign Oil Dependence

With locally produced biofuels, many countries have reduced their dependence on fossil fuels. It may not solve all problems in one stroke, but a nation can save billions of dollars by reducing their usage of foreign oil [214].

13.13.1.10 More Health Benefits

Air pollution cause more deaths and diseases than any other form of pollution. Pollutants from gasoline engines when released in the air, form smog and make thousands of people sick every year. Biodiesel produces less toxic pollutants than petroleum products [215].

13.13.2 Disadvantages of Biodiesel

13.13.2.1 Variation in Quality of Biodiesel

Biodiesel is made from a variety of biofuel crops. When the oil is extracted and converted to fuel using chemical process, the result can vary in ability to produce power. In short, not all biofuel crops are the same as the amount of vegetable oil produced may vary [216].

13.13.2.2 Not Suitable for Use at Low Temperatures

Biodiesel gels in cold weather, but the temperature at which it will gel depends on the oil or fat that was used to make it. The best way to use biodiesel during the colder months is to blend it with winterized diesel fuel [217].

13.13.2.3 Food Shortage

Since biofuels are made from animal and vegetable fat, a greater demand for these products may raise their prices and create food crises in some countries. For example, the production of biodiesel from corn may raise its demand and corn might become more expensive, which may deprive poor people from buying it [218].

13.13.2.4 Increased Use of Fertilizers

As more crops are grown to produce biofuels, more fertilizers will be used, which in turn can have a devastating effect on the environment. Excess use of fertilizers can result in soil erosion and can lead to land pollution [219].

13.13.2.5 Regional Suitability

Some regions are not suitable for oil-producing crops. The most productive crops cannot be produced everywhere, and they need to be transported to the plants, which increases the cost and the amount of emission associated with the production and transportation [220].

13.13.2.6 Water Shortage

The use of water to produce more crops can put pressure on local water resources. In an area with water scarcity, the production of crops to be used in making biofuels is not a wise idea [221].

13.13.2.7 Monoculture

Monoculture refers to the practice of producing the same crop over and over again rather than producing different crops. While this results in fetching best price for the farmer, it has some serious environmental drawbacks. When the same crop is grown repeatedly over large acres, the pest population may grow, and it may go beyond control. Without crop rotation, the nutrients of soil are not put back, which may result in soil erosion [222].

13.13.2.8 Fuel Distribution

Biodiesel is not distributed as widely as petrodiesel. The infrastructure still requires a boost so that it is adopted as the most preferred way to run engines [223].

13.13.2.9 Use of Petro-Diesel to Produce Biodiesel

It requires a great deal of energy to produce biodiesel fuel from soy crops, as energy is needed for sowing, fertilizing, and harvesting crops. Apart from that, raw material needs to be transported by trucks, which may consume additional fuel [224].

13.13.2.10 Slight Increase in Nitrogen Oxide Emissions

Biodiesel produces 10% higher NO_x than other petroleum products. Nitrogen oxide is one of the gases responsible for the formation of smog and ozone [225].

13.14 BIODIESEL AND DIESEL ENGINE EMISSIONS

The U.S. Environmental Protection Agency designates six criteria pollutants for determining air quality. These are CO, NO_x, SO_x, ground-level ozone (O_3), particulate matter (including particles like soot, dust, asbestos fibers, pesticides, and metals), and lead [226].

Petroleum-fueled vehicles, engines, and industrial processes directly release the vast majority of CO and NO_x into the atmosphere. They are also the principal source of gaseous HCs (also called volatile organic compounds), which combine with NO_x in sunlight to create O_3. Ozone, while important for blocking ultraviolet rays in the upper atmosphere, is also a key component of urban smog and creates human health problems when present in the lower atmosphere. SO_2 is a trace component of crude oil and can cause acid rain when released into the air by oil refineries or petroleum power plants. Particulate matter is directly emitted via vehicle exhaust and can also form from the reaction of exhaust gases with water vapor and sunlight. Finally, leaded gasoline is a huge contributor of lead in the atmosphere, and the use of unleaded gasoline has decreased lead concentrations dramatically. The Environmental Protection Agency is working to encourage the phaseout of leaded gasoline worldwide. Petroleum-fueled transportation and coal-burning power plants are considered the chief causes of global warming. Excess amounts of CO_2, methane, and NO_x, among other gases, trap heat in the atmosphere and create the greenhouse effect. CO_2 is a main constituent of petroleum fuel exhaust, even though it is not toxic and therefore not classified as a pollutant. About one-third of the CO_2 emitted into the atmosphere every year comes from vehicle exhaust. Methane, although usually associated with natural gas, is also emitted whenever crude oil is extracted, transported, refined, or stored.

13.14.1 Emissions

Petroleum products give off the following emissions when they are burned as fuel; these by-products have negative impacts on the environment and on human health:

1. Carbon dioxide (CO_2).
2. Carbon monoxide (CO).
3. Sulfur dioxide (SO_2).
4. Nitrogen oxides (NO_x) and volatile organic compounds (VOC).
5. Particulate matter (PM).
6. Lead and various air toxins such as benzene, formaldehyde, acetaldehyde, and 1,3-butadiene may also be emitted when some types of petroleum are burned.

Roy et al. [227] tested a direct injection (DI) diesel engine with biodiesel–petroleum diesel and *canola* oil–diesel blends for performance and emissions at high idling operations. Three fuel series were examined: pure canola biodiesel, used canola biodiesel, and pure *canola* oil. In all the series, fuels were blended with petroleum diesel 2–20 vol%. The emission contained CO, HC, nitric oxide (NO), NO_2, NO_x, CO_2, and others. Pure and used canola biodiesel blends showed very similar fuel properties in both engine performance and emissions. CO and HC emissions from biodiesel–diesel blends were significantly less than pure petroleum diesel fuel. Pure canola oil up to 5% in diesel fuel showed significantly less CO emissions than that of petroleum diesel fuel. Up to 5% biodiesel and canola oil in diesel fuel, either reduced or main-tained similar level of NO_x emissions to that of petroleum diesel fuel.

Lue et al. [228] developed a new domestic biodiesel production procedure using waste fryer vegetable oil by transesterification method and investigated its emission characteristics on a small direct injection diesel engine using biodiesel blends and die-sel fuels. The emission characteristics include smoke emissions, gaseous emissions (CO, HC, NO_x, and SO_2), particle size distributions, and number concentrations at a variety of steady-state engine speed points. The diesel engine fueled with biodiesel blends emitted more particle number concentrations than those with diesel fuel. Particle number concentration increased as biodiesel concentration increased. As for the smoke and gaseous emissions, such as CO, HC, NO_x, and SO_2, they favored biodiesel blends.

Valente et al. [229] tested the impacts on fuel consumption and exhaust emissions of a diesel power generator operating with fuel blends of 5%, 20%, 35%, 50%, and 85% of *soybean* biodiesel in petroleum diesel oil, and 5%, 20%, and 35% of *castor* oil biodiesel in petroleum diesel oil, using varying engine load from 9.6 to 35.7 kW. At low and moderate loads, CO emission increased by nearly 40% and over 80% when fuel blends containing 35% of *castor* or *soybean* oil biodiesel were used in comparison to unblended diesel oil.

Valente et al. [230] studied the exhaust emissions from a diesel power genera-tor operating with waste cooking oil biodiesel blends of 25%, 50%, and 75% of biodiesel concentration in diesel oil. The addition of biodiesel to the fuel increased NO_x, CO, CO_2, exhaust gas opacity, and HC emissions. A major increase of NO_x was observed at low loads, while CO and HCs mostly increased at high loads. Using 50% of fuel mixture increased CO_2, CO, HC, and NO_x throughout the load range by 8.5%, 20.1%, 23.5%, and 4.8%, respectively.

Shaheed and E. Swain [231] discussed the effects of fuel injection pressure on the exhaust emissions and BSFC of a direct injection diesel engine using biodiesel–diesel blends. The BSFC, CO_2, NO_x, and O_2 emission increased; smoke opacity, unburned HC, and CO emissions decreased due to the fuel properties and combustion charac-teristics of biodiesel. On the other hand, the increased injection pressure caused a decrease in the BSFC of high percentage biodiesel–diesel blends (such as B20, B50, and B100), smoke opacity, emissions of CO, and unburned HC and increased the emissions of CO_2, O_2, and NO_x.

Sayin et al. [232] investigated the effects of fuel injection timing on the exhaust emission characteristics of a single-cylinder, direct injection diesel engine, using *canola* oil methyl ester–diesel blends. The BSFC and CO_2 and NO_x emissions increased, but smoke opacity, HC, and CO emissions decreased because of the fuel properties and combustion characteristics of *canola* oil methyl ester. The effect of injection timing on the exhaust emissions of the engine exhibited similar trends for diesel fuel and canola oil methyl ester–diesel blends. The emissions of NO_x and CO_2 increased, and the smoke opacity and the emissions of HC and CO decreased for all test conditions.

Ozsezen et al. [233] discussed the performance and biodiesel combustion char-acteristics in a direct injection diesel engine obtained from methyl ester of *palm* oil and canola oil. The biodiesels caused reductions in CO and HC emissions and smoke opacity but also increased NO_x emissions.

Sharon et al. [234] tested biodiesel produced from *palm* oil blended with diesel by different volume proportions (25%, 50%, and 75%). Biodiesel and its blends were used in a direct injection diesel engine at constant speed with varying loads (between 20% and 100%). They analyzed its performance, emission, and combustion profile. Smoke densities of B100 and B75 were lower than petroleum diesel by 19% and 10% at full load, respectively. At full load, CO emission from B100 and B75 were 52.9% and 35.2%, and lower than diesel HC emission. NO_x emission was higher for all biodiesel blends. B75 showed a lower amount of emissions.

Rahman et al. [235] evaluated fuel consumption and emission parameters of diesel engines under high idling conditions using diesel blended with *Jatropha curcas* biodiesel. Compared to diesel fuel, biodiesel–diesel blends decreased CO and HC emissions and increased NO_x emissions.

Karabektas [236] investigated the effects of turbocharger on the performance of a diesel engine using diesel fuel and biodiesel in terms of CO and NO_x emissions. The author tested aspirated four-stroke direct injection diesel engine with diesel fuel and neat biodiesel, which is *rapeseed* oil methyl ester, at full load conditions at speeds between 1200 and 2400 rpm with intervals of 200 rpm. A turbocharger system was installed on the engine. The author showed that CO in the operations with biodiesel was lower than those in the operations with diesel fuel, whereas NO_x emission in biodiesel operation was higher.

Ghorban et al. [237] compared the combustion of B5, B10, B20, B50, B80, and B100 with petroleum diesel over wide input air flows at two energy levels in an experimental boiler, in terms of flue gas emissions (CO, CO_2, NO_x, and SO_2). The results showed that except for B10, biodiesel and other blends emitted less pollutant CO, SO_2, and CO_2 than diesel. B10 emitted lower CO_2 and NO_x, but emitted higher SO_2 than diesel.

Mohamed and Bhatti [238] measured the effect of different blends of a *soybean* methyl ester with diesel fuel on the cylinder pressure, HRR, CO, HC, NO_x, and smoke opacity using a single-cylinder, direct injection diesel engine. The use of biodiesel produced less smoke opacity of up to 48.23%, with 14.65% higher BSFC compared to diesel fuel. CO emissions of B20% and B100% of *soybean* methyl ester were 11.36% and 41.7% lower than diesel fuel. All blends of *soybean* methyl ester were emitting lower HC concentration compared to that of diesel. NO_x emissions are observed to be higher for all blends of *soybean* methyl ester.

Qi et al. [239] examined the effects of biodiesel addition to diesel fuel on the performance, emissions, and combustion characteristics of a naturally aspirated direct injection compression ignition engine using *soybean* biodiesel, produced from *soybean* crude oil by a method of alkaline-catalyzed transesterification. They showed significant improvement in the reduction of CO and smoke for biodiesel and its blends at high engine loads. HC had no evident variation for all tested fuels. NO_x were slightly higher for biodiesel and its blends.

Öztürk et al. [180] investigated performance, emissions, combustion, and injection characteristics of a diesel engine fueled with blends of diesel fuel No. 2 and a mixture of *canola* oil–*hazelnut* biodiesels, using a single-cylinder direct injection diesel engine. The O_2 content of B5 enhanced the combustion resulting in increased

NO_x emission and decreased total HC, CO, and smoke emissions. However, B10 fuel deteriorated the combustion due to higher density, viscosity, and surface tension. Therefore, for B10, total HC, CO, and smoke emissions increased, while NO_x emission decreased. CO_2 emissions for both blends were very similar to those of No. 2 diesel.

Dhar and Agarwal [240] investigated *karanja* biodiesel and its blends on particulate size-number distribution, size-surface area distribution, and total particle number concentration at various engine operating conditions. The peak number concentration of particulates increased with increasing engine speed. Smaller concentrations of *karanja* biodiesel (up to 20%) were effective in reducing the particle number emissions.

Cheung et al. [241] investigated the gaseous and particle emissions of a four-cylinder naturally aspirated direct injection diesel engine fueled with different mixed concentrations of *castor* oil biodiesel–diesel fuel B10 (diesel containing 10% vol. of biodiesel), B20, B30, pure biodiesel, and pure diesel. The biodiesel led to reduction of HC, CO, and particle mass concentrations and number concentrations but an increase in NO_x.

Uyumaz et al. [242] examined JP-8 aviation fuel and *sunflower* methyl ester blends at 7.5, 11.25, 15, and 18.75 Nm engine loads and at maximum torque speed (2200 min^{-1}) in a single-cylinder and direct injection diesel engine. For all test fuels, as the engine load increases, the specific fuel consumptions decrease, and NO_x emissions increased with the increase in the amount of biodiesel. CO emissions decreased as the amount of biodiesel fuel increased in the test fuels.

Rahman et al. [235] evaluated fuel consumption and emission parameters under high idling conditions when diesel blended with *Jatropha curcas* biodiesel is used to operate a diesel engine. They showed that biodiesel–diesel blends decreased CO and HC emissions, but increased NO_x emissions in high idling modes.

Utlu and Koçak [243] tested methyl ester obtained from waste frying oil in a diesel engine with turbocharged, four cylinders, and direct injection and compared results with No. 2 diesel fuel. They showed that the amount of emission such as CO, CO_2, NO_x, and smoke darkness of waste frying oils was less than No. 2 diesel fuel alone.

Gopal et al. [244] compared blends of biodiesel made from waste frying oil with diesel in direct injection diesel engines. The waste frying oil biodiesel had similar characteristics to that of petroleum diesel. The BSFC, CO, HC, and smoke opacity were lower in waste frying oil biodiesel blends than in diesel alone. Specific energy consumption and NO_x of waste frying oil biodiesel blends were found to be higher than diesel.

Selvam and Vadivel [245] investigated methyl esters of beef tallow as neat biodiesel (B100) and its blend (B5, B25, B50, and B75) with diesel fuel to determine exhaust emissions of CO, HCs, NO_x, and smoke density, at different loading conditions and at a constant engine speed of 1500 rpm. A drastic reduction in CO, unburned HC, and smoke density was observed for all the blended fuels as well as for neat biodiesel. However, in the case of NO_x, there was a slight increase for all the blended fuels and the neat biodiesel relative to diesel fuel.

Kalam et al. [246] evaluated emission and performance characteristics of a multicylinder diesel engine operating on waste cooking oil blends of 5% *palm* oil with 95% ordinary diesel fuel (P5) and 5% coconut oil with 95% ordinary diesel fuel (C5). Pure petroleum diesel was used for comparison purposes. The blended fuels reduced exhaust emissions such as unburned HCs, smoke, CO, and NO_x.

Nabi et al. [247] investigated exhaust emissions of neat diesel fuel and diesel–biodiesel blends in a four-stroke naturally aspirated direct injection diesel engine. They showed lower CO and smoke emissions but higher NO_x emission. However, compared to the diesel fuel, NO_x emissions with diesel–biodiesel blends were slightly reduced.

Saqib et al. [248] studied biodiesel production using *rapeseed* oil as feedstock and $NaOCH_3$ as a catalyst. The results showed that CO and particulate matter emissions of *rapeseed* oil methyl ester were lower than those of diesel fuel. NO_x emissions of *rapeseed* oil methyl ester were lower for B5, B20, B40, and B50, while higher for B80 and B100. These results show the environment benefits of biodiesel.

13.15   BIODIESEL TOXICITY

Toxicity studies [249] for biodiesel emissions are far from complete. Toxicity, like the biology of people, animals, and their many diseases, is a broad and complex subject. A wide array of toxicity tests is available to probe the ability of materials to cause or enhance many different diseases. The tests assess different properties of potential toxins, so it is not surprising that biodiesel emissions appear to be less toxic than diesel emissions in some assays and more toxic in others. Given the inherent variability of the results, it is too early to tell if biodiesel from any particular feedstock will be more toxic than biodiesel from other feedstocks. On the positive side, most (but not all) studies [250] have shown lower toxicity for particulate matter emissions from biodiesel compared with particulate matter emissions from petroleum diesel. The reduction in toxicity is modest however, in the range of 10%–40%. The average particle size from biodiesel (which affects lung deposition properties) is reported to increase modestly in some studies and decrease in others. It is most likely that there is little influence from changes in particle size on particle toxicity. On the other hand, several studies [251] report the gas-phase and semivolatile fractions of the emissions from biodiesel combustion may be more toxic than the same fraction from conventional diesel combustion. Semivolatiles are organics that can be in either the gas or particle phase depending primarily on the temperature. Volatile emissions from biodiesel, as expected, contain many more oxygenated compounds, including several carbonyls such as formaldehyde. Formaldehyde, a known air toxicant, is highly elevated in biodiesel exhaust compared to diesel exhaust. Biodiesel fuel itself also appears to be moderately less toxic than petroleum diesel fuel, again by 10%–40%. At this point, it is not clear if the lower or higher toxicity aspects of biodiesel emissions will dominate, but it is clear that from a toxicity point of view, biodiesel fuel is not a panacea. It is important to note that emissions from the combustion of unprocessed oils (raw feedstocks that have not been converted to biodiesel) exhibit a highly elevated toxicity relative to conventional diesel, often by a factor of eight or more [252]. Several researches and reviews were published describing comparisons between the toxicity of biodiesel, unprocessed oil, petroleum diesel, and biodiesel–petroleum diesel blends.

Lapinskienė et al. [253] compared diesel fuel to biodiesel fuel by determining the toxicity of analyzed materials and by quantitatively evaluating the microbial transformation of these materials in nonadapted aerated soil. The toxicity levels were determined by measuring the respiration of soil microorganisms and the activity of soil dehydrogenases. They found that diesel fuel has toxic properties in higher concentrations of 3% (w/w), while biodiesel up to a concentration of 12% (w/w). Diesel fuel is more resistant to biodegradation and produces more humus products.

Leme et al. [254] assessed the cytotoxic effects of water systems contaminated with neat biodiesel and its diesel blends by means of different procedures on human T-cell leukemia and human hepatocellular carcinoma cells (detection of changes in mitochondrial membrane potential using tetramethyl rhodamine ethyl ester). They observed the cytotoxic effects as a dose-dependent response only for water contaminated with pure diesel (D100) and diesel–biodiesel (B5) blend. They hypothesized that diesel accounts for those adverse effects observed and that biodiesel does not worsen the effects of diesel pollution.

Leme et al. [255] assessed the genotoxicity of water-soluble fractions (WSF) of biodiesel and its diesel blends using the *Salmonella* assay and the *in vitro* MicroFlow® kit (Litron) assay. They showed that care while using biodiesel should be taken to avoid harmful effects on living organisms in cases of water pollution.

Demirbas [224] reported that biodiesel fuel has better properties than petroleum diesel fuel. It is renewable, biodegradable, nontoxic, and essentially free of sulfur and aromatics. Biodiesel seems to be a realistic fuel for the future. It has become more attractive recently because of its environmental benefits. Biodiesel is an environmentally friendly fuel that can be used in any diesel engine without modification.

Pereira et al. [256] compared the water-soluble fractions from biodiesel and biodiesel/diesel blends to diesel in their sublethal toxicity to microalgae carried out in the water-soluble fractions (WSF). They showed positive correlation with increasing diesel concentrations (B100 < B5 < B3 < B2 < D). Biodiesel interacted with the aqueous matrix, methanol, which showed lower toxicity than the diesel contaminants in blends. Water-soluble fractions from biodiesel caused 50% culture growth inhibition (LC_{50}-96 h) at concentrations varying from 2.3% to 85.6%, depending on the tested fuels and species. LC_{50}, also called sublethal concentration, is the concentration of a compound that can kill 50% of the microorganisms in the medium. However, the same species sensitivity trend (*S. costatum* > *N. oculata* > *T. chuii* > *P. subcapitata*) was observed for all the tested fuels.

Fregolente et al. [257] conducted water absorbance of biodiesel and biodiesel–diesel fuel blends evaluating the temperature and blend ratio parameters. They showed that water in biodiesel ranged from 1500 to 1980 mg/kg in the temperature range of 283–323 K, which was 10–15 times higher than diesel. At constant relative humidity, biodiesel absorbed 6.5 times more moisture than diesel. The presence of free and/or emulsified water in biodiesel and blends was determined through turbidity experiments.

Leite et al. [258] determined the toxicity to two marine organisms of the water-soluble fractions of three different biodiesel fuels obtained by methanol transesterification of *castor* oil, *palm* oil, and waste cooking oil. They showed that the highest toxicity was for *castor* oil, followed by water-soluble fractions and the *palm* oil.

Methanol was the most prominent contaminant; its concentrations increased over time in samples of water-soluble fractions stored up to 120 days.

Khan et al. [259] evaluated the effect of neat biodiesel, biodiesel blends, and diesel on *Oncorhynchus mykiss* and *Daphnia magna* using acute toxicity testing. Static nonrenewal bioassays of freshwater organisms containing B100, B50, B20, B5, and conventional diesel fuel were used to compare the acute effects of biodiesel to diesel. The percent mortality and lethal concentration (LC_{50}) at different exposure times were determined from the acute toxicity tests performed. Trials were considered valid if the controls exhibited >90% survival. Based on the percentage of mortality and LC_{50} values, a toxicity ranking of fuels was developed.

Rodinger [260] estimated the toxic and ecotoxic hazards of biodiesel and diesel petroleum fuel by means of unispecies testing, testing a microcosm, and analyzing the biodegradability extent. In the first test, biodiesel and diesel were applied directly to animals or to the medium the animals or plants lived in. During the second test, biodiesel and diesel were evaluated and the residue was tested for toxic effects. Results showed that biodiesel is much better for the environment when compared to petroleum diesel fuel.

Batista et al. [261] analyzed *soybean* biodiesel stored at room temperature for 5 years and *soybean* biodiesel obtained by accelerated oxidation, by gas chromatography–mass spectrometry. Results showed that the main degradation products were 2-hexanal, 2-heptenal, and 2,4-decadienal, which increase the toxicity of biodiesel.

Dvm [262] reported that the relative health effects of biodiesel and petroleum diesel exhaust can best be estimated by comparing the dust mass produced, the mutagenicity of organic emissions, and the irritant potentials of the two materials when inhaled. Biodiesel contains less dust mass than petroleum diesel, and a greater portion of biodiesel dust is soluble. The soluble organic fraction of biodiesel dust is less mutagenic than that of petroleum diesel dust. These differences suggest that the carcinogenic potential of inhaled biodiesel exhaust is probably less than that of petroleum diesel exhaust. Depending on the fuel, engine, and operating conditions, irritating gases and vapors can be greater or lesser for biodiesel than for petroleum diesel. Overall, the presented information suggests that the health risks with biodiesel might be less than with petroleum diesel, and that the future of biodiesel fuels will be limited more by economic factors than by health concerns. Cruz et al. [263] investigated the biodegradation of contaminated soil with biodiesel, diesel, and petroleum by autochthonous soil microorganisms and also of soil enriched with *Bacillus subtilis* using colorimetric method. The phytotoxicity was evaluated in recently contaminated soil and after 240 days to ensure the decrease of toxicity by the use of *Lactuca sativa* seeds. The biodegradation assessment was carried out with redox 2,6-dichlorophenol indophenol indicator and by the extraction of the contaminant in the soil with hexane. The amount of contaminant extracted from recently contaminated soil was compared to the amount found in the buried samples for 240 days. They revealed that the autochthonous microorganisms were active on recently contaminated soil with biodiesel, because all the biodiesel was biodegraded. Hence, only 0.001 g of biodiesel was extracted, and the phytotoxicity decreased after 240 days. On the other hand, the contaminated soil with diesel and petroleum was less active in 2,6-dichlorophenol indophenol test, and consequently, there was a large amount of contaminant in soil after 240 days. Furthermore, petroleum and diesel were phytotoxic after biodegradation.

The complex composition of petroleum and diesel requires interactions of the microbial community able to biodegrade HCs and metabolites from biodegradation. The naturally present microorganisms in the soil were capable of degrading the pollutant as much as the samples enriched with *B. subtilis*. The 2,6-dichlorophenol indophenol test is a simple and inexpensive method to analyze the potential biodegradation of all microorganisms of the soil and if inoculation of the biodegrading microorganisms will be necessary. Therefore, it would be helpful in bioremediation strategies.

Borges et al. [264] produced different types of biodiesel by a homogeneous alkali transesterification reaction using *soybean* oil, *pork lard*, and *castor* bean oil as raw materials, to evaluate how their different compositions may affect biodegradability, namely, in the presence of benzene. Biodiesels were characterized according to the European standard EN 14214. The anaerobic biodegradation of the different types of biodiesel was examined as well as its influence on the biodegradation of benzene. They showed that methane production occurred from the anaerobic degradation of all biodiesel types. The differences between the degradation behaviors of these fuels were negligible, contrary to what was expected. However, the amount of methane produced was low due to nutrient limitations. This fact was confirmed by the organic acid analysis and by the addition of new media. Anaerobic benzene biodegradation was found to be negatively impacted by the presence of all biodiesel types on average. Therefore, the results of this study may impact the management of sites that contain biodiesel and fuel HC contamination.

Cruz et al. [265] evaluated the loss of the lysosomal membrane integrity in liver homogenate of juvenile *Tilapia* exposed to biodiesels–waste sunflower oils, through the increase in acid phosphatase activity, as an evidence of cytotoxicity. Differences in the enzyme activity levels (3.4, 2.3, and 0.8 mU/mg total protein over the control value, which was 1.6 mU/mg total proteins), found for *castor* oil, waste cooking oil, and *palm* oil biodiesels, respectively, were indicative of their toxicity according to this decreasing trend. Waste sunflower oil chromatograms suggest the cytotoxicity as related to methanol.

13.16 BIODIESEL QUALITY FROM DIFFERENT SOURCES AS COMPARED TO CONVENTIONAL PETROLEUM DIESEL FUEL

Several researches investigated the quality between the petroleum diesel fuel and the biodiesel fuel from different sources. The results of the studies varied depending on the studied properties.

Ávila and Sodré et al. [266] investigated the characteristics of *fodder radish*, a potential biodiesel source. The results showed that *fodder radish* biodiesel can meet physicochemical property specifications, although its acid number requires attention.

Silitonga et al. [267] investigated the opportunity of biodiesel characterization and production from *Ceiba pentandra* seed oil. The suggested biodiesel–diesel blending improved the properties of biodiesel such as viscosity, density, flash point, calorific value, and oxidation stability. It can be concluded that this feedstock can be considered as a potential source for biodiesel production.

Lin and Li [268] used the soap stock of a mixture of marine fish, refined fish oil transesterified to produce biodiesel. The fuel properties of the biodiesel were analyzed. Results showed greater kinematic viscosity, higher heating value, higher cetane index, more carbon residue, and a lower peroxide value, flash point, and distillation temperature than those of waste cooking oil biodiesel.

Gandure et al. [269] investigated fuel characteristics of biodiesel derived from *kernel* oils of *Sclerocarya birrea, Tylosema esculentum, Schiziophyton rautanenii,* and *Jatropha curcas* plants in comparison with petroleum diesel. Flash point, CP, kinematic viscosity, density, calorific value, acid value, and free fatty acids were measured using biodiesel standards such as ASTM D6751 and EN 14214. They found the heating values of *S. birrea* and *S. rautanenii* biodiesel fuels were 9.2% and 10.3% lower than that of petroleum diesel, while those of *T. esculentum* and *J. curcas* were both 9.7% lower. Other fuel properties analyzed demonstrated that biodiesel fuels produced from *kernel* oils of *S. birrea, T. esculentum, S. rautanenii,* and *J. curcas* plants have properties that were comparable to, and in some cases better than, those of petroleum diesel.

Lin and Fan [270] investigated the fuel properties of biodiesel produced from *Camellia oleifera* Abel oil through supercriticalmethanol transesterification with no catalyst. The fuel properties of the resulting biodiesel were analyzed and compared with those of a commercial biodiesel and with ASTM No. 2D diesel fuel. They found that the biodiesel produced from *C. oleifera* Abel oil had more favorable fuel properties than the commercial biodiesel produced from waste cooking oil, including a higher heat of combustion, higher flash point, and lower levels of kinematic viscosity, water content, and carbon residue. Moreover, the former appears to have much lower peroxide and acid values, and thus a much higher degree of oxidative stability than the latter.

Valente et al. [271] evaluated the physicochemical properties of the fuel blends of waste cooking oil biodiesel or *castor* oil biodiesel with petroleum diesel oil. The properties evaluated were fuel density, kinematic viscosity, cetane index, distillation temperatures, and sulfur content, which were measured according to standard test methods. The results were analyzed based on the present specifications for biodiesel fuel in Brazil, Europe, and the United States. They showed that density and viscosity increased with increasing biodiesel concentration, while fuel sulfur content was reduced. Cetane index decreased with high biodiesel content in diesel oil. The biodiesel blends distillation temperatures T10 and T50 were higher than those of diesel oil, while the distillation temperature T90 was lower. T10, T50, and T90 were designated for the temperature at which 10%, 50%, or 90% of the biodiesel volume evaporates during distillation. The maximum biodiesel concentration in diesel oil that meets the required characteristics for internal combustion engine application was evaluated, based on the results obtained.

13.17 EFFECT OF WASTEWATER IRRIGATION ON BIODIESEL QUALITY

Wastewater effluent can constitute an important source of irrigation water and nutrients for bioenergy crop cultivations, with minor adverse impacts on soil properties and seed yield. Plant species play an important role with regard to the changes in soil properties and to the related factors of seed and biodiesel yields.

Tsoutsos et al. [272] studied the use of wastewater as irrigation feedstock for cultivations of *sunflower* and *castor* crops and to track the effect of critical parameters on oil and biodiesel quality in terms of measured parameters such as oil yield, acid value, density, and viscosity. They observed that wastewater irrigation can have a positive impact on oil quality for biodiesel production.

Gandure et al. [273] investigated physicochemical properties, performance, and emission characteristics of *Jatropha* methyl ester and its blends with petroleum diesel. They showed that viscosity values for all fuels fall within specifications of ASTM, with a maximum variation of 21% observed between B0 and B100. Cold-flow properties of cloud and pour points indicate that *Jatropha* methyl ester and its blends can power the diesel engine without much difficulty in cold weather. The flash points of *Jatropha* methyl ester and its blends were found to be lower than the ASTM specification of a minimum of 130°C, implying that the fuels are highly flammable and need extreme handling precaution during transportation. Biofuels performed better in engine performance when compared to petroleum diesel in terms of brake power, BSFC, and BTE. BTE is defined as the brake power of a heat engine as a function of the thermal input from the fuel. It is used to evaluate how well an engine converts the heat from a fuel to mechanical energy. This is largely attributed to higher combustion efficiency due to extra inbound O_2. A higher combustion efficiency of biofuels led to the reduced production of HC, CO, and CO_2 emissions when compared to petroleum diesel. Petroleum diesel was also observed to produce the highest proportion of soot during combustion in the magnitude of approximately 3% per 3 mL of fuel.

Martínez et al. [274] measured the kinematic viscosity, density, heating value, oxidation stability, lubricity, water content, CFPP, and smoke point for different tire pyrolysis liquid/diesel blends. CFPP is the lowest temperature, expressed in degrees Celsius (°C), at which a given volume of diesel type of fuel still passes through a standardized filtration device in a specified time when cooled under certain conditions. The values for both pure fuels and blends meet the requirements established for diesel fuel (standard EN 590). Other tire pyrolysis liquid properties data reported in literature, determined the flash point, the distillation curve, the calculated cetane index, and the total acid number have also been determined for the pure fuels. These properties showed the useful window for the utilization of tire pyrolysis liquid as a fuel in diesel engines or other power systems. Mixing rules for calculating some key properties of tire pyrolysis liquid/diesel blends as a function of tire pyrolysis liquid content have also been used in order to model the blend behavior.

Wang et al. [275] conducted a comparative study of the composition, biodiesel production, and fuel properties of nonedible oils from *Euphorbia lathyris*, *Sapium sebiferum*, and *Jatropha curcas*. They showed that, under optimal conditions, the fatty acid methyl esters content and yield of the three oils were greater than 97.5 wt% and 84.0%, respectively. The best biodiesel was produced from *Euphorbia lathyris* due to its high monounsaturation (82.66 wt%), low polyunsaturation (6.49 wt%), and appropriate proportion of saturated components (8.78 wt%). Namely, *Euphorbia lathyris* biodiesel possessed a CN of 59.6, an oxidation stability of 10.4 h, and a CFPP of −11°C. However, the CN (40.2) and oxidative stability (0.8 h) of dewaxed *Sapium sebiferum kernel* oil biodiesel were low due to the high polyunsaturation

(72.79 wt%). In general, the results suggest that *E. lathyris* is a promising species for biodiesel feedstock.

Hu et al. [276] reported for the first time the detailed impacts of cultivation period on growth dynamics and biochemical composition of a microalga strain *Nannochloropsis gaditana*. They showed that the biomass accumulation, lipid content, neutral lipid content, monounsaturated fatty acid composition, or the favorable fatty acid profile of C_{16}–C_{18} increased along with the cultivation period extension. However, the lipid productivity displayed a decrease for those cultured after 16 days, with the highest value reaching 289.51 mg/L/day. Biodiesel properties of this microalga also changed with the cultivation period extension, with average unsaturated degree decreased from 1.3 to 0.6, CP increased from 3.3°C to 12.1°C, CN increased from 54.5 to 58.9, and iodine number reduced sharply from 105.1 to 56.4 g I_2/100 g biodiesel, all of which satisfied the specifications of biodiesel standard.

Johnson and Wen [277] subjected the biodiesel prepared via the direct transesterification of dry biomass to ASTM standard tests. Parameters such as free glycerol, total glycerol, acid number, soap content, corrosiveness to copper, flash point, viscosity, and particulate matter met the ASTM standards, while water and sediment content, as well as the sulfur content, did not pass the standard. Collectively, they indicate the alga *S. limacinum* is a suitable feedstock for producing biodiesel via the direct transesterification method.

The production of biodiesel by transesterification of crude *rice bran* oil was studied and compared according to fuel properties of *rice bran* oil biodiesel using ASTM D6751-02 and DIN V51606 standards for biodiesel [278]. The study showed that most fuel properties complied with the limits prescribed in the aforementioned standards. A subsequent engine test showed a similar power output compared to regular diesel, but the consumption rate was slightly higher. Emission tests showed a marked decrease in CO, HCs, and particulate materials, however, with a slight increase in NO_x.

Karmee and Chadha [279] prepared biodiesel from the nonedible oil of *Pongamia pinnata* by transesterification of the crude oil with methanol in the presence of KOH as a catalyst. Important fuel properties of methyl esters of *Pongamia* oil (biodiesel) compared well (viscosity = 4.8 cSt at 40°C and flash point = 150°C) with ASTM and German biodiesel standards.

Crabbe et al. [280] showed that the fuel properties of biodiesel and the biodiesel–(acetone–butanol–ethanol) mixture were comparable to that of No. 2 diesel, but their CNs and boiling points of the 90% fractions were higher. Therefore, they could serve as efficient No. 2 diesel substitutes. The biodiesel–(acetone–butanol–ethanol) mixture had the highest CN.

Ali et al. [281] considered aspects related to the production of biodiesel from *neem* oil and investigating the fuel properties of biodiesel obtained from *neem* oil. The fuel properties including density, kinematic viscosity, and calorific value were within the standards of biodiesel properties, especially ASTM D6751-02, where the recommended standard of kinematic viscosity lies between 1.90 to 6.0 cSt.

Silitonga et al. [40] showed that *S. oleosa* methyl ester exhibited a satisfactory oxidative stability of 7.23 h and high CN (50.6) compared to petroleum diesel (49.7). In addition, *S. oleosa* methyl ester has good pour and cloud points of −3.0°C and −1.0°C, respectively, due to its high unsaturated fatty chain.

Encinar et al. [282] showed that the viscosity, CN, and CFPP of *castor* oil methyl ester biodiesel do not attain the values established by the standard EN 14214. These circumstances would force to use *castor* biodiesel oil mixed with diesel fuel oil, or to add additives in order to improve its properties to the specified values.

One of the most important roles in obtaining oil from microalgae is the choice of species [283]. A total of fifteen microalgal isolates, obtained from brackish and fresh waters, were assayed at the laboratory for their ability for high biomass productivity and lipid content. Only three microalgae were selected as the most potent isolates for biomass and lipid production. They have been identified as *Chlorella vulgaris*, *Scenedesmus quadri*, and *Trachelomonas oblonga*. All of them were cultivated on BG11 media and harvested by centrifugation. The dry weight of the three isolates was recorded as 1.23, 1.09, and 0.9 g/L, while the lipid contents were 37%, 34%, and 29%, respectively. This can be considered as a promising biomass production and lipid content.

The densities of *castor* oil methyl esters and *castor* oil ethyl esters are higher than the limit defined by the standard EN 14214 [284]. The viscosities are more than twice higher than the limit value. The CNs are lower than defined by the standard EN 14214. For the remaining parameters, *castor* oil methyl esters and *castor* oil ethyl esters meet the standard EN 14214. The presence of the free hydroxyl group has virtually no effect on the values of such parameters as carbon residue, filterability at low temperatures, and oxidation stability.

Abdelmalik et al. [285] synthesized alkyl ester derivatives from laboratory puri-fied *palm kernel* oil. The steps in the synthesis involved transesterification of *palm kernel* oil to produce a methyl ester, followed by epoxidation and then the grafting of side chains by esterification with propionic and butyric anhydride. The melting point of the ester derivatives was reduced with side chain attachment, and antioxidant improved its thermal stability.

May et al. [286] synthesized ethyl and isopropyl esters of crude *palm* oil and crude *palm stearin* via chemical transesterification reactions and subsequently evaluated their fuel properties. Generally, these alkyl esters exhibited higher viscosity com-pared to that of petroleum diesel. However, compared to petroleum diesel, these alkyl esters exhibited an acceptable gross heat of combustion (39–41 MJ/kg). Originated from a renewable source, the low sulfur content in alkyl esters emits much lower SO_2. These alkyl esters are much safer than petroleum diesel in terms of safety for storage and transportation as they possess high flash points. They may find applica-tions in the fuel industry besides utilization as oleo-chemicals.

Jurcak et al. [287] compared the sublethal effects of biodiesel and crude oil expo-sure on chemically mediated behaviors in a freshwater keystone species of crayfish (*Orconectes rusticus*) for their ability to respond appropriately to a positive chemical stimulus within a Y-maze choice paradigm. They indicated negative impacts of both biodiesel and crude oil on the ability of crayfish to locate the food source. However, there were no significant differences between behavioral performances when cray-fish were exposed to crude oil compared with biodiesel. Thus, biodiesel and crude oil have equally negative effects on the chemosensory behavior of crayfish. These findings indicate that biodiesel has the potential to have similar negative ecological impacts as other fuel source toxins.

Wyatt et al. [288] prepared fatty acid methyl esters of *lard*, *beef tallow*, and chicken fat by base-catalyzed transesterification for use as biodiesel fuels. Properties of the neat fat-derived methyl esters (B100) meet ASTM standard. The cold-flow properties of the fat-based fuels were less desirable than those of soy-based biodiesel. The three animal fat–based B20 fuels had lower NO_x emission levels (3.2%–6.2%) than did soy-based B20 fuel. These biodiesels were comparable to or better than soy-based biodiesel.

Yahyaee et al. [289] determined the high heating value, viscosity, density, flash point, acidity index, moisture content, and ash content measurements as well as FTIR and GC-MS analyses to characterize the pyrolytic oil obtained during waste fish pyrolysis at 500°C. They revealed that the bio-oil properties indicated a good calorific value (~9391 kcal/kg) compared to European biodiesel specifications. In contrast, higher acidity (103 mg KOH/g sample) and viscosity (7 cSt) compared to conventional fuels were obtained. These properties limit the direct use of bio-oils as alternative fuel in a diesel engine. An efficient mixture with fossil fuel may be a promising solution to improve the fuel properties.

Yang et al. [121] studied the characteristics of intermediate pyrolysis oils derived from sewage sludge and de-inking sludge (a paper industry residue), with a view to their use as fuels in a diesel engine. The organic fraction of the oils was separated from the aqueous phase and characterized. This included elemental and compositional analysis, heating value, cetane index, density, viscosity, surface tension, flash point, total acid number, lubricity, copper corrosion, water, carbon residue, and ash content. Most of these results were compared with commercial diesel and biodiesel. They showed that both pyrolysis oils have high carbon and hydrogen contents and their higher heating values compare well with biodiesel. The water content of the pyrolysis oils is reasonable and the flash point is found to be high. Both pyrolysis oils have good lubricity, but show some corrosiveness. Cetane index is reduced, which may influence ignition. Also viscosity is increased, which may influence atomization quality.

13.18 TRIBOLOGICAL ISSUES WITH BIODIESEL

13.18.1 Favorable Tribological Effect of Biodiesel

The lubricity issue is significant, because the advent of low-sulfur petroleum diesel fuels, as required by regulations in the United States and Europe, has led to the failure of engine parts, such as fuel injectors and pumps, because they are lubricated by the fuel itself [166]. It was reported [290,291] that neat biodiesel possesses inherently greater lubricity than petroleum diesel, and that adding biodiesel at low blend levels (1%–2%) to low-sulfur petrodiesel or aviation fuel [292]. Such effectiveness was reported for even lower (1%) blend levels or higher (10%–20%) levels.

Mondal et al. [293] studied the effects of biodiesel blends on engine performance and lubricating oil. Two long duration endurance tests of 1000 h were conducted using petroleum diesel and 5% biodiesel blend (B5) on a new generation multicylinder engine. Straight fatty acid esters engine oil meeting API was used for both the tests.

The change in viscosity of engine oil with both the fuels is within the specified oil rejection limits. The minor difference in engine viscosity change by using B5 as fuel is within the repeatability values of engine test. The endurance test of B5 indicates Fe levels between 7 and 35 ppm, which is very well below the specified lubricant rejection limit of 150 ppm. The wear values of the Fe for petroleum diesel were in the range of 25–65 ppm. Moreover, other metals such as Cu, Al, and Si were within the specified limits for both the fuels. These results seem to imply that the alkyl esters that largely comprise biodiesel are responsible for this lubricity enhancement. Free fatty acids enhanced the boundary lubrication behavior of sunflower oil formulations [294]. The esters of vegetable oils with hydroxylated fatty acids such as *castor* and *Lesquerella* oils improved lubricity at lower levels than the esters of nonhydroxylated vegetable oils [295,296]. Oxidized biodiesel showed improved lubricity, compared to its nonoxidized counterpart [297].

13.18.2 SOME SERIOUS TRIBOLOGICAL PROBLEMS WITH BIODIESEL

Some researchers reported an adverse effect of biodiesel on the tribological performance of engine and requires further attention. Srinivasa and Gopalakrishnan [298] investigated the neat soybean oil–derived biodiesel (B100) and its 50 vol% blend with petroleum diesel (B50) on diesel passenger cars. Metals concentration in the lubricant was determined, in order to assess the biodiesel impact on wear. The analysis of the lubricating oil samples showed that the use of B50 and B100 may lead to increased wear in terms of higher amounts of metallic elements, originating from the different moving parts. Even though biodiesel demonstrates better lubricity properties compared to diesel fuel, the wear of various vital parts of the engine was higher during the test fuel application. Iron and copper content implies cylinder and bearing wear, which increased by 67% and 272%, respectively [299].

Batko et al. [300] investigated the aspect of lubricity in steel–aluminum alloys for regular petroleum diesel and biodiesel of methyl esters of *colza* oil, and their mixtures using a friction machine with a roller–ring friction couple. The results showed that a small supplement of methyl esters of *colza* oil to diesel fuel causes a sudden increase of the wear. The highest wear value was obtained with a 20% supplement of methyl esters of *colza* oil. A possible explanation of higher wear is that high biodiesel concentrations partly dissolve the lubricant. The friction coefficient of the engine's moving parts increases, resulting in higher wear. Moreover, it is possible that some acidic components are formed during the combustion process, and these can be dissolved in the lubricant. This may result in corrosive wear due to the higher total acidity of the lubricant.

13.19 LUBRICITY OF BIODIESEL RELATIVE TO PETROLEUM DIESEL

Diesel fuel injection equipments depend on diesel fuel as a lubricant. The lubricating properties of diesel fuel are important, especially for rotary and distributor type fuel injection pumps. In these pumps, moving parts are lubricated by the fuel itself as it moves through the pump. Other diesel fuel systems, which include unit injectors, injectors, unit pumps, and in-line pumps, are partially fuel lubricated.

Several years ago, the lubricity of diesel fuel was sufficient to provide the required protection to preserve sufficient performance. With the worldwide trend to reduce emission from diesel engines, ultralow-sulfur diesel has been introduced with the sulfur concentration of less than 10 ppm. Recent changes in the composition of diesel fuel, primarily the need to reduce fuel sulfur and aromatic levels, have inadvertently caused the removal of some of the compounds that provide lubricity to the fuel.

The mechanisms for lubrication vary with test methods and operating conditions. For instance, monolayers of the additive, usually carboxylic acids or methyl esters, form on the surface, thus preventing contact between the two metal surfaces and reducing the wear. Under other conditions, the lubrication occurred by the formation of organometallic polymers from carboxylic acids on metallic surfaces [301].

The lubricity of diesel fuel varies considerably depending on the crude oil source, the refining processes used to produce the fuel, and the inclusion of lubricity enhancing additives in the diesel fuel. Typically, No. 1 diesel fuel (kerosene), which is used in colder climates, has poorer lubricity than No. 2 diesel fuel. The lubricity of fuel plays an important role in the wear that occurs in an engine. Engines depend on fuel to provide lubricity for the metal components that are constantly in contact with each other [302].

Biodiesel is a much better lubricant compared with petroleum diesel due to the presence of esters. Tests have shown that the addition of a small amount of biodiesel to diesel can significantly increase the lubricity of the fuel in the short term [303]. However, over a longer period of time (2–4 years), biodiesel loses its lubricity [304]. This is due to enhanced corrosion over time due to oxidation of the unsaturated molecules or increased water content in biodiesel from moisture absorption [305].

Knothe and Steidley [166] showed that the addition of biodiesel, even in very small quantities, provides an increase in fuel lubricity using ball-on-cylinder lubricity evaluator (BOCLE) and the high-frequency reciprocating rig (HFRR). The BOCLE is commonly used to evaluate the lubricity of fuels or fuel blends but does a poor job of characterizing the lubricity of fuels containing lubricity additives. While, HFRR test is commonly used for both the neat fuels and fuels containing small amounts of lubricity enhancing additives. The lubricity of numerous fatty compounds was studied and compared to that of HC compounds found in petroleum diesel. Fatty compounds in biodiesel possess better lubricity than HCs, because of their polarity-imparting O_2 atoms. The study showed that the lubricity is improved by increasing the chain length of the fatty acid methyl esters and by the presence of double bonds. The order of oxygenated moieties enhancing the lubricity of the biodiesel was COOH > CHO > OH > COOCH$_3$ > C=O > C–O–C. Additives containing OH, NH$_2$, and SH groups show that O_2 enhances lubricity more than nitrogen and sulfur. The addition of polar compounds such as free fatty acids or monoglycerides improves the lubricity of low-level blends of esters in low-lubricity petroleum diesel.

Yaşar and coauthors [306] determined the lubricity of biodiesel fuels obtained from *canola* oil, waste frying oil, and *sunflower* oil by using the HFRR test. They reported that the lubricity of biodiesels obtained after the transesterification of the different oils is superior to petroleum diesel fuel. The *sunflower* oil–based biodiesel showed greater wear scar diameter over *canola* oil– and waste frying oil–based biodiesels.

Topaiboul and Chollacoop [307] evaluated the lubricity of ultralow-sulfur diesel containing *palm* oil and *Jatropha* oil biodiesels using a high-frequency reciprocating rig (HFRR). Particularly, it was found that a very small amount (less than 1%) of biodiesel significantly improves the lubricity in ultralow-sulfur diesel, and the biodiesel from *Jatropha* oil is a superior lubricity enhancer.

Moser and Vaughn [308] prepared *Camelina* oil methyl and ethyl esters in moderately high yields (89 and 84 wt%, respectively) from *Camelina* oil by homogeneous base-catalyzed transesterification. The oxidative stability of the biodiesel prepared from *Camelina* oil was unsatisfactory according to ASTM D6751 and EN 14214 as a result of its high polyunsaturated fatty acid content. The high iodine values of *Camelina* oil methyl and ethyl esters were also in excess of the limit prescribed by EN 14214. The low-temperature operability of *Camelina* oil methyl and ethyl esters was similar to biodiesel prepared from *canola* and *soybean* oils, and superior to petroleum diesel–biodiesel blends. Evaluation of *Camelina* oil methyl and ethyl esters as blend components in ultralow-sulfur diesel revealed that blends at the B20 level were not satisfactory with regard to the oxidative stability specification contained within ASTM D7467. *Camelina*–ultralow-sulfur diesel blends were identical to *soybean*–ultralow-sulfur diesel blends with regard to low temperature performance, kinematic viscosity, lubricity, and surface tension. *Camelina* oil alkyl esters, like *soybean* oil alkyl esters can be used as lubricity-enhancing additives in petroleum diesel.

13.20   CONCLUSIONS

1. Biofuel can be obtained from different oils, either edible oils such as corn oil, sunflower oil, and *cottonseeds* oil or nonedible oils such as *castor* oil, *Jatropha* oil, and neem oil.
2. Biofuel can be obtained by transesterification reaction of oils, or by cracking of the oils using different catalysts.
3. Biofuel has several advantages and also some disadvantages.
4. The increased use of biofuel is expected due to diminishing world petroleum reserves, severe SO_x and NO_x emissions, and climate change due to greenhouse effect, which occurred due to the use of petroleum fuel.
5. The use of biofuel is creating tribological challenges.
6. Different challenges related to tribological issues of three most common biofuels, namely, ethanol, biodiesel, and straight-chain vegetable oil, are discussed as reported in world literature over the last three decades.
7. Quality control of biofuels identified as a key factor for sustainable market growth of these fuels can lead to many tribological issues.
8. Biodiesel provides good lubricity for diesel engines due to its chemical structure.
9. Physical properties of biofuel have great effect on its atomization during combustion in diesel engines.
10. The atomization of biodiesel depends on its density, viscosity, surface tension, and chemical structure of the different fatty acid esters.

REFERENCES

1. R.M. Joshi and M.J. Pegg, Flow properties of biodiesel fuel blends at low temperatures, *Fuel*, 86, 143–151 (2007).
2. S. Som, D.E. Longman, A.I. Ramirez, and S.K. Aggarwal, A comparison of injector flow and spray characteristics of biodiesel with petrodiesel, *Fuel*, 89, 4014–4024 (2010).
3. C.E. Ejim, B.A. Fleck, and A. Amirfazli, Analytical study for atomization of biodiesels and their blends in a typical injector: Surface tension and viscosity effects, *Fuel*, 86, 1534–1544 (2007).
4. S.V.D. Freitas, M.B. Oliveira, A.J. Queimada, M.J. Pratas, Á.S. Lima, and J.A.P. Coutinho, Measurement and prediction of biodiesel surface tensions, *Energy Fuels*, 25, 4811–4817 (2011).
5. Q. Shu, J. Wang, B. Peng, D. Wang, and G. Wang, Predicting surface tension of biodiesel by a mixture topological index method, at 313 K, *Fuel*, 87, 3586–3590 (2008).
6. H.J. Kim, S.H. Park, H.K. Suh, and C.S. Lee, Atomization and evaporation characteristics of biodiesel and dimethyl ether compared to diesel fuel in a high pressure injection system, *Energy Fuels*, 23, 1734–1742 (2009).
7. C.D. Bolszo and V.G. Mcdonell, Emissions optimization of a biodiesel fired gas turbine, *Proc. Combust. Inst.*, 32, 2949–2956 (2009).
8. G. Knothe, Biodiesel and renewable diesel: A comparison, *Prog. Energy Combust. Sci.*, 36, 364–373 (2009).
9. S. Kim, J.W. Hwang, and C.S. Lee, Experiments and modeling on droplet motion and atomization of diesel and biodiesels fuels in cross-flowed air stream, *Int. J. Heat Fluid Flow*, 31, 667–679 (2010).
10. P.P. Deng, J.I. Rhodes, and P.J. Lammers, Conversion of waste cooking oil to biodiesel using ferric sulfate and supercritical methanol processes, *Fuel*, 89, 360–364 (2010).
11. A. Ninni, Policies to support biofuels in Europe: The changing landscape of instruments, *AgBioForum*, 13, 131–141 (2005).
12. G.W. Huber, P. O'Connor, and A. Corma, Processing biomass in conventional oil refineries: Production of high quality diesel by hydrotreating vegetable oils in heavy vacuum oil mixtures, *Appl. Catal. A Gen.*, 65, 120–129 (2007).
13. S.A. Khan, M.Z. Hussain, S. Prasad, and U.C. Banerjee, Prospects of biodiesel production from microalgae in India, *Renew. Sustain. Energy Rev.*, 13, 2361–2372 (2009).
14. S.S. Kapdi, V.K. Vijay, S.K. Rajesh, and R. Prasad, Biogas scrubbing, compression and storage: Perspective and prospectus in Indian context, *Renew. Energy*, 30, 1195–1202 (2005).
15. M. Lualdi, S. Lögdberg, F. Regali, M. Boutonnet, and S. Järås, Investigation of mixtures of a Co-based catalyst and a Cu-based catalyst for the Fischer–Tropsch synthesis with bio-syngas: The importance of indigenous water, *Top. Catal.*, 54, 977–985 (2011).
16. A.V. Bridgewater and M.L. Cottam, Opportunities for biomass pyrolysis liquids production and upgrading, *Energy Fuels*, 6, 113–120 (1992).
17. Q. Hu, M. Sommerfeld, E. Jarvis, M. Ghirardi, M. Posewitz, M. Seibert, and A. Darzins, Microalgal triacylglycerols as feedstocks for biofuel production: Perspectives and advances, *Plant J.*, 54, 621–639 (2008).
18. S.S. Acharya, Agricultural price policy and development: Some facts and emerging issues, *Indian J. Agric. Econ.*, 52, 1–47 (1997).
19. D.J. Murphy, Engineering oil production in *rapeseed* and other oil crops, *Trends Biotechnol.*, 14, 206–213 (1996).
20. R.D. O'Brian, *Characterization of Fats and Oils, in Fats and Oils: Formulating and Processing for Applications*, Second Edition. CRC Press, Boca Raton, 8–29, (2004).

21. A.E. Atabani, A.S. Silitonga, H.C. Ong, T.M.I. Mahlia, H.H. Masjuki, and H. Fayaz, Non-edible vegetable oils: A critical evaluation of oil extraction, fatty acid compositions, biodiesel production, characteristics, engine performance and emissions production, *Renew. Sustain. Energy Rev.*, 18, 211–245 (2013).

22. R. Beccles, Biofuel as the solution of alternative energy production, *Future Food J. Food Agric. Soc.*, 1, 23–29 (2013).

23. A. Fröhlich and B. Rice, Bio-diesel from waste cooking oil. In: *Proceedings of the Second Activity Meeting of the International Energy Agency*, Vienna, Austria, November 11–18 (1995).

24. A. Banerjee, R. Sharma, Y. Chisti, and U.C. Banerjee, *Botryococcus braunii*: A renewable source of hydrocarbons and other chemicals, *Crit. Rev. Biotechnol.*, 22, 245–279 (2002).

25. E.A. Ehimen, Z.F. Sun, C.G. Carrington, E.J. Birch, and J.J. Eaton-Rye, Anaerobic digestion of microalgae residues resulting from the biodiesel production process, *Appl. Energy*, 88, 3454–3463 (2010).

26. K.K. Sharma, H. Schuhmann, and P.M. Schenk, High lipid induction in microalgae for biodiesel production, *Energies*, 5, 1532–1553 (2012).

27. D.S. Dhiraj and M.M. Deshmukh, Biodiesel production from animal fats and its impact on the diesel engine with ethanol-diesel blends, *Int. J. Emerg. Technol. Adv. Eng.*, 2, 179–185 (2012).

28. D. Bolonio, A. Llamas, J. Fernández, A.M. Al-Lal, L. Canoira, M. Lapuerta, and L. Gómez, Estimation of cold flow performance and oxidation stability of fatty acid ethyl esters from lipids obtained from *Escherichia coli*, *Energy Fuels*, 29, 2493–2502 (2015).

29. C.H. Biradar, K.A. Subramanian, and M.G. Dastidar, Production and fuel quality upgradation of pyrolytic bio-oil from *Jatropha Curcas* de-oiled seed cake, *Fuel*, 119, 81–89 (2014).

30. H.C. Ong, H.H. Masjuki, T.M.I. Mahlia, A.S. Silitonga, W.T. Chong, and K.Y. Leong, Optimization of biodiesel production and engine performance from high free fatty acid *Calophyllum inophyllum* oil in CI diesel engine, *Energy Convers. Manag.*, 81, 30–40 (2014).

31. G. Sakthivel, G. Nagarajan, M. Ilangkumaran, and A.B. Gaikwad, Comparative analysis of performance, emission and combustion parameters of diesel engine fuelled with ethyl ester of fish oil and its diesel blends, *Fuel*, 132, 116–124 (2014).

32. S. Sinha, A.K. Agarwal, and S. Garg, Biodiesel development from rice bran oil: Transesterification process optimization and fuel characterization, *Energy Convers. Manag.*, 49, 1248–1257 (2008).

33. N. Usta, E. Öztürk, Ö. Can, E.S. Conkur, S. Nas, A.H. Çon, A.Ç. Can, and M. Topcu, Combustion of biodiesel fuel produced from hazelnut soapstock/waste *sunflower* oil mixture in a Diesel engine, *Energy Convers. Manag.*, 46, 741–755 (2005).

34. M.S. Ardebili, B. Ghobadian, G. Najafi, and A. Chegeni, Biodiesel production potential from edible oil seeds in Iran, *Renew. Sustain. Energy Rev.*, 15, 3041–3044 (2011).

35. K. Gautam, N.C. Gupta, and D.K. Sharma, Physical characterization and comparison of biodiesel produced from edible and non-edible oils of *Madhuca indica* (mahua), *Pongamia pinnata* (karanja), and *Sesamum indicum* (til) plant oilseeds, *Biomass Convers. Biorefin.*, 4, 193–200 (2013).

36. S. Kaul, R. Singhal, B. Behera, D. Bangwal, and M.O. Garg, Reactive extraction of non-edible oil seeds for biodiesel production, *J. Sci. Ind. Res.*, 73, 235–242 (2014).

37. A.E. Atabani and A.S. César, *Calophyllum inophyllum*—A prospective non-edible biodiesel feedstock: Study of biodiesel production, properties, fatty acid composition, blending and engine performance, *Renew. Sustain. Energy Rev.*, 37, 644–655 (2014).

38. M. Naik, L.C. Meher, S.N. Naik, and L.M. Das, Production of biodiesel from high free fatty acid Karanja (*Pongamia pinnata*) oil, *Biomass Bioenergy*, 32, 354–357 (2008).

39. R.R. Nasaruddin, M.Z. Alam, and M.S. Jami, Evaluation of solvent system for the enzymatic synthesis of ethanol-based biodiesel from sludge *palm* oil (SPO), *Bioresour. Technol.*, 154, 155–161 (2014).

40. A.S. Silitonga, H.H. Masjuki, T.M.I. Mahlia, H.C. Ong, F. Kusumo, and H.B. Aditiya, *Schleichera oleosa* L oil as feedstock for biodiesel production, *Fuel*, 156, 63–70 (2015).

41. R.S. Kumar, K. Sureshkumar, and R. Velraj, Optimization of biodiesel production from *Manilkara zapota* (L.) seed oil using Taguchi method, *Fuel*, 140, 90–96 (2015).

42. Z.M. Phoo, L.F. Razon, G. Knothe, Z. Ilham, F. Goembira, and C.F. Madrazo, Evaluation of Indian milkweed (*Calotropis gigantea*) seed oil as alternative feedstock for biodiesel, *Ind. Crops Prod.*, 54, 226–232 (2014).

43. A.S. Silitongaa, H.C. Onga, T.M. Mahliac, and H.H. Masjukia, Characterization and production of *Ceiba pentandra* biodiesel and its blends, *Fuel*, 108, 855–858 (2013).

44. A.B. Fadhil, Biodiesel production from beef tallow using alkali-catalyzed transesterification, *Arab. J. Sci. Eng.*, 38, 41–47 (2013).

45. F. Mohammadi, Utilization of waste coral for biodiesel production via transesterification of *soybean* oil, *Int. J. Environ. Sci. Technol.*, 11, 805–812 (2014).

46. T. Maneerung, S. Kawi, and C. Wanga, Biomass gasification bottom ash as a source of CaO catalyst for biodiesel production via transesterification of palm oil, *Energy Convers. Manag.*, 92, 234–243 (2015).

47. I. Sarantopoulos, E. Chatzisymeon, S. Foteinis, and T. Tsoutsos, Optimization of biodiesel production from waste lard by a two-step transesterification process under mild conditions, *Energy Sustain. Dev.*, 23, 110–114 (2015).

48. F.A. Yassin, F.Y. El Kady, H.S. Ahmed, L.K. Mohamed, S.A. Shaban, and A.K. Elfadaly, Highly effective ionic liquids for biodiesel production from waste vegetable oils, *Egypt. J. Petrol.*, 24, 103–111 (2015).

49. F.A. Dawodu, O.O. Ayodele, and T. Bolanle-Ojo, Biodiesel production from *Sesamum indicum* L. seed oil: An optimization study, *Egypt. J. Petrol.*, 23, 191–199 (2014).

50. Y. Zhang, Y. Li, X. Zhang, and T. Tan, Biodiesel production by direct transesterification of microalgal biomass with co-solvent, *Bioresour. Technol.*, 196, 712–715 (2015).

51. W. Ahmed, M.F. Nazar, S.D. Ali, U.A. Rana, and S.U. Khan, Detailed investigation of optimized alkali catalyzed transesterification of *Jatropha* oil for biodiesel production, *J. Energy Chem.*, 24, 331–336 (2015).

52. Q. Zhou, H. Zhang, F. Chang, H. Li, H. Pan, W. Xue, D. Hu, and S. Yang, Nano La$_2$O$_3$ as a heterogeneous catalyst for biodiesel synthesis by transesterification of *Jatropha curcas* L. oil, *J. Ind. Eng. Chem.*, 31, 385–392 (2015).

53. J. Liu, Y. Nan, R. Lin, and L. Tavlarides, Production of biodiesel from microalgae oil (*Chlorella protothecoides*) by non-catalytic transesterification in supercritical methanol and ethanol: Process optimization, *J. Supercrit. Fluids*, 97, 174–182 (2015).

54. A. Bilgin, M. Gülüm, İ. Koyuncuoglu, E. Nac, and A. Cakmak, Determination of transesterification reaction parameters giving the lowest viscosity waste cooking oil biodiesel, *Proc. Soc. Behav. Sci.*, 195, 2492–2500 (2015).

55. K. Ullah, M. Ahmad, and F. Qiuc, Assessing the experimental investigation of milk thistle oil for biodiesel production using base catalyzed transesterification, *Energy*, 89, 887–895 (2015).

56. O.L. Bernardes, J.V. Bevilaqua, M.C. Leal, D.M. Freire, and M.A. Langone, Biodiesel fuel production by the transesterification reaction of soybean oil using immobilized lipase, *Appl. Biochem. Biotechnol.*, 140, 105–114 (2007).

57. R. Chakraborty and A. Banerjee, Prediction of fuel properties of biodiesel produced by sequential esterification and transesterification of used frying soybean oil using statistical analysis, *Waste Biomass Valorization*, 1, 201–208 (2010).

58. G. Vicente, M. Martínez, and J. Aracil, Integrated biodiesel production: A comparison of different homogeneous catalysts systems, *Bioresour. Technol.*, 92, 297–305 (2004).

59. X. Liu, H. He, Y. Wang, S. Zhu, and X. Piao, Transesterification of *soybean* oil to biodiesel using CaO as a solid base catalyst, *Fuel*, 87, 216–221 (2008).

60. X. Liu, H. He, Y. Wang, and S. Zhu, Transesterification of *soybean* oil to biodiesel using SrO as a solid base catalyst, *Catal. Commun.*, 8, 1107–1111 (2007).

61. N.U. Soriano Jr, R. Venditti, and D.S. Argyropoulos, Biodiesel synthesis via homogeneous Lewis acid-catalyzed transesterification, *Fuel*, 88, 560–565 (2009).

62. L. Gu, W. Huang, S. Tang, S. Tian, and X. Zhang, A novel deep eutectic solvent for biodiesel preparation using a homogeneous base catalyst, *Chem. Eng. J.*, 259, 647–652 (2015).

63. F. Li, H. Li, L. Wang, and Y. Cao, Waste carbide slag as a solid base catalyst for effective synthesis of biodiesel via transesterification of *soybean* oil with methanol, *Fuel Process. Technol.*, 131, 421–429 (2015).

64. Z. Al-Hamamre and J. Yamin, Parametric study of the alkali catalyzed transesterification of waste frying oil for Biodiesel production, *Energy Convers. Manag.*, 79, 246–254 (2014).

65. H. Karabas, Biodiesel production from crude acorn (*Quercus frainetto* L.) *kernel* oil: An optimization process using the Taguchi method, *Renew. Energy*, 53, 384–388 (2013).

66. M. Jaimasith and S. Phiyanalinmat, Biodiesel synthesis from transesterification by clay-based catalyst, *Chiang Mai J. Sci.*, 34, 201–207 (2007).

67. H.N. Bhatti, M.A. Hanif, M. Qasim, and A. Rehman, Biodiesel production from waste tallow, *Fuel*, 87, 2961–2966 (2008).

68. S.M.P. Meneghetti, M.R. Meneghetti, C.R. Wolf, E.C. Silva, G.E.S. Lima, and L. de L Silva, Biodiesel from *castor* oil: A comparison of ethanolysis versus methanolysis, *Energy Fuels*, 20, 2262–2265 (2006).

69. E. Maleki, M.K. Aroua, and N.M. Sulaiman, *Castor* oil—A more suitable feedstock for enzymatic production of methyl esters, *Fuel Process. Technol.*, 112, 129–132 (2013).

70. G. Ciudad, L. Azócar, H.J. Heipieper, R. Muñoz, and R. Navia, Improving fatty acid methyl ester production yield in a lipase-catalyzed process using waste frying oils as feedstock, *J. Biosci. Bioeng.*, 109, 609–614 (2010).

71. L.L. Sousa, I.L. Lucena, and F.A.N. Fernandes, Transesterification of *castor* oil: Effect of the acid value and neutralization of the oil with glycerol, *Fuel Process. Technol.*, 91, 194–196 (2010).

72. J.M. Dias, J.M. Araújo, J.F. Costa, M.C.M. Alvim-Ferraz, and M.F. Almeida, Biodiesel production from raw *castor* oil, *Energy*, 53, 58–66 (2013).

73. N. Sánchez, R. Sánchez, J.M. Encinar, J.F. González, and G. Martínez, Complete analysis of *castor* oil methanolysis to obtain biodiesel, *Fuel*, 147, 95–99 (2015).

74. H. Bateni and K. Karimi, Biodiesel production from *castor* plant integrating ethanol production via a biorefinery approach, *Chem. Eng. Res. Des.*, 107, 4–12 (2016), doi:10.1016/j.cherd.2015.08.014.

75. J.K. Rodríguez-Guerrero, M.F. Rubens, and P.T.V. Ros, Production of biodiesel from *castor* oil using sub and supercritical ethanol: Effect of sodium hydroxide on the ethyl ester production, *J. Supercrit. Fluids*, 132, 83–124 (2013).

76. M. Kılıç, B.B. Uzuna, E. Pütünb, and A.E. Pütün, Optimization of biodiesel production from *castor* oil using factorial design, *Fuel Process. Technol.*, 111, 105–110 (2013).

77. H. Bateni, K. Karimi, A. Zamani, and F. Benakashani, *Castor* plant for biodiesel, biogas, and ethanol production with a biorefinery processing perspective, *Appl. Energy*, 136, 14–22 (2014).

78. F. Halek, A. Delavari, and A. Kavousi-rahim, Production of biodiesel as a renewable energy source from *castor* oil, *Clean Technol. Environ. Policy*, 15, 1063–1068 (2013).

79. R. Peña, R. Romero, S.L. Martínez, M.J. Ramos, A. Martínez, and R. Natividad, Transesterification of *castor* oil: Effect of catalyst and co-solvent, *Ind. Eng. Chem. Res.*, 48, 1186–1189 (2009).

80. O.U. Dairo, T.M. Olayanju, E.S. Ajisegiri, O.J. Alamu, and A.E. Adeleke, Optimization of in-situ biodiesel production from raw *castor* oil-*bean* seed, *J. Energy Technol.*, 3, 14–19 (2013).

81. B. Antizar-Ladislao and J.L. Turrion-Gomez, Second-generation biofuels and local bio-energy systems, *Biofuel Bioprod. Biorefin.*, 2, 455–469 (2008).

82. V.E. Efeovbokhan, V. Enontiemonria, and A. Ayodeji, The effects of trans-esterification of *castor* seed oil using ethanol, methanol and their blends on the properties, *Int. J. Eng. Technol.*, 2, 1734–1742 (2012).

83. G.M. Hincapié, S. Valange, J. Barrault, J.A. Moreno, and D.P. Lópezl, Effect of micro-wave-assisted system on transesterification of *castor* oil with ethanol, *Univ. Sci.*, 19, 193–200 (2014).

84. N. Da Silva, Ć.B. Batistella, R.M. Filho, and M.R. Maciel, Biodiesel production from *castor* oil: Optimization of alkaline ethanolysis, *Energy Fuels*, 23, 5636–5642 (2009).

85. G. Jeong and D. Park, Optimization of biodiesel production from *castor* oil using response surface, *Appl. Biochem. Biotechnol.*, 156, 1–11 (2008).

86. C. Cao and Y. Zhao, Transesterification of *castor* oil to biodiesel using KOH/NaY as solid base catalyst, *Int. J. Green Energy*, 10, 219–229 (2013).

87. G. Chen, M. Ying, and W. Li, Enzymatic conversion of waste cooking oils into alter-native fuel-biodiesel. In: *27th Symposium on Biotechnology for Fuels and Chemicals ABAB Symposium*, pp. 911–921, Denver, CO (2006).

88. K.A. Shah, J.K. Parikh, and K.C. Maheria, Optimization studies and chemical kinetics of silica sulfuric acid-catalyzed biodiesel synthesis from waste cooking oil, *BioEnergy Res.*, 7, 206–216 (2014).

89. X. Miao and Q. Wu, Biodiesel production from heterotrophic microalgal oil, *Bioresour. Technol.*, 97, 841–847 (2006).

90. E. Vivek and A.K. Gupta, Biodiesel production from *Karanja* oil, *J. Sci. Ind. Res.*, 63, 39–47 (2004).

91. F. Ataya, M.A. Dubé, and M. Ternan, Acid-catalyzed transesterification of Canola oil to biodiesel under single- and two-phase reaction conditions, *Energy Fuels*, 21, 2450–2459 (2007).

92. B. De and R.S. Panua, An experimental study on performance and emission characteris-tics of vegetable oil blends with diesel in a direct injection variable compression ignition engine, *Procedia Eng.*, 90, 431–438 (2014).

93. S. Nagaraja, K. Sooryaprakash, and R. Sudhakaran, Investigate the effect of compres-sion ratio over the performance and emission characteristics of variable compression ratio engine fueled with preheated palm oil-diesel blends, *Procedia Earth Planet. Sci.*, 11, 393–401 (2015).

94. R.D. Misra and M.S. Murthy, Performance, emission and combustion evaluation of soapnut oil–diesel blends in a compression ignition engine, *Fuel*, 90, 2514–2518 (2011).

95. M. Mbarawa, The effect of clove oil and diesel fuel blends on the engine performance and exhaust emissions of a compression-ignition engine, *Biomass Bioenergy*, 34, 1555–1561 (2010).

96. R. Vallinayagam, S. Vedharaj, W.M. Yang, C.G. Saravanan, P.S. Lee, K.J.E. Chua, and S.K. Chou, Impact of ignition promoting additives on the characteristics of a diesel engine powered by pine oil–diesel blend, *Fuel*, 117, 278–285 (2014).

97. T. Ozaktas, Compression ignition engine fuel properties of a used sunflower oil-diesel fuel blend, *Energy Sources*, 22, 377–382 (2000).

98. K. Pramanik, Properties and use of *Jatropha curcas* oil and diesel fuel blends in com-pression ignition engine, *Renew. Energy*, 28, 239–248 (2003).

99. F.K. Forson, E.K. Oduro, and E.H. Donkoh, Performance of *jatropha* oil blends in a diesel engine, *Renew. Energy*, 29, 1135–1145 (2004).

100. O.M.I. Nwafor and G. Rice, Performance of *rapeseed* oil blends in a diesel engine, *Appl. Energy*, 54, 345–354 (1996).

101. G. Labeckas and S. Slavinskas, Performance and exhaust emissions of direct-injection diesel engine operating on rapeseed oil and its blends with diesel fuel, *Transport*, 20, 186–194 (2005).

102. T. Aysu, Catalytic pyrolysis of *Alcea pallida* stems in a fixed-bed reactor for production of liquid bio-fuels, *Bioresour. Technol.*, 191, 253–262 (2015).

103. J. Xu, J. Jiang, J. Chen, and Y. Sun, Biofuel production from catalytic cracking of woody oils, *Bioresour. Technol.*, 101, 5586–5591 (2010).

104. F. Yu, L. Gao, W. Wang, G. Zhang, and J. Ji, Bio-fuel production from the catalytic pyrolysis of soybean oil over Me-Al-MCM-41 (Me = La, Ni or Fe) mesoporous materials, *J. Anal. Appl. Pyrolysis*, 104, 325–329 (2013).

105. O.Y. Sang, Biofuel production from catalytic cracking of palm oil, *Energy Sources A Recov. Utiliz. Environ. Effects*, 25, 859–869 (2003).

106. I.V. Babich, M. Hulst, L. Lefferts, J.A. Moulijn, P. O'Connor, and K. Seshan, Catalytic pyrolysis of microalgae to high-quality liquid bio-fuels, *Biomass Bioenergy*, 35, 3199–3207 (2011).

107. H. Li, P. Yu, and B. Shen, Biofuel potential production from cottonseed oil: A comparison of non-catalytic and catalytic pyrolysis on fixed-fluidized bed reactor, *Fuel Process. Technol.*, 90, 1087–1092 (2009).

108. H. Zhang, R. Xiao, and H. Huang, Comparison of non-catalytic and catalytic fast pyrolysis of corncob in a fluidized bed reactor, *Bioresour. Technol.*, 100, 1428–1434 (2009).

109. S.A.P. Da Mota, A.A. Mancio, D.E.L. Lhamas, D.H. de Abreu, M.S. da Silva, W.G. dos Santos, D.A.R. de Castro, R.M. de Oliveira, and M.E. Araújo, Production of green diesel by thermal catalytic cracking of crude palm oil (*Elaeis guineensis Jacq*) in a pilot plant, *J. Anal. Appl. Pyrolysis*, 110, 1–11 (2014).

110. S. Biswas and D.K. Sharma, Effect of different catalysts on the cracking of *Jatropha* oil, *J. Anal. Appl. Pyrolysis*, 110, 346–352 (2014).

111. Z.D. Yigezu and K. Muthukumar, Catalytic cracking of vegetable oil with metal oxides for biofuel production, *Energy Convers. Manag.*, 84, 326–333 (2014).

112. L. Li, K. Quan, J. Xu, F. Liu, S. Liu, S. Yu, C. Xie, B. Zhang, and X. Ge, Liquid hydrocarbon fuels from catalytic cracking of rubber seed oil using USY as catalyst, *Fuel*, 123, 189–193 (2014).

113. J. Xu, J. Jiang, Y. Sun, and J. Chen, Production of hydrocarbon fuels from pyrolysis of soybean oils using a basic catalyst, *Bioresour. Technol.*, 101, 9803–9806 (2010).

114. X. Tang and F. Wei, Waste edible oil fluid catalytic cracking in a Downer reactor, In *The 12th International Conference on Fluidization—New Horizons in Fluidization Engineering*, Agassiz, British Columbia, Canada (2007).

115. V.P. Doronin, O.V. Potapenko, P.V. Lipin, T.P. Sorokina, and L.A. Buluchevskaya, Catalytic cracking of vegetable oils for production of high-octane gasoline and petrochemical feedstock, *Petrol. Chem.*, 52, 392–400 (2012).

116. H. Tian, C. Li, C. Yang, and H. Shan, Alternative processing technology for converting vegetable oils and animal fats to clean fuels and light olefins, *Chin. J. Chem. Eng.*, 16, 394–400 (2008).

117. T. Kraiem, A.B. Trabelsi, S. Naoui, H. Belayouni, and M. Jeguirim, Characterization of the liquid products obtained from Tunisian waste fish fats using the pyrolysis process, *Fuel Process. Technol.*, 138, 404–412 (2015), doi:10.1016/j.fuproc.2015.05.007.

118. T. Karayildirim, J. Yanik, M. Yuksel, and H. Bockhorn, Characterization of products from pyrolysis of waste sludge, *Fuel*, 85, 1498–1508 (2006).

119. C.M.R. Prado and N.R. Filho, Production and characterization of the biofuels obtained by thermal cracking and thermal catalytic cracking of vegetable oils, *J. Anal. Appl. Pyrolysis*, 86, 338–347 (2009).

120. N. Özbay, E.A. Varol, B.B. Uzun, and A.E. Pütün, Characterization of bio-oil obtained from fruit pulp pyrolysis, *Energy*, 33, 1233–1240 (2008).
121. Y. Yang, J.G. Brammer, M. Ouadia, J. Samanya, A. Hornung, H.M. Xu, and Y. Li, Characterization of waste derived intermediate pyrolysis oils for use as diesel engine fuels, *Fuel*, 103, 247–257 (2013).
122. I.M.S. Correia, M.J.B. Souz, A.S. Araújo, and E.M. Sousa, Thermal stability during pyrolysis of sunflower oil produced in the northeast of Brazil, *J. Therm. Anal. Calorim.*, 109, 967–974 (2012).
123. S. Yorgun, S. Şensöz, and M. Koçkar, Characterization of the pyrolysis oil produced in the slow pyrolysis of sunflower-extracted bagasse, *Biomass Bioenergy*, 20, 141–148 (2011).
124. V.R. Wiggers, A. Wisniewski Jr, L.A.S. Madureira, A.A. Barros, and H.F. Meier, Biofuels from waste fish oil pyrolysis: Continuous production in a pilot plant, *Fuel*, 88, 2135–2141 (2009).
125. V.R. Wiggers, H.F. Meier, A. Wisniewski, A.A. Barros, and M.R. Maciel, Biofuels from continuous fast pyrolysis of soybean oil: A pilot plant study, *Bioresour. Technol.*, 100, 6570–6577 (2009).
126. G. Chen, X. Zhang, W. Ma, B. Yan, and Y. Li, Co-pyrolysis of corn-cob and waste cooking-oil in a fixed bed reactor with HY upgrading process, *Energy Procedia*, 61, 2363–2366 (2014).
127. V.R. Wiggers, G.R. Zonta, A.P. França, D.R. Scharf, E.L. Simionatto, L. Ender, and H.F. Meier, Challenges associated with choosing operational conditions for triglyceride thermal cracking aiming to improve biofuel quality, *Fuel*, 107, 601–608 (2013).
128. H.F. Meier, V.R. Wiggers, G.R. Zonta, D.R. Scharf, E.L. Simionatto, and L. Ender, A kinetic model for thermal cracking of waste cooking oil based on chemical lumps, *Fuel*, 144, 50–59 (2015).
129. H. Lappi and R. Alén, Pyrolysis of vegetable oil soaps—Palm, olive, rapeseed and castor oils, *J. Anal. Appl. Pyrolysis*, 91, 154–158 (2011).
130. N.K. Nayan, S. Kumar, and R.K. Singh, Production of the liquid fuel by thermal pyrolysis of neem seed, *Fuel*, 103, 437–443 (2013).
131. R.V. Silva, A. Casilli, A.L. Sampaio, B.M. Ávila, M.C. Veloso, D.A. Azevedo, and G.A. Romeiro, The analytical characterization of *castor* seed cake pyrolysis bio-oils by using comprehensive GC coupled to time of flight mass spectrometry, *J. Anal. Appl. Pyrolysis*, 106, 152–159 (2014).
132. M.H. Salmani, S. Rehman, K. Zaidi, and A.K. Hasan, Study of ignition characteristics of microemulsion of coconut oil under off diesel engine conditions, *Eng. Sci. Technol.*, 18, 318–324 (2015).
133. R. Prakash, R.K. Singh, and S. Murugan, Use of biodiesel and bio-oil emulsions as an alternative fuel for direct injection diesel engine, *Waste Biomass Valorization*, 4, 475–484 (2013).
134. Y. Xu, X. Zheng, Y. Yin, J. Huang, and X. Hu, Comparison and analysis of the influence of test conditions on the tribological properties of emulsified bio-oil, *Tribol. Lett.*, 55, 543–552 (2014).
135. N.A. Ishak, I.A. Raman, M.A. Yarmo, and W.M. Mahmood, Ternary phase behavior of water microemulsified diesel-palm biodiesel, *Front. Energy*, 9, 162–169 (2015).
136. N. Arpornpong, D.A. Sabatini, S. Khaodhiar, and A. Charoensaeng, Life cycle assessment of palm oil microemulsion-based biofuel, *Int. J. Life Cycle Assess.*, 20, 913–926 (2015).
137. Z. Chen, X. Wang, Y. Pei, C. Zhang, M. Xiao, and J. He, Experimental investigation of the performance and emissions of diesel engines by a novel emulsified diesel fuel, *Energy Convers. Manag.*, 95, 334–341 (2015).

138. H. Raheman and S. Kumari, Combustion characteristics and emissions of a compression ignition engine using emulsified *jatropha* biodiesel blend, *Biosyst. Eng.*, 123, 29–39 (2014).

139. Y. Lin and H. Lin, Spray characteristics of emulsified *castor* biodiesel on engine emissions and deposit formation, *Renew. Energy*, 36, 3507–3516 (2011).

140. C.H. Cheng, C.S. Cheung, T.L. Chan, S.C. Lee, C.D. Yao, and K.S. Tsang, Comparison of emissions of a direct injection diesel engine operating on biodiesel with emulsified and fumigated methanol, *Fuel*, 87, 1870–1879 (2008).

141. Y. Liu, W. Jiao, and G. Qi, Preparation and properties of methanol–diesel oil emulsified fuel under high-gravity environment, *Renew. Energy*, 36, 1463–1468 (2011).

142. P. Benjumea, J.R. Agudelo, and A.F. Agudelo, Effect of the degree of unsaturation of biodiesel fuels on engine performance, combustion characteristics, and emissions, *Energy Fuels*, 25, 77–85 (2011).

143. G.I. Martins, D. Secco, H.A. Rosa, R.A. Bariccatti, B.D. Dolci, and S.N. de Souza, Physical and chemical properties of fish oil biodiesel produced in Brazil, *Renew. Sustain. Energy Rev.*, 42, 154–157 (2015).

144. K. Abbott, Effect of the degree of unsaturation of biodiesel fuels on NO_x and particulate emissions, EP Patent 0,796,687 (2001).

145. Ş. Altun, Effect of the degree of unsaturation of biodiesel fuels on the exhaust emissions of a diesel power generator, *Fuel*, 117, 450–457 (2014).

146. Ş. Altun and M. Lapuert, Properties and emission indicators of biodiesel fuels obtained from waste oils from the Turkish industry, *Fuel*, 117, 288–295 (2014).

147. M. Salamanca, F. Mondragón, J. Agudelo, P. Benjumea, and A. Santamarí, Variations in the chemical composition and morphology of soot induced by the unsaturation degree of biodiesel and a biodiesel blend, *Combust. Flame*, 159, 1100–1108 (2012).

148. S. Pinzi, P. Rounce, J.M. Herreros, A. Tsolakis, and M.P. Dorado, The effect of biodiesel fatty acid composition on combustion and diesel engine exhaust emissions, *Fuel*, 104, 170–182 (2013).

149. M.D. Redel-Macías, S. Pinzi, D.E. Leiva-Candia, A.J. Cubero-Atienza, and M.P. Dorado, Influence of fatty acid unsaturation degree over exhaust and noise emissions through biodiesel combustion, *Fuel*, 109, 248–255 (2013).

150. M.A. Islam, R.J. Brown, P.R. Brooks, M.I. Jahirul, H. Bockhorn, and K. Heimann, Investigation of the effects of the fatty acid profile on fuel properties using a multi-criteria decision analysis, *Energy Convers. Manag.*, 98, 340–347 (2015).

151. G. Knothe, A.C. Matheaus, and T.W. Ryan, Cetane numbers of branched and straight-chain fatty esters determined in an ignition quality tester, *Fuel*, 82, 971–975 (2003).

152. G. Knothe, A comprehensive evaluation of the cetane numbers of fatty acid methyl esters, *Fuel*, 119, 6–13 (2014).

153. D. Tong, C. Hu, K. Jiang, and Y. Li, Cetane number prediction of biodiesel from the composition of the fatty acid methyl esters, *J. Am. Oil Chem. Soc.*, 88, 415–423 (2011).

154. L.F. Ramírez-Verduzco, J.E. Rodríguez-Rodríguez, and A.R. Jaramillo-Jacob, Predicting cetane number, kinematic viscosity, density and higher heating value of biodiesel from its fatty acid methyl ester composition, *Fuel*, 91, 102–111 (2012).

155. C.A.W. Allen, K.C. Watts, R.G. Ackman, and M.J. Pegg, Predicting the viscosity of biodiesel fuels from their fatty acid ester composition, *Fuel*, 78, 1319–1326 (1999).

156. G. Knothe and K.R. Steidley, Kinematic viscosity of biodiesel fuel components and related compounds: Influence of compound structure and comparison to petrodiesel fuel components, *Fuel*, 84, 1059–1065 (2005).

157. L. Wang, H. Yu, X. He, and R. Liu, Influence of fatty acid composition of woody biodiesel plants on the fuel properties, *J. Fuel Chem. Technol.*, 40, 397–404 (2012).

158. S.K. Hoekman, A. Broch, C. Robbins, E. Ceniceros, and M. Natarajan, Review of bio-diesel composition, properties, and specifications, *Renew. Sustain. Energy Rev.*, 16, 143–169 (2012).

159. M.A. Sokoto, L.G. Hassan, S.M. Dangoggo, H.G. Ahmad, and A. Uba, Influence of fatty acid methyl esters on fuel properties of biodiesel produced from the seeds oil of *Curcubita pepo*, *Nig. J. Basic Appl. Sci.*, 89, 1981–1986 (2012).

160. F. Levine, R.V. Kayea, R. Wexler, D.J. Sadvary, C. Melick, and J.L. Scala, Heats of com-bustion of fatty acids and fatty acid esters, *J. Am. Oil Chem. Soc.*, 91, 235–249 (2014).

161. S. Pinz, I.L. Garcia, F.J. Lopez-Gimenez, M.D. Luque de Castro, G. Dorado, and M.P. Dorado, The ideal vegetable oil-based biodiesel composition: A review of social eco-nomical and technical implications, *Energy Fuels*, 23, 2325–2341 (2009).

162. M.B. Dantas, A.R. Albuquerque, L.E. Soledade, N. Queiroz, A.S. Maia, I.M. Santos, A.L. Souza, E.H. Cavalcanti, A.K. Barro, and A.G. Souza, Biodiesel from soybean oil, *castor* oil and their blends, *J. Therm. Anal. Calorim.*, 106, 607–611 (2011).

163. R.L. McCormick, M.S. Graboski, T.L. Alleman, A.M. Herring, and K.S. Tyson, Impact of biodiesel source material and chemical structure on emissions of criteria pollutants from a heavy-duty engine, *Environ. Sci. Technol.*, 35, 1742–1747 (2001).

164. S. Arumugam and G. Sriram, Comparative study of engine oil tribology, wear and com-bustion characteristics of direct injection compression ignition engine fuelled with *cas-tor* oil biodiesel and diesel fuel, *Aust. J. Mech. Eng.*, 10, 105–112 (2012).

165. C. Leonor, B. Forero, and F. De Paula, Biodiesel from *castor* oil: A promising fuel for cold weather, *Power*, 12, 4–11 (2012).

166. G. Knothe and K.R. Steidley, Lubricity of components of biodiesel and petrodiesel: The origin of biodiesel lubricity, *Energy Fuels*, 19, 1192–1200 (2005).

167. B. Tesfa, R. Mishra, F. Gu, and N. Powles, Prediction models for density and viscosity of biodiesel and their effects on fuel supply system in CI engines, *Renew. Energy*, 35, 2752–2760 (2010).

168. R.M.C. Farias, M.M. Conceição, R.A. Candeia, M.C.D. Silva, V.J. Fernandes, and A.G. Souza, Evaluation of the thermal stability of biodiesel blends of *castor* oil and passion fruit, *J. Therm. Anal. Calorim.*, 106, 651–655 (2011).

169. C.D. Rakopoulos, K.A. Antonopoulos, D.C. Rakopoulos, D.T. Hountalas, and E.G. Giakoumis, Comparative performance and emissions study of a direct injection diesel engine using blends of diesel fuel with vegetable oils or bio-diesels of various origins, *Energy Convers. Manag.*, 47, 3272–3287 (2006).

170. P. Lisiecki, Ł. Chrzanowski, A. Szulc, Ł. Ławniczak, and W. Białas, Biodegradation of diesel/biodiesel blends in saturated sand microcosms, *Fuel*, 116, 321–327 (2014).

171. D.Y. Leung, B.C. Koo, and Y. Guo, Degradation of biodiesel under different storage conditions, *Bioresour. Technol.*, 97, 250–256 (2006).

172. Z. Yang, B.P. Hollebone, Z. Wang, C. Yang, C. Brown, and M. Landriault, Storage sta-bility of commercially available biodiesels and their blends under different storage con-ditions, *Fuel*, 115, 366–377 (2014).

173. C. Lin and C. Chiu, Burning characteristics of palm-oil biodiesel under long-term stor-age conditions, *Energy Convers. Manag.*, 51, 1464–1467 (2010).

174. Z. Yang, B.P. Hollebone, Z. Wang, C. Yang, and M. Landriault, Effect of storage period on the dominant weathering processes of biodiesel and its blends with diesel in ambient conditions, *Fuel*, 104, 342–350 (2013).

175. A. Khalid, N. Tamaldin, M. Jaat, M.F. Ali, B. Manshoor, and I. Zaman, Impacts of bio-diesel storage duration on fuel properties and emissions, *Procedia Eng.*, 68, 225–230 (2013).

176. T. Kivevele and Z. Huan, Influence of metal contaminants and antioxidant additives on storage stability of biodiesel produced from non-edible oils of Eastern Africa origin (*Croton megalocarpus* and *Moringa oleifera oils*), *Fuel*, 158, 530–537 (2015).

177. B. Tesfa, R. Mishra, C. Zhang, F. Gu, and A.D. Ball, Combustion and performance characteristics of CI (compression ignition) engine running with biodiesel, *Energy*, 51, 101–115 (2013).

178. S. Puhan, N. Saravanan, G. Nagarajan, and N. Vedaraman, Effect of biodiesel unsaturated fatty acid on combustion characteristics of a DI compression ignition engine, *Biomass Bioenergy*, 34, 1079–1088 (2010).

179. M. Shahabuddin, A.M. Liaquat, H.H. Masjuki, M.A. Kalam, and M. Mofijur, Ignition delay, combustion and emission characteristics of diesel engine fueled with biodiesel, *Renew. Sustain. Energy Rev.*, 21, 623–632 (2013).

180. E. Öztürk, Performance, emissions, combustion and injection characteristics of a diesel engine fuelled with *canola* oil–hazelnut soapstock biodiesel mixture, *Fuel Process. Technol.*, 129, 183–191 (2015).

181. C. Allen, E. Toulson, D. Tepe, H. Schock, D. Miller, and T. Lee, Characterization of the effect of fatty ester composition on the ignition behavior of biodiesel fuel sprays, *Fuel*, 111, 659–669 (2013).

182. K. Muralidharan, D. Vasudevan, and K.N. Sheeba, Performance, emission and combustion characteristics of biodiesel fuelled variable compression ratio engine, *Energy*, 36, 5385–5393 (2011).

183. N. Kumar and S.R. Chauhan, Performance and emission characteristics of biodiesel from different origins: A review, *Renew. Sustain. Energy Rev.*, 21, 633–658 (2013).

184. M. Canakci, Combustion characteristics of a turbocharged DI compression ignition engine fueled with petroleum diesel fuels and biodiesel, *Bioresour. Technol.*, 98, 1167–1175 (2007).

185. N. Zhang, Z. Huang, X. Wang, and B. Zheng, Combustion and emission characteristics of a turbo-charged common rail diesel engine fuelled with diesel-biodiesel-DEE blends, *Front. Energy*, 5, 104–114 (2011).

186. R. Arslan, Emission characteristics of a diesel engine using waste cooking oil as biodiesel fuel, *African J. Biotechnol.*, 10, 3790–3794 (2011).

187. D.H. Qi, L.M. Geng, H. Chen, Y.Z. Bian, J. Liu, and X.C. Ren, Combustion and performance evaluation of a diesel engine fueled with biodiesel produced from soybean crude oil, *Renew. Energy*, 34, 2706–2713 (2009).

188. M.M. Rahman, A.M. Pourkhesalian, M.I. Jahirul, S. Stevanovic, P.X. Pham, H. Wang, A.R. Masri, R.J. Brown, and Z.D. Ristovski, Particle emissions from biodiesels with different physical properties and chemical composition, *Fuel*, 134, 201–208 (2014).

189. V.H. Nguyen and P.X. Pham, Biodiesels: Oxidizing enhancers to improve CI engine performance and emission quality, *Fuel*, 154, 293–300 (2015).

190. H. Masjuki and M.Z. Abdulmunin, Investigations on preheated palm oil methyl esters in the diesel engine, *Proc. Inst. Mech. Eng.*, 210, 131–137 (1996).

191. C. Pattamaprom, W. Pakdee, and S. Ngamjaroen, Storage degradation of palm-derived biodiesels: Its effects on chemical properties and engine performance, *Renew. Energy*, 37, 412–418 (2015).

192. G. Karavalakis, S. Stournas, and E. Bakeas, Light vehicle regulated and unregulated emissions from different biodiesels, *Sci. Total Environ.*, 407, 3338–3346 (2009).

193. Y. Ra, R.D. Reitz, J. Mcfarlane, and C.S. Daw, Effects of fuel physical properties on diesel engine combustion using diesel and bio-diesel fuels, *Engineering*, 28, 776–790 (2008).

194. X. Wang, Z. Huang, O.A. Kuti, W. Zhang, and K. Nishid, Experimental and analytical study on biodiesel and diesel spray characteristics under ultra-high injection pressure, *Int. J. Heat Fluid Flow*, 31, 659–666 (2010).

195. Y. Gao, J. Deng, C. Li, F. Dang, Z. Liao, Z. Wu, and L. Li, Experimental study of the spray characteristics of biodiesel based on inedible, *Biotechnol. Adv.*, 27, 616–624 (2009).

196. S.H. Yoon, H.K. Suh, and C.S. Lee, Effect of spray and EGR rate on the combustion and emission characteristics of biodiesel fuel in a compression ignition engine, *Energy Fuels*, 23, 1486–1493 (2009).

197. H.K. Suh and C.S. Lee, Experimental and analytical study on the spray characteristics of dimethyl ether (DME) and diesel fuels within a common-rail injection system in a diesel engine, *Fuel*, 87, 925–932 (2008).

198. H.K. Suh, H.G. Roh, and C.S. Lee, Spray and combustion characteristics of biodiesel/diesel blended fuel in a direct injection common-rail diesel engine, *J. Eng. Gas Turbines Power*, 130, 732–807 (2008).

199. X. Wang, Z. Huang, O.A. Kuti, W. Zhang, and K. Nishida, An experimental investigation on spray, ignition and combustion characteristics of biodiesels, *Proc. Combust. Inst.*, 33, 2071–2077 (2011).

200. H.J. Kim, S.H. Park, and C.S. Lee, A study on the macroscopic spray behavior and atomization characteristics of biodiesel and dimethyl ether sprays under increased ambient pressure, *Fuel Process. Technol.*, 91, 354–363 (2010).

201. M. Gumus, C. Sayin, and M. Canakci, The impact of fuel injection pressure on the exhaust emissions of a direct injection diesel engine fueled with biodiesel-diesel fuel blends, *Fuel*, 95, 486–494 (2012).

202. S.H. Park, H.J. Kim, H.K. Suh, and C.S. Lee, A study on the fuel injection and atomization characteristics of soybean oil methyl ester (SME), *Int. J. Heat Fluid Flow*, 30, 108–116 (2009).

203. P.C. Chen, W.C. Wang, W.L. Roberts, and T. Fang, Spray and atomization of diesel fuel and its alternatives from a single-hole injector using a common rail fuel injection system, *Fuel*, 103, 850–861 (2013).

204. B. Kegl and S. Pehan, Influence of biodiesel on injection, fuel spray, and engine characteristics, *Therm. Sci.*, 12, 171–182 (2008).

205. M.H. Shojaeefard, M.M. Etgahni, F. Meisami, and A. Barari, Experimental investigation on performance and exhaust emissions of *castor* oil biodiesel from a diesel engine, *Environ. Technol.*, 34, 2019–2026 (2013).

206. J. Liu, X. Zhang, T. Wang, J. Zhang, and H. Wang, Experimental and numerical study of the pollution formation in a diesel/CNG dual fuel engine, *Fuel*, 159, 418–429 (2015).

207. Z. Wang, Q. Li, Z. Lin, R. Whiddon, K. Qiu, M. Kuang, and K. Cen, Transformation of nitrogen and sulphur impurities during hydrothermal upgrading of low quality coals, *Fuel*, 164, 254–261 (2016).

208. A. Fernandez, J. Wendt, and M.L. Witten, Health effects engineering of coal and biomass combustion particulates: Influence of zinc, sulfur and process changes on potential lung injury from inhaled ash, *Fuel*, 84, 1320–1327 (2005).

209. J.M. Encinar, A. Pardal, and N. Sánchez, An improvement to the transesterification process by the use of co-solvents to produce biodiesel, *Fuel*, 166, 51–58 (2016).

210. M. Koncar, Criteria for the development and selection of low cost and high quality technologies for biodiesel. In: *Proceedings of the Second European Motor Biofuels Forum*, Joanneum Research, Graz, Austria (1996).

211. W. Rodinger, Ecotoxicological risks of biodiesel compared with diesel based on mineral oil. In: *Minutes of the Second Activity Meeting—Workshop*, Vienna, Austria (1995).

212. A. Demirbas, Biofuels sources, biofuel policy, biofuel economy and global biofuel projections, *Energy Convers. Manag.*, 49, 2106–2116 (2008).

213. J. Hill, E. Nelson, D. Tilman, S. Polasky, and D. Tiffany, Environmental, economic, and energetic costs and benefits of biodiesel and ethanol biofuels, *Proc. Natl. Acad. Sci. USA*, 103, 11206–11210 (2008).

214. G. Hochman, D. Rajagopal, and D. Zilberman, The effect of biofuels on the international oil market, *Appl. Econ. Perspect. Policy*, 33, 402–427 (2012).

215. R. Ballesteros, J. Guillén-Flores, and J. Barba, Environmental and health impact assessment from a heavy-duty diesel engine under different injection strategies fueled with a bioethanol–diesel blend, *Fuel*, 157, 191–201 (2015).

216. J.C.L. Alves and R.J. Poppi, Quantification of conventional and advanced biofuels contents in diesel fuel blends using near-infrared spectroscopy and multivariate calibration, *Fuel*, 165, 379–388 (2016).

217. K. Krisnangkura, T. Yimsuwan, and R. Pairintra, An empirical approach in predicting biodiesel viscosity at various temperatures, *Fuel*, 85, 107–113 (2006).

218. S. Kim and B.E. Dale, Global potential of bioethanol production from wasted crops and crop residues, *Biomass Bioenergy*, 26, 361–375 (2004).

219. A.N. Giri, M.N. Deshmukh, and S.B. Gore, Effect of cultural and integrated methods of weed control on cotton, intercrop yield and weed-control efficiency in cotton-based cropping system, *Ind. J. Agron.*, 51, 34–36 (2006).

220. J. Hill, E. Nelson, D. Tilman, S. Polasky, and D. Tiffany, Environmental, economic and energetic costs and benefits of biodiesel and ethanol biofuels, *Proc. Natl. Acad. Sci. USA*, 103, 11206–11210 (2006).

221. A.Y. Hoekstra and A.K. Chapagain, *Globalization of Water: Sharing the Planet's Freshwater Resources*, Blackwell Publishing, Oxford, U.K. (2008).

222. A.H. Kassam and K.R. Stockinger, Growth and nitrogen uptake of sorghum and millet in 83 mixed cropping, *Samaru Agric. Newslett.*, 15, 28–35 (1973).

223. J.B. Woerfel, Harvest, storage, handling, and trading of soybeans. In: *Practical Handbook of Soybean Processing and Utilization* (D.R. Erickson, ed.), AOCS Press, Champlain, IL (1995).

224. A. Demirbas, Progress and recent trends in biodiesel fuels, *Energy Convers. Manag.*, 50, 14–34 (2009).

225. F. Millo, B.K. Debnath, T. Vlachos, C. Ciaravino, L. Postrioti, and G. Buitoni, Effects of different biofuels blends on performance and emissions of an automotive diesel engine, *Fuel*, 159, 614–627 (2015).

226. P. Emberger, D. Hebecker, P. Pickel, E. Remmele, and K. Thuneke, Emission behaviour of vegetable oil fuel compatible tractors fuelled with different pure vegetable oils, *Fuel*, 167, 257–270 (2016).

227. M.M. Roy, W. Wang, and J. Bujold, Biodiesel production and comparison of emissions of a DI diesel engine fueled by biodiesel-diesel and canola oil-diesel blends at high idling operations, *Appl. Energy*, 106, 198–208 (2013).

228. Y. Lue, Y. Yeh, and C. Wu, Emission characteristics of a small D.I. diesel engine using biodiesel blended fuels, *J. Environ. Sci. Health*, 36, 845–859 (2001).

229. O.S. Valente, M.J. Da Silva, V.M. Pasa, and C.R. Belchior, Fuel consumption and emissions from a diesel power generator fuelled with *castor* oil and soybean biodiesel, *Fuel*, 89, 3637–3642 (2010).

230. O.S. Valente, V.M. Pasa, and C.R. Belchior, Exhaust emissions from a diesel power generator fuelled by waste cooking oil biodiesel, *Sci. Total Environ.*, 431, 57–61 (2012).

231. A. Shaheed and E. Swain, Combustion analysis of coconut oil and its methyl esters in a diesel engine, *Proc. Inst. Mech. Eng.*, 213, 417–425 (1999).

232. C. Sayin, M. Gumus, and M. Canakci, Effect of fuel injection timing on the emissions of a direct-injection (DI) diesel engine fueled with *Canola* oil methyl ester–diesel fuel blends, *Energy Fuels*, 24, 2675–2682 (2010).

233. A.N. Ozsezen, M. Canakci, A. Turkcan, and C. Sayin, Performance and combustion characteristics of a DI diesel engine fueled with waste *Palm* oil and *Canola* oil methyl esters, *Fuel*, 88, 629–636 (2009).

234. H. Sharon, K. Karuppasamy, D.R. Kumar, and A. Sundaresan, A test on DI diesel engine fueled with methyl esters of used *Palm* oil, *Renew. Energy*, 47, 160–166 (2012).

235. S.M. Rahman, H.H. Masjuki, M.A. Kalam, M.J. Abedin, A. Sanjid, and S. Imtenan, Effect of idling on fuel consumption and emissions of a diesel engine fueled by *Jatropha* biodiesel blends, *J. Clean. Prod.*, 69, 208–215 (2014).

236. M. Karabektas, The effects of turbocharger on the performance and exhaust emissions of a diesel engine fuelled with biodiesel, *Renew. Energy*, 34, 989–993 (2009).

237. A. Ghorban, B. Bazooyar, A. Shariati, S.M. Jokar, H. Ajami, and A. Naderi, A comparative study of combustion performance and emission of biodiesel blends and diesel in an experimental boiler, *Appl. Energy*, 88, 4725–4732 (2011).

238. M.F. Mohamed and S.K. Bhatti, Experimental and computational investigations for combustion, performance and emission parameters of a diesel engine fueled with *Soybean* biodiesel-diesel blends, *Energy Procedia*, 52, 421–430 (2014).

239. D.H. Qi, H. Chen, L.M. Geng, and Y.Z. Bian, Experimental studies on the combustion characteristics and performance of a direct injection engine fueled with biodiesel/diesel blends, *Energy Convers. Manag.*, 51, 2985–2992 (2010).

240. A. Dhar and A.K. Agarwal, Effect of *Karanja* biodiesel blends on particulate emissions from a transportation engine, *Fuel*, 141, 154–163 (2015).

241. C.S. Cheung, X.J. Man, K.W. Fong, and O.K. Tsang, Effect of waste cooking oil biodiesel on the emissions of a diesel engine, *Energy Procedia*, 66, 93–96 (2015).

242. A. Uyumaz, H. Solmaz, E. Yılmaz, H. Yamık, and S. Polat, Experimental examination of the effects of military aviation fuel JP-8 and biodiesel fuel blends on the engine performance, exhaust emissions and combustion in a direct injection engine, *Fuel Process. Technol.*, 128, 158–165 (2014).

243. Z. Utlu and M.S. Koçak, The effect of biodiesel fuel obtained from waste frying oil on direct injection diesel engine performance and exhaust emissions, *Renew. Energy*, 33, 1936–1941 (2008).

244. K.N. Gopal, A. Pal, S. Sharma, C. Samanchi, and K. Sathyanarayanana, Investigation of emissions and combustion characteristics of a CI engine fueled with waste cooking oil methyl ester and diesel blends, *Alex. Eng. J.*, 53, 281–287 (2014).

245. D. Selvam and K. Vadivel, Performance and emission analysis of DI diesel engine fuelled with methyl esters of beef tallow and diesel blends. *Procedia Eng.*, 38, 342–358 (2012).

246. M.A. Kalam, H.H. Masjuki, M.H. Jayed, and A.M. Liaquat, Emission and performance characteristics of an indirect ignition diesel engine fuelled with waste cooking oil, *Energy*, 36, 397–402 (2011).

247. M. Nabi, M. Akhter, and M. Shahadat, Improvement of engine emissions with conventional diesel fuel and diesel–biodiesel blends, *Bioresour. Technol.*, 97, 372–378 (2006).

248. M. Saqib, M.W. Mumtaz, A. Mahmood, and M.I. Abdullah, Optimized biodiesel production and environmental assessment of produced biodiesel, *Biotechnol. Bioprocess Eng.*, 17, 617–623 (2012).

249. J.J. Bergh, I.J. Cronjé, J. Dekker, T.G. Dekker, L.M. Gerritsma, and L.J. Mienie, Noncatalytic oxidation of water-slurried coal with oxygen: Identification of fulvic acids and acute toxicity, *Fuel*, 76, 149–154 (1997).

250. W. Yinhui, Z. Rong, Q. Yanhong, P. Jianfei, L. Mengren, L. Jianrong, W. Yusheng, H. Min, and S. Shijin, The impact of fuel compositions on the particulate emissions of direct injection gasoline engine, *Fuel*, 166, 543–552 (2016).

251. S. Manzetti and O. Andersen. A review of emission products from bioethanol and its blends with gasoline. Background for new guidelines for emission control, *Fuel*, 140, 293–301 (2015).

252. O. Özener, L. Yüksek, A.T. Ergenç, and M. Özkan, Effects of soybean biodiesel on a DI diesel engine performance, emission and combustion characteristics, *Fuel*, 115, 875–883 (2014).

253. A. Lapinskienė, P. Martinkus, and V. Rėbždaitė, Eco-toxicological studies of diesel and biodiesel fuels in aerated soil, *Environ. Pollut.*, 142, 432–437 (2006).

254. D. Leme, T. Grummt, R. Heinze, A. Sehr, and M. Skerswetat, Cytotoxicity of water-soluble fraction from biodiesel and its diesel blends to human cell lines, *Ecotoxicol. Environ. Saf.*, 74, 2148–2155 (2011).

255. D. Leme, T. Grummt, D. Oliveira, S. Andrea, S. Renz, S. Reinel, E.R.A. Ferraz et al., Genotoxicity assessment of water soluble fractions of biodiesel and its diesel blends using the *Salmonella* assay and the in vitro MicroFlow® kit (Litron) assay, *Chemosphere*, 86, 512–520 (2012).

256. S. Pereira, V.Q. Araújo, M.V. Reboucas, and F. Vieira, Toxicity of biodiesel, diesel and biodiesel/diesel blends: Comparative sub-lethal effects of water-soluble fractions to microalgae species, *Bull. Environ. Contamin. Toxicol.*, 88, 234–238 (2012).

257. P. Fregolente, L. Fregolente, and M. MacIel, Water content in biodiesel, diesel, and biodiesel-diesel blends, *J. Chem. Eng. Data*, 57, 1817–1821 (2012).

258. M. Leite, M. Araújo, and I. Nascimento, Toxicity of water-soluble fractions of biodiesel fuels derived from *Castor* oil, *Palm* oil, and waste cooking oil, *Environ. Toxicol. Chem.*, 30, 893–897 (2011).

259. N. Khan, M.A. Warith, and G. Luk, A comparison of acute toxicity of biodiesel, biodiesel blends, and diesel on aquatic organisms, *J. Air Waste Manag. Assoc.*, 57, 286–296 (2007).

260. W. Rodinger, Toxicology and ecotoxicology of biodiesel fuel. In: *Plant Oils as Fuels* (N. Martini, and J.S. Schell, eds.), Springer, Berlin, Germany, pp. 161–180 (1998).

261. L. Batista, V. Da Silva, É. Pissurno, T. Soares, and M. Jesus, Formation of toxic hexanal, 2-heptenal and 2,4-decadienal during biodiesel storage and oxidation, *Environ. Chem. Lett.*, 13, 353–358 (2015).

262. J.L. Dvm, Health issues concerning inhalation of petroleum diesel and biodiesel exhaust. In: *Plant Oils as Fuels* (N. Martini, and J.S. Schell, eds.), Springer, Berlin, Germany, pp. 92–103 (1998).

263. J.M. Cruz, I.S. Tamada, P.R. Lopes, R.N. Montagnolli, and E.D. Bidoia, Biodegradation and phytotoxicity of biodiesel, diesel, and petroleum in soil, *Water Air Soil Pollut.*, 225, 1962–1971 (2014).

264. J.M. Borges, J.M. Dias, and A.S. Danko, Influence of the anaerobic biodegradation of different types of biodiesel on the natural attenuation of benzene, *Bull. Environ. Contamin. Toxicol.*, 225, 2146–2152 (2014).

265. A.C. Cruz, M.N. Leite, L.E. Rodrigues, and I.A. Nascimento, Estimation of biodiesel cytotoxicity by using acid phosphatase as a biomarker of lysosomal integrity, *Bull. Environ. Contamin. Toxicol.*, 89, 219–224 (2012).

266. R.N. Ávila and J.R. Sodré, Physical–chemical properties and thermal behavior of fodder radish crude oil and biodiesel, *Ind. Crops Prod.*, 38, 54–57 (2012).

267. A.S. Silitonga, H.H. Masjuki, T.M. Mahlia, H.C. Ong, and A.E. Atabani, A global comparative review of biodiesel production from *jatropha curcas* using different homogeneous acid and alkaline catalysts: Study of physical and chemical properties, *Renew. Sustain. Energy Rev.*, 24, 514–533 (2013).

268. C.Y. Lin and R.J. Li, Fuel properties of biodiesel produced from the crude fish oil from the soapstock of marine fish, *Fuel Process. Technol.*, 90, 130–136 (2009).

269. J. Gandure, C. Ketlogetswe, and A. Temu, Fuel properties of biodiesel produced from selected plant kernel oils indigenous to Botswana: A comparative analysis, *Renew. Energy*, 68, 414–420 (2014).

270. C.Y. Lin and C.L. Fan, Fuel properties of biodiesel produced from *Camellia oleifera Abel* oil through supercritical-methanol transesterification, *Fuel*, 90, 2240–2244 (2011).

271. O.S. Valente, V.M. Pasa, C.R. Belchior, and J.R. Sodré, Physical–chemical properties of waste cooking oil biodiesel and *castor* oil biodiesel blends, *Fuel*, 90, 1700–1702 (2011).

272. T. Tsoutsos, M. Chatzakis, I. Sarantopoulos, and A. Nikologiannis, Effect of wastewater irrigation on biodiesel quality and productivity from *castor* and sunflower oil seeds, *Renew. Energy*, 57, 211–215 (2013).

273. J. Gandure, C. Ketlogetswe, and A. Temu, Fuel properties of *jatropha* methyl ester and its blends with petroleum diesel, *J. Eng. Appl. Sci.*, 8, 900–908 (2013).

274. J.D. Martínez, M. Lapuerta, R. Contreras, R. Murillo, and T. García, Fuel properties of tire pyrolysis liquid and its blends with diesel fuel, *Energy Fuels*, 27, 3296–3305 (2013).

275. R. Wang, M.A. Hanna, W. Zhou, P.S. Bhadury, and Q. Chen, Production and selected fuel properties of biodiesel from promising non-edible oils: *Euphorbia lathyris* L., *Sapium sebiferum* L. and *Jatropha curcas* L., *Bioresour. Technol.*, 102, 1194–1199 (2011).

276. Q. Hu, W. Xiang, S. Dai, T. Li, F. Yanga, Q. Jia, G. Wang, and H. Wu, The influence of cultivation period on growth and biodiesel properties of microalga *Nannochloropsis gaditana*, *Bioresour. Technol.*, 192, 157–164 (2015).

277. M.B. Johnson and Z. Wen, Production of biodiesel fuel from the microalga *Schizochytrium limacinum* by direct transesterification of algal biomass, *Energy Fuels*, 23, 5179–5183 (2009).

278. L. Lin, D. Ying, S. Chaitep, and S. Vittayapadung, Biodiesel production from crude rice bran oil and properties as fuel, *Appl. Energy*, 86, 681–688 (2009).

279. S.K. Karmee and A. Chadha, Preparation of biodiesel from crude oil of *Pongamia pinnata*, *Bioresour. Technol.*, 96, 1425–1429 (2005).

280. E. Crabbe, C. Nolasco-Hipolito, G. Kobayashi, K. Sonomoto, and A. Ishizaki, Biodiesel production from crude palm oil and evaluation of butanol extraction and fuel properties, *Process Biochem.*, 37, 65–71 (2001).

281. M.H. Ali, M. Mashud, M.R. Rubel, and R.H. Ahmad, Biodiesel from *Neem* oil as an alternative fuel for Diesel engine, *Procedia Eng.*, 56, 625–630 (2013).

282. J.M. Encinar, J.F. González, and A. Pardal, Transesterification of *castor* oil under ultrasonic irradiation conditions, *Fuel Process. Technol.*, 103, 9–15 (2012).

283. E.A. Mahmoud, L.A. Farahat, Z.K. Abdel Aziz, N.A. Fatthallah, and R.A. Salah El Din, Evaluation of the potential for some isolated microalgae to produce biodiesel, *Egypt. J. Petrol.*, 24, 97–101 (2015).

284. J. Cvengros, J. Paligova, and Z. Cvengrošova, Properties of alkyl esters base on *castor* oil, *Eur. J. Lipid Sci. Technol.*, 108, 629–635 (2006).

285. A.A. Abdelmalik, J.C. Fothergill, S.J. Dodd, A.P. Abbott, and R.C. Harris, Effect of side chains on the dielectric properties of alkyl esters derived from *palm kernel* oil. In: *Proceedings of IEEE International Conference on Dielectric Liquids*, Bled, Slovenia (2011).

286. C.Y. May, Y.C. Liang, C.S. Foon, and M.A. Ngan, Key fuel properties of palm oil alkyl esters, *Fuel*, 84, 1717–1720 (2005).

287. A.M. Jurcak, S.J. Gauthier, and P.A. Moore, The effects of biodiesel and crude oil on the foraging behavior of rusty crayfish, *Orconectes rusticus*, *Arch. Environ. Contamin. Toxicol.*, 69, 557–565 (2015), doi:10.1007/s00244-015-0181-4.

288. V.T. Wyatt, M.A. Hess, R.O. Dunn, T.A. Foglia, and M.J. William, Fuel properties and nitrogen oxide emission levels of biodiesel produced from animal fats, *J. Am. Oil Chem.*, 82, 585–591 (2005).

289. R. Yahyaee, B. Ghobadian, and G. Najafi, Waste fish oil biodiesel as a source of renewable fuel in Iran, *Renew. Sustain. Energy Rev.*, 17, 312–319 (2013).

290. L. Schumacher, Biodiesel lubricity. In: *The Biodiesel Handbook* (G. Knothe, J. Krahl, and J. Van Gerpen, eds.), AOCS Press, Champaign, IL, pp. 137–144 (2005).

291. P.I. Lacey and S.R. Westbrook, Lubricity requirement of low sulfur diesel fuels, SAE Paper 950248, Society of Automotive Engineers: Washington, DC (1995).

292. G. Anastopoulos, E. Lois, F. Zannikos, S. Kalligeros, and C. Teas, HFRR lubricity response of an additized aviation kerosene for use in CI engines, *Tribol. Int.*, 35, 599–604 (2005).

293. P. Mondal, M. Basu, and N. Balasubramanian, Direct use of vegetable oil and animal fat as alternative fuel in internal combustion engine, *Biofuels Bioprod. Biorefin.*, 2, 155–174 (2008).

294. N.J. Fox, B. Tyrer, and G.W. Stachowiak, Boundary lubrication performance of free fatty acids in sunflower oil, *Tribol. Lett.*, 16, 275–281 (2004).
295. J.W. Goodrum and D.P. Geller, Influence of fatty acid methyl esters from hydroxylated vegetable oils on diesel fuel lubricity, *Bioresour. Technol.*, 96, 851–855 (2008).
296. D.C. Drown, K. Harper, and E. Frame, Screening vegetable oil alcohol esters as fuel lubricity enhancers, *J. Am. Oil Chem. Soc.*, 78, 579–584 (2001).
297. K.S. Wain and J.M. Perez, Oxidation of biodiesel fuel for improved lubricity, *ICE*, 38, 27–34 (2002).
298. R.P. Srinivasa and V.K. Gopalakrishnan, Vegetable oils and their methyl esters as fuels for diesel engines, *Ind. J. Technol.*, 129, 292–297 (1991).
299. A. Agarwal, Biofuels (alcohols and biodiesel) applications as fuels for internal combustion engines, *Prog. Energy Combust. Sci.*, 33, 233–271 (2007).
300. B. Batko, T.K. Dobek, and A. Koniuszy, Evaluation of vegetable and petroleum based diesel fuels in the aspect of lubricity in steel–aluminium association, *J. Int. Agrophys.*, 22, 31–34 (2008).
301. F.P. Bowden and D. Tabor, *Friction and Lubrication of Solids*, Clarendo Press, Gloucestershire, U.K. (1950).
302. M.A. Fazal, A.S.M. Haseeb, and H.H. Masiuki, Biodiesel feasibility study: An evaluation of material compatibility; performance; emission and engine durability, *Renew. Sustain. Energy Rev.*, 15, 1314–1324 (2011).
303. H.H. Masjuki and M.A. Maleque, The effect of palm oil diesel fuel contaminated lubricant on sliding wear of cast irons against mild steel, *Wear*, 198, 293–299 (1996).
304. S.J. Clark, L. Wagner, M.D. Schrock, and P.G. Piennaar, Methyl and ethyl soybean esters as renewable fuels for diesel engines, *J. Am. Oil Chem. Soc.*, 61, 1632–1638 (1984).
305. A. Monyem and J. Van Gerpen, The effect of biodiesel oxidation on engine performance and emissions, *Biomass Bioenergy*, 20, 317–325 (2001).
306. F. Yaşar, H. Adin, S. Aslan, and Ş. Altun, Comparing the lubricity of different biodiesel fuels. In: *Sixth International Advanced Technologies Symposium (IATS'11)*, Elazığ, Turkey (2011).
307. S. Topaiboul and N. Chollacoop, Biodiesel as a lubricity additive for ultra low sulfur diesel, *Songklanakarin J. Sci. Technol.*, 32, 153–156 (2010).
308. B.R. Moser and S.F. Vaughn, Evaluation of alkyl esters from Camelina sativa oil as biodiesel and as blend components in ultra low-sulfur diesel fuel, *Bioresour. Technol.*, 101, 646–653 (2010).

14 Field Pennycress
A New Oilseed Crop for the Production of Biofuels, Lubricants, and High-Quality Proteins

*Roque L. Evangelista, Steven C. Cermak,
Milagros P. Hojilla-Evangelista,
Bryan R. Moser, and Terry A. Isbell*

CONTENTS

14.1 INTRODUCTION

Field pennycress or pennycress (*Thlaspi arvense* L., Brassicaceae) is a native of Eurasia and widely naturalized in temperate and subarctic regions in North America [1,2]. The name thlaspi was derived from the Greek word *thlao*, to flatten, and *aspis*, shield, which described the shape of the fruit. Arvense is from the Latin word arvens, meaning "of the field" [3]. Other names for pennycress include stinkweed, fanweed, Frenchweed, and wild garlic. Pennycress can be found in a wide variety of habitats such as croplands, fallow fields, roadsides and railroads, gardens, meadows, and pastures. It is considered a major agricultural weed causing serious yield losses in field crops, reduced crop quality, and contaminated hay and grain feed [1,2]. Growing pennycress as a source of oil for industrial use was suggested by Clopton and Triebold [4] in the early 1940s. Test plantings of pennycress produced 1680 kg of seeds/hectare on irrigated land.

Renewed interest in developing alternative sources of feedstock for biofuels in the past few years has led researchers at United States Department of Agriculture-Agricultural Research Service (USDA-ARS) in Peoria, Illinois, to develop pennycress as an oilseed crop. The combination of exceptional winter hardiness and early harvest date of pennycress compared to other winter annual oilseed crops makes it suitable as an off-season crop between corn and soybeans production in most of the upper Midwestern United States [5]. This allows for the production of oil for industrial applications without disrupting the current crop rotation for growing commodity crops for food use. This rotation scheme also offers a distinct advantage to farmers by providing additional farm income from an otherwise fallow season with little impact on subsequent soybean production [6]. Planting pennycress as a cover crop could also reduce soil erosion, decrease nutrient loss, and improve nitrogen management [7,8]. Pennycress has the potential to provide late fall and early spring weed suppression. The early canopy development in pennycress can reduce light quality and light quantity, resulting in changes in the morphology of weeds. Pennycress begins flowering in April and early May; thus, pennycress can serve as a food source for honeybees and native pollinators in early spring when no other plants are in bloom [9]. In addition, pennycress may also provide overwintering habitat for predatory insects, increasing the potential for the biocontrol of pests [10–12].

The pennycress breeding program was initiated by Western Illinois University (WIU) [13] and USDA-ARS in 2009. The goals were to develop varieties of pennycress with early maturity, higher oil content, reduced seed dormancy, and uniform stand establishment. Three lines currently under commercial development are derived from wild selections: W-12, a line from WIU containing 36% oil, and two lines from USDA-ARS, Patton containing 34% oil and Beecher containing 36% oil. Katelyn (Reg. No. GP-35, PI 673443) pennycress, publicly released by the USDA-ARS in 2014, resulted from selections from the Beecher line [14]. Katelyn has a postharvest germination rate of 91% for seeds kept in the dark and 81% under 12 h light/dark 27.5°C/11.5°C conditions. This new cultivar will improve stand establishment and help alleviate the potential of pennycress to develop a seed bank in the soil by promoting the germination of seed spilled at harvest to grow and die under the canopy of the subsequent soybean crop where it should not be capable of competing.

14.2 PLANT DESCRIPTION

The morphology of pennycress is described in great detail by Best and McEntyre [1] and Warwick et al. [2]. Pennycress can be either a winter or spring annual propagating by seeds. Seeds that germinate in the fall grow into winter hardy rosettes (Figure 14.1a). As warm weather returns in early spring, more leaves emerge from the basal rosette and the central stem with several side stems develops rapidly. The central and side stems terminate in erect racemes of small white flowers with four small petals (Figure 14.1b). Flowers are self-pollinated but are visited by small flies and bees. Mature plants grow to about 80 cm in height. The seed pods are borne on slender upward curving stalks. The pods are round and flat, about 1 cm in diameter, broadly winged, and notched at the tip (Figure 14.1c). The seed pods containing 4–16 seeds are bright green but become greenish orange or light yellow at maturity (Figure 14.1d). Mature seeds are dark reddish brown to black and are 1.5–2 mm long and obovate to ovate in shape. The seed surface has concentric ridges resembling a fingerprint (Figure 14.1e).

14.3 SEED PRODUCTION

Small-scale production of pennycress in Central Illinois has been conducted recently using Beecher seeds harvested from a wild stand. Planting was done around the third week of September after corn was harvested. Seeds are planted over corn stubble using a grain drill to a depth of no more than 6.4 mm at the rate of 11 kg seeds/ha. In small plot studies, planting when the soil temperature was above 15°C resulted in a high seed yield (1568 kg/ha). Seeds planted when soil temperature was below 10°C did not germinate until spring [15]. Most emergence occurs 7 days after planting. No nitrogen fertilizer was used in field trials, although, in small plot

FIGURE 14.1 Photograph of field pennycress (a) rosette, (b) at flowering stage (c) green seed pods, (d) mature seed pod, and (e) seeds.

experiments [16], 23 kg/ha, divided into fall and spring applications, appeared to be the optimal nitrogen rate in pennycress production. No herbicides or pesticides are currently used in field trials.

Harvest takes place in early June using a traditional combine with grain heads when seed moisture is around 12%. In 2008 field trials, the average combine yield was 1080 kg/ha, while the yield from hand-harvested 1 m² random sampling of the field was 1140 kg/ha. Hand-harvested seeds from 2009 experimental plots ranged from 1075 to 2534 kg/ha [17]. Pennycress has to be harvested as soon as the pods are amenable to threshing to minimize seed loss due to shattering. Estimates of seed loss by combine harvesting of 400 m² field plots with plant densities of 34.6k–666.9k plants per hectare ranged from 30–109 kg/ha, with the highest plant densities yielding the highest loss rates. A qualitative evaluation of the plots prior to harvest indicated that more seeds were dispersed on the ground than were collected in the weigh boats after mechanical harvesting, indicating that weather-induced shattering could be a problem [18].

The seeds must be dried immediately or at least aerated after harvest to prevent mold growth. Prescreening the seeds before drying may be needed depending on the amount of trash from spring weeds that contains high moisture. After drying to 8%–10% moisture content (MC), the seeds are cleaned to remove the remaining light trash and fine dirt before storage. Seeds for planting must be stored near constant room temperature to maintain high germination rates [18].

Field operations costs to produce pennycress will be similar to those for winter wheat and rye. A preliminary estimate for a commercial production is $268/ha [19]. This cost includes seeding, fertilizer, combine harvesting, and hauling. The estimated cost for seeding camelina by drilling is $39.50/ha [19], which is similar to that of aerial seeding.

14.4 PROPERTIES AND PROXIMATE ANALYSIS OF PENNYCRESS SEEDS

The physical properties of Beecher pennycress seeds harvested from a wild stand near Hanna City, IL, are similar to those reported in the literature. The seeds were 1.9 mm long, 1.4 mm wide, and about 1 mm thick. At 9.5% MC, 1000 seeds weighed 0.97 g and had a bulk density of 583 kg/m³ [20].

Pennycress seed contained, on a dry basis, 30.9% crude oil, 23.2% crude protein, 10.6% crude fiber, 5.6% ash, and 29.7% carbohydrates (Table 14.1) [20–22]. The pennycress seed crude protein content was slightly lower than the 36%–44% (dry and fat-free basis) protein reported for rapeseed [23,24], but very close to the 30% reported for lesquerella seed [25]. Defatted pennycress seed meals contained 39.7 and 36.7 mg sinigrin/g sample [26,27]. Sinigrin (allyl-glucosinolate) is the only glucosinolate detected in pennycress seed [28].

Pennycress oil contains a diverse set of fatty acids with 14–24 carbon chains, with erucic, linoleic, linolenic, and oleic as the major fatty acids (Table 14.2). Most of the fatty acids are either monounsaturated (60%) or polyunsaturated (36%). The total saturated fatty acid is 3.9%, of which palmitic acid accounts for 69%. The erucic acid content of pennycress oil is lower than that of a typical high erucic acid rapeseed. The amounts of palmitic, oleic, and linoleic acids in pennycress oil are less than half of that found in soybean oil.

TABLE 14.1
Proximate Composition of Pennycress Seeds

Component	Amount % (w/w) Dry Basis
Crude oil	30.9
Crude protein	23.2
Crude fiber	10.6
Ash	5.6
Carbohydrates	29.7

Sources: Evangelista, R.L. et al., *Ind. Crops Prod.*, 37(1), 76, 2012; Hojilla-Evangelista, M.P. et al., *Ind. Crops Prod.*, 45(1), 223, 2013; Hojilla-Evangelista, M.P. et al., *J. Am. Oil Chem. Soc.*, 92, 905, 2015.

TABLE 14.2
Fatty Acid Composition of Field Pennycress, Rapeseed, and Soybean Oils

Fatty Acids	Trivial Name	% (w/w)		
		Pennycress[a]	Rapeseed[b]	Soybean[c]
C14:0	Myristic	0.1	0.5	0.1
C16:0	Palmitic	2.7	3.5	10.6
C16:1 c9	Palmitoleic	0.2	—	0.1
C18:0	Stearic	0.4	1.0	4.0
C18:1 c9	Oleic	11.5	13.0	23.3
C18:1 c11	Vaccenic	1.5	—	—
C18:2 c9, 12	Linoleic	20.6	14.0	53.7
C18:3 c9, 12, 15	Linolenic	12.3	9.0	7.6
C20:0	Arachidic	0.4	1.0	0.3
C20:1 c11	Gondoic	8.7	7.5	—
C20:2 c11, 14	Eicosadienoic	1.7	1.0	—
C22:0	Behenic	0.3	0.5	0.3
C22:1 c13	Erucic	35.3	47.5	—
C22:2 c13, 16	Docosadienoic	0.7	1.0	—
C22:3 c13, 16, 19	Dihomo-γ-linolenic	0.6	—	—
C24:1 c15	Nervonic	3.4	—	—
$\sum$ SFA[d]		3.9	6.6	15.3
$\sum$ MUFA[d]		60.6	68.0	23.4
$\sum$ PUFA[d]		35.9	25.0	61.3

[a] Average of values reported in literature [29–32].

[b] Traditional rapeseed oil [33].

[c] O'Brien [34].

[d] Total saturated fatty acids (SFA), monounsaturated fatty acids (MUFA), and polyunsaturated fatty acids (PUFA).

14.5   OIL EXTRACTION

The preparation of pennycress seeds for oil extraction follows the same protocol as rapeseed and other oilseeds with glucosinolates. When the seeds are crushed, the myrosinase (a β-thioglucosidase enzyme) can hydrolyze glucosinolates to produce nitriles, isothiocyanates, thiocyanates, and other sulfur-containing compounds. Overcooking the seeds also produces these sulfur-containing compounds due to thermal degradation of the glucosinolates [35]. Some of these sulfur-containing compounds end up in the oil during extraction. In rapeseed oil, isothiocyanates were found to poison the catalyst used in hydrogenation [36,37]. Therefore, seed cooking must be carried out in such a way that myrosinase is inactivated but not in conditions severe enough to cause the thermal decomposition of glucosinolates.

Seed MC has a strong influence on inactivating myrosinase during cooking. Myrosinase in pennycress seeds with 4.5% MC remained active even when cooked at 116°C for 30 min. For seeds with 9.2% MC, myrosinase was inactivated after 30 min of cooking at 82°C or after 20 min at 93°C. It took only 10 and 5 min of cooking at 82°C to inactivate myrosinase when seed MCs were 13% and 18%, respectively [20]. Seed cooking enhances oil expression by rupturing oil cells, reducing oil viscosity, and coagulating meal protein. Cooking also reduces the seed moisture, which increases friction as the seed goes through the screw press [38]. Furthermore, cooking inactivates enzymes in the seed that may affect oil quality.

Oil extraction studies were conducted by the USDA-ARS in Peoria, Illinois, using seeds harvested from wild stands as well as from field trials. On a laboratory scale, unheated and cooked pennycress seeds (2.5 kg/batch) were pressed using a 2.2 kW Scott Tech radial screw press (Model ERT60II, Scott Tech Equipment, Vinhedo, Brazil). The seeds were cooked using a fluidized bed dryer (Model FBD2000, Endecotts, Ltd., London, U.K.). The heater was initially set at 148°C, the same temperature as the steam-heated cooker. If the seed was subjected to an additional drying time, the heater temperature was lowered to maintain the target seed temperature. The oil yields, expressed as the percent of total oil in the seed, were calculated from the oil contents of the starting seed and the press cake. This calculation assumed that the pressed oil is free of solids (foots), which is the case when the solids are separated and recycled to the press. The amounts of phospholipids (measured as phosphorus, P) and sulfur-containing compounds (measured as sulfur, S) in the crude oil were analyzed using an inductively coupled plasma-optical emission spectrometer.

Both the unheated seeds and seeds cooked to 60°C, the upper limit for cold pressing, had oil yields of around 70% (Table 14.3). The resulting crude oil contained low levels of phospholipids and sulfur-containing compounds. The oil yield increased by about 6% when the seeds were heated to 82°C and dried to <4% MC. The P and S contents of the crude oil were below detection limits (<5 ppm). The residual oils in the press cake were lower (~10%) when the seeds were cooked and dried compared with the cold-pressed seeds (~12%).

One-pass pressing (59 kg seeds/batch) was also conducted using a 14.9 kW heavy-duty screw press (Model L250, French Oil Mill Machinery Company, Piqua, OH). Screw pressing trials were conducted to determine the optimum screw arrangement and screen bar spacing as described by Evangelista et al. [20]. Seed cooking was

TABLE 14.3

Results of Oil Extraction from Pennycress Seeds by Screw Pressing

Seed Cooker and Screw Press Used	Temperature (°C)	Seed			Press Cake				
		Time to Reach Temp. (min)	MC[a] % (w/w)	Oil % (w/w, db[b])	MC % (w/w)	Oil % (w/w, db)	Oil Yield % (w/w) of Total Oil	P[c] (ppm)	S[d] (ppm)
Fluidized bed dryer and Scott Tech ERT60II	Not heated	—	9.0	30.6	10.3	11.6	70.2	<5	16
	60	9	7.0		7.5	12.1	68.6	8	12
	82	18	3.8		4	10.3	73.8	<5	<5
	82	(38)[e]	2.5		2.5	9.8	75.3	<5	<5
French 324 seed cooker and French L250 press	Not heated	—	9.4	31.0	9.2	9.7	76.1	16	<5
	60	5	9.1		8.7	11.6	70.6	14	<5
	82	19	7.1		7.6	8.3	79.8	31	14
	82–93	(50)[f]	5.2		6.0	4.9	88.5	125	144

[a] Moisture content.
[b] Dry basis.
[c] Phosphorus.
[d] Sulfur.
[e] Includes 20 min hold time after reaching target temperature.
[f] Includes 30 min hold time to dry the seed to around 5% MC.

done using a steam-heated 3-deck cooker (Model 324, French Oil Mill Machinery Company). The seed cooker was preheated using steam (148°C) for at least an hour before use. The oil yields from cold-pressed unheated seeds and seeds cooked to 60°C were 76% and 71%, respectively (Table 14.3). Compared to the oils obtained from the Scott Tech press, the crude oils obtained from the French L250 press have a slightly higher P content, but the amount of S was below the detection limit (<5 ppm). The highest oil yield was obtained when the seeds were dried to 5% MC where 88% of the total oil was recovered. However, the P and S levels were an order of magnitude higher than that of the seed heated to only 82°C. The overall good quality of crude oil in the laboratory-scale pressing can be attributed to the shorter heating and drying times in the fluidized bed dryer and to the less aggressive pressing (as indicated by the ≥10% oil content of the press cakes) by the Scott Tech ERT60II screw press.

14.6   PENNYCRESS OIL AS BIOFUEL

14.6.1   PENNYCRESS FATTY ACID METHYL ESTERS

The pennycress fatty acid methyl ester (PME) was prepared from crude penny-cress oil by employing the classic alkali-catalyzed transesterification process [39]. Methanol was added to pennycress oil at a 6:1 mol ratio. After heating the oil and methanol mixture to 60°C, 0.50 wt% sodium methoxide (25 wt% in methanol) was added and stirred for 1.5 h. The reaction mixture was cooled to room temperature and then transferred to a separatory funnel. After allowing it to settle for >2 h, the heavier glycerol layer was drained. The excess methanol in the methyl ester was removed under reduced pressure (20 mbar; 30°C) using a rotary evaporator. The product was rinsed with distilled water until neutral pH was obtained and then dried over $MgSO_4$.

The properties of pennycress oil and PME (Table 14.4) were determined following standard methods and compared to the biodiesel (B100) standards ASTM D6751 [40] and EN 14214 [41]. The acid value (AV, mg KOH/g) was determined as described in the American Oil Chemists' Society (AOCS) official method Cd 3d-63 [42] using a Metrohm 836 Titrando (Metrohm USA, Inc., Riverview, FL) equipped with a Solvotrode electrode. The amount of free glycerol and total glycerol in PME were determined by gas chromatography (GC) according to ASTM D6584 [43]. The Agilent 7890A Gas Chromatograph (Agilent, Santa Clara, CA) employed was equipped with a flame ionization detector and an Agilent D8-5HT column (15 m × 0.32 mm i.d., 0.10 μm film thickness). The oxidative stability (induction period, IP, h) was determined at 110°C using Metrohm 743 Rancimat following standard method EN 14112 [44].

The cloud point (CP, °C) and pour point (PP, °C) were determined following ASTM D5773 [45] and D5949 [46], respectively, using PSA-70S Phase Technology Analyzer (Richmond, BC, Canada). The cold filter plugging point (CFPP, °C) was measured using FPP 5Gs ISL Automatic CFPP Analyzer (PAC L.P., Houston, TX) according to ASTM D6371 [47]. Kinematic viscosity (mm^2/s) was measured at 40°C using a Canon Fenske viscometer (Cannon Instrument Co., State College, PA) following ASTM D445 [48].

TABLE 14.4
Properties of Pennycress Oil and Pennycress Methyl Ester (PME) Relative to Biodiesel Standard

Property	Pennycress Oil	PME	Biodiesel Standard ASTM D6751	EN 14214
Acid value (mg KOH/g)	0.61	0.04	0.50 maximum	0.50 maximum
Free glycerol (mass %)	—	0.005	0.020 maximum	0.020 maximum
Total glycerol (mass %)	—	0.041	0.240 maximum	0.250 maximum
Cloud point (°C)	−25	−10	Report	—
Pour point (°C)	−28	−18	—	—
Cold filter plugging point (°C)	21	−17	—	a
Oxidative stability (110°C) (h)	5.0	4.4	3 minimum	6 minimum
Kinematic viscosity (40°C) (mm²/s)	40.97	5.24	1.9–6.0	3.5–5.0
Wear scar diameter (HFRR, 60°C) (μm)	125	125	≤520[b]	≤460[b]
Sulfur (ppm)	—	7	15 maximum	10 maximum
Phosphorus (mass %)	—	0.000	0.001 maximum	0.001 maximum
Cetane number	—	59.8[c]	47 minimum	51 minimum

Source: Moser, B.R. et al., *Energy Fuels*, 23, 4149, 2009.

[a] Variable by location and time of year.

[b] Wear scar value specified for petrodiesel under ASTM D975 [56] and EN 590 [57].

[c] Estimated using derived cetane number.

Lubricity, as indicated by the wear scar on the ball rubbing against a disk submerge in the fuel, was measured using a high frequency reciprocating rig (HFRR) lubricity tester (PCS Instruments, London, England) in accordance with ASTM D6079 [49]. The wear scar (μm) is the average of the maximum length along the x- and y-axis of the wear scar measured with the aid of a microscope (Epimat M4000, Prior Scientific, Rockland, MA).

The sulfur (ppm) and phosphorus (wt%) content of pennycress oil and PME were obtained according to ASTM D5453 [50] and D4951 [51], respectively. The cetane number (CN) of PME was determined as a derived cetane number (DCN) using an Ignition Quality Tester (Advanced Engine Technology Ltd., Ottawa, Ontario, Canada) in accordance with ASTM D6890 [52]. The DCN obtained by ASTM D6890 correlates with CN determined by ASTM D613 [53].

As shown in Table 14.4, the PME prepared for property evaluation had acid value and free and total glycerol contents well below the maximum limits required for biodiesel. The PME had CP, PP, and CFPP of −10°C, −18°C, and −17°C, respectively. These excellent low-temperature properties were attributed to the preponderance of monounsaturated (erucate and oleate) and polyunsaturated (linoleate and linolenate) methyl esters, which have low melting points, and to the low amounts of saturated fatty acid methyl esters (FAME) (4%) in pennycress oil (Table 14.2). The PME's CN (59.8) was much higher than the minimum required by ASTM 6751 and EN 14214. The high CN of PME was attributed to the high concentration of methyl erucate and methyl gondoate, which had been reported to have CN of 74.2 and 73.2, respectively [54].

The oxidative stability of the PME (4.4 h) passed the minimum standard (3 h) for ASTM D6751 but did not meet the specification (6 h) required by EN 14214. PME contains 36% polyunsaturated methyl esters, a third of which are methyl linolenate. The oxidative stability of PME, however, is higher than the major unsaturated methyl esters (erucate, linoleate, linolenate, and oleate) in its composition. This could be due to the presence of residual native tocopherols (644 ppm) acting as natural antioxidant in the PME [39]. The kinematic viscosity of PME at 40°C (5.24 mm²/s) was within the ASTM D6751 (1.5–6.0 mm²/s) limit but higher than the EN 14214 requirement (3.5–5.0 mm²/s). The high kinematic viscosity of PME was due to the high proportion of long-chain fatty acid methyl ester (FAME) in its composition. About half of PME are $\geq$C20 long FAMEs, mostly methyl erucate and methyl gondoate, which have kinematic viscosities of 7.33 and 5.77 mm²/s, respectively [55]. Like biodiesel produced from other oils, PME also offers excellent lubricity. The wear scar diameter from HFRR measurement (60°C) was 125 μm, much smaller than the maximum wear scar diameter for petrodiesel standards under ASTM D975 (520 μm) [56] and EN 590 (460 μm) [57].

14.6.2 Pennycress Oil Blend with Gasoline

Some U.S. farmers have been successfully using straight vegetable oil (VO)–gasoline blends for several years as fuel for unmodified diesel engines [58]. Using gasoline as a blending component has several benefits: it is readily available, has high energy content, is inexpensive, and is completely miscible and stable with VO. In addition, compared to biodiesel, producing a VO–gasoline blend is fast, has low energy inputs, does not create waste products, and does not require a catalyst [59,60].

Peer-reviewed literatures evaluating VO and gasoline blends as fuels for diesel engines (referred to as triglyceride blends, TGB) were only available recently [32,58,60]. Pennycress oil was one of the VOs evaluated in a TGB [32,60]. The crude pennycress oil was prefiltered through a 10 μm polypropylene filter before mixing it with E10 gasoline (contains 10% v/v ethanol) at 3:1 (v/v) pennycress oil:E10 gasoline ratio. The blended fuel was further filtered using a 1 μm filter.

Fuel properties of pennycress oil, TGB, PME, and No. 2 diesel (Table 14.5) were evaluated following ASTM standards [32]. The density (g/cm³) was measured at 20°C using Anton Paar DSM 5000 (Anton Paar USA, Inc., Ashland, VA) density meter according to ASTM D4052 [61]. The kinematic viscosity (mm²/s) was determined using Anton Paar's Stabinger Viscometer SVM3000 following ASTM D445 [48]. The fuel's high heating value (HHV, J/g), also known as gross caloric value, was quantified using an IKA C200 calorimeter (IKA Works, Inc., Wilmington, NC) according to ASTM D240 [62]. The CFPP (°C) was determined using an automated CFPP analyzer (Lawler DR4-14, Lawler Manufacturing Corp., Edison, NJ) in conformity with ASTM D6371 [47].

Details of the test engine and the methods used in emissions testing can be found in Drenth et al. [60]. Engine performance and emissions were determined using a 4.5 L, 175 hp (130.5 kW) John Deere 4045 test engine. Engine performance and emissions

TABLE 14.5

Properties, Fuel Consumption, and Exhaust Gas Emissions of Pennycress Triglyceride Blends with E10 Gasoline (Pennycress TGB) and Other Fuels

Property	Pennycress Oil	PME	Pennycress TGB	No. 2 Diesel
Density at 20°C (g/cm³)	0.905	0.894	0.880	0.841
Kinematic viscosity at 40°C (mm²/s)	43.7	7.5	12.0	2.3
High heating value (J/g)	41,478	41,906	42,074	47,566
Cold filter plugging point[a] (°C)	21	−18	−17	−26
Brake-specific fuel consumption[a] (g/kW-h)	—[b]	248	239	223
Brake-specific emissions[a] (g/kW-h)				
CO	—	0.83	1.22	1.16
NOx	—	7.14	6.31	6.55
NMHC	—	0.219	0.318	0.209
Particulate matter	—	0.131	0.108	0.080
VOC (ppm)[a]	—	21	32	27
CH₂O (ppm)[a]	—	1.7	4.1	2.0

Sources: Drenth, A.C. et al., *Fuel*, 153, 19, 2015; American Society of Testing Materials, Standard test method for kinematic viscosity of transparent and opaque liquids (and calculation of dynamic viscosity), ASTM D445-06, in *ASTM Annual Book of Standards*, American Society of Testing Materials, West Conshohocken, PA, 2006.

[a] Values were obtained from bar graphs.
[b] Not determined.

data were collected at 50% load (250 N-m) and intermediate speed (1700 rpm) corresponding to mode 7 of ISO 8178 Non-Road Steady Cycle. Constant engine speed and load were maintained during each test using an eddy current dynamometer and a controller. The engine's fuel consumption was measured using a Coriolis meter and verified gravimetrically using a precision balance. The engine's exhaust was sampled and analyzed for pollutants using a Rosemount 5-gas emissions analysis system. A Fourier transform infrared spectrometer (Nicolet 6700, Thermo Fisher Scientific, Madison, WI) was also used to measure the volatile organic compounds (VOCs). The particulate matter was collected using a separate probe to sample a small portion of the exhaust stream. The sample exhaust stream passed through a PM10 cyclone to remove particulates bigger than 10 μm. The remaining smaller particles were collected using a Whatman PLC 7592-104 Teflon filter. The amount of particulate matter trapped in the filter was determined gravimetrically.

Density is an important property of diesel performance because fuel injection is based on volume metering. The addition of 25% v/v E10 gasoline to pennycress oil resulted in a TGB with a density (0.880 g/cm³) that was still higher than that of No. 2 diesel fuel (0.841 g/cm³) but similar to PME (0.894 g/cm³) (Table 14.5). The density of No. 2 diesel may range from 0.820 to 0.845 g/cm³ (at 15°C) [57], while biodiesel (B100) falls between 0.86 and 0.90 g/cm³ [41]. Aside from the volumetric blending of VO and E10 gasoline, farmers also use a hydrometer to produce TGBs with a specific gravity of 0.865 [58].

Pennycress oil had the highest kinematic viscosity (43.7 mm²/s) among the VOs evaluated by Drenth et al. [32]. Blending E10 gasoline significantly reduced the kinematic viscosity of pennycress oil, but its value (12.0 mm²/s) (Table 14.5) was above the acceptable range of 1.9–4.1 mm²/s for No. 2 diesel [56]. Pennycress TGB may be more suitable as fuel for low- to medium-speed heavy-duty diesel engines, which use No. 4 diesel fuel. No. 4 diesel fuel has a kinematic viscosity of 5.5–24.0 mm²/s [56].

The E10 gasoline also lowered the CFPP of pennycress oil to −17°C, which is comparable to that of PME B100 (−18°C) (Table 14.5). TGBs from other VOs also had much lower CFPP than their corresponding biodiesel. This suggests that TGBs would be widely adaptable most of the year in areas like the Midwestern United States [32].

E10 gasoline, which has a higher energy content than pennycress oil, resulted in a TGB with higher high heating value (HHV, 42,074 J/g) than PME B100 (41,906 J/g) but still much lower than that of No. 2 diesel (47,566 J/g) (Table 14.5). The brake-specific fuel consumption (BSFC), a measure of the amount of fuel consumed per unit of power produced by an engine, obtained from each type of fuel is consistent with their HHV. The BSFC of pennycress TGB (239 g/kW-h) is a little less than that of PME B100 (248 g/kW-h) but more than that of No. 2 diesel (223 g/kW-h) (Table 14.5). However, factoring in their densities, the differences in their fuel consumption are reduced.

Pennycress TGB generated more carbon monoxide (CO) and nonmethane hydrocarbon (NMHC) but lower total oxides of nitrogen (NOx) than PME B100 and No. 2 diesel (Table 14.5). Pennycress TGB emission also contained more VOCs and formaldehyde (CH₂O) than those of PME B100 and No. 2 diesel.

14.7 PENNYCRESS ESTOLIDES

VO-based lubricants and their derivatives have excellent lubricity and biodegradability. However, the combination of poor cold temperature performance and low resistance to thermal oxidation is a problem for lubricant applications [63]. The development of estolides has solved these undesirable physical properties by improving performance in both of these areas, thus reducing the need for expensive additive packages, which can sacrifice biodegradability and increase toxicity and cost.

Estolides are a class of esters based on VOs [64] made by the formation of a carbocation at the site of unsaturation that can undergo nucleophilic addition by another fatty acid (FA), with or without carbocation migration along the length of the chain, to form an ester linkage (Figure 14.2). The secondary ester linkages of the estolide are more resistant to hydrolysis than those of the triglycerides (TGs), and the unique structure of the estolide results in materials that have far superior physical properties (such as PP, CP, and wear) for lubricant applications than vegetable and mineral oils [65,66]. Estolides have been developed from both FAs and directly from the VO or TG (Figure 14.3).

Three different types of estolides have been synthesized at the USDA-ARS in Peoria, Illinois: (1) oleic FA-based estolides, (2) hydroxy-based estolides from

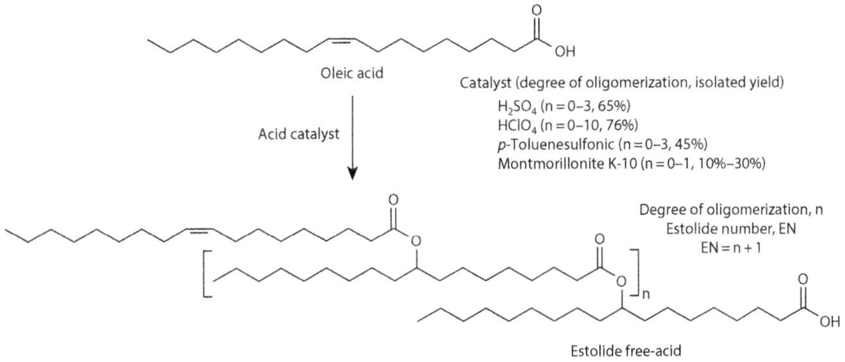

FIGURE 14.2 General scheme for oleic estolide free-acid synthesis. Degree of oligomerization (n) and isolated yield (%) are dependent on the acid catalyst used for the synthesis. (From Isbell, T.A. et al., *J. Am. Oil Chem. Soc.*, 71, 169, 1994.)

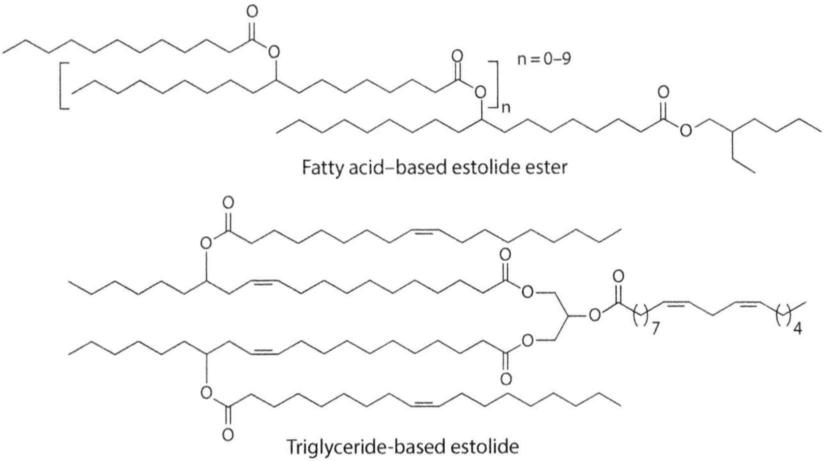

FIGURE 14.3 Structures of fatty acid estolide 2-ethylhexyl ester and triglyceride estolide. (From Isbell, T.A., *Grasas Aceites*, 62, 8, 2011.)

TG, and (3) hydroxy-based estolides from FA. The simple estolide structure (type a) has been easily modified by the addition of a saturated FA using this same technology [67,68]. A FA-based estolide similar to type (1) was produced from pennycress oil.

14.7.1 Synthesis of Pennycress Estolides

Estolides are formed from the cationic homo-oligomerization of unsaturated fatty acids resulting from the addition of a fatty acid carboxyl moiety across the olefin [69]. This condensation can continue, resulting in oligomeric compounds where

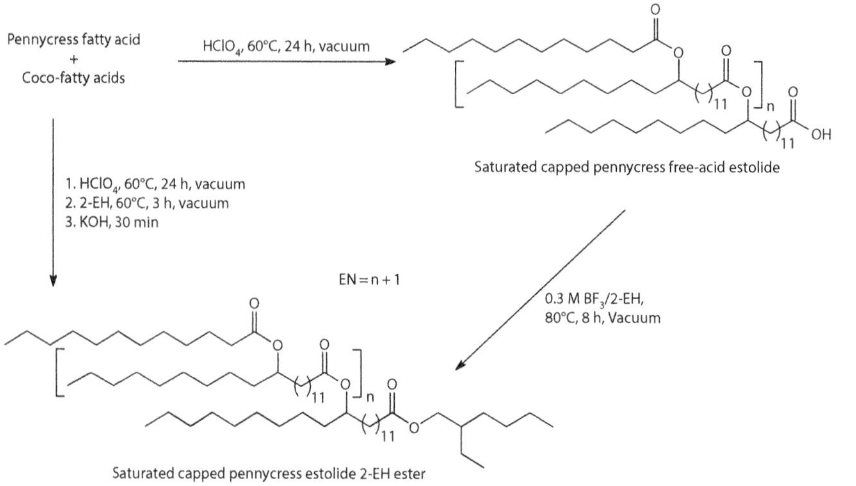

FIGURE 14.4 Synthetic routes for saturated capped pennycress estolides. (From Cermak, S.C. et al., *Ind. Crops Prod.*, 67, 179, 2015.)

the average extent of oligomerization (n) is related to the estolide number (EN) as follows (Figure 14.4) [70]:

$$EN = n + 1 \tag{14.1}$$

When saturated fatty acids are added to the reaction mixture, the oligomerization terminates upon the addition of the saturated fatty acid to the olefin since the saturate provides no additional reaction site to further the oligomerization. Consequently, the estolide is stopped at this point from further growth; thus, the estolide is referred to as "capped" [67].

Pennycress estolide was produced and evaluated as described by Cermak et al. [31]. Crude pennycress oil was hydrolyzed and distilled to yield a mixture of pennycress fatty acids. A series of pennycress-based free-acid estolides (Table 14.6) were synthesized by an acid-catalyzed condensation reaction conducted under vacuum at 60°C. Varying chain lengths (C) of saturated and unsaturated (U) capping fatty acids were used while all other reaction parameters were held constant. Vacuum distillation was used to remove excess fatty acids and other by-products. In most cases, the new pennycress-based free-acid estolides have a pennycress backbone with a terminal saturated capping group (Figure 14.4) [31].

The low-temperature properties of estolides are greatly improved when the free acids are esterified with 2-EH alcohol [63,64]. Estolide esters were synthesized with a 2:1 excess of pennycress fatty acid to capping fatty acid followed by esterification with 0.3 M 2-EH/BF$_3$ at 80°C under vacuum for 8 h (Figure 14.4). The samples were vacuum distilled to remove any remaining 2-EH, providing neat estolide 2-EH esters [31].

TABLE 14.6

Effect of Capping Fatty Acid Structure on the Physical Properties of Pennycress Free-Acid Estolides[a]

| Estolide | Capping Fatty Acid | | | Capped[d] (%) | Pour Point (°C) | Cloud Point (°C) | Vis. at 40°C (cSt) | Vis. at 100°C (cSt) | Vis. Index | Gardner Color |
	Name	C:U[b]	EN[c]							
A	Pennycress	22:1[e]	1.11	6.3	-9	-3	870.5	75.3	163	14+
B	Caprylic	8:0	1.18	23.8	-15	-9	642.7	54.4	146	12+
C	Capric	10:0	1.18	40.8	-12	4	494.4	44.9	144	15-
D	Lauric	12:0	1.28	37.7	-15	-11	608.1	53.6	149	14-
E	Myristic	14:0	1.33	53.0	-12	-4	682.7	57.0	146	16
F	Palmitic	16:0	1.22	51.4	-6	2	763.8	65.0	154	18
G	Oleic	18:1	1.49	6.5	-15	-11	692.3	61.5	156	13-
H	Coco	12:0[f]	1.16	48.2	-9	6	497.1	42.2	134	17+

Source: Cermak, S.C. et al., *Ind. Crops Prod.*, 67, 179, 2015.

[a] Ratio of pennycress fatty acids to capping fatty acid was 1:0.5.

[b] Capping fatty acid chain length (C) and unsaturation (U).

[c] Estolide number—see Equation 14.1.

[d] Percent of estolide saturated capped.

[e] Mainly C22:1—see Table 14.2.

[f] Mainly C12:0.

14.7.2 Pennycress Estolides Lubricant Properties

Estolides have certain physical characteristics that could help eliminate common problems, such as low resistance to thermal oxidative stability [71] and poor low-temperature properties [72], associated with VOs as functional fluids. Simple oleic estolide esters, when formulated with a small amount of oxidative stability package, show better oxidative stability than both petroleum and VO-based fluids [63], but there is still room for improvement.

The physical properties of the estolides (Tables 14.6 and 14.7) were determined employing standard methods. PPs (°C) were measured according to ASTM D97 [73] to an accuracy of ±3°C. The PPs were determined by placing a test jar with 50 mL of the sample into a cylinder submerged in a cooling medium. The sample temperature was reduced in 3°C increments at the top of the sample until the material stopped pouring. The sample no longer poured when the material in the test jar did not flow when held in a horizontal position for 5 s. The PP is the coldest temperature at which the sample still poured. CPs (°C) were determined following ASTM D2500 [74] to an accuracy of ±1°C. A test jar with 50 mL of the sample was immersed into a cylinder submerged in a cooling medium. The sample temperature was reduced in 1°C increments until any cloudiness was observed at the bottom of the test jar.

Viscosity measurements were made using calibrated Cannon-Fenske viscometer tubes (Cannon Instrument Co., State College, PA) and a Temp-Trol (Precision Scientific, Chicago, IL) viscometer bath set at 40.0°C and 100.0°C. Viscosity and viscosity index (VI) were calculated using ASTM Methods D445 [48] and D2270 [75], respectively.

One of the most important physical properties to a consumer is the color of the oil or oil-based products. As a potential hydraulic fluid, the estolides need to meet the color requirements of currently used hydraulic fluids. Gardner color was measured on a Lovibond 3-Field Comparator (Tintometer Ltd., Salisbury, England) according to AOCS Official Method Td 1a-64 [76]. The sample is compared against 18 glass standards, which range from lightest yellow (1) to brownish red (18). A (+) or (−) notation was employed to indicate samples that did not match one particular Gardner color.

The pennycress-based free-acid estolides have a pennycress backbone with primarily a terminal saturated capping group (Figure 14.4). The saturated capped % values in Tables 14.6 and 14.7 were obtained from GC analysis of the corresponding estolides, which were saponified and esterified with methanol. The components from GC analysis were classified as one of the following: unsaturated, saturated, and hydroxy fatty acids. The percent saturated capped, C, was calculated as follows:

$$C(\%) = \left[\frac{S}{100 - H} \right] \times 100 \qquad (14.2)$$

where
  C is the saturated capped (%)
  S is the saturated fatty acids (% mole)
  H is the hydroxyl fatty acids (% mole)

TABLE 14.7

Effect of Capping Fatty Acid Structure on the Physical Properties of Pennycress Estolide 2-Ethylhexyl Esters[a]

Estolide	Capping Fatty Acid			Capped[d] (%)	Pour Point (°C)	Cloud Point (°C)	Vis. at 40°C (cSt)	Vis. at 100°C (cSt)	Vis. Index	Gardner Color
	Name	C:U[b]	EN[c]							
A-2EH	Pennycress	22:1[e]	1.00	6.2	−18	—[f]	245.8	33.6	183	17−
B-2EH	Caprylic	8:0	1.69	27.7	−21	—	138.1	20.4	171	16−
C-2EH	Capric	10:0	1.38	43.1	−21	—	131.5	19.5	169	18+
D-2EH	Lauric	12:0	1.51	41.1	−21	—	124.4	19.0	173	15+
E-2EH	Myristic	14:0	1.40	50.4	−18	—	142.7	21.1	173	18+
F-2EH	Palmitic	16:0	1.20	52.3	−12	—	167.7	24.6	179	18+
G-2EH	Oleic	18:1	2.23	7.4	−24	−17	191.7	27.1	178	15
H-2EH	Coco	12:0[g]	1.13	47.9	−18	—	116.3	18.2	175	18

a   Ratio of pennycress fatty acids to capping fatty acid was 1:0.5.

b   Capping fatty acid chain length (C) and unsaturation (U).

c   Estolide number—see Equation 14.1.

d   Percent of estolide saturated capped.

e   Mainly C22:1—see Table 14.2.

f   Dash indicates sample too dark for cloud point determination.

g   Mainly C12:0.

Generally, as the chain length increased, the percentage of saturated capped estolide (Table 14.6) also increased, which is the same trend that was observed with the coriander-based estolides [77]. Caprylic, C8:0, exhibited the least amount of saturated capped material with only 23.8% while myristic, C14:0, had the highest at 53.0%.

In terms of low-temperature properties, the pennycress free-acid estolides (Table 14.6) have higher PP and CP values (less desirable) than the estolide 2-EH esters (Table 14.7). Of the new pennycress free-acid estolides, the lauric and oleic capped pennycress estolides had the best low-temperature properties with a PP of −15°C and CP of −11°C (Estolide D and G, Table 14.6). The remaining pennycress free-acid estolides had PPs that ranged from −9°C to −12°C; thus, the free-acid estolides generally have relatively poor low-temperature properties.

Of these new pennycress estolide 2-EH esters, the oleic capped pennycress estolide 2-EH ester had the best low-temperature properties with a PP of −24°C and CP of −17°C (Estolide G-2EH, Table 14.7). With the exception of palmitic acid, there was very little variation in the low-temperature properties of the remaining capped estolide esters, as seen by PPs ranging from −18°C to −21°C and CPs (where available) of about −14°C. A general trend of increasing PP and CP, or less desirable low-temperature properties, was observed as the chain length increased for the saturated capped 2-EH estolides (estolide F-2EH, Table 14.7, palmitic capped 2-EH estolide, PP −12°C).

The pennycress-based free-acid estolides and estolide 2-EH esters produced have a noticeable viscosity increase over previously synthesized oleic-based estolides. Oleic-based estolides have viscosities at 40°C and 100°C in the range of 80 and 13 cSt [78]. The viscosities of the pennycress-based materials are much higher due to longer-chain fatty acids and the position of the estolide linkage (Figure 14.5). As the estolide position shifts on the fatty acid backbone, the physical properties of the materials change. Viscosities (Table 14.7) of the pennycress estolide 2-EH esters ranged from 116.3 to 245.8 cSt at 40°C and 18.2 to 33.6 cSt at 100°C with VI from 169 to 183. Due to the hydrogen bonding of the carboxylate functionality, as expected, the free-acid estolides displayed higher viscosity than their 2-EH ester counterparts. Viscosities (Table 14.6) of the free-acid estolides ranged from 494.4 to 870.5 cSt at 40°C and 53.6 to 75.3 cSt at 100°C with VI from 134 to 163.

There was no significant trend in the Gardner colors of the estolides in Tables 14.6 and 14.7. The free-acid estolides, in general, have a darker color than any previously produced estolides [63,79]. This is most likely due to additional color bodies in the fatty acids derived from crude pennycress oil. This could easily be improved by a second distillation of the initial pennycress fatty acids prior to

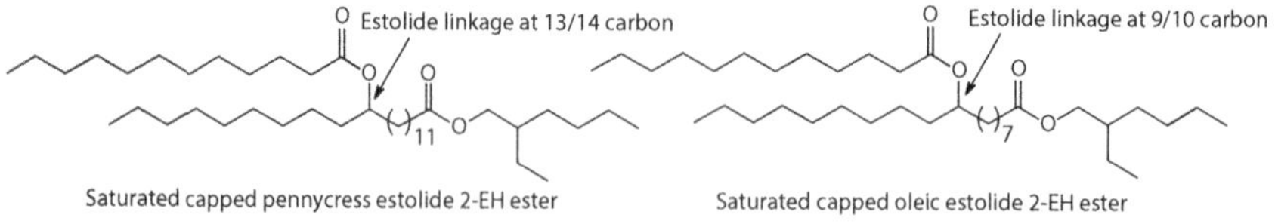

Estolide linkage at 13/14 carbon

Estolide linkage at 9/10 carbon

Saturated capped pennycress estolide 2-EH ester   Saturated capped oleic estolide 2-EH ester

FIGURE 14.5 Estolide linkage position. (From Cermak, S.C. et al., *Ind. Crops Prod.*, 67, 179, 2015.)

estolide synthesis or by further refining of the crude oil. Finally, the estolide 2-EH esters incurred an expected increase in Gardner color over that of the free-acid estolides, which resulted from the second acid treatment necessary for the esterification reaction.

Previous developments of a new synthetic method [31,66,68,77,80,81] have made it possible to eliminate the second acid treatment. This involves a one-pot perchloric acid–catalyzed process, followed by an *in situ* esterification of the free-acid estolide, followed by a double distillation of the pennycress fatty acids prior to estolide formation. A final method of color reduction involves an additional distillation step, which proved to be too costly. This second distillation step can improve the Gardner color of 2-EH estolide esters from 10 to 11 range to a 6 [78].

14.8 PENNYCRESS SEED PROTEINS

Pennycress seed protein is anticipated to be a major coproduct from oil production. Substantial work has already been done by USDA-ARS in Peoria, Illinois, on the extraction and characterization of pennycress seed proteins. The effects of seed cooking before oil extraction and the process of protein extraction method on the functional properties of the protein have been determined.

Pennycress seed protein consists primarily of the water-soluble (albumins) and NaCl-soluble (globulins) fractions, which account for 19 w/w% and 23 w/w% of total protein, respectively. Only minimal amounts (2%–5%) of alkali-soluble proteins were found and no ethanol-soluble protein was detected [21,82]. In contrast, canola seed and lesquerella seed (both also in the Brassicaceae family) have the NaOH- and NaCl-soluble fractions as the major protein classes [83,84]. In pennycress press cake from cooked seeds, the amounts of albumins and globulins were substantially reduced (to 14% and 17% of total protein, respectively). This decrease was not observed in pennycress oil cake from cold pressing [21], which clearly showed that the albumins and globulins in pennycress seed were susceptible to thermal treatment during seed cooking.

Sodium dodecyl sulfate-polyacrylamide gel electrophoresis revealed eleven polypeptide bands in pennycress seed protein, with molecular weight (MW) range 6.5–100 kDa, but the eight dominant polypeptide bands resolved at lower MW (6.5–41 kDa). Pennycress press cake protein showed band patterns that were similar to those of unprocessed ground seed, with an additional faint band visible between 107 and 198 kDa [21,22].

Amino acid compositions of pennycress seed and press cake are very similar, with glutamic acid, aspartic acid, alanine, arginine, leucine, and glycine having the greatest amounts (Table 14.8) [21,82]. Pennycress seed protein had adequate amounts of the essential amino acids to meet the nutritional needs of children 10–12 years old and adults, but was deficient in leucine, methionine + cysteine-cystine, lysine, and tryptophan for infants, according to FAO-suggested requirements [85]. The amino acid scoring patterns (sum of essential amino acids less tyrosine [85]) for pennycress seed and press cake were calculated to be 39.9 and 39.1 g/100 g protein, respectively, which were greater than the 37.4 g/100 g protein for soybean meal [86] and 36.0 g/100 g protein for canola–rapeseed meal [23].

TABLE 14.8

Amino Acid Compositions (g/100 g protein) of Defatted Pennycress Seed,
Cooked Seed-Press Cake, Saline-Extracted (SE) or Acid Precipitated (AP)
Protein Isolates (PI), Soybean Meal, and Soybean Protein Isolate[a]

Amino Acid	Pennycress Seed			Pennycress Press Cake			Soybean	
	Seed	SE PI	AP PI	Press Cake	SE PI	AP PI	Meal	PI[a]
Histidine[b]	2.8	2.8	2.8	2.7	2.6	2.6	2.6	2.7
Isoleucine	4.4	4.1	4.7	4.2	3.9	4.0	4.6	4.9
Leucine	7.8	7.2	8.8	7.8	7.5	7.8	7.8	7.7
Lysine	6.0	5.2	3.6	5.5	4.7	4.6	6.4	6.4
Methionine	1.5	1.4	1.6	1.6	1.2	1.6	1.1	1.1
Phenylalanine	5.0	4.9	5.2	5.0	4.9	5.2	5.0	5.4
Threonine	4.8	4.4	5.2	4.8	4.4	4.6	3.9	3.7
Tryptophan	1.3	1.7	1.2	1.6	1.7	1.7	1.4	1.4
Tyrosine	3.3	3.0	4.2	3.2	2.8	2.9	3.8	3.7
Valine[b]	6.3	5.7	6.4	5.8	5.6	6.1	4.6	4.8
Alanine	5.1	4.7	5.6	5.2	4.6	4.8	4.3	3.9
Arginine	7.8	8.2	8.3	7.9	8.0	8.3	7.3	7.8
Aspartic acid + asparagine	8.8	8.9	9.0	8.9	9.0	8.8	11.8	11.9
Cystine	2.0	2.2	1.6	2.0	2.1	2.2	1.4	1.3
Glutamic acid + glutamine	16.3	19.1	15.1	16.7	20.7	18.7	18.6	20.5
Glycine	6.8	6.2	6.2	7.0	6.4	6.2	4.3	4.1
Proline	6.0	7.2	6.2	6.0	6.2	6.2	5.5	5.3
Serine	3.8	3.4	4.3	4.0	3.7	3.8	5.5	5.5
AA scoring pattern[c]	39.9	37.4	39.5	39.1	36.4	38.1	38.8	38.1

Note: Values for pennycress seed, press cake, and protein isolates are means of duplicate
        determinations.

[a] Data for ground pennycress seed and press cake from Hojilla-Evangelista et al. [22,27]; soybean meal
     and protein isolate from Wolf [66].
[b] Histidine through valine are essential amino acids.
[c] Amino acid scoring pattern is the sum of all essential amino acids less tyrosine.

14.8.1 PRODUCTION AND COMPOSITION OF PENNYCRESS PROTEIN ISOLATES

Hojilla-Evangelista and coworkers [22,27] used two methods to produce protein isolates from ground pennycress seed and press cake: saline extraction (SE), which was
based on the major soluble proteins in pennycress, and conventional alkali solubili-
zation-acid precipitation (AP), which was adapted from the commercial process for
soybean protein isolates [87]. In SE, the main steps include protein extraction for
2 h using 0.1 M NaCl at 50°C and ratio of 10 mL: 1 g (two stages), centrifugation to
separate the supernatant, ultrafiltration-diafiltration, and freeze-drying. On the other
hand, the key steps in AP were alkali solubilization for 1.5 h using water and 1 M

NaOH (solvent pH = 10) at 50°C and a ratio of 10 mL: 1 g, centrifugation to recover the supernatant, precipitation of the soluble proteins in the supernatant by the addition of 1 M HCl until pH 4.5, redispersion of the recovered precipitate in pH 7 water, dialysis against distilled water of the soluble protein, and then freeze-drying.

For both pennycress seed and press cake, SE generally had greater protein recovery than did AP, with values of 40% versus 35% using seed as starting material and 45% versus 23% when press cake was used [22,27]. SE also produced from ground seed a protein isolate with higher purity (97% crude protein) than did AP (90% crude protein). However, when press cake was used as the protein source, the recovered product from SE contained considerably less protein (67% crude protein) than the one produced by AP (86.5% crude protein) [22,27]. Previous work that isolated protein from Brassicaceae seed meals, such as canola and rapeseed, using similar methods reported protein recoveries of 33%–65% [88–90] and protein contents of 76%–99% for the isolates [90–92].

Proteins isolated from press cake by both SE and AP had little residual oil ($\leq$0.5%) and no detectable crude fiber. However, the ash and carbohydrate contents of SE protein (15% and 18%, respectively) were much greater than the amounts of AP protein (3.5% and 9.5%) [22]. Sinigrin in protein isolates from ground pennycress seed and press cake was either absent or detected in trace amounts (0.1–1.7 mg sinigrin/g sample), which is a significant decrease when compared with sinigrin contents of defatted meal (Table 14.1) [22,27]. Membrane filtration techniques, which were used in the production of pennycress protein isolates, have been reported to be particularly effective in reducing glucosinate contents because of the lower MWs of glucosinolates compared to proteins [91,93].

The amino acid composition of pennycress protein isolates had only minor differences with those of their starting seed or press cake, regardless of the extraction method used [22,27]. SE seed protein isolate contained more tryptophan, glutamic acid + glutamine, and proline than ground seed but had less lysine, alanine, and glycine (Table 14.8). AP seed protein isolate had the greatest amount of serine but the least amounts of lysine, cysteine, and glutamic acid + glutamine compared with ground seed and SE seed protein isolate. With regard to the press cake protein isolates, they contained markedly more glutamic acid/glutamine and tryptophan but less lysine and tyrosine compared with ground press cake (Table 14.8). In general, pennycress seed or press cake protein isolates had amino acid contents that compared favorably with those reported for soybean protein isolate [86] (Table 14.8).

As with the amino acid contents, pennycress seed protein isolates showed polypeptide band patterns that were virtually identical to those of the ground seed, with the seven major polypeptides resolving between 6.5 and 41 kDa but appearing as darker and wider bands [27], which indicated their greater amounts. SE seed protein isolate also had more bands between 50 and 100 kDa than did AP seed protein isolate, with such bands corresponding to those detected in the water- and NaCl-soluble protein fractions in ground seed [21]. In contrast, both SE and AP press cake protein extracts showed only five dominant polypeptide bands, although they also resolved between 6.5 and 45 kDa [22]. It is assumed that the polypeptides with MW less than 6.5 kDa and those with MW 17–45 kDa are related to napin and cruciferin, respectively (major storage proteins of Brassica oilseeds) [94].

14.8.2 Functional Properties of Pennycress Protein Isolates

The protein isolates from seed meal and press cake were evaluated for their solubility, heat stability, foaming capacity, foam stability, emulsifying activity, and water holding capacity (WHC). Protein solubility at different pHs was determined using aqueous dispersions containing 1% (w/v) protein [95]. The pH was adjusted to 2, 4, 5.5, 7, 8.5, and 10 with dilute HCl or NaOH. After centrifugation, the protein content of the supernatant was determined spectrophotometrically and quantified with a standard curve developed using bovine serum albumin. Protein solubility is expressed as the percentage of protein in the sample that remained in the supernatant.

Heat stability was evaluated using aqueous dispersions containing 50 mg protein/mL. The pH was adjusted to 2.0, 7.0, or 10.0 with dilute HCl or NaOH and centrifuged at $10,000 \times g$ for 30 min. Twenty milliliters of the supernatant was heated for 30 min in a 90°C–100°C water bath, cooled to ambient temperature, centrifuged, and then poured through a Whatman No. 2 filter paper. The protein content of the filtrate and that of the unheated supernatant were determined by combustion according to AOCS Official method Ba 4e-93 [96]. Heat stability is the decrease in protein solubility (expressed in %) after heating.

The emulsifying activities of the protein isolates were evaluated by measuring their emulsifying activity index (EAI) and emulsion stability index (ESI) at pH 2, 7, and 10 using the method described by Wu et al. [97]. Six milliliters of sample (1 mg protein/mL) and 2 mL corn oil were emulsified using a homogenizer (Fisher PowerGen 35, Fisher Scientific, Pittsburgh, PA) set at high speed (30,000 rpm) for 1 min. The absorbance readings of the emulsified samples were taken at 500 nm and used to calculate the EAI (m²/g) and ESI (min.) as follows:

$$\text{EAI}\left(m^2/g\right) = 2T\left(A_0 \times \text{dilution factor}/C \times \Phi \times 10{,}000\right) \qquad (14.3)$$

where

T = 2.303

A_0 is the absorbance measured immediately after emulsion formation

dilution factor = 100

C is the concentration of protein (mg/mL) in the aqueous phase before emulsification

Φ is the oil volume fraction of the emulsion; and

$$\text{ESI}\left(\text{min}\right) = A_0 \times \frac{\Delta t}{\Delta A} \qquad (14.4)$$

where

Δt = 10 min

$\Delta A = A_0 - A_{10}$

A_{10} is the absorbance of the emulsion taken after 10 min.

The foaming capacity and foam stability of protein samples (10 mg protein/mL) were determined at pH 2, 7, and 10. The foaming apparatus was a graduated column fitted with a coarse fritted disk at the bottom. Air flowed in through the stem at a rate of 100 mL/min at 138 kPa (20 psi). The volume (mL) of foam produced in 1 min represented foaming capacity, while the portion of foam left (%) after standing for 15 min represented foam stability [95].

The WHC was measured using the method described by Balmaceda et al. [98] for insoluble or partly soluble materials with some modification. One gram of protein sample and 30 mL of distilled water were loaded into preweighed centrifuge tubes and were placed on a platform shaker and mixed for 15 min. The sample pH was adjusted to 2.0, 7.0, or 10.0 by adding dilute HCl or NaOH. The tubes were then heated for 30 min in a water bath set at 60°C followed by cooling in tap water for 30 min. After cooling, the samples were centrifuged at $18,000 \times g$ for 10 min and the supernatant was decanted. The weight of the tube and its contents was then obtained. The WHC (g water/g protein) was calculated as follows:

$$\text{WHC}\left(\text{g water/g protein}\right) = \left[W_f - W_e - \left(100 - A\right)/100\right]/1\,\text{g sample} \quad (14.5)$$

where
  W_f is the final weight of the tube (g)
  W_e is the empty weight of the tube (g)
  A is the % solubility × % protein in dry sample

As shown in Table 14.9, pennycress protein isolate from press cake produced by the SE method was generally more soluble than the protein isolate from ground seed or produced by AP. SE seed protein isolate showed the greatest amounts of soluble protein (74%–91%) at pH ≥ 7 and at pH 2 (68%), while the least amount of soluble protein was detected at pH 4. This behavior appears to follow that of SE canola protein isolate that was found to have high solubility (52%–97%) from pH 3 to pH 9 [99]. In contrast, AP seed protein isolate showed a higher isoelectric point at pH 5.5 but a maximum solubility of only 50% at pH 2 and 10. The solubility profile of AP pennycress seed protein isolate was similar to that observed for AP canola protein isolate, that is, less than 60% solubility from pH 2 to 10 [83]. AP press cake protein isolate also showed the least amount of soluble protein at pH 5.5 (isoelectric pH) but was far more soluble than the AP seed protein isolate at the other pH points tested, especially pH 2 and 10 where its solubility was almost 100% [22]. Seed cooking during oil pressing likely contributed to the enhanced solubility of the press cake protein. Damodaran [100] explained that thermal treatment can induce reversible structural changes that consequently improve solubility and other functional properties. SE press cake protein extract had a near-constant extraordinarily high solubility (76%–95%) from pH 2 to 10 [22], with the least amount of soluble protein (76%) determined at pH 4.0, which was the same isoelectric pH as that of the SE seed protein isolate. Albumins and globulins, the likely dominant fractions in saline-extracted protein, are usually highly soluble in aqueous media and could contribute to increased protein solubility.

Most proteins are susceptible to the detrimental effects of heating, which usually manifest through a substantial and irreversible reduction in solubility brought about

TABLE 14.9

**Saline-Extracted (SE) or Acid Precipitated (AP) Protein Isolates (PI)
from Pennycress Seed and Press Cake: Solubility, Heat Stability,
Foaming Properties, Emulsion Activity Index, Emulsion Stability Index,
and Water Holding Capacity**

Functional Property	pH	Pennycress Seed		Pennycress Press Cake	
		SE PI	AP PI	SE PI	AP PI
Solubility (% soluble protein)	2.0	68	50	95	97
	4.0	17	18	76	40
	5.5	27	3	86	4
	7.0	74	34	93	62
	8.5	85	46	88	79
	10.0	91	54	87	97
Heat stability (% loss of protein	2.0	12	0	5	3
solubility)	7.0	62	4	73	23
	10.0	64	0	79	2
Foaming capacity (mL)	2.0	107	127	109	115
	7.0	102	127	107	119
	10.0	101	128	104	122
Foam stability (% remaining foam	2.0	95	92	82	96
after 15 min)	7.0	39	98	95	97
	10.0	69	95	96	95
Emulsifying activity index (m²/g)	2.0	141	88	226	24
	7.0	192	105	265	91
	10.0	208	178	412	136
Emulsion stability index (min)	2.0	14	12	14	24
	7.0	21	14	12	35
	10.0	34	16	17	31
Water holding capacity	2.0	6.6	7.5	8.8	Not done
(g water/g protein)	7.0	2.5	4.3	10.7	10.1
	10.0	6.2	4.5	12.5	Not done

Sources: Hojilla-Evangelista, M.P. et al., *J. Am. Oil Chem. Soc.*, 92, 905, 2015; Hojilla-Evangelista, M.P.
et al., *Ind. Crops Prod.*, 55, 173, 2014.

by the aggregation of unfolded protein molecules [101]. Pennycress protein isolates
produced by acid precipitation from either ground seed or press cake were more stable to heating than SE protein isolates, as shown by the lower loss of solubility at all
pH levels tested (Table 14.9). In contrast, when solutions of SE protein isolates (pH 7
or 10) were heated at 100°C, substantial loss of solubility was noted (60%–79% loss
of solubility) (Table 14.9). The heat-induced loss of solubility was especially severe
for SE press cake protein isolates. This behavior may be related to the substantial
presence of albumins and globulins, the amounts of which were reduced significantly

when seeds were subjected to cooking at 82°C before oil pressing [21] or when protein extraction was done at 77°C [82].

Pennycress seed and press cake protein isolates were observed to produce substantial foam volumes (>100 mL) and very stable foams (Table 14.9). Their foaming properties did not appear to be affected by the pH of testing. Foaming capacity and foam stability of pennycress seed and press cake protein isolates were very similar to values reported (131 mL, 95% remaining foam) for AP soybean protein [102]. SE seed protein isolate had a lower foaming capacity than AP seed protein isolate and its most stable foam was obtained at pH 2. Tan et al. [93] attributed poor foam stabilities to weak protein–protein interactions at the air–water interface that, in turn, hinder the formation of interfacial membranes. Foam volumes produced by SE press cake protein isolate were almost identical to those produced by SE seed protein isolate and were likewise less than that of the AP press cake protein isolate (Table 14.9). However, both press cake protein isolates produced equally stable foams. In comparison, earlier work on canola protein isolate produced similarly by AP or SE reported that the foaming capacity and stability of the isolates were less than those of its meal and decreased further as pH was raised from 10 to 12 [92,93,103].

Pennycress protein isolates produced from seed or press cake had outstanding emulsifying activities that increased as the pH of the analysis moved toward the alkaline range. The emulsifying capacity of SE protein isolates (based on EAI values 141–412 m^2/g protein), particularly those from press cake, was evidently superior to those of the AP protein isolates (Table 14.9) and to that of soybean protein extracted by similar methods (56–99 m^2/g protein; [102]). The greater solubility of SE pennycress protein isolates may have been a positive influence in emulsion formation [93,104]. Emulsions formed by the SE seed protein isolates were more stable than those formed by the AP seed protein isolates (Table 14.9), especially at pH 7 and 10. The ESI for SE seed protein isolates at pH 7 (21 min) was greater than the 15 min reported for soybean protein at the same pH [102]. On the other hand, AP press cake protein isolate formed considerably more stable emulsions than those formed by SE press cake protein (Table 14.9). AP press cake protein emulsions can remain stable for approximately two times longer than those formed by the other pennycress protein isolates and soybean protein isolate [22,102].

Pennycress press cake protein isolates produced by SE or AP showed unusually high WHC (>8.0 g water/g protein), which increased with pH, in the case of SE press cake protein isolate only (Table 14.9). On the other hand, WHC values for the seed protein isolates were lower than those of the press cake protein isolates and appeared to be affected by pH and the extraction method. SE seed protein isolate showed the highest WHC at pH 2 and 10, while AP seed protein isolate had maximum WHC at pH 2, and the value was greater than that of the SE seed protein isolate at the same pH. WHC values for pennycress protein isolates were similar to those of lesquerella meal and press cake (6–8 g water/g protein) [25]. Given the range of WHC values determined for pennycress seed and press cake protein isolates, possible applications may be in comminuted meat products, baked goods, and gel-type materials [100].

14.9 CONCLUSION

Field pennycress has numerous positive attributes that make it a very promising industrial oilseed crop. Its short growing season makes it suitable as an off-season crop between corn and soybean production in most of the upper Midwestern United States. The planting of pennycress may also serve as a cover crop over winter and as food source for honeybees in early spring when very few or no other plants are in bloom. The mechanical oil extraction of pennycress seeds can be accomplished with minimal seed conditioning, and good oil yields are achieved in a single pass through the screw press. The superior cold flow properties of pennycress biodiesel and pennycress oil–E10 gasoline blend make them adaptable most of the year in areas with colder climates. Pennycress estolide esters capped with saturated fatty acids have modest cold temperature properties, which could fill a specialty niche as a high-viscosity industrial lubricant. Pennycress seed proteins have excellent nutritional qualities and the isolates have outstanding solubility, foaming, emulsifying, and water holding properties.

ACKNOWLEDGMENTS

The authors are extremely grateful to Ms. Benetria Banks, Mr. Billy Deadmond, Ms. Amber Durham, Mr. Jeff Forrester, Mr. Ben Lowery, Mr. Gary Grose, Mr. Ray Holloway, Ms. Mardell Schaer, Mr. Kevin Steidley, and Ms. Kelly Utt for their assistance in the sample preparation and data collection. We also extend our appreciation to Dr. Winthrop B. Phippen for his work on pennycress crop development and to Dr. Aaron C. Drenth and coworkers for their work on TGB fuel.

DISCLAIMER

The mention of trade names or commercial products in this publication is solely for the purpose of providing specific information and does not imply recommendation or endorsement by the U.S. Department of Agriculture (USDA). The USDA is an equal opportunity provider and employer.

REFERENCES

1. K.F. Best and G.I. McEntyre, The biology of Canadian weeds 9, *Thlaspi arvense* L., *Can. J. Plant Sci.*, 55, 279–292 (1975).
2. S.I. Warwick, A. Francis, and D.J. Susko, The biology of Canadian Weeds 9, *Thlaspi arvens* L. (Updated), *Can. J. Plant Sci.*, 82, 803–823 (2002).
3. L.W. Mitch, Field pennycress (*Thlaspi arvense* L.) -The stinkweed, *Weed Technol.*, 10, 675–678 (1996).
4. J.R. Clopton and H.O. Triebold, Fanweed seed oil: Potential substitute for rapeseed oil, *Ind. Eng. Chem.*, 36(3), 218–219 (1944).
5. T.A. Isbell, US effort in the development of new crops (Lesquerella, Pennycress, Coriander and Cuphea), Oleagineux, Corps Gras, *Lipides*, 16, 205–210 (2009).
6. W.B. Phippen and M.E. Phippen, Soybean seed yield and quality as a response to field pennycress residue, *Crop Sci.*, 52, 2767–2773 (2012).

7. E.J. Kladivko, T.C. Kaspar, D.B. Jaynes, R.W. Malone, J. Singer, M.K. Morin, and T. Searchinger, Cover crops in the upper midwestern United States: Potential adoption and reduction of nitrate leaching in the Mississippi river basin, *J. Soil Water Conservat.*, 69, 279–291 (2014).

8. T.C. Kaspar and J.W. Singer, The use of cover crops to manage soil, *USDA-ARS/UNL faculty paper*, 1382, 321–336 (2011). http://digitalcommons.unl.edu/usdaarsfacpub/1382 (accessed September 8, 2015).

9. C.E. Eberle, M.D. Thom, K.T. Nemec, F. Forcella, J.G. Lundgren, R.W. Gesh, W.E. Riedell et al., Using pennycress, camelina, and canola cash cover crops to provision pollinators, *Ind. Crops Prod.*, 75, 20–25 (2015).

10. J.H. Groeneveld, H.P. Luhrs, and A.M. Klein, Pennycress double-cropping does not negatively impact spider diversity, *Agric. Forest Entomol.*, 17, 247–257 (2015).

11. J.H. Groeneveld and A.M. Klein, Pennycress-corn double-cropping increases ground beetle diversity, *Biomass Bioenerg.*, 77, 16–25 (2015).

12. G.A. Johnson, M.B. Kantar, K.J. Betts, and D.L. Wyse, Field pennycress production and weed control in a double crop system with soybean in minnesota, *Agron. J.*, 107(2), 532–540 (2015).

13. Pennycress Resource Network 2010. http://www.wiu.edu/pennycress/ (accessed September 4, 2015).

14. T.A. Isbell, S.C. Cermak, D.A. Dierig, F.J. Eller, and L. Marek, Registration of Katelyn *Thlaspi arvense* L. (Pennycress) with improved nondormant traits, *J. Plant Register.*, 9, 212–215, (2015).

15. W.B. Phippen, B. John, and M.E. Phippen, Planting date, herbicide, and soybean rotation studies with field pennycress (*Thlaspi arvense* L.), in: *22nd Annual Meeting of the Association for the Advancement of Industrial Crops: Program and Abstracts*, T.A. Isbell and D.A. Dierig (eds.), Association for the Advancement of Industrial Crops, Fort Collins, CO (September 19–22, 2010).

16. H. Rukavina, D.C. Sahm, L.K. Manthey, and W.B. Phippen, The effect of nitrogen rate on field pennycress seed yield and oil content, in: *23rd Annual Meeting of the Association for the Advancement of Industrial Crops: Program and Abstracts*, B.L. Johnson and M.T. Berti (eds.), Association for the Advancement of Industrial Crops, Fargo, ND (September 11–14, 2011).

17. T.A. Isbell and S.C. Cermak, *Thlaspi arvense* (Pennycress) germination, development and yield potential, in: *22nd Annual Meeting of the Association for the Advancement of Industrial Crops: Program and Abstracts*, T.A. Isbell and D.A. Dierig (eds.), Association for the Advancement of Industrial Crops, Fort Collins, CO (September 19–22, 2010).

18. T.A. Isbell, S.C. Cermak, and R.L. Evangelista, *Thlaspi arvense* (Pennycress) germination, bolting, and mechanical harvest seed loss, in: *23rd Annual Meeting of the Association for the Advancement of Industrial Crops: Program and Abstracts*, B.L. Johnson and M.T. Berti (eds.), Association for the Advancement of Industrial Crops, Fargo, ND (September 11–14, 2011).

19. M. Smith, Field pennycress, Agricultural marketing resource center (updated April 2015). http://www.agmrc.org/commodities-products/grains-oilseeds/pennycress/ (accessed September 2, 2015).

20. R.L. Evangelista, T.A. Isbell, and S.C. Cermak, Extraction of pennycress (*Thlaspi arvense* L.) seed oil by full pressing, *Ind. Crops Prod.*, 37(1), 76–81 (2012).

21. M.P. Hojilla-Evangelista, R.L. Evangelista, T.A. Isbell, and G.W. Selling, Effects of cold-pressing and seed cooking on functional properties of protein in pennycress (*Thlaspi arvense* L.) seed and press cakes, *Ind. Crops Prod.*, 45(1), 223–229 (2013).

22. M.P. Hojilla-Evangelista, G.W. Selling, M.A. Berhow, and R.L. Evangelista, Extraction, composition and functional properties of pennycress (*Thlaspi arvense* L.) press cake protein, *J. Am. Oil Chem. Soc.*, 92, 905–914 (2015).

23. F. Shahidi and M. Naczk, Removal of glucosinolates and other antinutrients from canola and rapeseed by methanol/ammonia processing, in: *Canola and Rapeseed: Production, Chemistry, Nutrition, and Processing Technology*, F. Shahidi (ed.), pp. 291–306, AVI-Van Nostrand Reinhold, New York (1990).

24. D.K. Salunkhe, J.K. Chavan, R.N. Adsule, and S.S. Kadam, *World Oilseeds: Chemistry, Technology, and Utilization*, p. 62, AVI-Van Nostrand Reinhold, New York (1992).

25. M.P. Hojilla-Evangelista and R.L. Evangelista, Functional properties of protein from *Lesquerella fendleri* seed and press cake from oil processing, *Ind. Crops Prod.*, 29, 466–472 (2009).

26. S.F. Vaughn, T.A. Isbell, D. Weisleder, and M.A. Berhow, Biofumigant compounds released by field pennycress (*Thlaspi arvense* L.) seedmeal, *J. Chem. Ecol.*, 31, 167–177 (2005).

27. M.P. Hojilla-Evangelista, G.W. Selling, M.A. Berhow, and R.L. Evangelista, Preparation, composition and functional properties of pennycress (*Thlaspi arvense* L.) seed protein isolates, *Ind. Crops Prod.*, 55, 173–179 (2014).

28. S.F. Vaughn, D.E. Palmquist, S.M. Duval, and M.A. Berhow, Herbicidal activity of glucosinolate-containing seedmeals, *Weed Sci.*, 54, 743–748 (2006).

29. B.R. Moser, S.N. Shah, J.K. Winkler-Moser, S.F. Vaughn, and R.L. Evangelista, Composition and physical properties of cress (*Lepidium sativum* L.) and field pennycress (*Thlaspi arvense* L.) oils, *Ind. Crops Prod.*, 30, 199–205 2009.

30. S.C. Cermak, G. Biresaw, T.A. Isbell, R.L. Evangelista, S.F. Vaughn, and R.E. Murray, New crop oils—Properties as potential lubricants, *Ind. Crops Prod.* 44, 232–239 (2013).

31. S.C. Cermak, A.L. Durham, T.A. Isbell, R.L. Evangelista, and R.E. Murray, Synthesis and physical properties of pennycress estolides and esters, *Ind. Crops Prod.*, 67, 179–184 (2015).

32. A.C. Drenth, D.B. Olsen, and K. Denef, Fuel property of triglyceride blends with emphasis on industrial oilseeds camelina, carinata, and pennycress, *Fuel*, 153, 19–20 (2015).

33. H.B.W. Patterson, *Handling and Storage of Oilseeds, Oils, Fats, and Meal*, Elsevier Science Publishing Co., Inc., New York (1989).

34. R.D. O'Brien, *Fats and Oils: Formulating and Processing for Applications*, 2nd edn., CRC Press, Boca Raton, FL (2003).

35. A.J. MacLeod, S.S. Panesar, and V. Gil, Thermal degradation of glucosinolates, *Phytochemistry*, 20, 977–980 (1981).

36. B. Drozdowski and M. Zajac, Effect of the concentration of some nickel catalyst poisons in oils on the course of hydrogenation, *J. Am. Oil Chem. Soc.*, 54, 595–599 (1977).

37. H. Klimmek, Influence of various catalyst poisons and other impurities on fatty acid hydrogenation, *J. Am. Chem. Soc.*, 61, 200–204 (1984).

38. L.M. Khan and M.A. Hanna, Expression of oil from oilseeds—A review, *J. Agric. Eng. Res.*, 28, 495–503, 1983.

39. B.R. Moser, G.H. Knothe, S.F. Vaughn, and T.A. Isbell, Production and evaluation of biodiesel from field pennycress (*Thlaspi arvense* L.) oil, *Energy Fuels*, 23, 4149–4155 (2009).

40. ASTM D6751-11b, Standard specification for biodiesel fuel (B-100) blend stock for middle distillate fuels, in: *Annual Book of ASTM Standards*, pp. 1003–1091, American Society of Testing Materials, West Conshohocken, PA (2012).

41. European Committee for Standardization (CEN), Automotive fuels—Fatty acid methyl esters (FAME) for diesel engines—Requirements and test methods; EN 14214, European Committee for Standardization (CEN), Brussels, Belgium (2003).

42. American Oil Chemists' Society, AOCS official method Cd 3d-63, acid value, in: *Official Methods and Recommended Practices of the American Oil Chemists' Society*, 5th edn., D. Firestone (ed.), American Oil Chemists' Society, Champaign, IL, 1–2 (1999).

43. ASTM D6584-10a, Standard test method for determination of total monoglycerides, total diglycerides, total triglycerides, and free and total glycerin in B-100 biodiesel methyl esters by gas chromatography, in: *Annual Book of ASTM Standards*, pp. 501–509, American Society of Testing Materials, West Conshohocken, PA (2012).

44. European Committee for Standardization (CEN), Fat and oil derivatives. Fatty acid methyl esters (FAME). Determination of oxidative stability (accelerated oxidation test); EN 14112, European Committee for Standardization (CEN), Brussels, Belgium (2003).

45. ASTM D5773-10, Standard test method for cloud point of petroleum products (constant cooling rate method), in: *Annual Book of ASTM Standards*, pp. 1290–1296, American Society of Testing Materials, West Conshohocken, PA (2012).

46. ASTM D5949-10, Standard test method for pour point of petroleum products (automatic pressure pulsing method), in: *Annual Book of ASTM Standards*, pp. 1402–1406, American Society of Testing Materials, West Conshohocken, PA (2012).

47. ASTM D6371-99, Standard test method for cold filter plugging point of diesel and heating fuels, in: *Annual Book of ASTM Standards*, pp. 599–605, American Society of Testing Materials, West Conshohocken, PA (2000).

48. ASTM D445-97, Standard test method for kinematic viscosity of transparent and opaque liquids (and calculation of dynamic viscosity), in: *Annual Book of ASTM Standards*, pp. 188–196, American Society of Testing Materials, West Conshohocken, PA (2000).

49. ASTM D6079-97, Standard test method for evaluating lubricity of diesel fuels by high frequency reciprocating rig (HFRR), in: *Annual Book of ASTM Standards*, pp. 191–196, American Society of Testing Materials, West Conshohocken, PA (2000).

50. ASTM D5453-09, Standard test method for determination of total sulfur in light hydrocarbons, spark ignition engine fuels, diesel engine fuels, and engine oils by ultraviolet fluorescence, in: *Annual Book of ASTM Standards*, pp. 1051–1060, American Society of Testing Materials, West Conshohocken, PA (2012).

51. ASTM D4951-09, Standard test method for determination of additive elements in lubricating oils by inductively coupled plasma atomic emission spectroscopy, in: *Annual Book of ASTM Standards*, pp. 780–787, American Society of Testing Materials, West Conshohocken, PA (2012).

52. ASTM D6890-13a, Standard test method for determination of ignition delay and derived cetane number (DCN) of diesel fuel oils by combustion in a constant volume chamber, in: *Annual Book of ASTM Standards*, pp. 1–17, American Society of Testing Materials, West Conshohocken, PA (2013).

53. ASTM D613-10a, Standard test method for cetane number of diesel fuel oil, in: *Annual Book of ASTM Standards*, pp. 3–14, American Society of Testing Materials, West Conshohocken, PA (2012).

54. G. Knothe, A.C. Matheaus, and T.W. Ryan III, Cetane numbers of branched and straight-chain fatty esters determined in an ignition quality tester, *Fuel*, 82, 971–975 (2003).

55. G. Knothe and K.R. Steidley, Kinematic viscosity of biodiesel fuel components and related compounds. Influence of compound structure and comparison to petrodiesel fuel components, *Fuel*, 84, 1059–1065 (2005).

56. ASTM D975-98b, Standard specification for diesel fuel oils, in: *Annual Book of ASTM Standards*, pp. 345–362, American Society of Testing Materials, West Conshohocken, PA (2000).

57. European Committee for Standardization (CEN), Automotive fuels—Diesel—Requirements and test methods; EN 590, European Committee for Standardization (CEN), Brussels, Belgium (2009).

58. A. Lakshminarayan, D.B. Olsen, and P.E. Cabot, Performance and emission evaluation of triglyceride-gasoline blends in agricultural compression ignition engines, *Appl. Eng. Agric.*, 30, 523–534 (2014).

59. R.O. Dunn and M.O. Bagby, Low temperature phase behavior of vegetable oil/co-solvent blends as alternative diesel fuel, *J. Am. Oil Chem. Soc.*, 77, 11315–11322 (2000).

60. A.C. Drenth, D.B. Olsen, P.E. Cabot, and J.J. Johnson, Compression ignition engine performance and emission evaluation of industrial oilseed biofuel feedstock, camelina, carinata, and pennycress across three fuel pathways, *Fuel*, 136, 143–155 (2014).

61. ASTM D4052-96, Standard test method for density, relative density, and API gravity of liquids by digital density meter, in: *Annual Book of ASTM Standards*, pp. 680–683, American Society of Testing Materials, West Conshohocken, PA (2000).

62. ASTM D240-92, Standard test method for heat of combustion of liquid hydrocarbon fuels by bomb calorimeter, in: *Annual Book of ASTM Standards*, pp. 146–154, American Society of Testing Materials, West Conshohocken, PA (2000).

63. S.C. Cermak and T.A. Isbell, Improved oxidative stability of estolide esters, *Ind. Crops Prod.*, 18, 223–230 (2003).

64. S.C. Cermak and T.A. Isbell, Physical properties of saturated estolides and their 2-ethylhexyl esters, *Ind. Crops Prod.*, 16, 119–127 (2002).

65. T.A. Isbell, M.R. Edgcomb, and B.A. Lowery, Physical properties of estolides and their ester derivatives, *Ind. Crops Prod.*, 13, 11–20 (2001).

66. S.C. Cermak and T.A. Isbell, Biodegradable oleic estolide ester having saturated fatty acid end group useful as lubricant base stock, U.S. Patent 6,316,649 B1 (2000).

67. S.C. Cermak and T.A. Isbell, Synthesis of estolides from oleic and saturated fatty acids, *J. Am. Oil Chem. Soc.*, 78, 557–565 (2001).

68. T.A. Isbell, Chemistry and physical properties of estolides, *Grasas Aceites*, 62, 8–10 (2011).

69. T.A. Isbell, R. Kleiman, and B.A. Plattner, Acid-catalyzed condensation of oleic acid into estolides and polyestolides, *J. Am. Oil Chem. Soc.*, 71, 169–174 (1994).

70. T.A. Isbell and R. Kleiman, Characterization of estolides produced from the acid-catalyzed condensation of oleic acid, *J. Am. Oil Chem. Soc.*, 71, 379–383 (1994).

71. R. Becker and A. Knorr, An evaluation of antioxidants for vegetable oils at elevated temperatures, *Lubr. Sci.*, 8, 95–117 (1996).

72. S. Asadauskas and S.Z. Erhan, Depression of pour points of vegetable oils by blending with diluents used for biodegradable lubricants, *J. Am. Oil Chem. Soc.*, 76, 313–316 (1999).

73. ASTM D97-96a, Standard test method for pour point of petroleum products, in: *Annual Book of ASTM Standards*, pp. 87–94, American Society of Testing Materials, West Conshohocken, PA (2000).

74. ASTM D2500-98a, Standard test method for cloud point of petroleum products, in: *Annual Book of ASTM Standards*, pp. 847–849, American Society of Testing Materials, West Conshohocken, PA (2000).

75. ASTM D2270-93, Standard practice for calculating viscosity index from kinematic viscosity at 40 and 100 °C, in: *Annual Book of ASTM Standards*, pp. 755–760, American Society of Testing Materials, West Conshohocken, PA (2000).

76. American Oil Chemists' Society, AOCS official method Td 1a-64, color—Gardner 1963 (glass standards), in: *Official Methods and Recommended Practices of the American Oil Chemists' Society*, 5th edn., D. Firestone (ed.), American Oil Chemists' Society, Champaign, IL, 1–1 (1999).

77. S.C. Cermak, T.A. Isbell, R.L. Evangelista, and B.L. Johnson, Synthesis and physical properties of petroselinic based estolide esters, *Ind. Crops Prod.*, 33, 132–139 (2011).

78. S.C. Cermak, A.L. Skender, A.B. Deppe, and T.A. Isbell, Synthesis and physical properties of tallow-oleic estolide 2-ethylhexyl esters, *J. Am. Oil Chem. Soc.*, 84, 449–456 (2007).

79. T.A. Isbell and S.C. Cermak, Purification of meadowfoam monoestolide from polyestolide, *Ind. Crops Prod.*, 19, 113–118 (2004).

80. S.C. Cermak and T.A. Isbell, Synthesis and physical properties of cuphea-oleic estolides and esters, *J. Am. Oil Chem. Soc.*, 81, 297–303 (2004).

81. T.A. Isbell, T.P. Abbott, S. Asadauskas, and J.E. Lohr, Biodegradable oleic estolide ester base stocks and lubricants, U.S. Patent 6,018,063 (2000).

82. G.W. Selling, M.P. Hojilla-Evangelista, R.L. Evangelista, T.A. Isbell, N. Price, and K.M. Doll, Extraction of proteins from pennycress seeds and press cake, *Ind. Crops Prod.*, 41(1), 113–119 (2013).

83. D.M. Klockeman, R. Toledo, and K.A. Sims, Isolation and characterization of defatted canola meal protein, *J. Agric. Food Chem.*, 45, 3867–3870 (1997).

84. Y.V. Wu and M.P. Hojilla-Evangelista, *Lesquerella fendleri* protein fractionation and characterization, *J. Am. Oil Chem. Soc.*, 82, 53–56 (2005).

85. Food and Agriculture Organization/World Health Organization/United Nations University (FAO/WHO/UNU), Energy and protein requirements, WHO Technical Report Series No. 724, World Health Organization, Geneva, Switzerland (1985).

86. W.J. Wolf, Edible soybean protein products, in: *CRC Handbook of Processing and Utilization in Agriculture*, Vol. 2, Part 2, Plant Products, I.A. Wolff (ed.), pp. 23–55, CRC Press, New York (1983).

87. E.W. Lusas and K.C. Rhee, Soybean processing and utilization, in: *Practical Handbook of Soybean Processing and Utilization*, D.R. Erickson (ed.), pp. 117–160, AOCS Press, Champaign, IL (1997).

88. L. Gillberg and B. Tornell, Preparation of rapeseed protein isolates: Dissolution and precipitation behavior of rapeseed proteins, *J. Food Sci.*, 41, 1063–1069 (1976).

89. L.L. Diosady, Y.-M. Tzeng, and L.J. Rubin, Preparation of rapeseed protein concentrates and isolates using ultrafiltration, *J. Food Sci.*, 49, 768–770, 776 (1984).

90. L. Xu and L.L. Diosady, Functional properties of chinese rapeseed protein isolates, *J. Food Sci.*, 59, 1127–1130 (1994).

91. L.J. Rubin, L.L. Diosady, and Y.-M. Tzeng, Ultrafiltration in rapeseed processing, in: *Canola and Rapeseed: Production, Chemistry, Nutrition, and Processing Technology*, F. Shahidi (ed.), pp. 307–330, AVI-Van Nostrand Reinhold, New York (1990).

92. R.E. Aluko, T. McIntosh, and F. Katepa-Mupondwa, Comparative study of the polypeptide profiles and functional properties of *Sinapis alba* and *Brassica juncea* seed meals and protein concentrates, *J. Sci. Food Agric.*, 85, 1931–1937 (2005).

93. S.H. Tan, R.J. Mailer, C.L. Blanchard, and S.O. Agboola, Canola proteins for human consumption: Extraction, profile, and functional properties, *J. Food Sci.*, 76, R16–R28 (2011).

94. J.P.D. Wanasundara, S.J. Abeysekara, T.C. McIntosh, and K.C. Falk, Solubility differences of major storage proteins in brassicaceae oilseeds, *J. Am. Oil Chem. Soc.*, 89, 869–881 (2012).

95. D.J. Myers, M.P. Hojilla-Evangelista, and L.A. Johnson, Functional properties of protein extracted from flaked, defatted, whole corn by ethanol/alkali during sequential extraction processing, *J. Am. Oil Chem. Soc.*, 71, 1201–1204 (1994).

96. American Oil Chemists' Society, AOCS official method Ba 4e-93, generic combustion method for determination of crude protein, in: *Official Methods and Recommended Practices of the American Oil Chemists' Society*, 5th edn., D. Firestone (ed.), American Oil Chemists' Society, Champaign, IL, 1–4 (1999).

97. W.U. Wu, N.S. Hettiarachchy, and M. Qi, Hydrophobicity, solubility, and emulsifying properties of soy protein peptides prepared by papain modification and ultrafiltration, *J. Am. Oil Chem. Soc.*, 75, 845–850 (1998).

98. E.A. Balmaceda, M.K. Kim, R. Franzen, B. Mardones, and J.C. Lugay (1984) Protein functionality methodology—standard tests, in: *Food Protein Chemistry*, J.M. Regenstein and C.E. Regenstein (eds.), pp. 278–291, Academic Press, New York.

99. Y. Yoshie-Stark, Y. Wada, and A. Wasche, Chemical composition, functional properties, and bioactives of rapeseed protein isolates, *Food Chem.*, 107, 32–39 (2008).

100. S. Damodaran, Amino acids, peptides, and proteins, in: *Food Chemistry*, 3rd edn., O.R. Fennema (ed.), pp. 356–359, Marcel Dekker, Inc., New York (1996).

101. J.E. Kinsella, Functional properties of soy proteins, *J. Am. Oil Chem. Soc.*, 56, 242–258 (1976).

102. M.P. Hojilla-Evangelista and D.J. Sessa, Functional properties of soybean and lupin protein concentrates produced by ultrafiltration-diafiltration, *J. Am. Oil Chem. Soc.*, 81, 1153–1157 (2004).

103. J. Pedroche, M.M. Yust, H. Lqari, J. Giron-Calle, M. Alaiz, J. Vioque, and F. Millan, *Brassica carinata* protein isolates: Chemical composition, protein characterization and improvement of functional properties by protein hydrolysis, *Food Chem.* 88, 337–346 (2004).

104. J.F. Kinsella, S. Damodaran, and J.B. German, Physicochemical and functional properties of oilseed proteins with emphasis on soy proteins, in: *New Protein Foods: Seed Proteins*, A. Altshul and H. Wilcke (eds.), pp. 107–179, Academic Press, London, U.K. (1985).

15 Biobased Lubricant Additives

Girma Biresaw, Grigor B. Bantchev,
Zengshe Liu, David L. Compton,
Kervin O. Evans, and Rex E. Murray

CONTENTS

15.1 INTRODUCTION

The basic ingredients of a lubricant formulation are the base oil and one or more additives [1–3]. The base oil could be a single component (e.g., refined petroleum of certain distillation range) or a blend of several base oils of varying viscosity grades and chemistry. Each component is carefully selected so that the lubricant can properly perform its primary and secondary functions for the intended length of its service life.

Among the primary functions of a lubricant are to provide the appropriate friction and wear for the specific tribological process it is intended for. Other primary functions include removing heat (cooling) from the friction zone and protecting the friction components from damage, rust, and corrosion [3]. The lubricant has numerous secondary functions that must be properly carried out for it to be applied in the lubrication process. The lubricant must be easy to handle under all weather conditions; must maintain its physical and chemical properties, that is, must not break down or polymerize during use; must be safe to exposed workers and the environment; must not produce or emit toxic components; must be easy to process and dispose after use; etc.

The lubricant formulations can be applied as an oil-based or water-based fluid. In oil-based fluids, the lubricant formulation, comprising base oil and additives, is used as formulated or is diluted with the appropriate base oil prior to application. In water-based fluids, the lubricant formulation is dispersed in water to form oil–water mixtures (emulsions, microemulsions, micellar solutions) of various degrees of stability. Water-based fluids provide several benefits over oil-based fluids including excellent cooling, low cost, and low greenhouse gas emission. An important additive in water-based formulations is the emulsifier (also referred to as the surfactant or detergent), which allows mixing of the oil formulation with the water and also allows to control the stability of the resulting emulsion.

The selection of ingredients for lubricant formulation begins with the base oil, since it comprises the largest proportion of the formulation (60%–99%) [1–3]. The most important property of the base oil is its viscosity at the temperature of its application, which must match the viscosity requirement for the process. The viscosity of the base oil affects the lubricant film thickness that will be formed during the tribological process [4], and thereby affects the outcome of the lubrication process. Other critical properties of base oils include viscosity index (VI), pour point, cloud point, flash point, etc. An example of lubricant base oil specification, detailing these and other properties, is given in Table 15.1 [5]. The data in Table 15.1 are for a biobased base oil obtained from estolide technology [6], and similar data are supplied by all manufacturers of Group I through V base oils [7].

In many instances, the selected base oil or base oil blend will not fully satisfy the requirements for formulating a lubricant that adequately performs its primary and secondary functions while at the same time being cost competitive. In such cases,

TABLE 15.1

An Example of a Typical Specification of Lubricant Base Oil Supplied by the Manufacturer

Property	Unit	Method	Typical Result
kVisc at 100°C	mm²/s	ASTM D-445	63
kVisc at 40°C	mm²/s	ASTM D-445	30.3
Viscosity index		ASTM D-2270	164
Pour point	°C	ASTM D-97	−18
Cloud point	°C	ASTM D-2500	−15
Flash point	°C	ASTM D-92	280
Fire point	°C	ASTM D-92	300
Iodine value	mg I₂/100 g	NMR	<0.5
Specific gravity (15.5°C)		ASTM D-4052	0.898
Evaporative loss (NOACK)	% w/w	ASTM D-5800	2
Acid value	mg KOH/g	ASTM D-664	≤0.10

Data are for biobased base oil SE7B (estolide) from biosynthetic technologies from Reference 5.

the formulation is optimized with the help of additives. Additives perform two broad categories of tasks. The first task is to enhance the properties of the base oil to the value required for the formulation. Examples of such properties include viscosity, VI, pour point, and others listed in Table 15.1. The second task is to provide the formulation with critical properties that cannot be provided by the base oil alone. Examples of such properties include friction, wear, corrosion resistance, bioresistance, etc. The number of additives required for optimizing the lubricant formulation depends on the type (oil- or water-based) and application of the lubricant formulation. Simple formulations contain one to three additives, whereas complex formulations such as those for engine oils can have more than ten additives [8]. Examples of additives used in lubricant formulations are listed in Table 15.2 [1,2].

The current global demand for additives is estimated at 4.05 million metric tons, of which 60% or 2.4 million metric tons are used to formulate heavy-duty and passenger car engine oils [8]. Figure 15.1 shows the share of global additive consumption by the lubricant function as well as by the additive function [8]. It is clear from the data that lubricants for engine/automotive-related applications account for up to 80% of the global additive demand. The data also show that three additives, namely, dispersants, viscosity index improvers (VIIs), and detergents, account for 70% of the global additive demand (Figure 15.1b).

Lubricant additives currently in the market are almost exclusively petroleum based, with biobased additives accounting for close to 0% of the market share [9,10]. As a result, biobased lubricants currently in the market (e.g., biobased hydraulic and gear oils) are routinely formulated by blending biobased base oils (such as vegetable oils and/or their derivatives) exclusively with petroleum-based additives [9–12]. Thus, there is a need to develop, commercialize, and substantially grow the market share of

TABLE 15.2

Examples of Additives Used in Lubricant Formulations

Antifoams/defoamers

Anti-mist

Antioxidants/metal deactivators

Antiwears

Biocides/fungicides

Corrosion inhibitors/rust inhibitors

Couplers

Demulsifiers

Detergents/emulsifiers/surfactants

Dispersants

Extreme pressure

Friction modifiers

pH buffers

Pour point depressants

Seal swell

Tackifiers

Viscosity modifiers/viscosity index improvers

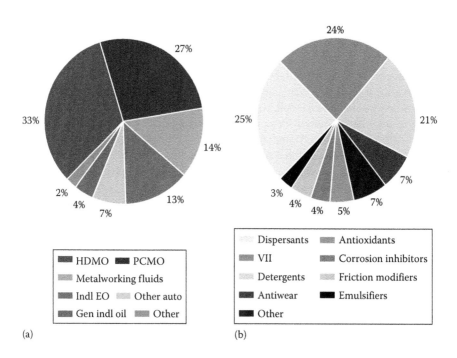

(a)   (b)

FIGURE 15.1 Share of global additive demand by lubricant function (a), and additive function (b). *Abbreviations*: HDMO, heavy duty motor oil; PCMO, passenger car motor oil; Indl EO, industrial engine oil; Gen Indl oil, general industrial oil; VII, viscosity index improver. (From Anon., *A Supplement to Lubes'N'Greases*, 21(8 Suppl.), 1, 2015.)

biobased additives in the global market. Such effort will enhance the sustainability of biobased lubricants and also reduce their greenhouse gas footprint. Currently, several groups are engaged in the synthesis and evaluation of various types of biobased additives. This chapter is a review of such efforts in the authors' laboratory.

15.2 BIOBASED FRICTION MODIFIERS

Most lubricants currently in the market are formulated using petroleum-based base oils. These base oils can be either Group I through III petroleum distillates or Group IV polyalphaolefins synthesized from petroleum-based feedstocks [7]. All these base oils are highly nonpolar hydrocarbons, manufactured to be free of elements such as N, O, and S that can render them some degree of polarity. As a result, while these base oils can form lubricant films under hydrodynamic conditions, they are unable to bear load or reduce friction under boundary or extreme pressure (EP) conditions [3]. Thus, the use of these base oils in lubricant applications will require incorporating friction modifiers into the formulation. Examples of hydrocarbon structures, 1–5, typically found in Group I through IV base oils are given in Figure 15.2.

Friction modifiers are organic molecules whose structure includes both nonpolar hydrocarbon and polar functional groups such as carboxylic acids, alcohols, esters, etc. [1,13]. Such structure provides the molecules with amphiphilic properties [14], allowing it to interact with both polar and nonpolar materials. As a result, friction modifiers, while they are very soluble in the highly nonpolar hydrocarbon base oils, can also adsorb on the highly polar metallic friction surfaces. The thin films formed by the adsorption of friction modifiers can separate friction surfaces and thereby reduce both friction and wear during a tribological process.

Biobased oils such as vegetable oils, 6, and their derivatives (fatty acids, 7; fatty acid methyl esters, 8; estolides, 9; etc.) are amphiphilic because their structures comprise both nonpolar hydrocarbon chains and polar functional groups (Figure 15.3). As a result, biobased oils can be used as friction modifiers in lubricant formulations. The friction modifier properties of various biobased oils have been investigated

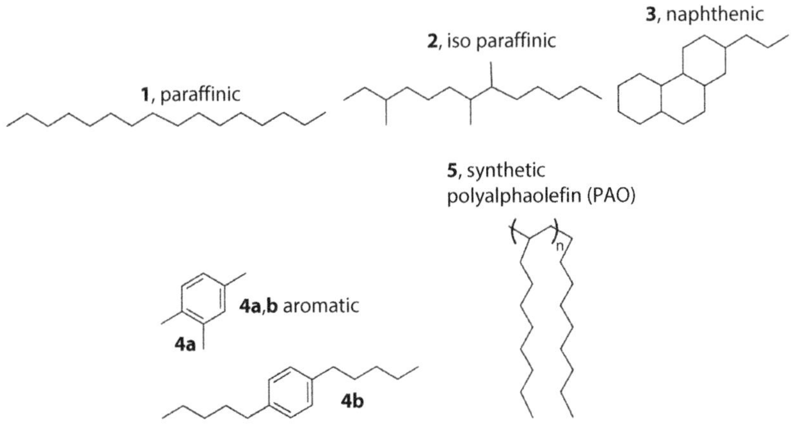

FIGURE 15.2 Examples of hydrocarbon structures found in Group I through IV base oils.

using a variety of experimental and theoretical methods [15–21]. The studies were conducted using highly purified (>99.9% anhydrous) hexadecane (structure **1** in Figure 15.2) as the model petroleum base oil. Table 15.3 shows the list of vegetable oils, **6**; fatty acid methyl esters, **8**; and estolides, **9**, used in the investigations. The structures of some of these biobased oils are given in Figure 15.3.

The friction modifiers investigated were selected to have the variations in their structure that will allow for investigating the effect of chemical structure on additive properties. Thus, vegetable oils with triglyceride, **6a** and **6b**, and waxy ester, **6c**, structures and a wide range of fatty acid compositions [22] were studied. Similarly, methyl and 2-ethylhexyl esters of fatty acids, **8**, with different chain lengths and degrees of saturations were also used (Table 15.3). The various approaches to friction modifier investigations are discussed next.

15.2.1 Tribological Investigation of Biobased Friction Modifiers

The friction properties of hexadecane with various concentrations of friction modifiers (0.0–0.6 M) were investigated under boundary lubrication conditions

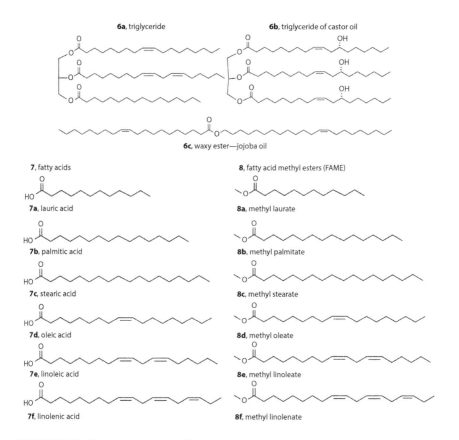

FIGURE 15.3   Structures of vegetable oils and derivatives.   *(Continued)*

9, estolides

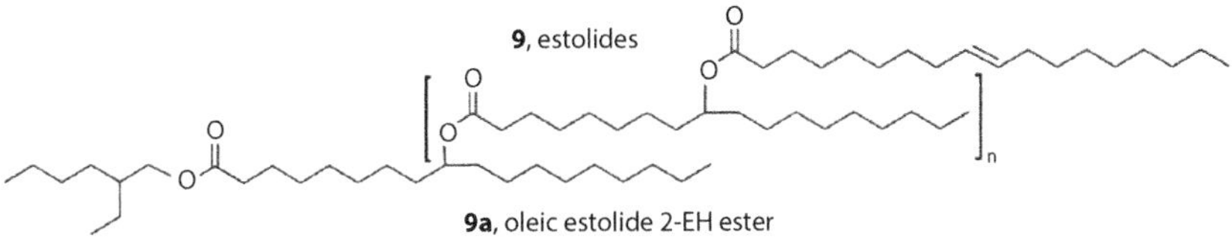

9a, oleic estolide 2-EH ester

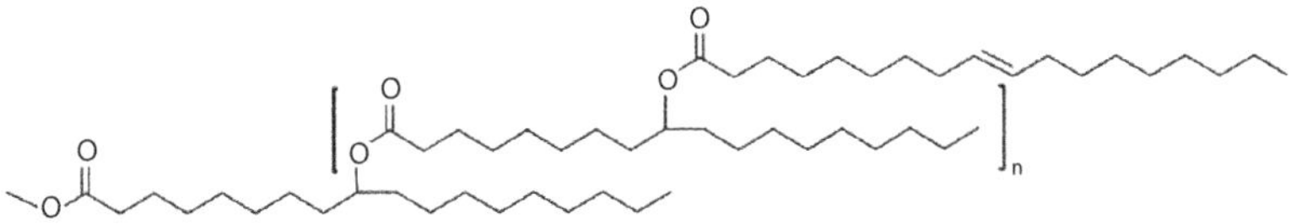

9b, oleic estolide methyl ester

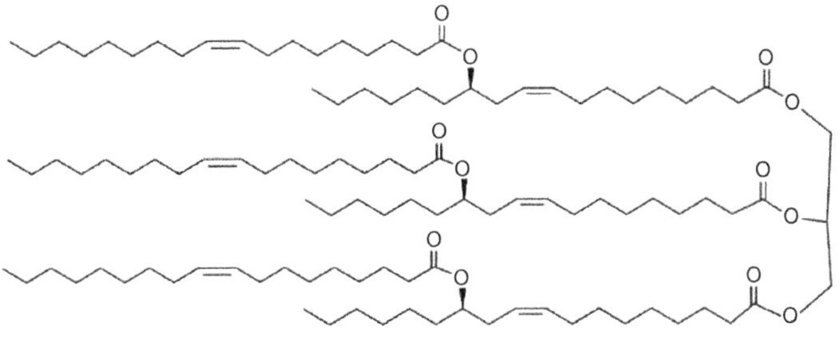

9c, castor oil-oleic estolide

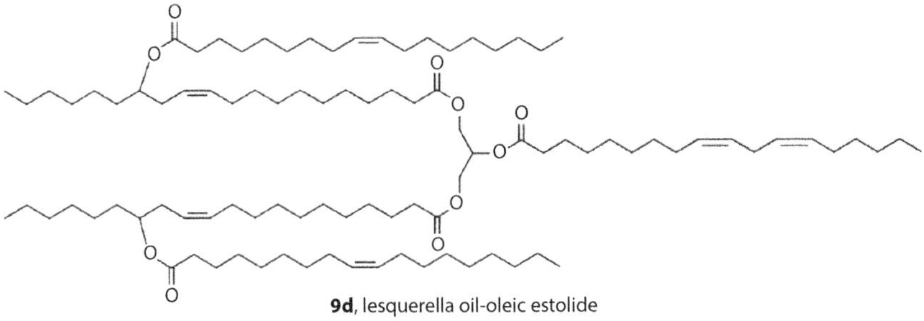

9d, lesquerella oil-oleic estolide

FIGURE 15.3 (*Continued*) Structures of vegetable oils and derivatives.

[15,16,18–20]. Tests were conducted on a Multi-Specimen Tribometer from Falex Corporation (Sugar Grove, IL) setup in a ball-on-disk configuration as depicted in Figure 15.4. Test conditions were as follows: test duration, 15 min; sliding speed, 6.22 mm/s (5 rpm); applied load, 1778 N (181.4 kgf, 400 lbf); and temperature of specimen and test fluid maintained at room temperature (25°C ± 2°C).

A typical coefficient of friction (COF) vs. time trace from a ball-on-disk test is shown in Figure 15.5. The COF slowly increases with time in the first 5–10 min and levels off to a steady-state value for the rest of the test duration. The average COF in the steady-state region is reported. Each lubricant blend was used in at least two friction tests, and the average values of multiple tests were used in data analysis.

The results from testing a set of blends of a friction modifier in hexadecane are summarized as COF vs. concentration data, similar to that depicted in Figure 15.6

TABLE 15.3
Biobased Friction Modifiers Investigated in Hexadecane Base Oil

Friction Modifier		Method—Instrument	References
Type	Name		
Vegetable oil, 6	Canola, 6a	Adsorption model	[15–17,19,21]
	Cottonseed, 6a	Boundary friction—ball-on-disk	
	High-oleic safflower, 6a	tribometer	
		Gravimetry—QCM-D[a]	
	High-oleic sunflower, 6a	Interfacial tension—ADSA[b]	
	Jojoba, 6c		
	Meadowfoam		
	Olive, 6a		
	Safflower, 6a		
	Soybean, 6a		
Fatty acid alkyl	2-ethylhexyl oleate	Adsorption model	[15–18,20,21]
ester, 8	Methyl laurate, 8a	Boundary friction—ball-on-disk	
		tribometer	
	Methyl palmitate, 8b	QCM-D[a]	
	Methyl stearate, 8c	Interfacial tension—ADSA[b]	
	Methyl oleate, 8d		
Estolide, 9	Oleic estolide 2-ethylhexyl	Adsorption model	[20]
	ester, 9a		
	Oleic estolide methyl ester, 9b	Boundary friction—ball-on-disk	
		tribometer	

[a] Quartz crystal microbalance with dissipation (QCM-D).
[b] Axisymmetric drop shape analysis (ADSA).

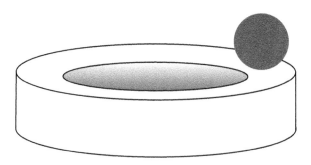

FIGURE 15.4 Schematic of ball-on-disk tribotest configuration.

for soybean oil [15]. The data in Figure 15.6 show the typical profile observed from such tests of all friction modifiers and have the following four common features: a very high COF (~0.5) without the friction modifier (pure base oil, which in the current investigation was hexadecane, **1** [15]); an initial sharp decrease of COF with increasing friction modifier concentration; a moderate decrease of COF with further

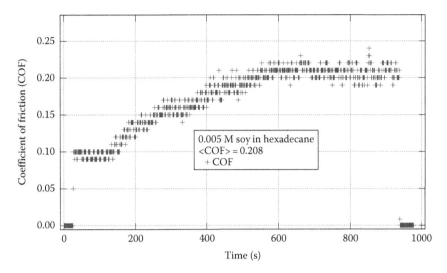

FIGURE 15.5 Typical trace from a ball-on-disk test. (From Biresaw, G. et al., *J. Am. Oil Chem. Soc.*, 79, 53, 2002.)

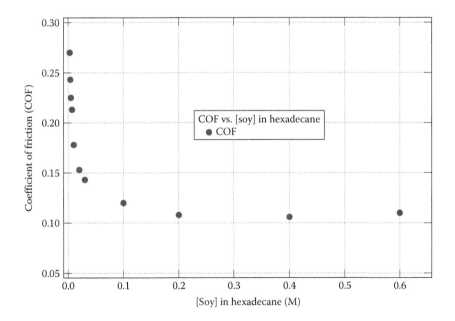

FIGURE 15.6 Effect of friction modifier concentration on COF of hexadecane. (From Biresaw, G. et al., *J. Am. Oil Chem. Soc.*, 79, 53, 2002.)

TABLE 15.4
Minimum Steel/Steel Steady-State Coefficient
of Friction of Biobased Friction Modifiers
in Hexadecane Base Oil

Biobased Friction Modifier	Coefficient of Friction
None (pure hexadecane)	0.50[a]
Safflower	0.11[b]
High-oleic safflower	0.10[b]
Jojoba	0.10[b]
Methyl oleate	0.13[b]
Methyl palmitate	0.11[b]

[a] Data from Reference 15.
[b] Data from Reference 16.

increase in friction modifier concentration; and, at high friction modifier concentrations, a region of steady-state low COF (~0.1) that is independent of friction modifier concentration. All friction modifiers display these four regions and provide slightly different values of the steady-state low COF as shown in Table 15.4 [15,16]. Thus, it is clear from Figure 15.6 and Table 15.4 that the friction modifier can produce a fivefold decrease in the COF of the base oil, which is a significant modification of the friction properties of the base oil.

15.2.2 INTERFACIAL INVESTIGATION OF BIOBASED FRICTION MODIFIERS

As mentioned before, friction modifiers reduce friction because of their amphiphilic properties, which allows them to adsorb on polar metallic surfaces and also to dissolve in nonpolar oils. The same property also allows friction modifiers to adsorb on polar liquid surfaces such as water and to lower the water–oil interfacial tension [14]. Thus, friction modifiers can assist in the mixing of oil and water and the formation of oil–water emulsions. Emulsions are an important class of lubricants used in various applications, including in water-based metal forming and hydraulic fluids [2,23].

In this investigation, the dynamic interfacial tension between highly purified water (deionized water purified to a conductivity of 18.2 MΩ-cm on a Barnstead EASYpure UV/UF water purification system model #D8611, EASYpure UVIUF; Barnstead International, Dubuque, IA) and hexadecane, **1**, with various concentrations of friction modifier, was measured by the method of axisymmetric drop shape analysis (ADSA) [24,25] using the FTA-200 (First Ten Angstroms, Portsmouth, VA) tensiometer. In the ADSA method, the surface tension or interfacial tension is measured from the analysis of the shape of a liquid pendant drop in air or in another liquid [24,25]. In a typical procedure, a predetermined volume of water is automatically pumped through a blunt needle into the hexadecane solution contained in a glass cuvette to form a pendant drop, similar to that depicted in Figure 15.7a. Figure 15.7b shows the main components of the FTA-200 tensiometer [17]. The instrument is

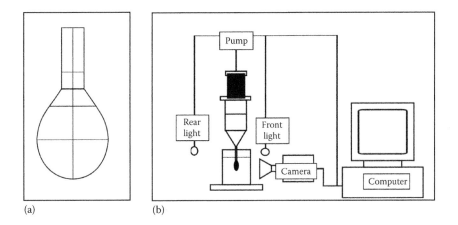

(a)   (b)

FIGURE 15.7 Schematics of pendant drop (a) and the FTA 200 tensiometer (b). (From Biresaw, G., *J. Am. Oil Chem. Soc.*, 82, 285, 2005.)

programmed to begin acquiring and saving the image of the pendant drop at a set rate (e.g., 0.067 s/image) for a set period of time (835.6 s), as soon as pumping is stopped. At the end of the image acquisition period, the instrument automatically analyzes each image and displays the dynamic interfacial tension data vs. time elapsed as a plot or spreadsheet on the computer monitor. An example of such data, from repeat measurement of interfacial tension between water and hexadecane, 1, with 0.004 M methyl palmitate, **8b**, friction modifier, is given in Figure 15.8. As shown in Figure 15.8, the interfacial tension starts very high, decreases sharply with

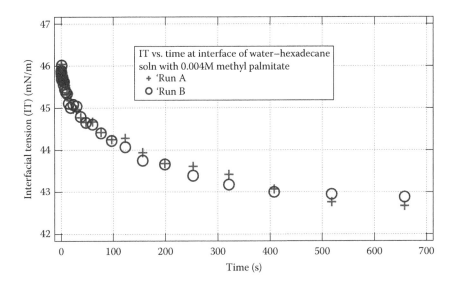

FIGURE 15.8 Dynamic interfacial tension between water and hexadecane with 0.004M of methyl palmitate, **8b**. (From Biresaw, G., *J. Am. Oil Chem. Soc.*, 82, 285, 2005.)

increasing time initially, then decreases gradually with further increase of time, and levels off to a constant value after a very long period of time.

The minimum interfacial tension, which changes little with further increase of time, is called the equilibrium interfacial tension and is a function of the friction modifier concentration in the hexadecane. The equilibrium interfacial tensions (which will be referred to as interfacial tension from here on) from dynamic interfacial tension measurements on the various blends are determined and analyzed as a function of the friction modifier concentration in hexadecane. An example of such analysis for blends of jojoba (vegetable oil **6c**) friction modifier in hexadecane is given in Figure 15.9 [17].

The interfacial tension vs. jojoba oil concentration data in Figure 15.9 have a similar profile as COF vs. soybean oil concentration data in Figure 15.6 discussed earlier. In both cases, four distinct regions are observed: a region of maximum value (of COF or interfacial tension) with no added friction modifier (pure hexadecane), followed by a region of sharp decrease with increasing friction modifier concentration, a third region of gradual decrease with further increase of friction modifier concentration, and the final region where no change in COF or interfacial tension occurs with further increase of friction modifier concentration.

It is assumed that in the fourth region, all sites at the interface have been occupied by the friction modifier and that further increase in friction modifier concentration will not change the value (COF or interfacial tension). Thus, the interfacial tension in the fourth region of Figure 15.9 is the minimum value possible at the interface of pure hexadecane with the pure friction modifier. This value corresponds to the interfacial energy between the two pure materials [26] and is a function of the chemical structures of both materials. This is illustrated in Table 15.5 where one of the components, hexadecane, was held constant, while the second component was varied

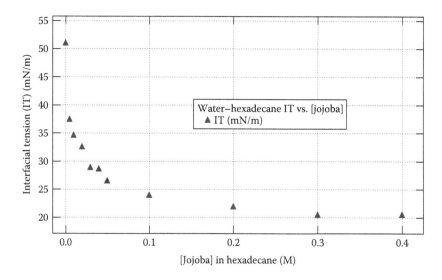

FIGURE 15.9 Effect of friction modifier (jojoba vegetable oil, **6c**) concentration on hexadecane-water interfacial tension. (From Biresaw, G., *J. Am. Oil Chem. Soc.*, 82, 285, 2005.)

TABLE 15.5

Interfacial Tension between Hexadecane and Various Friction Modifiers[a]

Hexadecane (HX)—Friction Modifier	Interfacial Tension (mN/m)
Hexadecane—water[a]	50.7[b]
Hexadecane—methyl palmitate	29.4[c]
Hexadecane—safflower oil	21.3[c]
Hexadecane—jojoba	20.5[c]

[a] Data from Reference 17.
[b] Corresponds to the equilibrium interfacial tension of water with neat hexadecane (hexadecane without solubilized friction modifier).
[c] Corresponds to the equilibrium interfacial tension of water with hexadecane containing friction modifier at a concentration that provides full coverage of the interface by the friction modifier.

from the very polar pure water to the very nonpolar vegetable oils, with the fatty acid methyl ester of intermediate polarity in the middle. The interfacial tension results in Table 15.5 follow the general rule of low interfacial tension (high compatibility) between materials of similar polarity or vice versa, that is, high interfacial tension (poor compatibility) between materials of dissimilar polarity [27]. In the example shown in Table 15.5, the interfacial tension with the highly nonpolar hexadecane increased with increasing polarity of the second material in the order: safflower oil ~ jojoba < methyl palmitate ≪ water.

15.2.3 QUARTZ CRYSTAL MICROBALANCE INVESTIGATION OF BIOBASED FRICTION MODIFIERS

The adsorption of two biobased oils, soybean oil, **6a**, and methyl oleate, **8d**, on steel surface was investigated using quartz crystal microbalance with dissipation (QCM-D) method. These two oils are commercially available and have structures similar to those depicted in Figure 15.3 for vegetable oil, **6a**, and fatty acid methyl esters (FAME), **8d**, respectively. Methyl oleate is commercially produced from soybean and other vegetable oils by a transesterification process commonly known as the biodiesel process [28].

A QCM-D monitors the film development through material adsorption onto surfaces using the piezoelectric property of quartz and is sensitive to nanogram deposits made in vacuum [29] and liquids [30]. An alternating voltage is applied across the quartz, inducing resonance. Adsorbed material causes a frequency shift in the MHz range of the quartz's resonance frequency. Sauerbrey analysis approximates a linear relationship between the frequency shift and adsorbed mass [31]. This linear relationship holds true based on three assumptions: (1) the adhering layer is small and thin relative to the crystal; (2) the adhering material binds rigidly to the surface without slippage; and (3) the material adheres uniformly over the surface. QCM-D measures the frequency shifts (Δf) of the quartz and changes in the oscillations' dissipation (ΔD) simultaneously [32]. The dissipation changes of the oscillations are due to frictional interactions on the molecular level causing energy losses. These dissipation changes are measured by temporarily disconnecting the alternating voltage applied across the quartz and monitoring the decay of the resonance frequency. The decaying resonance frequency relates directly to the viscoelastic properties of the adhered material.

QCM-D depends on the total adhered mass (including solvent) that oscillates with the vibrating quartz. The measured frequencies (fundamental and several overtones) will decrease as mass increases on the surface. It should be noted that the mass measured due to the frequency changes includes all material coupled to the vibrating quartz, including trapped or adsorbed solvent.

Dissipation changes (ΔDs) are affected by any loss of energy. Any coupling between the bulk medium and vibrating surface will alter dissipation and in turn will affect the viscoelastic properties of the layer adsorbed on the surface. Dissipation changes, measured as $\Delta D \times 10^6$, are typically less than 1 for rigidly adsorbed layers; dissipation changes >1 indicate a highly viscoelastic system.

Soybean oil was purchased from a local supermarket. Methyl oleate, **8d** ($\geq 99\%$ purity), and hexadecane, **1** ($\geq 99\%$ pure, anhydrous), were purchased from Sigma-Aldrich Chemical Company (St. Louis, MO). The hexadecane, **1**, was further purified by passing down a packed silica bead column to remove all trace impurities ($\leq 1\%$), which were found previously to affect experimental results. Ethanol (200 proof) was purchased from VWR (Batavia, IL) and used for initial cleaning of stainless steel sensor crystals. Clear glass vials, methanol, Alconox, and chloroform were all purchased from Fisher Scientific (Suwannee, GA). Purified (distilled, deionized, and UV-treated) water of 18.2 MΩ-cm resistance from a Barnstead Nanopure Diamond water purification system (model D11911; Thermo Fisher Scientific, Inc., Waltham, MA) was used. Stainless steel sensor crystals (5 MHz) were purchased from Q-Sense AB (Västra Frölunda, Sweden). (The stainless steel coating was prepared by sputtering onto gold-coated quartz crystals according to the supply brochure.) High-purity nitrogen and argon gases used for drying the QCM-D were purchased locally (Airgas, Inc., East Peoria, IL). Hellmanex II cleaning solution, purchased from Hellma GmbH & Co. KG (Müllheim, Germany), was used. Chemicals and supplies were used as received unless otherwise stated.

Soybean oil, methyl oleate, **8d**, and hexadecane, **1**, were degassed by stirring under vacuum for 1 h minimum. Clear glass vials were rinsed with diluted Alconox detergent, purified water, methanol, and chloroform, sequentially. Appropriate amounts of degassed soybean oil or methyl oleate were thoroughly mixed with degassed hexadecane to a final volume of 8 mL. Methyl oleate samples were continuously protected from light at all times.

A single-chamber QCM-D system with an axial flow sample chamber (Q-Sense Inc., Glen Burnie, MD) was used to record the adsorption of soybean oil and methyl oleate onto stainless steel Q-Sense crystals. Data were recorded and analyzed using the software provided with the instrument. All measurements were conducted at 25°C $\pm$ 0.05°C after signal stabilization. Frequency and dissipation changes were recorded for the fundamental (5 MHz) and three overtones (15, 25, and 35 MHz;

third, fifth, and seventh overtones). Stainless steel sensor crystals were cleaned sequentially in nanopure water and ethanol, dried under a gentle nitrogen stream, and exposed to UV/ozone (BioForce Nanosciences, Ames, IA) for 10 min [33]. The crystals were then cleaned in 4% (v/v) Hellmanex (Hellma GmbH & Co. KG, Müllheim, Germany) for 1 h, rinsed with water and ethanol, and exposed to UV/ozone. Crystals were stored under argon until used in experiments.

Stainless steel sensor crystals were exposed to preheated hexadecane, **1**, followed by hexadecane containing either soybean oil or methyl oleate at various concentrations, and then rinsed with hexadecane, **1**. Atomic force microscope (AFM) imaging was conducted on Q-Sense stainless steel crystals, highly ordered pyrolytic graphite (HOPG), and mica in hexadecane using a multimode system equipped with a NonoScope IV controller and E-scanner (Veeco Instruments, Inc. Santa Barbara, CA).

The exposure of stainless steel sensor crystals to solutions of soybean oil or methyl oleate at varying concentrations resulted in final frequency and dissipation values that indicated molecules were deposited onto the surface. The measured frequency and dissipation values for the two materials, as a function of their concentrations in hexadecane, are presented in Table 15.6. The final frequency values show that the

TABLE 15.6
Final Frequency (Δf_3) and Final Dissipation Change (ΔD_3) for Soybean Oil and Methyl Oleate in Hexadecane Measured Using QCM-D

[Oil] (mM), in Hexadecane	Final Δf_3 (Hz)	Final $\Delta D_3 \times 10^6$
Soybean oil		
1	−1.38	0.037
2	−1.11	0.134
5	−2.71	0.091
10	−2.30	−0.098
25	−4.16	0.086
50	−6.77	0.392
100	−6.99	0.247
200	−12.6	2.46
Methyl oleate		
1	−0.550	0.109
2	−1.22	0.160
5	−1.84	0.309
10	−3.63	0.671
25	−2.94	0.947
50	−3.69	0.393
100	−4.16	0.476
200	−3.99	0.710
400	−3.26	0.440

mass was deposited in a concentration-dependent manner. The dissipation values for soybean oil and methyl oleate at almost all concentrations studied (1–400 mM) were relatively small ($\Delta D \leq 1 \times 10^{-6}$), indicating that the mass adhered as a rigid, thin layer.

The exception is 200 mM soybean oil, which gave ΔD of 2.46×10^{-6}. The frequency changes in Table 15.6 were analyzed using the Sauerbrey equation (Equation 15.1) to determine the deposited mass [34]:

$$\Delta m = K \frac{\Delta f}{n} \tag{15.1}$$

where

Δm is the deposited mass in ng/cm^2

$K = 17.7$ ng/cm^2 Hz is the mass sensitivity constant for quartz

Δf is the frequency change in Hz

n is the fundamental frequency or third, fifth, etc., overtone

Voight analysis [35], which incorporated frequency changes and dissipation changes into a viscoelastic model, was used to determine the film thickness from the data in Table 15.6. The resulting deposited mass and film thickness for methyl oleate and soybean oil as a function of concentration in hexadecane are summarized in Figure 15.10.

Frequency changes of the adsorbed soybean oil and methyl oleate indicated full surface coverage at 100 mM for soybean oil and 10 mM for methyl oleate. Additional soybean oil and methyl oleate was not adsorbed above 100 and 10 mM, respectively. This agrees well with concentrations shown to provide a maximum surface coverage in friction studies [15,36]. The maximum adsorbed mass for soybean oil was approximately 2.5 times that for methyl oleate, roughly equal to the 3-to-1 ratio of ester groups and the 2.9-to-1 ratio of their molecular weights. Further analysis shows that the maximum adhered mass translates into a film thickness of approximately 15 and 7 Å, respectively, for soybean oil and methyl oleate. Molecular modeling indicates that both soybean oil and methyl oleate are 20 Å long, each being 18 carbons in length. QCM-D analysis suggests that both adhered as films less than 20 Å thick. Either soybean oil and methyl oleate adsorbed as truncated hydrocarbons, or both adsorbed with their acyl chains forming less than a 90° angle relative to the surface (or a tilt angle of more than 0° from normal to the surface). The nonzero tilt angle is more likely because hydrophobic interactions would occur between adsorbed molecules and the surrounding solvent since both are hydrophobic. This would account for the greater molecular extension into the bulk solvent. This also suggests that soybean oil molecules would align as fully extended bristles on a brush and methyl oleate molecules would align nearly parallel to the surface.

AFM imagery and analysis were able to confirm that methyl oleate did deposit onto steel (indicated by change in surface roughness when only hexadecane was present versus hexadecane containing 100 mM methyl oleate). Using highly oriented pyrolytic graphite (HOPG) and mica, which have proven to be good surfaces for

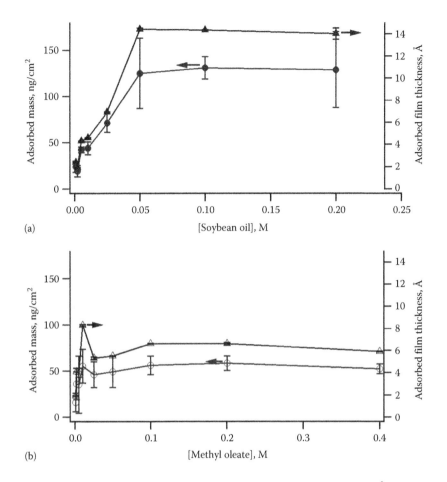

FIGURE 15.10 Adsorbed mass (triangles, ng/cm²) and film thickness (circles, Å) of soybean oil (a) and methyl oleate (b) on steel surface, as a function of concentration in hexadecane, estimated from QCM-D data shown in Table 15.6.

imaging fatty acids and long-chain alkanes [37,38], it was determined that methyl oleate adsorbed as a thin film on HOPG with some periodicity in alignment and a thickness of approximately 1 nm that agreed well with QCM-D measurements. AFM imaging also showed that methyl oleate formed a thin film on mica of thickness in agreement with the QCM-D values. AFM imaging proved difficult for soybean oil as the AFM tip could not maintain contact with the surface under the conditions explored in this work.

Comparing these soybean oil and methyl oleate adsorption studies to those of other hydrocarbons adsorbed onto metal surfaces shows that tilt angle analysis [39] puts soybean oil a tilt angle of approximately 41°, which is in good agreement with tilt angles found for other hydrocarbons adsorbed onto metal surfaces [39–41]. Such a high tilt angle suggests that soybean oil packs in a disordered liquid-like state [42]. Methyl oleate appears to have a tilt angle of about 69°, which is well beyond any reported for organic molecules adsorbed onto metal surfaces [41]. However, if it is assumed that methyl oleate is organized in an alternating head-to-head/tail-to-tail fashion on stainless steel as seen for its isomeric form, trans-oleic (elaidic) acid on HOPG [37], then a tilt angle of 69° is reasonable. The periodicity of methyl oleate on HOPG may further indicate a head-to-head/tail-to-tail arrangement. It should be noted that the rapid adsorption of both soybean oil and methyl oleate onto stainless steel also suggests that structural arrangement was rapid.

15.2.4 ADSORPTION PROPERTIES OF BIOBASED FRICTION MODIFIERS

Friction modifiers change the interfacial properties (COF, interfacial tension, etc.) by diffusing from the solution and adsorbing at the interface (liquid–liquid, liquid–gas, solid–solid, solid–gas, etc.). This is illustrated in the dynamic interfacial tension data shown in Figure 15.8. Initially, all the friction modifier is in the hexadecane solution and none at the interface, resulting in the high interfacial tension expected between pure water and hexadecane [43]. With time, diffusion of the friction modifier to the interface occurs, causing the interfacial tension to decrease. The interfacial tension continues to decrease with increasing time as more friction modifier adsorbs at the interface and its surface concentration grows. After a while, the surface concentration of the friction modifier reaches its maximum equilibrium value, and the interfacial tension attains its minimum constant value, and does not further change with time.

The minimum or equilibrium (interfacial tension or COF) value is a function of the concentration of the friction modifier in hexadecane. This is illustrated in Figures 15.6 and 15.9, where the minimum values are plotted as a function of the friction modifier concentration in the hexadecane. It is clear from these data that the minimum value decreases with increasing the friction modifier concentration up to a point beyond which it becomes constant and independent of the friction modifier concentration in hexadecane. This is because full coverage at the interface has been achieved and further increase of friction modifier concentration in hexadecane will not produce further adsorption at the interface, and the minimum or equilibrium (interfacial tension or COF) value becomes independent of the friction modifier concentration.

The concentration of the friction modifier adsorbed at the interface, $[A_I]$, expressed in moles/unit surface area, is in thermodynamic equilibrium with the concentration of the friction modifier in solution, that is, in hexadecane, $[A_H]$, expressed in moles/unit volume. Usually, such equilibrium is written in terms of the concentration of interface sites occupied by the additive, $[A_I]$, by the solvent (hexadecane), $[H_I]$, and the concentration of additive in the solvent, $[A_H]$, as follows:

$$A_H + H_I \overset{K_0}{\rightleftarrows} A_I \tag{15.2}$$

The concentration of adsorbed species at an interface is usually expressed in terms of fractional surface coverage, θ, defined as follows:

$$\theta = \frac{A_I}{A_T} \tag{15.3}$$

where $A_T = H_I + A_I$, is the total concentration of interface sites where adsorption is possible.

The value of the fractional surface coverage, θ, is calculated from measured COF or interfacial tension data as follows:

$$\theta = \frac{X_H - X_I}{X_H - X_T} \tag{15.4}$$

where

X_H is the value for pure hexadecane, that is, without friction modifier additive
X_T is the value at full surface coverage, that is, the minimum equilibrium values observed at very high friction modifier concentrations in Figures 15.6 and 15.9
X_I is the values at friction modifier concentrations before full surface coverage is observed

The relationship between the concentration of the additive in solution, $[A_H]$, and that on the surface, θ, is called an adsorption isotherm. An example of an adsorption isotherm, calculated from the interfacial tension data of jojoba oil shown in Figure 15.9 is given in Figure 15.11. As can be seen in Figure 15.11, the profile of the adsorption

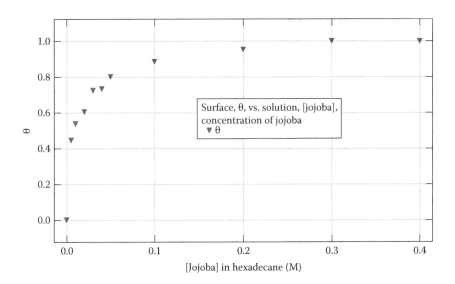

FIGURE 15.11 Adsorption isotherm: surface (expressed in terms of fractional surface coverage, θ) vs. solution (in hexadecane) concentrations of jojoba oil (**6c**) from interfacial tension with purified water. (From Kurth, T.L. et al., *J. Am. Oil Chem. Soc.*, 82, 293, 2005.)

isotherm is a mirror image of the interfacial data, since full coverage corresponds to minimum interfacial tension, while zero coverage corresponds to maximum interfacial tension between water and hexadecane.

The adsorption isotherm data such as those shown in Figure 15.11 can be analyzed to determine the free energy of adsorption, ΔG_{ads}, of the friction modifier from hexadecane to the interface. A generalized equation for calculating ΔG_{ads} is as follows [24]:

$$\Delta G_{ads} = \Delta G_0 + \alpha\theta; \quad (kcal/mol) \tag{15.5}$$

where

ΔG_0 is the free energy of adsorption due to primary interactions between the friction modifier and the surface
α is the free energy of adsorption due to lateral interactions, which could be attractive ($\alpha < 0$), repulsive ($\alpha > 0$), or inconsequential ($\alpha = 0$)
θ is the fractional surface coverage

The exact form of Equation 15.5 used for calculating ΔG_{ads} depends on the model selected to analyze the adsorption isotherm. The simplest model is the Langmuir model [44], which assumes no lateral interactions ($\alpha = 0$), and Equation 15.5 simplifies to

$$\Delta G_{ads} = \Delta G_0 = -RT\ln(K_0); \quad (kcal/mol) \tag{15.6}$$

where K_0 is estimated from the adsorption isotherm using the following relationships (derived from Equations 15.3 and 15.4):

$$\frac{1}{\theta} = 1 + \frac{1}{K_0[A_H]} \tag{15.7}$$

Thus, K_0 is obtained from a plot of $1/\theta$ vs. $1/[A_H]$ and used in Equation 15.6 to calculate ΔG_{ads}. An example of such analysis for friction-derived adsorption isotherm of high-oleic safflower oil friction modifier is illustrated in Figure 15.12.

The Temkin model assumes repulsive lateral interactions ($\alpha > 0$) and gives the following relationship between fractional surface coverage and concentration in solution [45]:

$$\theta = \frac{RT}{\alpha}\ln\left(\frac{[A_H]}{K_0}\right) \tag{15.8}$$

Values of α and K_0 are obtained from a plot of θ vs. $\ln[A_H]$ and used in Equations 15.5 and 15.6 to calculate ΔG_{ads}.

The Frumkin–Fowler–Guggenheim model assumes attractive lateral interactions ($\alpha < 0$) [46].

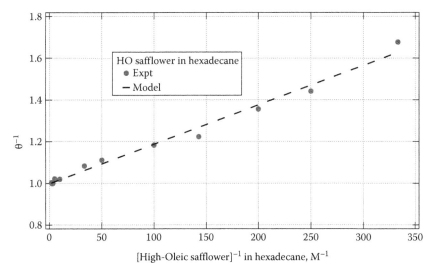

FIGURE 15.12 Langmuir analysis of fractional surface coverage, θ, vs. solution concentration data (friction derived adsorption isotherm) for high oleic safflower oil.

Free energies of adsorption, ΔG_{ads}, of friction modifiers, estimated from the analysis of friction and interfacial data using the Langmuir and Temkin models, are summarized in Table 15.7 [15–17,19,47]. The data in Table 15.7 show that friction and interfacial methods result in similar Langmuir ΔG_{ads} values for the friction modifiers, which were within ±0.25 kcal/mol. This is an indication that the polarity of the steel surface and water are similar and, hence, result in similar adsorption of the friction modifiers on these two surfaces [17].

The data in Table 15.7 also show that ΔG_{ads} values were highly dependent on the chemical structure of the friction modifier as well as on the model used to analyze the adsorption isotherm. Thus, for example, the ΔG_{ads} values estimated from the Langmuir model were generally lower for vegetable oils (<−3.0 kcal/mol) than for the fatty acid methyl esters (>−3.0 kcal/mol), indicating a stronger adsorption of the vegetable oils on the polar surfaces than the fatty acid methyl esters. This is consistent with the ability of the vegetable oils to engage in multiple bonding to the surface due to the multiple ester groups present in their structures. The fatty acid methyl esters on the other hand can participate in single bonding only and, hence, will display weaker bonding. Among the vegetable oils, jojoba oil, **6c**, displayed the highest ΔG_{ads}, which corresponds to the weakest adsorption and is a reflection of its inability to participate in multiple bonding since it has only one ester group in its structure.

Table 15.7 shows that the Temkin model predicts higher ΔG_{ads} values, that is, weaker adsorption, than the Langmuir model for all the friction modifiers investigated. This result is consistent with the assumptions used in these two models. While the Langmuir model assumes no lateral interactions ($\alpha = 0$), the Temkin model assumes repulsive lateral interaction ($\alpha > 0$). Thus, in the Temkin model, repulsive

TABLE 15.7

Free Energy of Adsorption, ΔG_{ads} (kcal/mol), at Hexadecane–Water Interface, of Biobased Friction Modifiers, Obtained from Friction and Interfacial Tension–Derived Adsorption Isotherms Analyzed Using Different Adsorption Models

Friction Modifier	Expt. Method	ΔG_{ads}, kcal/mol		References
		Langmuir	Temkin	
Vegetable oils				
Canola	Friction	−3.81	−2.04	[19,32]
Cottonseed	Friction	−3.71	−2.16	[19,32]
Jojoba	Interfacial	−3.00		[17]
Jojoba	Friction	−3.27	−1.31	[16]
Meadowfoam	Friction	−3.57	−2.28	[19,32]
Olive	Friction	−3.98		[19]
Safflower	Interfacial	−3.70		[17]
Safflower	Friction	−3.65	−2.01	[16]
High-oleic safflower	Friction	−3.71	−2.53	[16]
Soybean	Friction	−3.60	−2.1	[15]
High-oleic soybean	Friction	−3.7	−2.1	[15]
Fatty acid methyl esters				
Methyl laurate	Friction	−1.90	−0.6	[15]
Methyl oleate	Friction	−2.91	−1.02	[16]
Methyl palmitate	Friction	−2.70	−1.63	[16]
Methyl palmitate	Interfacial	−3.06		[17]

lateral interaction will oppose the primary interaction and will result in a weaker net adsorption and, hence, a higher ΔG_{ads}.

15.3 BIOSURFACTANTS

Surfactants play critical roles in lubricant formulations, especially in water-based lubricants used in metalworking, hydraulic, and other fluids [1–3]. Surfactants allow for the mixing of the oil-based formulation with water to form an emulsion. These also allow for control of the stability of the emulsion to that desired for the specific lubrication application.

Surfactants accomplish this task because of their amphiphilic structure, which is comprised of a lipophilic hydrocarbon portion that is compatible with oils and a hydrophilic polar portion that is compatible with polar solvents such as water. The hydrophilic portion of the surfactant molecule can have a variety of structures that can be grouped under one of the following categories: anionic, cationic, zwitterionic, or nonionic [48]. Examples of anionic surfactants include soaps and polysoaps with metallic or amine counterions [48]. In the work described here, two types of polymeric polysoap biosurfactants were synthesized and their surface properties investigated [49–51,53].

15.3.1 SYNTHESIS AND CHARACTERIZATION

The first polysoap was obtained by ring-opening polymerization of epoxidized soybean oil, which was then hydrolyzed into the free fatty acid, and finally, the polycarboxylic acid was converted to the polysoap by reaction with various bases. The schematic of the ring-opening synthesis is shown in Figure 15.13a [49].

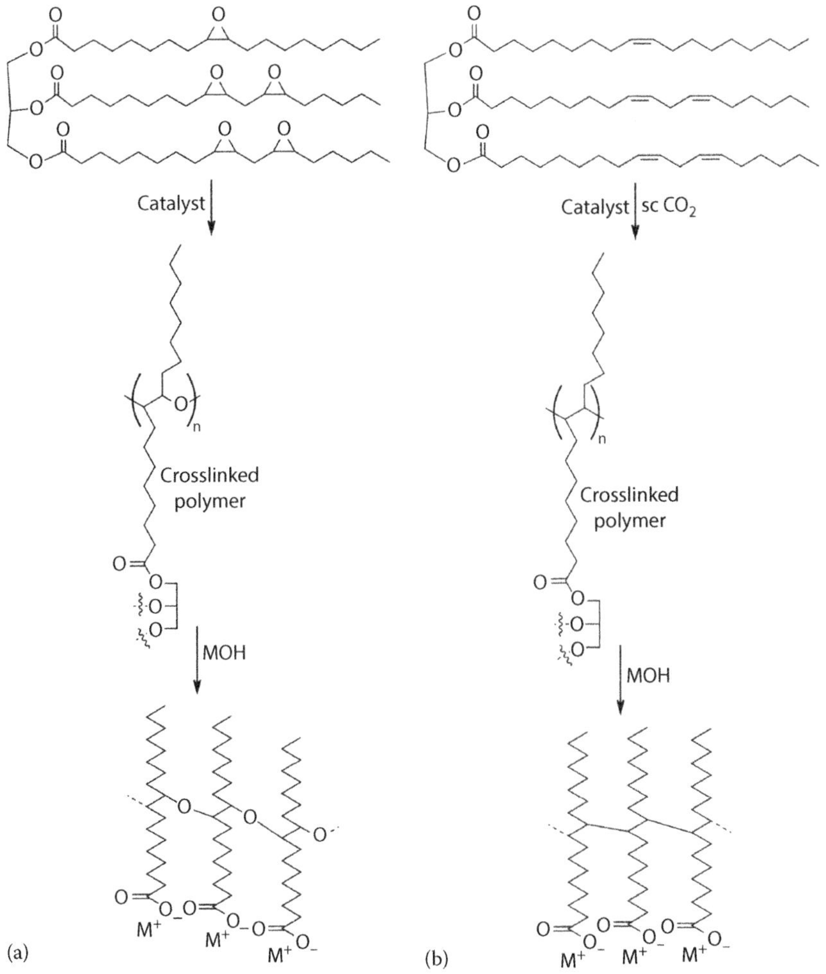

FIGURE 15.13 Synthesis of polysoaps from: (a) epoxidized soybean oil; (b) soybean oil. M^+ is Na^+, K^+, or tri-ethanol ammonium (TEA^+). (a: From Biresaw, G. et al., *J. Appl. Polym. Sci.*, 108, 1976, 2008; b: From Liu, Z. and Biresaw, G., *J. Agric. Food. Chem.* 59, 1909, 2011.)

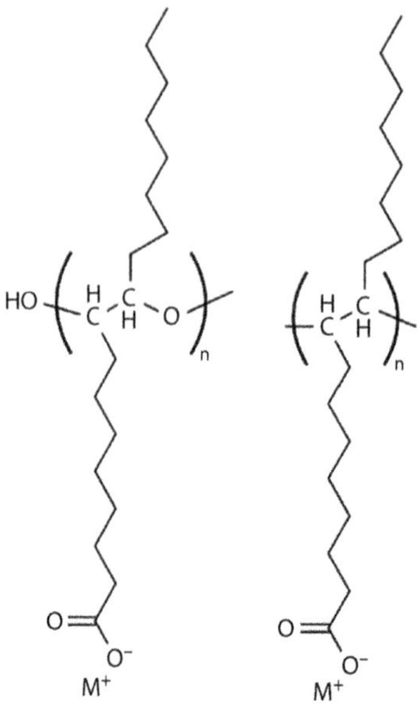

FIGURE 15.14   Polysoap repeat unit structures synthesized from epoxidized soybean oil (a), and soybean oil (b). M⁺ is Na⁺, K⁺, or tri-ethanol ammonium (TEA⁺).

The second was obtained by direct polymerization of soybean oil in supercritical carbon dioxide medium in the presence of boron trifluoride diethyl etherate catalyst. The resulting polymer was then hydrolyzed to the polycarboxylic acid, which was then converted to the polysoaps by saponification with selected bases. A schematic of the second synthesis is depicted in Figure 15.13b [50]. The polysoaps were properly characterized using FTIR, ^{1}H-NMR, and ^{13}CNMR. A gel permeation chromatography (GPC) analysis gave molecular weights of 2.6–3.2 kg/mol for epoxidized soybean oil–based polysoap (ESOP, 10) and 6.3 kg/mol for the soybean oil–based polysoap (SOP, 11) [49,50].

It should be noted that these two groups of polysoaps differ in the structures of their backbones, which are compared in Figure 15.14 [50]. The ESOP, 10, contains an ether linkage in its repeat unit, whereas the SOP, 11, does not. These structural differences, along with variations in counterion (Na⁺, K⁺, triethanol ammonium [TEA⁺]), were investigated using surface and interfacial tension measurements.

15.3.2   SURFACE AND INTERFACIAL PROPERTIES

The dynamic surface tensions of aqueous soap solutions were measured as a function of concentration using the ADSA method described earlier (Figure 15.7). The equilibrium surface tensions from such measurements were plotted as a function of concentration and used to estimate the surface energy of the polysoaps as described earlier for friction modifiers. An example of such a plot for SOP is given in Figure 15.15.

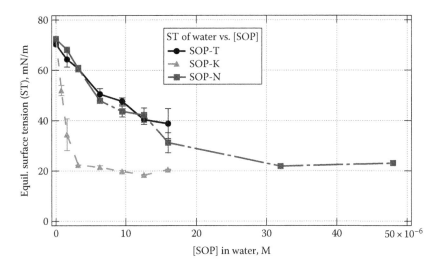

FIGURE 15.15 Effect of concentration of soybean oil-based polysoap, [SOP], on equilibrium surface tension of water. (From Liu, Z.S. and Biresaw, G., *J. Agric. Food. Chem.*, 59, 1909, 2011.)

Similarly, the dynamic interfacial tensions between various concentrations of aqueous polysoaps and hexadecane were measured as described earlier for friction modifiers. From plots of concentration vs. equilibrium interfacial tension data, similar to that shown in Figure 15.16 for SOP, the measured interfacial energy (γ_{sh}) between polysoap and hexadecane was estimated.

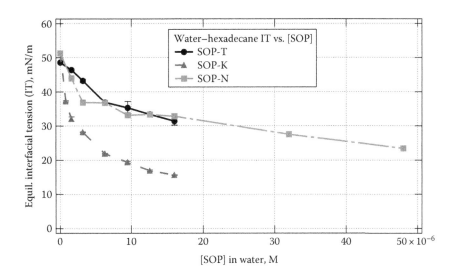

FIGURE 15.16 Effect of [SOP] on equilibrium interfacial tension.

15.3.3 Surface Energy

The surface energy of polysoaps (γ_s) comprise polar $\left(\gamma_s^p\right)$ and dispersive $\left(\gamma_s^d\right)$ components related as follows [52,54]:

$$\gamma_s = \gamma_s^p + \gamma_s^d \tag{15.9}$$

Equation 15.9 can be rearranged to express the polar and dispersive components in terms of fractional polarity (x^p) and fractional dispersity (x^d) terms as follows:

$$\gamma_s = \gamma_s^p + \gamma_s^d = \gamma_s\left(x^p\right) + \gamma_s\left(x^d\right) \tag{15.10}$$

where
$$x^p = \gamma_s^p / \gamma_s$$
$$x^d = \gamma_s^d / \gamma_s$$
$$x^p + x^d = 1$$

The fractional dispersity and fractional polarity of a material are highly dependent on the chemical structure of the material. These values can be estimated from the surface energy of the two materials (γ_s and γ_h for polysoap and hexadecane, respectively) and the interfacial energy between the materials (γ_{sh}) using the harmonic mean (HM) or geometric mean (GM) methods as follows [52,54]:

$$\gamma_{sh} = \gamma_s + \gamma_h - \frac{4\gamma_s^d\gamma_h^d}{\gamma_s^d + \gamma_h^d} - \frac{4\gamma_s^p\gamma_h^p}{\gamma_s^p + \gamma_h^p} \tag{15.11}$$

$$\gamma_{sh} = \gamma_s + \gamma_h - 2\sqrt{\gamma_s^d\gamma_h^d} - 2\sqrt{\gamma_s^p\gamma_h^p} \tag{15.12}$$

Hexadecane is a hydrocarbon, and its surface energy does not have a polar component, that is, $\gamma_h^p = 0$, which leads to

$$\gamma_h = \gamma_h^p + \gamma_h^d = \gamma_h^d \tag{15.13}$$

Combining Equations 15.9 through 15.13 gives the following expressions for calculating fractional dispersity and polarity values using the HM and GM methods, respectively:

$$x^d = \frac{A\gamma_h}{4\gamma_s\gamma_h - A\gamma_s} \tag{15.14}$$

$$x^d = \frac{A^2}{4\gamma_s\gamma_h} \tag{15.15}$$

where
$$A = \gamma_s + \gamma_h - \gamma_{sh}$$
$$x^p = 1 - x^d$$

15.3.4 Biosurfactant Structure and Surface Energy

Equations 15.14 and 15.15 were used to analyze the surface energy and interfacial energy (with hexadecane) data generated on polysoaps from epoxidized soybean (ESOP, **10**) and soybean (SOP, **11**), with sodium (N), potassium (K), and triethanol ammonium (T) counterions. The result of the analysis is summarized in Table 15.8. The data in Table 15.8 shows that the estimated x^d values varied depending on the structure of the polysoap, counterion, and even the model used to analyze the data. In general, the GM method gave a slightly lower x^d (slightly higher x^p) value than the HM methods. However, both methods gave similar trends on the effect of a counterion on x^d, which decreased in the order $K^+ > Na^+ > TEA^+$ on both ESOP and SOP polysoaps. Please note that the opposite trend is predicted for x^p.

Both models predict higher x^d (lower x^p) for ESOP than for SOP polysoaps. The average x^d values of polysoaps with different counterions were in the range 0.53 to 0.61 ± 0.05 and 0.35 to 0.43 ± 0.12 for ESOP and SOP, respectively. The corresponding x^p values were 0.39 to 0.47 ± 0.05 and 0.57 to 0.65 ± 0.12 for ESOP and SOP, respectively. The two polysoaps differ in their molecular weight as well as in the chemical structure of their backbone linkages. The molecular weight of ESOP is about half that of SOP. ESOP also has ether linkages in its backbone, whereas SOP has only hydrocarbon linkages in its backbone. It is not clear why these differences in molecular weight and backbone chemical structure can cause the ESOP polysoap to produce a more hydrophobic surface (larger dispersive component) than the SOP polysoap.

TABLE 15.8

Calculated Fractional Dispersive Energy (x^d) for Polysoaps Estimated Using the Harmonic Mean (x^d-HM, Equation 15.14) and Geometric Mean (x^d-GM, Equation 15.15) Methods

Material	MW (g/mol)	γ_h (dyn/cm)	γ_s (dyn/cm)	γ_{sh} (dyn/cm)	x^d-HM	x^d-GM
Hexadecane	224	27.5				
SOP-K	6300	27.5	20.5	15.6	0.56	0.47
SOP-N	6300	27.5	22.5	23.4	0.39	0.29
SOP-T	6300	27.5	39.6	31.4	0.33	0.29
ESOP-K	2615	27.5	19.9	11.9	0.66	0.58
ESOP-K	3219	27.5	19.9	12.7	0.64	0.55
ESOP-N	3219	27.5	21.6	13.0	0.62	0.55
ESOP-T	2615	27.5	22.9	14.2	0.59	0.52
ESOP-T	3219	27.5	23.9	16.9	0.53	0.45

Abbreviations: ESOP and SOP (structures **10** and **11**, respectively, in Figure 15.14) are polysoaps from soybean oil and epoxidized soybean oil, respectively; K, N, and T correspond to polysoaps with potassium, sodium, and triethanol ammonium counterions, respectively; γ_h and γ_s are the measured surface energy of hexadecane and polysoap, respectively [49,50]; γ_{sh} is the measured polysoap–hexadecane interfacial energy [49,50].

15.4 BIOBASED ANTIWEAR ADDITIVES

Antiwear additives control the fourth largest market share of the more than 4 million metric tons/year global additive market (Figure 15.1) [8]. They are the critical components of lubricants used in many applications, including in metalworking, hydraulic, and engine oils [1,51,53]. Antiwear additives protect tools used in metalworking from wear and damage and extend their useful lives. They also protect pistons, cylinders, and other components of engines from wear and provide long life to engines. Similarly, antiwear additives used in hydraulic lubricants protect the components of small and large hydraulic systems from wear, increase equipment reliability, and reduce repair and maintenance costs.

The details of how antiwear additives provide wear protection are not fully understood. The consensus is that, under certain conditions of pressure and temperature in the friction zone, the antiwear additives participate in an *in situ* tribochemical reaction with the metallic friction surfaces. The reaction results in the formation of new materials, with highly reduced wear rates, on the surface of the friction materials. Such mechanism is considered to be responsible for the effectiveness of the most commonly used antiwear additive in the market, zinc dialkyldithiophosphate (ZDDP), **12** (Figure 15.17).

Since phosphorous is one of the elements known to provide antiwear characteristics, we synthesized phosphonate containing biobased additives using free radical initiators (structure **13** in Figure 15.18) and investigated their physical and antiwear properties [52,54–57]. A brief description of the synthesis and property investigation is given next.

15.4.1 SYNTHESIS OF BIOBASED PHOSPHONATES

Two types of biobased precursors were used in the synthesis. The first was methyl oleate, **8d** [52,54–56], which can be readily obtained from transesterification of fats and vegetable oils, whose fatty acid profile is rich in oleic acid. The second is limonene, **16** [57], which is a major by-product of the citrus industry [58]. In both cases, the synthesis involved the free radical addition of dialkyl phosphite, **14**, to the precursor double bonds, as depicted in Figure 15.18. Three different dialkyl phosphites, namely, dimethyl (**14a**), diethyl (**14b**), and di-*n*-butyl (**14c**), were used in the synthesis to produce three different phosphonate structures (**15, 17–19**). The synthesis was conducted for 10–24 h, under inert atmosphere, at a temperature (64°C–125°C) appropriate for the selected free radical initiator, **13**. The initiators used included *tert*-butyl perbenzoate, **13a**; di-*tert*-butyl peroxide, **13b**; dilauroyl peroxide, **13c**; and 2,2′-azobis(2-methylpropionitrile), **13d**.

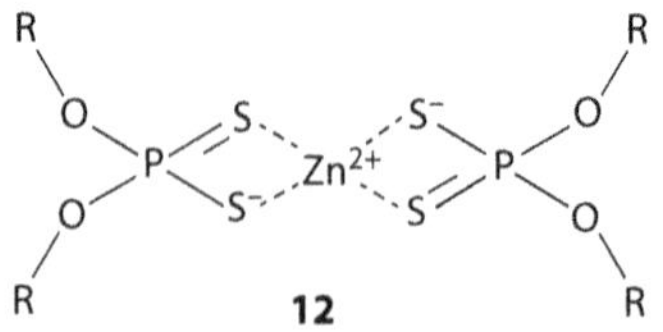

FIGURE 15.17   Structure of zinc dialkyldithiophosphate (ZDDP) anti-wear additive.

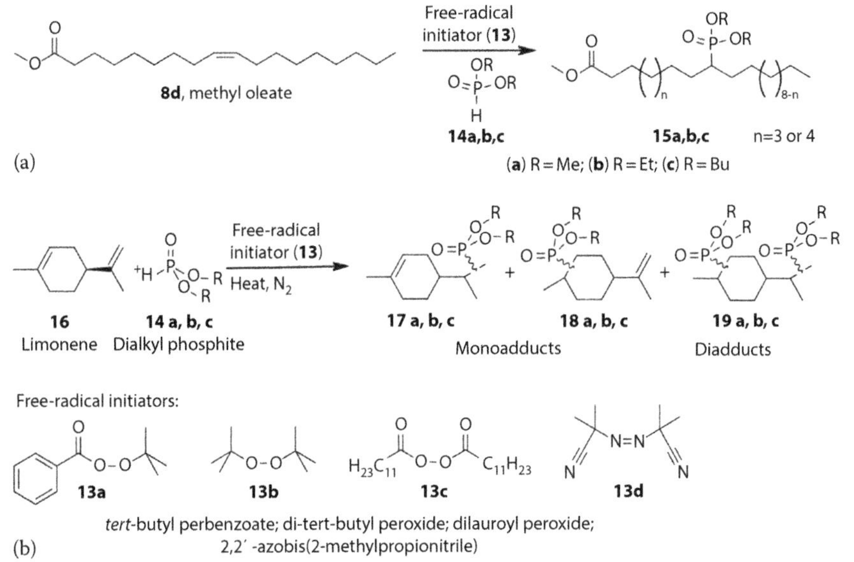

FIGURE 15.18 Schematics of phosphonate synthesis from (a) methyl oleate; and (b) limonene. (From Bantchev, G.B. et al., *Spectrochim. Acta Part A*, 110, 81, 2013; Biresaw, G. and Bantchev, G.B., *J. Am. Oil Chem. Soc.*, 90, 891, 2013; Bantchev, G.B. et al., *J. Am. Oil Chem. Soc.*, 93, 859; Biresaw, G. and Bantchev, G.B., *Tribol. Lett.*, 60, 11, 2015, doi: 10.1007/s11249-015-0578-2.)

As shown in Figure 15.18, the product from the synthesis with methyl oleate is a monoadduct (**15**) while that from limonene can contain both mono- (**17, 18**) and diadducts (**19**). However, the synthesis with limonene was allowed to proceed until both double bonds reacted to produce only the diadduct product (**19**) [57].

It should be noted that the location of the phosphonate branch is a chiral center and its position in both the mono- and diadduct products can be different (Figure 15.18). As a result, the products from both syntheses are mixtures of several positional and enantiomeric structures.

The biobased phosphonate products were obtained in 94%–95% isolated yields and positively identified using a combination of gas chromatography–mass spectrometry, nuclear magnetic resonance (NMR), and Fourier transform infrared (FTIR) spectroscopy [54–57].

15.4.2 Physical Property of Biobased Phosphonates

Tables 15.9 and 15.10 compare selected properties of dialkylphosphonate derivatives with each other and with the starting biobased feedstock. The property changes observed in the dialkyl phosphonate derivatives can be attributed to the insertion of heavier and polar atoms into their structures. Similar trends were observed for both the monoadducts of phosphonates from methyl oleate and the diadducts of phosphonates from limonene.

TABLE 15.9

Selected Physical Properties of Dialkylphosphonates Derived from Methyl Oleate[a]

| Property[a] | Method | Methyl Oleate, 13 | Dialkylphosphonate, 15 | | |
			Dimethyl, 15a	Diethyl, 15b	di(n-butyl), 15c
Density (g/mL)	D-7402				
40°C		0.8595	0.9735	0.9509	0.9397
100°C		0.8162	0.9290	0.9066	0.8963
kVisc (mm²/s)	D-7402				
40°C		4.5	20.2	16.5	19.0
100°C		1.8	4.2	3.8	4.4
VI	D-2270	N/A	111	129	145
Oxidation stability					
PDSC-OT (°C)	D-6186	172.7 ± 0.5	200.5 ± 1.4	199.4 ± 1.5	201.7 ± 0.2
PDSC-PT (°C)	D-6186	187.6 ± 0.1	217.7 ± 1.06	216.7 ± 0.6	220.8 ± 0.9
Cold flow					
Pour point (°C)	D-5949	−21	−33 ± 1		−41 ± 1
Cloud point (°C)	D-5773	−15	<−75		<−75

[a] Data from Reference 55 and references therein. See Figure 15.18a for structures.

Abbreviations: kVisc, kinematic viscosity; PDSC, pressurized differential scanning calorimetry; OT, onset temperature; PT, peak temperature.

Compared to the starting biobased feedstocks, the dialkylphosphonate displayed poorer solubility in base oils, both petroleum based (polyalphaolefin, PAO-6) and biobased (high-oleic sunflower oil, HOSuO) (Table 15.10). This was attributed to the increased polarity of the phosphonates relative to the starting materials due to the presence of oxygen and phosphorous groups in their structures. As a result, the solubility in the nonpolar PAO-6 base oil was much lower than in the polar HOSuO. Also, the solubility in both base oils improved with increasing chain length of the alkyl groups, attributed to decreased polarity with increasing chain length.

The dialkylphosphonates displayed a higher density than the starting materials, which was attributed to the presence of heavier elements (PO_3) in their structures (Tables 15.9 and 15.10). The effect was more pronounced with the diadduct or bis(dialkylphosphonate) derivatives, which displayed a density of >1.0 g/mL at 40°C and 100°C for all alkyl derivatives. The density of the dialkylphosphonates from both derivatives was a function of the alkyl chain length and decreased with increasing chain length in the order: dimethyl > diethyl > di-n-butyl (Tables 15.9 and 15.10).

Phosphonation also resulted in derivatives with a sharply higher viscosity than the starting materials, and the effect was much higher with the diadduct derivatives (Tables 15.9 and 15.10). Viscosity was also a function of the phosphonate alkyl chain length, and generally decreased with increasing chain length.

TABLE 15.10

Selected Physical Properties of Limonene Bis(dialkylphosphonates)[a]

			Limonene Bis(dialkylphosphonates), 19[b]		
Property	Method	Limonene, 16	Dimethyl, 19a	Diethyl, 19b	di(n-butyl), 19c
RT solubility in					
PAO-6 (%, w/w)		>50	2.15	4.21	6.96
HOSuO (%, w/w)		>50	7.06	7.15	10.17
Density (g/mL)	D-7042				
40°C		0.8267	1.1633	1.0809	1.0172
100°C		0.7794	1.1154	1.0344	0.9722
kVisc (mm²/s)	D-7042				
40°C		0.83 ± 0.00	299 ± 16	68.41 ± 0.02	63.80 ± 0.01
100°C		0.47 ± 0.00	15.30 ± 0.46	7.53 ± 0.01	8.17 ± 0.00
VI	D-2270	96	5	60	95
Oxidation stability					
PDSC-OT (°C)	D-6186	140.8 ± 1.4	199.9 ± 1.1	183.9 ± 1.9	191.5 ± 2.5
PDSC-PT (°C)	D-6186	154.3 ± 2.0	219.3 ± 0.7	210.4 ± 1.4	217.8 ± 1.4

[a] Data from Reference 57.

[b] See Figure 15.18b for structures.

Abbreviations: RT, room temperature; kVisc, kinematic viscosity; PDSC, pressurized differential scanning calorimetry; OT, onset temperature; PT, peak temperature; PAO-6, polyalphaolefin base oil (**5** in Figure 15.2) with kinematic viscosity of 6 mm²/s at 100°C, supplied as Durasyn 166 by Ineos Oligomers (League City, TX); HOSuO, high-oleic sunflower oil **6a** (81% oleic acid) obtained from Columbus Foods Company (Des Plaines, IL).

The viscosity indices (VIs) of the dialkylphosphonates were generally lower than those of the starting materials and increased with increasing alkyl chain length (Tables 15.9 and 15.10). Thus, for limonene derivatives, the VI increased in the order: dimethylphosphonate < diethylphosphonate < di-n-butylphosphonate ~ limonene (Table 15.10).

As shown in Figure 15.18, phosphonylation resulted in a major change in the structure of the starting biobased feedstock that profoundly impacted its oxidation stability. It has eliminated one or more double bonds, which are known to impart poor oxidation stability to these materials. As shown in Tables 15.9 and 15.10, pressurized differential scanning calorimetry (PDSC) results indicate low OT and PT values for the starting materials, which were lowest for limonene because of the multiple double bonds in its structure. However, a jump in OT and PT values was observed for the dialkylphosphonate derivatives from both feedstocks, indicating the elimination of the double bonds that caused poor oxidation stability.

In addition to eliminating unsaturation, phosphonylation also introduced branching into the structure of the starting biobased feedstocks (Figure 15.18), which profoundly altered the cold flow properties of the dialkylphosphonates. Branching interferes with crystallization and, therefore, depresses the pour point and cloud point of the oils. This was demonstrated for methyl oleate and its dialkylphospho-nates (Table 15.9). Thus, considerable improvement in cold flow properties, that is, significant lowerings in pour point and cloud point values, was observed after the conversion of methyl oleate, **8d**, into the dialkylphosphonate products, **15**.

Thus, phosphonylation has provided biobased dialkylphosphonate products with higher density and viscosity that was a function of the number of dialkylphosphonate adducts and the alkyl chain length; lower VI; lower solubility in petroleum-based and biobased base oils, which was also a function of the number of dialkylphosphonate adducts and alkyl chain length; and significantly improved oxidation stability and cold flow properties.

15.4.3 ANTIWEAR PROPERTIES

Antiwear investigations were conducted on a four-ball tribometer (model KTR-30L; Koehler Instrument Company, Bohemia, NY) according to ASTM standard test method D-4172 [59]. In the four-ball tribometer (Figure 15.19), a test is conducted between three stationary steel balls secured at the bottom of a pot and a rotating fourth steel ball mounted on the top ball drive and pressed against the three balls by a specified load. The pot also contains the lubricant to be tested, which is maintained at the specified temperature using a thermocouple connected to a heating element.

The steel balls used in the test were obtained from Falex Corporation (Aurora, IL) and have the following specifications: material, chrome steel alloy made from AISI E52100 standard steel; hardness, 64–66 Rc; diameter, 12.7 mm; and finish, grade 25 extra polish. The friction test conditions as specified in ASTM D-4172 four-ball antiwear test were as follows: load, 392 N; speed, 1200 rpm; lubricant temperature, 75°C; and test duration, 60 min.

Blends of biobased and petroleum-based base oils, with 0%–10% (w/w) biobased antiwear additives, were prepared and tested. Similar blends with ZDDP (Elco 108;

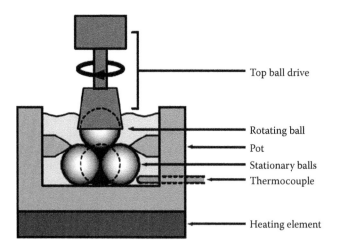

FIGURE 15.19 Schematic of 4-ball tribometer.

Elco Corporation, Cleveland, OH) were also prepared and tested. The base oils used in the evaluation included polyalphaolefin with a kinematic viscosity of 6 mm²/s at 100°C (PAO-6, Durasyn 166; Ineos Oligomers, League City, TX), soybean oil (Pioneer Hi-Breed International, Des Moines, IA), and high-oleic sunflower oil (HOSuO, 81% oleic acid; Columbus Foods Company, Des Plaines, IL).

At the end of each test, the COF for the lubricant was calculated from the measured torque data and corresponding load according to the ASTM D-5183 procedure [60]. The wear scar diameters (WSD), along and across the wear direction of each of the three balls from the test, were measured according to ASTM standard test method D-4172 [59], using a wear scar measurement system (ScarView; Koehler Instrument Company, Inc., Bohemia, NY), and averaged. Each test lubricant was used in at least two measurements, and average COF and WSD were used in the data analysis.

As discussed earlier, the ASTM D-4172 [59] antiwear test produces two sets of data: COF and WSD. In the discussion to follow, the COF results are discussed first followed by the WSD results.

Figures 15.20 and 15.21 summarize the COF results as a function of biobased antiwear additive concentration in biobased (soybean or HOSuO) and petroleum-based (PAO-6) base oils. Data for ZDDP in these base oils are also included for comparison. The data in Figure 15.20 are for the dialkyl phosphonate biobased antiwear additives derived from methyl oleate, **15 a,b,c** (Figure 15.18a); and the data in Figure 15.21 are for bis(dialkyl phosphonate), **19 a,b,c**, biobased antiwear additives derived from limonene (Figure 15.18b). The data in Figures 15.20 and 15.21 show, with some minor exceptions, the antiwear additives increased the COF of the biobased and the petroleum-based base oils. For most additives, including for ZDDP, the COF increased with increasing concentration until it reached a steady-state constant value above 2% (w/w) concentration. With minor exceptions, the steady-state COF value was higher than the COF of the neat base oil without the antiwear additive. This confirms the general observation that antiwear additives are not effective at reducing friction.

The effect of antiwear additive concentration in biobased (soybean or HOSuO) and petroleum-based (PAO-6) base oils on WSD is summarized in Figures 15.22 and 15.23. In almost all cases, the presence of the antiwear additive in the formulation resulted in significant reductions in the wear properties of the biobased and the petroleum-based base oils. As shown in Figures 15.22 and 15.23, the WSD decreased with increasing concentration until it reached a minimum steady-state value that was independent of concentration. The exceptions to this profile were in situations where the antiwear additive has poor solubility in the base oil. This is exemplified in Figure 15.23, where the limonene-based bis(dialkyl phosphonate) antiwear additives, **19**, which have poor solubility in PAO-6 (Table 15.10), displayed increased WSD with increasing concentration in PAO-6. A comparison of the WSD of the neat base oils (Figures 15.22 and 15.23) indicates a much lower value for the biobased than for the petroleum based. This is due to the polarity of the biobased base oils, which allows it to adsorb on friction surfaces and prevents direct contact between friction surfaces. It should be noted that, even though the neat biobased base oils have low WSD, further reduction of WSD was attained by addition of the biobased and ZDDP antiwear additives.

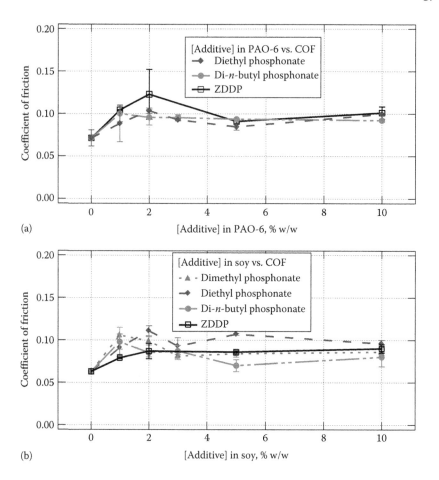

(a)

(b)

FIGURE 15.20 Effect of [dialkyl phosphonate], **15**, derived from methyl oleate on coefficient of friction: (a) in PAO-6 base oils; (b) in soybean oil base oil.

The data in Figures 15.22 and 15.23 also underscore the importance of the antiwear additive chemical structure on WSD. As mentioned earlier, poor solubility, due to vast differences in polarity between base oil and bis(dialkyl phosphonate) antiwear additive (Figure 15.23a), resulted in an increased WSD with increasing concentration of antiwear additive. In general, the dimethyl phosphonates (**15a** and **19a**), which are the most polar of all the structures investigated, displayed the smallest decline in the WSD of the neat base oil (Figures 15.22 and 15.23). On the other hand, the di-*n*-butyl phosphonates (**15c** and **19c**) displayed the largest reduction in WSD, which, in some cases, was almost as good as the values obtained for ZDDP, **12** (Figures 15.22 and 15.23).

This investigation showed that a minimum WSD of 0.4–0.5 mm is attained for biobased base oils with 1%–2% (w/w) of biobased antiwear additive concentration (Figures 15.22 and 15.23). The minimum WSD values were dependent on the chemical structure of the antiwear additive, namely, the chain length of

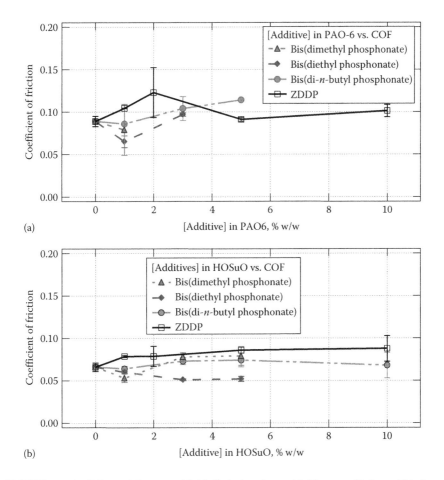

FIGURE 15.21   Effect of [limonene bis(dialkyl phosphonate)], **19**, on coefficient of friction: (a) in PAO-6 base oils; (b) in soybean oil base oil.

the dialkyl on the phosphonate group of the additives. The WSD increased with decreasing alkyl chain length in the order ZDDP $\leq$ di-n-butyl $\leq$ diethyl < dimethyl (Figures 15.22 and 15.23).

15.5   BIOBASED VISCOSITY INDEX IMPROVERS

Oligomers of lipoate ester (**23** in Figure 15.24) were found to display viscosity index improver (VII) properties in lubricant blends with the biobased base oil high-oleic sunflower oil (HOSuO) [61]. These oligomers were by-products comprising 3%–20% of the product mixture obtained from the condensation reaction of lipoic acid, **20**, with a variety of alcohols, **21a–g** (Figure 15.24).

Lipoic acid, **20**, is a natural product found in plants and animals [62]. It is a C_8 carboxylic acid with cyclic disulfide structure and is solid at room temperature. In the work described here, lipoic acid was reacted with various alcohols, **21a–g**, in

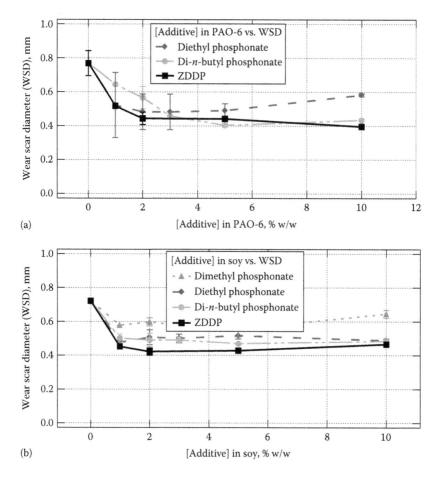

FIGURE 15.22 Effect of [dialkyl phosphonate] derived from methyl oleate, **15**, on wear scar diameter: (a) in PAO-6 base oils; (b) in soybean oil base oil.

the presence of a solid catalyst (Amberlyst-15; Sigma-Aldrich, St. Louis, MO), as depicted in Figure 15.24. The reaction was conducted without solvent, at 70°C, for 96 h. The optimized synthesis procedure gave 80% yield of purified lipoate ester product mixture.

As shown in Figure 15.24, the alcohols selected for this study, **21a–g**, varied in their chain length (from 2 to 18) and degree of branching (none, simple, complex). The product mixtures of the lipoate esters were identified and thoroughly characterized using ^{1}H NMR and GPC. All the biobased esters (lipoate esters) synthesized were liquid at room temperature and displayed different degrees of solubility in base oils. Lipoate ester solubility was highly dependent on the structure of the alcohol group and the polarity of the biobased (HOSuO) and petroleum-based (PAO-6) base oil [61]. Thus, lipoate esters with $\geq C_8$ and $\geq C_{10}$ alcohols displayed good solubility ($\geq$10% w/w) in HOSuO and PAO-6, respectively (Table 15.11).

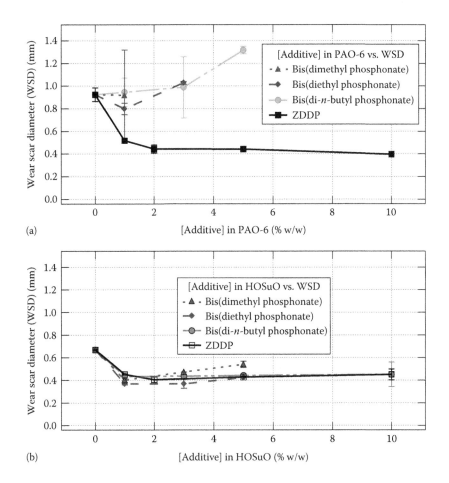

FIGURE 15.23 Effect of [limonene bis(dialkyl phosphonate)], **19**, on wear scar diameter: (a) in PAO-6 base oils; (b) in soybean oil base oil.

   The product mixture of the lipoate esters that displayed acceptable solubility in the two base oils were selected and further evaluated for their physical and other properties using a variety of methods. The selected lipoate esters were based on the following alcohols: 2-ethylhexanol (2-EH, **21d**), *n*-octanol (**21c**), *n*-dodecanol (**21f**), and isostearyl alcohol (**21g**). One common feature of these lipoate esters was that they contain a small quantity of polylipoate esters (**23c,d,f,g**), obtained by ring-opening polymerization of the corresponding lipoate esters [63]. The molecular weight and concentration of the polymeric by-product in these four lipoate ester product mixtures are summarized in Table 15.11. The data in Table 15.11 show that the presence of the polymeric by-products had a profound effect on the viscosity and VI of the 2-ethylhexyl, *n*-octyl, and *n*-dodecyl lipoates. The data show that the effect was a function of the polymer concentration in the product mixture. As a result, the presence of the polymer in the isostearyl lipoate had no effect on these two properties because of its low concentration in the product mixture.

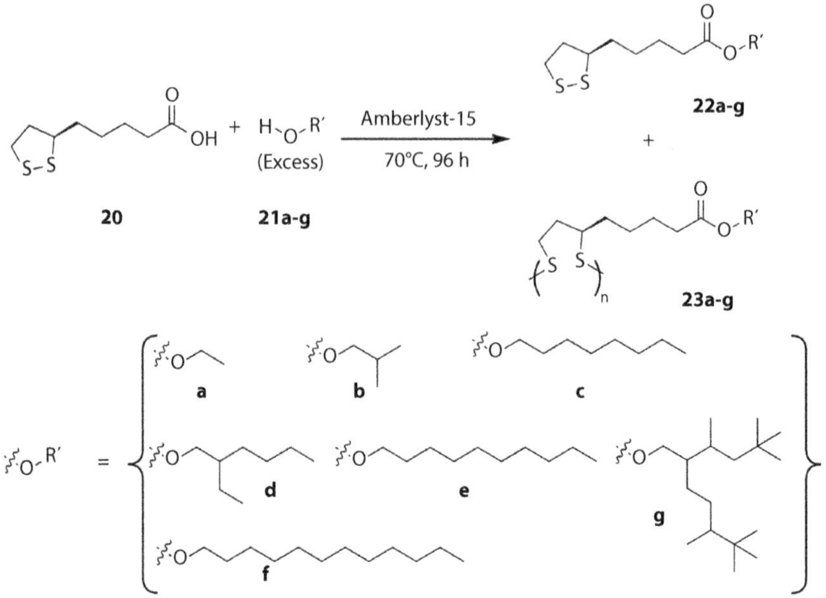

FIGURE 15.24 Schematic of lipoate ester synthesis. (From Biresaw, G. et al., *Ind. Eng. Chem. Res.*, 55, 373, 2016.)

The lipoate esters with the polymer by-products were further investigated for their additive properties as viscosity modifiers. Blends with 1% and 5% lipoate ester product mixture in HOSuO were prepared and their kinematic viscosity at 40°C and 100°C determined. The data were then used to determine the VI of the blend as per ASTM D-2270 [64]. The results, analyzed as a function of the concentration of the polylipoate ester (rather than lipoate ester product mixture) in the blend, are shown in Figure 15.25. At 40°C, all four lipoate ester product mixtures were effective at increasing the kinematic viscosity of HOSuO, which increased with increasing concentration of the polylipoate ester (Figure 15.25a). Similar results were observed at 100°C for three of the four polylipoate esters. The exception was the isostearyl polymer (**23g**), which displayed a decrease in the kinematic viscosity of HOSuO with increasing polymer concentration at 100°C.

Figure 15.25b compares the VI of the blends as a function of the polylipoate ester concentration. The data show a similar trend as that of the kinematic viscosity data at 100°C. Thus, the VI of HOSuO increased with increasing concentration of the polylipoate ester in the blend for three of the four polymers investigated. The exception was again the polylipoate ester from the isostearyl alcohol (**23g**), which displayed a reduction in VI with increasing polymer concentration in the blend. The result indicates that polylipoate esters synthesized from linear and long-chain alcohols are effective VIIs. Since lipoic acid and the long/linear alcohols can be sourced from renewable feedstocks, the VIIs synthesized from them will be biobased and provide numerous environmental, health, economic, and other benefits. The benefits are particularly large since VIIs comprise almost a quarter of the more than 4 million

TABLE 15.11

Properties of Selected Neat Lipoate Ester Product Mixtures[a]

	Method	Lipoate Ester Product Mixture (22/23)[b]			
		2-Ethylhexyl 22d/23d	n-Octyl 22c/23c	n-Dodecyl 22e/23e	Isostearyl 22g/23g
Oligomer (% w/w)	GPC	6.6	5.7	19.8	3.1
Oligomer MW (kD)	GPC	3435	3128	2,253	97
Density at 40°C (g/cm³)	D-7042	1.0041 ± 0.0000	1.0003 ± 0.0000	0.9774 ± 0.000	0.9477 ± 0.0000
Density at 100°C (g/cm³)	D-7042	0.9583 ± 0.0004	0.9564 ± 0.0003	0.9349 ± 0.0009	0.9075 ± 0.0002
kVisc at 40°C (mm²/s)	D-7042	219.5 ± 0.7	147.2 ± 0.3	53,700 ± 416	74.0 ± 0.5
kVisc at 100°C (mm²/s)	D-7042	96.2 ± 3.2	47.0 ± 0.6	10,869.8 ± 82.5	8.2 ± 0.0
VI	D-2270	447	355	567	70
RT solubility (% w/w) in	Visual				
HOSuO		>10	>10	>10	>10
PAO-6		<2	~2	>10	>10

[a] Data from Reference 61.

[b] See Figure 15.24 for structures.

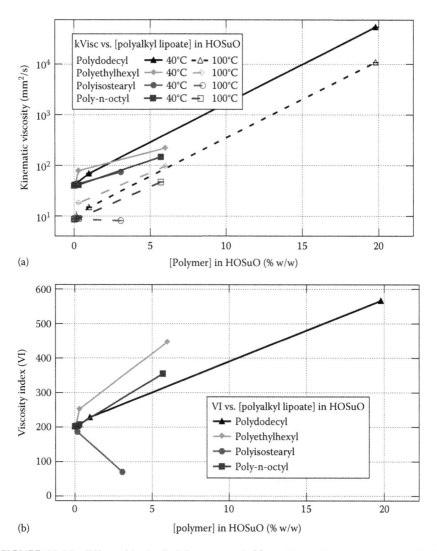

FIGURE 15.25   Effect of [polyalkyl lipoate ester], **23**, on kinematic viscosity (a), and VI (b) of HOSuO.

metric tons/year global additive market. Currently, this market is completely dominated by petroleum-based VII additives, whose replacement by polylipoate ester or
other biobased VII additives will result in a huge reduction of greenhouse gases and
many other environmental, health, and economic benefits.

15.6   BIOBASED ANTIOXIDANTS

Some organic compounds containing sulfur in their structures are known to resist
oxidation, while others display antioxidant properties and can be used as additives in
lubricant formulations [65,66]. Most such compounds currently in use are petroleum

based. Investigations were conducted to synthesize sulfur containing biobased compounds and evaluate their antioxidant properties. The following three types of compounds were synthesized and investigated: sulfide-modified (SM) vegetable oils, high-oleic sunflower-based lipoyl triglycerides, and lipoate esters of various alcohols. Each of these investigations will be discussed separately.

15.6.1 SULFIDE-MODIFIED VEGETABLE OIL BIOBASED ANTIOXIDANTS

SM vegetable oils (**26** in Figure 15.26) were synthesized by free radical addition of butanethiol, **24**, to the double bonds in corn and canola vegetable oils, **25**, as depicted in Figure 15.26 [67]. The reactions were initiated using ultraviolet light, with or without added photoinitiator (2,2-dimethoxy-2-phenylacetophenone, 99%; Acros Organics, Fair Lawn, NJ). Up to sixfold excess butanethiol, relative to moles of double bonds in the vegetable oils, were used. Optimized synthesis procedure was conducted at −78°C, for 2–8 h, resulting in 97% conversion of the double bonds and 61% isolated yield of the SM vegetable oil **26** [67].

The properties of the SM vegetable oils were investigated using a variety of methods [68]. Table 15.12 compares some properties of neat corn and canola oils to those of the corresponding neat SM products. The difference in the properties of the oils before and after addition of the butanethiol is consistent with the expectations based on the differences in their structures. Thus, the structure of the product after the addition of butanethiol will have no double bonds, will be branched, and will comprise a heavy atom (sulfur) in its structure. As a result, the new product will have higher density, higher viscosity, improved cold flow properties (lower cloud point and lower pour point), and improved oxidation stability (Table 15.12). The SM products also showed lower VI, which could be attributed to the effect of the various isomers in the product mixture on viscosity.

The antioxidant properties of the SM oils were evaluated in PAO-6, corn oil, and HOSuO base oils. Blends containing 0%–10% of the SM vegetable oils were tested for their oxidation stability using the rotating pressurized vessel oxidation test (RPVOT) (ASTM D-2272) [69] and PDSC (ASTM D-6186) [70] methods. The RPVOT results showed no change in the oxidation stability of corn oil with the addition of 6.2%–6.3% of SM corn or SM canola oils (Table 15.13). However,

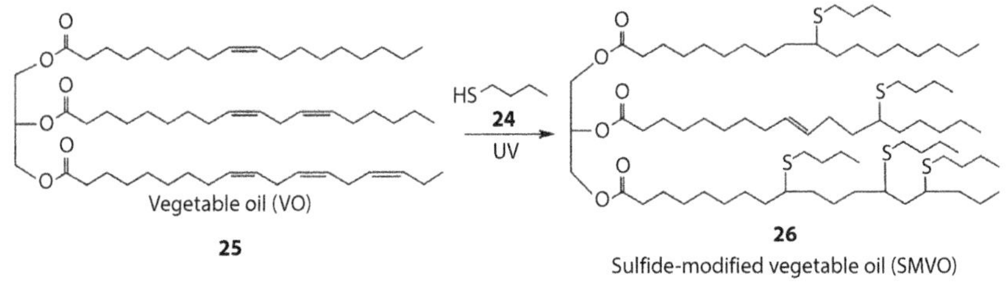

Vegetable oil (VO)

25

Sulfide-modified vegetable oil (SMVO)

26

FIGURE 15.26 Free radical reaction of vegetable oil with butanethiol. (From Bantchev, G.B. et al., *J. Agric. Food Chem.*, 57, 1282, 2009.)

TABLE 15.12

Properties of Neat Sulfide-Modified (SM) Vegetable Oils and Their Precursors[a]

	Method	Corn, 25[b]	SM-Corn, 26	Canola, 25	SM-Canola, 26
Density (g/cm³)	D-7042				
40°C				0.9052	0.9362
100°C				0.8655	0.8972
kVisc (mm²/s)[a]	D-7042				
40°C		32.4	124	36.7	179.5
100°C		7.7	17.7	8.3	23
VI	D-2270	220	158	212	155
Cold flow					
Cloud point (°C)[a]	D-2500	−1	0	−7	<−33
Pour point (°C)[a]	D-97	−12	−21	−21	−33
Oxidation stability					
RPVOT (min)[a]	D-2272	12.5 ± 0.7	100 ± 7	14.5 ± 0.7	204
PDSC-OT (°C)	D-6186	169.4 ± 0.2	210.0 ± 1.6	144.7 ± 1.0	200.0 ± 1.0
PDSC-PT (°C)	D-6186	179.4 ± 0.6	228.3 ± 1.6	162.9 ± 0.9	224.6 ± 1.6

[a] Data from Reference 68.
[b] See Figure 15.26 for structures.

an improved oxidation stability was observed in PAO-6 base oil, which showed an increase in the RPVOT time from 22 to 311 min with the addition of 6.2 % w/w SM corn oil. Similarly, addition of 6.3% w/w SM canola to PAO-6 increased the RPVOT time from 22 to 249 min (Table 15.13)

PDSC evaluations were conducted in blends of HOSuO base oil with 0%–10% SM corn, canola, and high-oleic sunflower oils. As shown in Table 15.13, all blends showed improvement in oxidation stability (higher OT and PT temperatures) relative to the base oils. Similar improvements were observed for blends with 1%–5% w/w of the additives, while a much higher improvement occurred by all the additives at 10% w/w concentrations. Thus, based on this investigation, the SM vegetable oils can be used as biobased antioxidant additives in lubricant formulation.

15.6.2 LIPOYL GLYCERIDE BIOBASED ANTIOXIDANTS

Lipoyl glycerides (LG, **29–32** in Figure 15.27) are biobased oils with cyclic disulfide moiety in their structure, which can potentially provide them with antioxidant properties. Lipoyl glycerides were synthesized by transesterification reaction between high-oleic sunflower oil (**27**, HOSuO; Columbus Foods, Des Plaines, IL) and lipoic acid (**20**; Sigma-Aldrich, St. Louis, MO) in 2-methyl-2-butanol solvent (Sigma-Aldrich, St. Louis, MO) as depicted in Figure 15.27 [71]. The synthesis was conducted in the presence of a lipase enzyme catalyst (Novozym 435, *Candida antarctica* lipase B immobilized on acrylic beads; Novozymes North America Inc., Franklinton, NC, USA), at 60°C for 24 h. After isolation and purification, the synthesis resulted in 89% yield of the

TABLE 15.13

Evaluation of the Antioxidant Properties of Sulfide-Modified (SM) Vegetable Oils, 26, in PAO-6, Corn Oil, and HOSuO Base Oils Using RPVOT[a] and PDSC[b] Methods

	RPVOT Results[c]			
Additive	SM-Corn[d]	SM-Corn	SM-Canola	SM-Canola
Base Oil	Corn Oil	PAO-6	Corn Oil	PAO-6
[Additive], % w/w	RPVOT (min)[c]			
0	12.5 ± 0.7	22.0 ± 0.0	12.5 ± 0.7	22.0 ± 0.0
6.2	13.0 ± 0.0	311 ± 44		
6.3			13	249
100	100 ± 7	100 ± 7	204	204

	PDSC Results			
Additive		SM-Corn	SM-Canola	SM-HOSuO
Base Oil		HOSuO	HOSuO	HOSuO
[Additive], % w/w		OT and PT (°C)		
0	OT, °C	181.6 ± 0.9	181.6 ± 0.9	181.6 ± 0.9
	PT, °C	195.3 ± 0.4	195.3 ± 0.4	195.3 ± 0.4
1	OT, °C	184.7 ± 1.2	185.3 ± 2.7	186.1 ± 1.3
	PT, °C	195.3 ± 0.4	198.1 ± 1.8	199.1 ± 0.8
3	OT, °C	187.8 ± 0.3	188.6 ± 1.4	185.8 ± 0.1
	PT, °C	200.7 ± 0.8	200.9 ± 1.9	198.6 ± 0.1
5	OT, °C	188.9 ± 1.8	189.0 ± 0.4	187.5 ± 0.3
	PT, °C	200.4 ± 2.8	200.9 ± 0.4	200.6 ± 0.1
10	OT, °C	191.6 ± 0.2	192.4 ± 0.6	191.5 ± 2.1
	PT, °C	200.3 ± 0.3	202.1 ± 0.8	202.9 ± 2.3
100	OT, °C	210.0 ± 1.6	200.0 ± 1.0	230.7 ± 0.2
	PT, °C	228.3 ± 1.6	224.6 ± 1.6	245.0 ± 0.5

[a] RPVOT, ASTM D-2272.
[b] PDSC, ASTM D-6186.
[c] RPVOT data from References 68.
[d] See Figure 15.26 for structures of SM vegetable oils.

crude lipoyl glyceride (LGc) product mixture, which, in addition to several mono- and di-lipoate glycerol products, also contained unreacted HOSuO (**27**), unreacted lipoic acid (**20**), and free fatty acids (**28**). The crude product mixture was further purified by removing the polar components (lipoic acid and free fatty acids) by extraction with methanol to produce a purified lipoyl glyceride fraction (LGp) in 61% isolated yield. The components in the product mixtures LGc and LGp were positively identified by liquid chromatography–high-resolution mass spectrometry (LC-HRMS) analysis [71].

The properties of neat lipoyl glycerides, crude (LGc) and purified (LGp) product mixtures, are compared in Table 15.14. The top part of Table 15.14 lists the

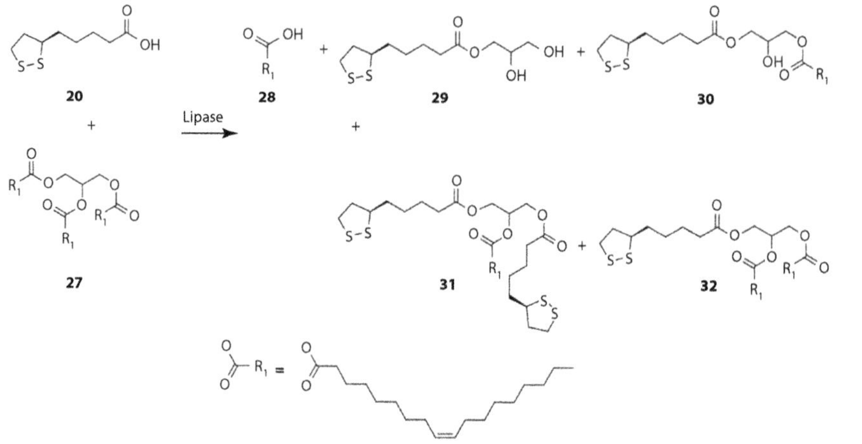

FIGURE 15.27 Enzymatic transesterification between lipoic acid and HOSuO.

components of the product mixtures, their molecular weights and relative quantities (% w/w) determined using LC-HRMS analysis. The composition of the two product mixtures has many common features. The major component in both product mixtures is unreacted HOSuO, **27** (45%–50% w/w), followed by lipoyl dioleoylglycerol, **32** (23%–41% w/w). They both contain small quantities (1%–6% w/w) of three different lipoate ester structures (**29–31**) from transesterification of lipoic acid with glycerol. The major difference in the composition between LGc and LGp is the presence of a large quantity (total of 22% w/w) of two polar compounds, lipoic acid (**20**) and free fatty acid (**28**), in LGc but none in LGp, which causes LGc to be more polar than LGp.

The bottom half of Table 15.14 compares the physical properties of the neat lipoyl glyceride product mixtures LGc and LGp, to each other and to the parent HOSuO. As expected, the polar product mixture LGc displayed poor solubility, whereas LGp displayed good solubility, in the nonpolar base oil PAO-6. Both product mixtures displayed a higher density at 40°C and 100°C than HOSuO, which can be attributed to the presence of disulfide in their structures. At 40°C, both product mixtures displayed a slightly higher viscosity than HOSuO, whereas at 100°C, LGc displayed lower viscosity, while LGp displayed higher viscosity than HOSuO. Both product mixtures displayed lower VI than HOSuO, which can be attributed to their complex composition relative to HOSuO. They also displayed lower pour point than HOSuO, which can be attributed to the replacement of one or more of the long-chain (C$_{18}$) oleate in HOSuO by the shorter (C$_8$) lipoate in the product mixture. In addition, the complex mixture in LGc and LGp will favor the depression of freezing point. The PDSC data in Table 15.14 show that LGc and LGp display much greater oxidation stability than HOSuO. Neat LGc and LGp product mixtures displayed similar onset temperature (OT) and peak temperature (PT) values that were 30°C to 40°C above the values for the neat HOSuO.

The antioxidant additive properties of lipoyl glycerides were investigated in HOSuO base oil. Blends with 0%–10% w/w LGc and LGp in HOSuO were prepared and their oxidation stability determined using PDSC according to ASTM D-6186

TABLE 15.14

Properties of Crude (LGc) and Purified (LGp) Neat Lipoyl Glycerides[a]

Product Mixture Composition (LC-HRMS)[b]

Component[c]	Mol. Wt. (g/mol)	Composition (% w/w)	
		LGc	LGp
Lipoic acid, **20**	206	4.0	0.0
Free fatty acids, **28**	282	18.0	0.0
Lipoyl glycerol, **29**	280	1.3	1.0
Lipoyl monooleoylglycerol, **30**	544	5.0	2.0
Dilipoyl oleoylglycerol, **31**	733	4.0	6.0
Lipoyl dioleoylglycerol, **32**	809	22.8	41.0
HO sunflower oil, **27**	885	45.0	50.0

Physical Properties

	Method	HOSuO	LGc	LGp
RT solubility in PAO-6, % w/w	Visual		<0.4	>11.0
Density at 40°C (g/cm³)	D-7042	0.8994	0.9284	0.9274
Density at 100°C (g/cm³)	D-7042	0.8683	0.8902	0.8875
kVisc at 40°C (mm²/s)	D-7042	40.77	42.18	45.96
kVisc at 100°C (mm²/s)	D-7042	8.72	8.35	9.12
VI	D-2270	200	178.5	185.0
Cloud point (°C)	D-2500		-3.5 ± 0.3	-1.4 ± 0.2
Pour point (°C)	D-97	-12^d	-17.7 ± 0.6	-16.3 ± 0.6
PDSC-OT (°C)	D-6186	181.6 ± 0.9	223.8 ± 0.8	222.4 ± 0.7
PDSC-PT (°C)	D-6186	195.3 ± 0.4	242.7 ± 0.7	237.7 ± 3.2

[a] Data from Reference 71 unless stated otherwise.

[b] LC-HRMS, liquid chromatography–high-resolution mass spectrometry.

[c] See Figure 15.27 for structures.

[d] Data from Reference 22.

procedure [70]. The resulting OT and PT are summarized in Table 15.15. The addition of LGc and LGp to HOSuO increased both OT and PT by more than 10°C–15°C relative to neat HOSuO. Both OT and PT displayed a slight increase with a further increase of LGc and LGp concentration to 10% w/w and were 20°C–30°C below the value observed for neat LGc and LGp. This indicates that further improvement in oxidation stability of HOSuO can be attained by increasing the concentrations of LGc and LGp above the 10% w/w values. This work indicates that crude and purified lipoyl glycerides are effective as antioxidant additives in lubricant formulations.

15.6.3 LIPOATE ESTER BIOBASED ANTIOXIDANTS

Lipoate esters were obtained by a condensation reaction of lipoic acid with a variety of alcohols, without solvent and in the presence of a solid catalyst as depicted in

TABLE 15.15

PDSC Evaluation (ASTM D-6186) of the Antioxidant Properties of Crude (LGc) and Purified (LGp) Lipoyl Glycerides in HOSuO Base Oils (°C)[a]

	Base Oil Additive[b]	HOSuO LGc (20, 27–32)	HOSuO LGp (27, 29–32)
[Additive], % w/w			
0	OT, °C	181.6 ± 0.9	181.6 ± 0.9
	PT, °C	195.3 ± 0.4	195.3 ± 0.4
5	OT, °C	199.2 ± 0.0	197.3 ± 0.0
	PT, °C	215.6 ± 0.8	214.9 ± 0.0
10	OT, °C	201.9 ± 0.2	200.3 ± 0.0
	PT, °C	216.3 ± 0.4	217.4 ± 0.0
100	OT, °C	223.8 ± 0.8	222.4 ± 0.7
	PT, °C	242.7 ± 0.7	237.7 ± 3.2

[a] Data from Reference 71.
[b] See Figure 15.27 for structure of additive components.

Figure 15.24 [61]. In some cases, the synthesis resulted in the production of a small quantity of polymeric by-products, which had a profound impact on the viscosity and VI of the materials (Table 15.11). Details of the synthesis, purification, and identification procedures are given in Section 15.5.

Four of the lipoate ester mixtures (**22c,d,e,g** and **23c,d,e,g**) that displayed acceptable solubility in biobased (HOSuO) and petroleum-based (PAO-6) base oils were further characterized for the physical property of the neat product mixture, and the resulting data are summarized in Table 15.11. Further investigation into the additive property showed that three of the four lipoate esters investigated had good VII properties in HOSuO base oil, which was attributed to the effect of the high molecular weight by-products in the product mixtures (Figure 15.25). The neat lipoate ester product mixtures were also evaluated for their oxidation stability by PDSC (ASTM D-6186), and the results, relative to neat HOSuO, are presented in Table 15.16. It is clear from the data in Table 15.16 that the lipoate ester product mixtures have much greater oxidation stability than HOSuO, with OT and PT values 40°C–50°C higher than neat HOSuO. The higher oxidation stability of lipoate esters relative to HOSuO can be attributed to the presence of sulfur in their structures, which is known to provide antioxidant properties [65,66]. Another reason for the large difference in the oxidation stability between HOSuO and lipoate esters is the poor oxidation stability of HOSuO due to the presence of double bonds in its structure. The allylic and bis-allylic protons in the HOSuO structure are highly reactive and cause it to display poor oxidation stability.

The four lipoate ester product mixtures were further evaluated for their antioxidant additive properties in HOSuO base oil using PDSC. Blends of HOSuO with

TABLE 15.16

Comparison of the Onset (OT) and Peak (PT) Oxidation Temperatures of Lipoate Esters and HOSuO Determined Using PDSC (ASTM D-6186)[a]

	PDSC Oxidation Temperatures (°C)	
Oil[b]	OT	PT
HOSuO, **27**	187.2 ± 0.6	200.0 ± 0.2
2-Ethylhexyl lipoate, **22d, 23d**	226.0 ± 0.8	247.3 ± 0.6
Octyl lipoate, **22e, 23e**	226.1 ± 0.4	247.9 ± 0.1
Dodecyl lipoate, **22f, 23f**	227.8 ± 0.2	253.2 ± 0.8
Isostearyl lipoate, **22g, 23g**	228.5 ± 0.5	249.8 ± 0.6

[a] Data from Reference 61.

[b] See Figure 15.24 for structure of lipoate esters.

0%–20% w/w lipoate esters were prepared, and their onset (OT) and peak (PT) oxidation temperatures measured. The results, summarized in Figure 15.28, show that the presence of any of the four lipoate ester product mixtures in the blend increased the OT and PT values of HOSuO. Furthermore, the OT and PT values increased with increasing concentration of the lipoate ester product mixtures. Thus, it can be concluded that lipoate esters are effective antioxidant additives in lubricant formulations.

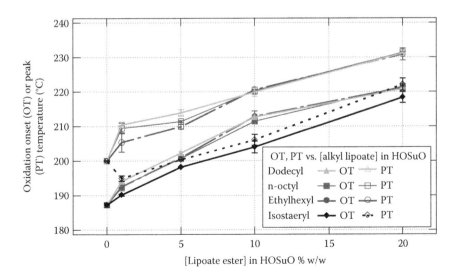

FIGURE 15.28 Onset (OT) and peak (PT) oxidation temperatures from evaluation of the antioxidant additive properties of alkyl lipoate esters in HOSuO blends using PDSC (ASTM D-6186). (From Biresaw, G. et al., *Ind. Eng. Chem. Res.*, 55, 373, 2016.)

15.7   BIOBASED EXTREME PRESSURE ADDITIVES

Extreme pressure (EP) additives are essential to allow the lubricant to be used in applications that occur in extreme conditions of load, speed, and temperatures [1–3]. Conducting such extreme processes without EP additives in the lubricant can lead to process failure, equipment damage, and possibly also injury to operators.

EP additives are organic and organometallic compounds containing elements such as sulfur, phosphorous, and halogens in their structures [1–3]. It is generally assumed that, under the severe conditions of the tribological process, these elements undergo tribochemical reactions with the friction surfaces and produce new tribomaterials. These new materials, produced *in situ* in the friction zone, provide acceptable friction and wear performance for the tribological process to proceed without causing excessive wear, friction, or damage to equipment.

Current EP additives in the market are almost exclusively petroleum based. As a result, biobased lubricant formulations for hydraulic and other applications are formulated with petroleum-based EP additives. Thus, there is a need and a vast untapped market for biobased EP additives in the lubricant economy. In order to tackle this challenge, our group has synthesized and investigated a number of biobased EP additives with phosphorous and sulfur in their structures. The investigations into structure of each biobased EP additive are separately discussed in the following text.

EP properties were investigated using a four-ball tribometer (Model KTR-30L; Koehler Instruments, Bohemia, NY) in accordance with ASTM D-2783 [72]. In this procedure, the test lubricant is used in a series of 10-s four-ball tests conducted at 1760 ± 40 rpm, room temperature, and progressively increasing load until welding of the four balls occurs (see Figure 15.19). Each 10-second test is conducted using fresh test lubricant and fresh clean balls secured in a clean ball pot. The load at which welding occurs is reported as the characteristic EP weld point for the test lubricant. EP additives that provide high weld point are preferred.

15.7.1   Methyl Oleate–Based Dialkyl Phosphonate Biobased EP Additives

These phosphonate derivatives (**15a, b, c** in Figure 15.18) were obtained by reacting methyl oleate (**8d** in Figure 15.18) with dialkyl phosphites). Dimethyl, diethyl, and di-*n*-butyl phosphites (**14a, b,** and **c** in Figure 15.18) were used in the synthesis. The general synthesis scheme and the structures of the dialkyl phosphonates derived from methyl oleate are shown in Figure 15.18. Details of the experimental procedure are given in Section 15.4.1. Selected physical properties of these materials are summarized in Table 15.9. Further investigation showed that these dialkyl phosphonates displayed antiwear properties that were comparable to ZDDP, **12**, in biobased base oils (Figures 15.22 and 15.23).

The results of EP investigation of the dialkyl phosphonate derivatives, along with those of the precursor methyl oleate and various base oils, are summarized in Table 15.17. As shown in Table 15.17, neat methyl oleate and its neat bisphosphonate derivatives displayed a weld point of 120 kgf, which was similar to the value reported for the neat biobased and petroleum-based base oils [22,55,73]. Thus,

TABLE 15.17

Four-Ball Extreme Pressure Weld Point (ASTM D-2783), in kgf, of Neat Dialkyl Phosphonate Derivatives of Methyl Oleate, and Blends in Biobased and Petroleum-Based Base Oils[a]

				Methyl Oleate	Dialkyl Phosphonate of Methyl Oleate			
	Base Oils							
	PAO-6	150 N[c]	Soy[e]	8d	Dimethyl, 15a	Diethyl, 15b	Di-*n*-butyl, 15c	ZDDP[b]
Neat oils	120	120	120	120	120	120	120	
Blends, 5 % w/w in								
PAO-6						120	120	130
150 N[c]								150[d]
Soy[e]						140	140	250[d]

[a] Weld point data from Reference 55 unless stated otherwise. See Figure 15.18a for the structure of dialkyl phosphonates.

[b] ZDDP, **12**, zinc dialkyldithiophosphate.

[c] 150 N, solvent-refined highly paraffinic mineral oil with a kinematic viscosity of 30 mm^2/s at 40°C (Reference 73).

[d] Weld point data from Reference 73.

[e] Soy, RBD soybean oil with a kinematic viscosity of 31 mm^2/s at 40 °C (Reference 22).

insertion of phosphorous atom into the structure of methyl oleate had no effect on the EP property of the neat oils. Dialkyl phosphonates were also investigated for their EP additive property in various base oils. The EP property of blends with 5% w/w dialkyl phosphonates in PAO-6 and soybean oil was investigated. The results, which are summarized in Table 15.17, show no change in weld point for PAO-6 blends but a slight increase of weld point for soybean blends. So, it appears that dialkyl phosphonates can have an EP additive property in the polar biobased base oil but not in the nonpolar petroleum based. Such a positive effect of polar base oils on the weld point of EP additives has been reported before [73] and can be attributed to improved compatibility between the polar dialkyl phosphonates additives and the polar base oil.

Table 15.17 shows that blends with 5% w/w ZDDP displayed a slightly higher weld point than those with the 5% w/w dialkyl phosphonates. The data also show that the weld point of blends with 5% w/w ZDDP increased with increased polarity of the base oils. Thus, the weld point of 5% w/w ZDDP blends and the polarity of the base oils increased in the order: PAO-6 < 150 N < soybean.

Based on the data in Table 15.17, it can be concluded that dialkyl phosphonates of methyl oleate show some EP additive property in biobased polar base oils. However, they show no EP properties in neat or in blends with nonpolar petroleum-based base oils.

15.7.2   LIMONENE BIS(DIALKYL PHOSPHONATES) AS BIOBASED EP ADDITIVES

Bis(dialkyl phosphonates) of limonene, **19**, were obtained by free radical reaction of limonene, **16**, with dialkyl phosphites, **14**, as outlined in Figure 15.18b [57]. Phosphites of methyl, ethyl, and *n*-butyl were used in the synthesis. Since limonene has two double bonds, the reaction is capable of providing a product mixture comprising mono- (**17, 18**) and diadduct (**19**) products. However, as detailed in Section 15.4.1, the reaction was optimized to produce only the diadduct bis(dialkyl phosphonate) product (**19a,b,c** in Figure 15.18b) in high yield. Table 15.10 provides a summary of the physical properties of the three limonene bis(dialkyl phosphonate) products. Tribological investigation showed that bis(dialkyl phosphonates) display comparable antiwear properties to ZDDP in biobased base oils (Figure 15.23b).

The EP properties of neat limonene bis(dialkyl phosphonates) were investigated using a four-ball tribometer (ASTM D-2783), and the results are summarized in Table 15.18. The investigation showed that the bis(dialkyl phosphonates) of limonene do display some EP property that is dependent on the nature of the alkyl group. Thus, the highest weld point was observed for the neat bis(dimethylphosphonate), **19a**, whereas lower but similar weld point was observed for the diethyl (**19b**) and di-*n*-butyl (**19c**) materials.

TABLE 15.18

Four-Ball Extreme Pressure Weld Point (ASTM D-2783), in kgf, of Neat Bis(dialkyl phosphonate) Derivatives of Limonene, and Blends in Biobased and Petroleum-Based Base Oils[a]

		Weld Point (kgf)
Neat base oils	PAO-6	120[a]
	150 N	120[b]
	HOSuO	120[c]
	Soybean	120[b]
Neat phosphonate additives[d]	Dimethyl, **19a**	260[e]
	Diethyl, **19b**	200[e]
	Di-*n*-butyl, **19c**	200[e]
5% w/w blends of ZDDP in	PAO-6	130[a]
	150 N	150[b]
	Soybean	250[b]

[a] Data from Reference 57.
[b] Data from Reference 73 in soybean base oil.
[c] Data from Reference 71.
[d] See Figure 15.18b for the structure of limonene bis(dialkyl phosphonate).
[e] Data from Reference 57.

15.7.3 SULFIDE-MODIFIED VEGETABLE OILS AS BIOBASED SULFURIZED EP ADDITIVES

SM vegetable oils, **26**, were obtained by free radical reaction of butanethiol, **24**, with vegetable oils (**6, 25**). The schematics of the synthesis and the structure of the products are given in Figure 15.26. SM corn and SM canola oils were synthesized in good yield and their properties investigated. SM corn and SM canola oils displayed highly improved oxidation and cold flow properties relative to the unreacted vegetable oils (Tables 15.12 and 15.13).

Results from a four-ball EP investigation (ASTM D-2783 [68]) of the SM vegetable oils are summarized in Table 15.19. Neat SM corn and SM canola oils displayed a slightly higher weld point than the biobased corn oil and the petroleum-based base oil PAO-6. SM corn oil was further investigated as an EP additive in corn and PAO-6 base oils. Blending SM corn oil resulted in an increase of the weld points for both base oils. Weld points were higher for corn oil blends than for PAO-6 blends, and increased with increasing SM corn in both blends. Table 15.19 also compares EP weld point data for a commercial pentasulfide EP additive TPS-32, **33** in Table 15.21 (di-*t*-dodecyl pentasulfide, obtained from Arkema Inc., Philadelphia, PA). The data in Table 15.19 show that the commercial polysulfide EP additive displayed a weld point that was almost six-fold higher than that observed for SM vegetable oils. Thus, more work is needed to bring the performance of the biobased SM EP additives to be competitive with the petroleum-based commercial products currently in the market.

TABLE 15.19
Four-Ball Extreme Pressure Weld Point (ASTM D-2783), in kgf, of Neat Sulfide-Modified Corn (SM-Corn), Canola (SM-Canola), and Blends in Corn Oil and PAO-6 Base Oils[a]

| | | Four-Ball Extreme Pressure Weld Point, kgf | | |
| | | Sulfide-Modified Vegetable Oils[b], 26 | | |
[Additive] (w/w %)	Base Oil	SM-Corn	SM-Canola	TPS-32[c], 33
0 (neat base oil)	PAO-6	120	120	120
0 (neat base oil)	Corn	120	120	120
2	PAO-6	130		740
2	Corn	140		800
6.2	PAO-6	140		
6.2	Corn	150		
100 (neat additive)	None	140	140	

[a] Data from Reference 68 unless stated otherwise.

[b] See Figure 15.25 for structures of SM vegetable oils.

[c] TPS-32 is a commercial sulfurized EP additive, di-*t*-dodecyl pentasulfide, obtained from Arkema Inc., Philadelphia, PA.

15.7.4 Lipoyl Glycerides Biobased Sulfurized EP Additives

Product mixtures containing lipoyl glycerides (LGs) (**20, 27–32** in Figure 15.27) were obtained via enzymatic transesterification reaction between lipoic acid (**20**) and high-oleic sunflower oil (**27**). The synthesis scheme along with the structures of the products from the synthesis is shown in Figure 15.27. The reaction results in a crude product mixture (LGc) of at least seven distinct compounds (Table 15.14), whose structures contain one or more of the following: lipoic acid, **20**; fatty acid (**28**, almost exclusively oleic acid, **7b**); and glycerol (Figure 15.27 and Table 15.14). A purified product mixture (LGp), with at least five different compounds, was also obtained by removing the polar components in LGc, namely, the free fatty acids, **28**, and unreacted lipoic acid, **20**. LGc and LGp were thoroughly characterized and used in tribological evaluations. Tables 15.14 and 15.15 provide a summary of the neat properties of LGc and LGp and their additive properties in HOSuO base oil.

The four-ball EP properties of neat LGc and LGp were evaluated. In addition, LGc was investigated as an EP additive in HOSuO, soybean, and PAO-6 base oils. The resulting weld point values are summarized in Table 15.20 along with data for the commercial EP additive TPS-32 (**33** in Figure 15.21) [68,74]. Neat LGc and LGp displayed weld points that were much higher than those observed with the phosphonate or SM additives discussed earlier. LGc was found to be a more effective product mixture than LGp and gave a weld point value of 590 kgf, as opposed to 400 kgf for the LGp product mixture (Table 15.20).

As an additive, LGc displayed weld points that were slightly higher than the value for the pure base oils, and much lower than the values obtained with TPS-32. The observed weld points of LGc and TPS-32 blends were functions of base oil polarity. Thus, the weld points of LGc and TPS-32 were lower for the blends in the petroleum-based PAO-6 and 150 N base oils than in the biobased soybean or HOSuO base oils. In addition, blends of LGc and TPS-32 in soybean or HOSuO base oils displayed weld point values that increased with increasing concentration of EP additive in the blend.

Even though LGc and TPS-32 displayed a similar effect as a function of concentration and base oil polarity, more work is needed to improve the EP performance of the biobased LGc to the level of the petroleum-based TPS-32 EP additive.

15.7.5 Lipoate Ester Biobased Sulfurized EP Additives

Lipoate esters with polymeric residues (**22a–g, 23a–g** in Figure 15.24) were obtained by the direct esterification of lipoic acid with a wide range of alcohol structures in the presence of a solid catalyst. The schematics of the synthesis procedure and the structures of the lipoate esters synthesized are shown in Figure 15.24. The lipoate esters were obtained in good yield, and in some cases, the product mixture contained a small amount of polymeric by-product (**23a–g**) that had a profound effect on the viscosity and VI of the product mixture (Table 15.11 and Figure 15.25). Details of the synthesis, identification, and characterization of the lipoate esters are given in Section 15.5. Some properties of the lipoate esters are summarized in Tables 15.11 and 15.16 and in Figures 15.25 and 15.28.

TABLE 15.20

Four-Ball Extreme Pressure Weld Point (ASTM D-2783), in kgf, of Crude (LGc) and Purified (LGp) Lipoyl Glyceride of HOSuO, and Their Blends in Soybean, HOSuO, and PAO-6 Base Oils[a]

| | | Four-Ball Extreme Pressure Weld Point, kgf | | |
| | | Lipoyl Glyceride Product Mixture (20, 27–32)[f] | | |
[Additive] (w/w %)	Base Oil	Crude (LGc)	Purified (LGp)	TPS-32[b], 33
0 (neat base oil)	Soybean	120	120	120
0 (neat base oil)	HOSuO	120	120	120
0 (neat base oil)	PAO-6	120	120	120
2	Corn			800[c]
2	PAO-6			740[c]
5	Soybean	200		942[d]
5	HOSuO	180		
5	PAO-6	170		
5	150 N[e]			590[d]
10	Soybean	210		
10	HOSuO	210		
20	Soybean			981[d]
20	150 N[e]			500[d]
100 (neat additive)	None	590	400	

[a] Data from Reference 71 unless stated otherwise.

[b] TPS-32 is a commercial sulfurized EP additive, di-t-dodecyl pentasulfide, obtained from Arkema Inc., Philadelphia, PA.

[c] Data from Reference 68.

[d] Data from Reference 74.

[e] 150 N, solvent-refined highly paraffinic mineral oil with a kinematic viscosity of 30 mm^2/s at 40°C (Reference 73).

[f] See Figure 15.27 for the structure of LG components.

Selected lipoate esters that displayed acceptable solubility in HOSuO (>10% w/w) were investigated for their EP additive properties using the four-ball EP procedure (ASTM D-2783 [72]). The results of the investigation are summarized in Figure 15.29, along with data for the commercial polysulfide EP additive TPS-32 (**33** in Table 15.21). The data in Figure 15.29 show that lipoate esters have EP additive properties and improve the weld point of HOSuO. For all four lipoate esters investigated, the weld point increased with increasing concentration of lipoate ester, to up to four-fold the weld point of the neat HOSuO at 3.4% w/w concentration. These values, however, were about half of what has been observed with the commercial polysulfide EP additive TPS-32 [68,74]. Thus, more work is needed to bring the performance of biobased EP additives to the level currently possible with the commercial petroleum-based EP additives.

TABLE 15.21

Representative Structures of Sulfurized EP Additives Investigated in this Work

Name-1	Name-2	x in S_x	Representative Structure
Vegetable oil	Soybean Corn HOSuO	0	**6a**
SM-Veg oil (SM, sulfide modified)	SM-corn SM-canola	1	**26**
Lipoyl glycerides (LG)	LGc LGp	2	**31**
Lipoate ester	2-ethylhexyl lipoate n-octyl lipoate n-dodecyl lipoate i-octadecyl lipoate	2	**23d**
TPS-32	di-t-dodecyl pentasulfide	5	**33** $(t\text{-}C_{12}H_{25})\text{-}S_5\text{-}(t\text{-}C_{12}H_{25})$

15.7.6 STRUCTURE–PROPERTY CONSIDERATIONS IN SULFURIZED EXTREME PRESSURE ADDITIVES

The base oils and sulfurized EP additives discussed in this section differ from each other in various structural aspects such as molecular weight, type, number of functional groups, degree and location of unsaturation, extent and complexity of branching, etc. They also differ in the composition of the sulfur atom in their structures, which is considered to be responsible for their EP properties. Thus, for the purpose of correlating EP properties, it is possible to ignore all aspects of structural differences except those related to sulfur. These materials can be categorized based on the structure of the sulfur atoms in the molecules, which can be represented by a generalized structure as follows:

$$R - S_x - R'$$ (15.16)

where x = 0, 1, 2, ...

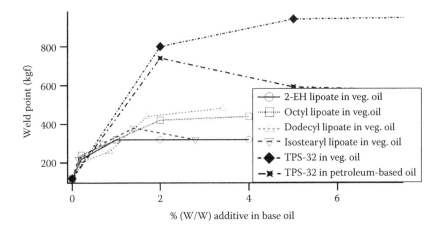

FIGURE 15.29 Four-ball extreme pressure weld point (ASTM D-2783), in kgf, of blends of biobased (lipoate esters) vs. petroleum based (TPS-32), sulfurized extreme pressure additives in vegetable oil and petroleum base oils. (From Biresaw, G. et al., *Ind. Eng. Chem. Res.*, 55, 373, 2016; Biresaw, G. et al., *Tribol. Lett.*, 43, 17, 2011; Biresaw, G. et al., *Ind. Eng. Chem. Res.*, 51, 262, 2012.)

The structure with x = 0 corresponds to the base oil, which contains no sulfur in its structure. The other structures investigated contain 1, 2, and 5 consecutive sulfur atoms in their structures. Table 15.21 is a summary of the structures of the sulfurized EP additives discussed in this work.

An examination of the weld point data for the neat EP additive structures listed in Table 15.21 shows some correlation between weld point and the number of consecutive sulfur atoms (x) in their structures. This correlation is illustrated in Figure 15.30, where the number of consecutive sulfur atoms in the structure, x, which varies in the range of 0–5, is plotted against the weld point of the neat EP additives. The points in Figure 15.30 are the measured weld point data, whereas the line corresponds to the best fit value obtained from linear regression analysis of the measured data. It is clear from Figure 15.30 that good correlation exists (correlation coefficient = 0.96) between the weld point and the number of consecutive sulfur atoms in the structure. This suggests that the mere presence of sulfur atoms is not sufficient to produce good EP property in the molecule. The sulfur atoms must have a specific structure which, in this case, is that they must all be connected to each other to produce the best EP outcome. If the sulfur atoms were scattered at different points in the molecule, poor EP property would be observed. In addition, the data strongly suggest that improvement in EP property, that is, larger weld point, is attained as the number of consecutive sulfur atoms in the structures increases.

EP property is related to the tribochemical reaction of the EP additive with the metallic friction surfaces to produce *in situ* new materials with good lubrication property. This process involves bond breaking in the EP additive to avail the sulfur atoms for reaction with the friction surfaces. The bond dissociation energy of the S—S

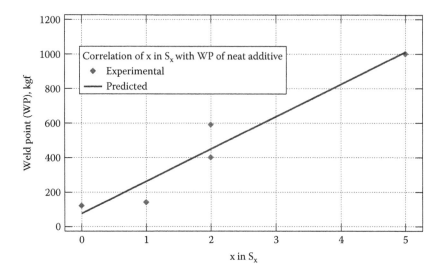

FIGURE 15.30 Effect of EP additive sulfur composition on weld point.

is lower than C—S in both diatomic and polyatomic molecules [75] (Table 15.22). Thus, during a tribochemical reaction, the S—S bond is easier to break and participate in the tribochemical process than the C—S bond. This suggests that EP additives with multiple S—S bonds can easily make the sulfur atoms available for tribochemical reactions, thereby producing superior EP performance. This can explain the correlation between weld point and the number of S—S bonds in an EP additive molecule

TABLE 15.22

Reported Bond Dissociation Energies (BDE) between Atoms in Diatomic and Polyatomic Molecules with EP Property (kJ/mol)[a]

	BDE (kJ/mol) in Molecules That Are			BDE (kJ/mol) in Molecules That Are	
Bond	**Diatomic[b]**	**Polyatomic[c]**	**Bond**	**Diatomic[b]**	**Polyatomic[c]**
C—S	713	301–312	C—C	618	323–377
S—S	425	273–277	C—Cl	395	300–360
S-Zn	225		C—Br	318	293–300
S—P	442		C—I	253	227–237

[a] Data from Reference 75.
[b] Data from Table 15.1 of Reference 75. Bond dissociation energies of diatomic molecules.
[c] Data for simple polyatomic hydrocarbons (without multiple bonds, aromatics, functional groups, halogens, etc.) from Table 15.3 of Reference 75. Bond dissociation energies of polyatomic molecules.

observed in Figure 15.30. Lower bond dissociation energy relative to C—C is also responsible for the effective EP characteristic of halogenated paraffin and other EP additives (Table 15.22).

15.8   SUMMARY AND CONCLUSIONS

The formulation of fully biobased lubricants requires using biobased base oils and biobased additives. While the base oil is selected mainly to provide an appropriate film thickness during lubrication, the additive is selected mainly to provide one of many specific properties, for example, corrosion resistance, to the formulation. Thus, several additives will be required to ensure the formulation has acceptable properties in many critical areas (e.g., oxidation, cold flow, bioresistance, etc.). Thus, depending on the intended application of the lubricant, it might be necessary to blend 10 or more different additives in the formulation.

Currently, there are several biobased base oil options in the market. These include base oils derived from vegetable oils (regular or genetically modified), synthetic derivatives of vegetable oils, and enzymatically produced base oils from sugar-based feedstock. However, currently there are no biobased additives in the market that can be used with the biobased base oils to formulate fully biobased lubricants. As a result, lubricant formulators are left with no choice but to use the commercially available petroleum-based additives and produce partially biobased formulations.

In order to remedy this problem, and to help with the formulation of fully biobased lubricants, a number of groups are engaged in the development of various biobased additives. This chapter is a summary of the work in the authors' group related to the synthesis and investigation of a number of biobased additives including friction modifiers, biosurfactants, antiwear additives, VIIs, antioxidants, and EP additives. The following is a summary of the results from these investigations.

- *Biobased fiction modifiers.* It was demonstrated that vegetable oils and their derivatives can be used as effective biobased friction modifiers in lubricant formulations. Vegetable oils and their derivatives have amphiphilic properties and can adsorb on friction surfaces and interfaces to reduce surface tension, interfacial tension, boundary friction, and wear. The free energy of adsorption of vegetable oils and their derivatives was quantified and correlated with their structures.
- *Biosurfactants.* Polysoaps were synthesized from soybean oil and epoxidized soybean and their surface and interfacial (oil–water) properties investigated. In addition, the surface energy of the polysoaps with the various counterions was quantified and related to their structures. Polysoaps with sodium, potassium, and triethanol ammonium counterions were found to be effective biosurfactants for the solubilization of oil in water and thereby allow for the formulation of water-based biobased lubricants for use in metalworking, hydraulic, and other applications.
- *Biobased antiwear additives.* Dialkyl phosphonates of methyl oleate and bis(dialkyl phosphonates) of limonene were found to be effective antiwear

additives in biobased base oils such as high-oleic sunflower oil. The antiwear performance of the biobased phosphonates in biobased base oils was
comparable to that of the commercial antiwear additive zinc dialkyldithiophosphate. The phosphonates were synthesized, their structures positively
identified using a combination of several analytical methods, and their
physical and tribological properties investigated.

- *Biobased viscosity index improvers.* Oligomers of lipoate esters were
  found to be effective VIIs in the biobased base oil high-oleic sunflower oil
  (HOSuO). These oligomers were by-products from the synthesis of lipoate
  esters from lipoic acid and several alcohols, comprising 3%–20% of the
  product mixture. The presence of the oligomers resulted in sharp increases
  in viscosity (>1000-fold) and VI (>2-fold) of the neat product mixtures. The
  product mixture was also an effective VII additive in blends with high-oleic
  sunflower oil, resulting in the increase of the blend viscosity and VI.
- *Biobased antioxidants.* Several biobased compounds, synthesized with sulfur in their structures, were found to display good oxidation resistance and
  also act as antioxidant additives in blends with biobased (corn oil, HOSuO)
  and petroleum-based (PAO-6) base oils. The materials were SM vegetable
  oils, lipoyl glycerides (LGs), and lipoate esters. The oxidation properties
  were evaluated using RPVOT and PDSC methods.
- *Biobased extreme pressure additives.* All the biobased products synthesized in this work, with phosphorous and sulfur atoms in their structures,
  displayed various degrees of EP properties in biobased base oils (soybean
  or HOSuO). The products, neat or in blends, were evaluated for their EP
  weld point according to the four-ball EP test (ASTM D-2783), relative to
  the pure base oil as well as to blends containing commercial EP additives
  (ZDDP, TPS-32). The general conclusions from the evaluation were the
  following:
  - Blends of 5% w/w dialkyl phosphonate of methyl oleate in soybean dis-
    played a higher weld point than the neat phosphonate or neat soybean
    oil, but lower weld point than 5% blend of ZDDP in soybean.
  - Neat limonene bis(dialkyl phosphonate) displayed a higher weld point
    than biobased (HOSuO, soybean) or petroleum-based (PAO-6, 150 N)
    base oils.
  - Neat SM corn and canola oils displayed higher weld points than bio-
    based and petroleum-based base oils, but much lower than that observed
    with 2% of the commercial polysulfide EP additive TPS-32 in the bio-
    based and petroleum-based base oils.
  - Neat LGs displayed much higher weld points than those observed by
    the above phosphonates or SM vegetable oils.
  - Blends of LGs in biobased and petroleum-based base oils displayed
    weld points that were higher than those for the neat base oils and
    increased with increasing LG concentration.
  - The weld points of LG blends were much lower than the values observed
    for a similar concentration of the commercial polysulfide EP additive
    TPS-32.

- Blends of alkyl lipoate esters in HOSuO displayed a higher weld point than the neat base oil; blend weld points increased with increasing alkyl lipoate ester concentration.
- The weld points of alkyl lipoate ester blends were about half the value observed for a similar concentration of the commercial polysulfide EP additive TPS-32.
- The weld point results of the biobased and petroleum-based sulfurized EP additives indicate that the structure of the sulfur atoms rather than the mere presence of sulfur in the molecule is critical for EP performance of sulfurized additives.

The progress in the development of biobased additives discussed earlier indicates that products competitive to petroleum-based additives currently in the market can be introduced in a very short time in the areas of friction modifiers, biosurfactants, and antiwear additives. However, more time will be required to develop competitive products in the areas of VIIs, antioxidants, and EP additives. In addition, there are many other categories of lubricant additives for which the work to develop competitive biobased products has yet to begin.

ACKNOWLEDGMENTS

The authors acknowledge the following people for help with the work described in this chapter: Ms. Judy Blackburn, Ms. Linda Cao, Mr. Daniel Knetzer, Mr. Benjamin Lowery, Ms. Linda Manthey, Ms. Leslie Smith, Dr. Karl Vermillion, and Dr. Kristen Woods.

REFERENCES

1. Rudnick, L.R. (ed.) *Lubricant Additives Chemistry and Applications*, 2nd edn. CRC Press, Boca Raton, FL (2010).
2. Byers, J.P. *Metalworking Fluids*. Marcel Decker, Inc., New York, 487pp. (1994).
3. Schey, J.A. *Tribology in Metalworking Friction, Lubrication and Wear*. American Society of Metals, Metals Park, OH (1983).
4. Hamrock, B.J., Dowson, D. Isothermal elastohydrodynamic lubrication of point contacts. III—Fully flooded results. *Trans. ASME J. Lubr. Technol.* 99, 264–276 (1977).
5. http://biosynthetic.com/biosynthetic-base-oil/lubricant-technical-performance (accessed December 22, 2015).
6. Isbell, T.A., Abbott, T.P., Asadauskas, S., Lohr, J.E. Jr. Biodegradable oleic estolide esterase stocks and lubricants. US Patent No. 6, 018,063 (2000).
7. Brown, F. Base oil groups: Manufacture, properties and performance. *Tribol. Lubr. Technol.* 71(4), 32–35 (2015).
8. Anon. Lubricants industry factbook. A Supplement to *Lubes'N'Greases*, 21(8, Suppl.), 1–52 (2015).
9. Schneider, M.P. Plant-oil-based lubricants and hydraulic fluids. *J. Sci. Food Agric.* 86, 1769–1780 (2006).
10. Bremmer, B.J., Plonsker, L. *Bio-Based Lubricants: A Market Opportunity Study Update*. Omni-Tech International, Midland, MI (2008).
11. Arca, M., Sharma, B.K., Perez, J.M., Doll, K.M. Gear oil formulation designed to meet biopreferred criteria as well as give high performance. *Int. J. Sustain. Eng.* 6, 326–331 (2013).

12. Cermak, S.C., Isbell, T.A. Estolides—The next generation biobased fluids. *Inform* 15, 515–517 (2004).
13. Spikes, H.A. Friction modifier additives. *Tribol. Lett.* 60(5): 1–26 (2015), DOI 10.1007/s11249-015-0589-z.
14. Hiemenz, P.C. *Principles of Colloid and Surface Chemistry*, 2nd edn. Marcel Dekker, New York (1986).
15. Biresaw, G., Adhvaryu, A., Erhan, S.Z., Carriere, C.J. Friction and adsorption properties of normal and high oleic soybean oils. *J. Am. Oil Chem. Soc.* 79, 53–58 (2002).
16. Biresaw, G., Adharyu, A., Erhan, S.Z. Friction properties of vegetable oils. *J. Am. Oil Chem. Soc.* 80, 697–704 (2003).
17. Biresaw, G. Adsorption of amphiphiles at an oil-water vs. an oil-metal interface. *J. Am. Oil Chem. Soc.* 82, 285–292 (2005).
18. Kurth, T.L., Biresaw, G., Adhvaryu, A. Cooperative adsorption behavior of fatty acid methyl esters from hexadecane *via* coefficient of friction measurements. *J. Am. Oil Chem. Soc.* 82, 293–299 (2005).
19. Adhvaryu, A., Biresaw, G., Sharma, B.K., Erhan, S.Z. Friction behavior of some seed oils: Bio-based lubricant applications. *Ind. Eng. Chem. Res.* 45, 3735–3740 (2006).
20. Kurth, T.L., Byars, J.A., Cermak, S.C., Sharma, B.K., Biresaw, G. Non-linear adsorption modeling of fatty esters and oleic estolides via boundary lubrication coefficient of friction measurements. *Wear* 262, 536–544 (2007).
21. Evans, K.O., Biresaw, G. Quartz crystal microbalance investigation of the structure of adsorbed soybean oil and methyl oleate onto steel surface. *Thin Solid Films* 519, 900–905 (2010).
22. Lawate, S.S., Lal, K., Huang, C. Vegetable oils—Structure and performance, in: Booser, E.R. (ed.) *Tribology Data Handbook*, pp. 103–116. CRC Press, New York (1997).
23. Papay, A.G., Rudnick, L.R. Hydraulics, in: Rudnick, L.R., Shubin, R.L. (eds.), *Synthetic Lubricants and High Performance Fluids*, 2nd edn., pp. 595–624. Marcel Dekker, New York (1999).
24. Adamson, A.P., Gast, A.W. *Physical Chemistry of Surfaces*. John Wiley & Sons, New York (1997).
25. Rotenberg, Y., Boruvka, L., Neumann, A.W. Determination of surface tension and contact angle from the shapes of axisymmetric fluid interfaces. *J. Colloid Interface Sci.* 93, 169–183 (1983).
26. van Oss, C.J. *Interfacial Forces in Aqueous Media*. Marcel Dekker, New York (1994).
27. Hansen, C.M. *Hansen Solubility Parameters: A User's Handbook*. CRC Press, New York (2000).
28. Knothe, G., Dunn, R.O. Biofuels derived from vegetable oils and fats, in: Gunstone, F.D., Hamilton, R.J. (eds.), *Oleochemical Manufacture and Applications*, pp. 106–163. Sheffield Academic Press, Sheffield, U.K. (2001).
29. Ayad, M.M., El-Hefnawey, G., Torad, N.L. A sensor of alcohol vapours based on thin polyaniline base film and quartz crystal microbalance. *J. Hazard. Mat.* 168, 85–88 (2009).
30. Arnau, A. A review of interface electronic systems for AT-cut quartz crystal microbalance applications in liquids. *Sensors* 8, 370–411 (2008).
31. Hook, F., Kasemo, B., Nylander, T., Fant, C., Scott, K., Elwing, H. Variations in coupled water, viscoelastic properties, and film thickness of a Mefp-1 protein film during adsorption and cross-linking: A quartz crystal microbalance with dissipation monitoring, ellipsometry, and surface plasmon resonance study. *Anal. Chem.* 73, 5796–5804 (2001).
32. Rodahl, M., Kasemo, B. A simple setup to simultaneously measure the resonant frequency and the absolute dissipation factor of a quartz crystal microbalance. *Rev. Sci. Instrum.* 67, 3238–3241 (1996).

33. Krozer, A., Rodahl, M. X-ray photoemission spectroscopy study of UV/ozone oxidation of Au under ultrahigh vacuum conditions. *J. Vac. Sci. Technol. A* 15, 1704–1709 (1997).

34. Larsson, C., Rodahl, M., Hook, F. Characterization of DNA immobilization and subsequent hybridization on a 2D arrangement of streptavidin on a biotin-modified lipid bilayer supported on SiO_2. *Anal. Chem.* 75, 5080–5087 (2003).

35. Voinova, M.V., Rodahl, M., Jonson, M., Kasemo, B. Viscoelastic acoustic response of layered polymer films at fluid-solid interfaces: Continuum mechanics approach. *Phys. Scr.* 59, 391–396 (1999).

36. Jahanmir, S., Beltzer, M. An adsorption model for friction in boundary lubrication. *ASLE Trans.* 29, 423–430 (1986).

37. Miao, X., Chen, C., Zhou, J., Deng, W. Influence of hydrogen bonds and double bonds on the alkane and alkene derivatives self-assembled monolayers on HOPG surface: STM observation and computer simulation. *Appl. Surf. Sci.* 256, 4647–4655 (2010).

38. Drummond, C., Alcantar, N.A., Israelachvili, J. Shear alignment of confined hydrocarbon liquid films. *Phys. Rev. E* 66, 011705/1–011705/6 (2002).

39. Knoben, W., Brongersma, S.H., Crego-Calama, M. Preparation and characterization of octadecanethiol self-assembled monolayers on indium arsenide (100). *J. Phys. Chem. C* 113, 18331–18340 (2009).

40. Chan, Y.-H., Schuckman, A.E., Perez, L.M., Vinodu, M., Drain, C.M., Batteas, J.D. Synthesis and characterization of a thiol-tethered tripyridyl porphyrin on Au(111). *J. Phys. Chem. C* 112, 6110–6118 (2008).

41. Love, J.C., Estroff, L.A., Kriebel, J.K., Nuzzo, R.G., Whitesides, G.M. Self-assembled monolayers of thiolates on metals as a form of nanotechnology. *Chem. Rev.* 105, 1103–1170 (2005).

42. Aswal, D.K., Lenfant, S., Guerin, D., Yakhmi, J.V., Vuillaume, D. Self assembled monolayers on silicon for molecular electronics: Analy. *Chim. Acta* 568, 84–108 (2006).

43. van Oss, C.J., Chaudhury, M.K., Good, R.J. Monopolar surfaces. *Adv. Colloid Interface Sci.* 28, 35–64 (1987).

44. Langmuir, I. The adsorption of gases on plane surfaces of glass, mica and platinum. *J. Am. Chem. Soc.* 40, 1361–1402 (1918).

45. Temkin, M.I. Adsorption equilibrium and the kinetics of processes on nonhomogeneous surfaces and in the interaction between adsorbed molecules. *J. Phys. Chem. (USSR)* 15, 296–332 (1941).

46. Butt, H.J., Graf, K., Kappl, M. *Physics and Chemistry of Interfaces*. Wiley-VCH, Weinheim, Germany (2003).

47. Biresaw, G. Tribological properties of ag-based amphiphiles, in: Biresaw, G., Mittal, K.L. (eds.), *Surfactants in Tribology*, Vol. 1, pp. 259–290. CRC Press, Boca Raton, FL (2008).

48. Rosen, M.J. *Surfactants and Interfacial Phenomenon*, 3rd edn. John Wiley & Sons, Hoboken, NJ (2004).

49. Biresaw, G., Liu, Z.S., Erhan, S.Z. Investigation of the surface properties of polymeric soaps obtained by ring opening polymerization of epoxidized soybean oil. *J. Appl. Polym. Sci.* 108, 1976–1985 (2008).

50. Liu, Z.S., Biresaw, G. Synthesis of soybean oil based polymeric surfactants in supercritical carbon dioxide and investigation of its surface properties. *J. Agric. Food. Chem.* 59, 1909–1917 (2011).

51. Liu, Z.S., Biresaw, G. Soy-based polymeric surfactants prepared in carbon dioxide media and influence of structure on their surface properties, in: Biresaw, G., Mittal, K.L. (eds.), *Surfactants in Tribology*, Vol. 4. pp. 419–442. CRC Press, Boca Raton, FL (2014).

52. Wu, S. *Polymer Interface and Adhesion*. Marcel Dekker, New York (1982).

53. Farng, L.O. Anti-wear additives and extreme pressure additives, in: Rudnick, L.R. (ed.), *Lubricant Additives Chemistry and Applications*, 2nd edn., Chapter 8, pp. 213–249. CRC Press, Boca Raton, FL (2009).

54. Bantchev, G.B., Biresaw G., Vermillion, K.E., Appell, M. Synthesis and spectral characterization of methyl 9(10)-dialkylphosphonostearates. *Spectrochim. Acta Part A* 110, 81–91 (2013).

55. Biresaw, G., Bantchev, G.B. Tribological properties of biobased ester phosphonates. *J. Am. Oil Chem. Soc.* 90, 891–902 (2013).

56. Bantchev, G.B., Biresaw, G., Palmquist, D.E., Murray, R.E. Radical-initiated reaction of methyl linoleate with dialkyl phosphites. *J. Am. Oil Chem. Soc.* 93, 859–868.

57. Biresaw, G., Bantchev, G.B. (2016) Tribological properties of limonene bisphosphonates. *Tribol. Lett.* 60, 11–25 (2015).

58. Walsh, M. Challenges impacting the North American orange oil industry, in: *Proceedings of IFEAT International Conference*, Montreal, Quebec, Canada, pp. 115–122 (2008).

59. ASTM D 4172-94. Standard test method for wear preventive characteristics of lubricating fluid (four-ball method), in: *Annual Book of ASTM Standards*, pp. 752–756. American Society for Testing and Materials, West Conshohocken, PA, 05.02 (2002).

60. ASTM D 5183-95. Standard test method for determination of the coefficient of friction of lubricants using the Four-Ball wear test machine, in: *Annual Book of ASTM Standards*, pp. 165–169. American Society for Testing and Materials, West Conshohocken, PA, 05.03 (2002).

61. Biresaw, G., Compton, D., Evans, K., Bantchev, G.B. Lipoate ester multi-functional lubricant additives. *Ind. Eng. Chem. Res.* 55, 373–383 (2016).

62. Gorąca, A., Huk-Kolega, H., Piechota, A., Kleniewska, P., Ciejka, E., Skibska, B. Lipoic acid—Biological activity and therapeutic potential. *Pharmacol. Rep.* 63, 849–858 (2011).

63. Kisanuki, A., Kimpara, Y., Oikado, Y., Kado, N., Matsumoto, M., Endo, K. Ring-opening polymerization of lipoic acid and characterization of the polymer. *J. Polym. Sci. Part A Polym. Chem.* 48, 5247–5253 (2010).

64. ASTM D2270-93. Standard practice for calculating viscosity index from kinematic viscosity at 40 and 100 °C. in: *Annual Book of ASTM Standards*, pp. 849–854. American Society for Testing and Materials, West Conshohocken, PA, 05.01 (2012).

65. Shelton, J.R. Organic sulfur compounds as preventive antioxidants, in: G. Scott (ed.), *Developments in Polymer Stabilization*, pp. 23–69. Applied Science Publishers, London, U.K. (1981).

66. Dong, J., Migdal, C.A.: Antioxidants, in: Rudnick, L.R. (ed.), *Lubricant Additives Chemistry and Applications*, 2nd edn., Chapter 1, pp. 3–50. CRC Press, Boca Raton, FL (2009).

67. Bantchev, G.B., Kenar, J.A., Biresaw, G., Han, M.G., Free radical addition of butanethiol to vegetable oil double bonds. *J. Agric. Food Chem.* 57, 1282–1290 (2009).

68. Biresaw, G., Bantchev, G.B., Cermak, S.C. Tribological properties of sulfur modified vegetable oils. *Tribol. Lett.* 43, 17–32 (2011).

69. ASTM D 2272-11 Standard test method for oxidation stability of steam turbine oils by rotating pressure vessel. in: *Annual Book of ASTM Standards*. American Society for Testing Materials, West Conshohocken, PA, 0.5.01, pp. 855–874 (2012).

70. ASTM D 6186-08 Standard test method for oxidation induction time of lubricating oils by pressure differential scanning calorimetry (PDSC). in: *Annual Book of ASTM Standards*, pp. 52–56. American Society for Testing and Materials, West Conshohocken, PA, 05.03 (2012).

71. Biresaw, G., Laszlo, J.A., Evans, K.O., Compton, D.L., Bantchev, G.B. Synthesis and tribological investigation of lipoyl glycerides. *J. Agric. Food Chem.* 62, 2233–2243 (2014).

72. ASTM D2783-88. Standard test method for measurement of extreme-pressure properties of lubricating fluids (Four-Ball Method), in: *Annual Book of ASTM Standards*, pp. 130–137. American Society for Testing and Materials, West Conshohocken, PA 05.02 (2002).

73. Asadauskas, S.J., Biresaw, G., McClure, T.G. Effects of chlorinated paraffin and ZDDP concentrations on boundary lubrication properties of mineral and soybean oils. *Tribol. Lett.* 37, 111–121 (2010).

74. Biresaw, G., Asadauskas, S.J., McClure, T.G. Polysulfide and bio-based extreme pressure additive performance in vegetable vs. paraffinic base oils. *Ind. Eng. Chem. Res.* 51, 262–273 (2012).

75. Luo, Y.-R. Bond dissociation energies, in: Haynes, W.M. editor in chief, *Handbook of Chemistry and Physics*. 91st edn., pp. 9-65–9-98 (2010–2011).

16 Assessment of Agricultural Wastes as Biosorbents for Heavy Metal Ions Removal from Wastewater

Nabel A. Negm, Hassan H.H. Hefni, and Ali A. Abd-Elaal

CONTENTS

16.1   INTRODUCTION

Heavy metals are a type of pollutant that are found as natural constituents of the Earth's crust. Serious biological hazards are caused by metal toxicity and cannot be controlled. A high degree of industrialization and urbanization has substantially enhanced the degradation of aquatic environments through the discharge of industrial wastewater and domestic wastes. This has resulted in significant amounts of heavy metals being deposited into natural aquatic and global ecosystems. In view of this, there is an urgent need for the removal of heavy metal contaminants from wastewater, either domestic or industrial.

Generally, a low-cost sorbent can be defined as one that requires little processing and is abundant in nature. In this context, agricultural by-products and industrial wastes can be seen as having great potential for the development of low-cost sorbents. The use of these materials could be beneficial not only to the environment in solving the solid waste disposal problem but also wisely help the economy. A literature survey revealed that numerous biological materials have been utilized as adsorbents. These include durian peel [1], rice husk [2], orange peels [3], spent tea leaves [4], and wheat straw [5].

Conventional methods for heavy metal removal from aqueous solution include chemical precipitation, electrolytic recovery, ion exchange, chelation, solvent extraction, and liquid membrane separation [6,7].

These methods are cost prohibitive and have inadequate efficiencies at low metal concentrations, particularly in the range of 1–100 mg/L [8]. Some of these methods generate toxic sludge such as biosorbents loaded with high concentrations of heavy metals, the disposal of which is an additional load on the techno-economic feasibility of treatment procedures. These constraints have prompted the search for alternative methods that would be efficient for metal sequestering. Such a possibility offers a method that uses sorbents of biological origin [9–11] for the removal of heavy metals from diluted aqueous solutions. The most frequently studied biosorbents are bacteria, fungi, and algae [12–16].

Conventional treatment technologies for the removal of these heavy metals are not economical and further generate a huge quantity of toxic chemical sludge. Biosorption is emerging as a potential alternative to the existing conventional technologies for the removal or recovery of metal ions from their aqueous solutions. The major advantages of biosorption over conventional treatment methods include low cost, high efficiency, minimization of chemical or biological sludge, regeneration of biosorbents, and the possibility of metal recovery. Cellulosic agricultural waste materials are abundant sources for significant metal biosorption.

The purpose of this chapter is to consolidate the available information on various biosorbents from the agricultural waste materials for heavy metal removal. Agricultural waste material being highly efficient, low cost, and a renewable source of biomass can be exploited for heavy metal remediation. Further, these biosorbents can be modified for higher efficiency and multiple reuses to enhance their applicability at an industrial scale.

16.2   ADSORPTION ISOTHERMS OF BIOSORBENTS FOR HEAVY METALS

Biosorption of heavy metals is a passive nonmetabolically mediated process including metal binding by biosorbent. Agricultural waste and its industrial by-products—bacteria, yeasts, fungi, and algae—can function as biosorbents for heavy metals. Biosorption is considered to be a fast physical/chemical process, and its rate is governed by the type of the process. In another sense, it can also be defined as a collective term for a number of passive accumulation processes, which in any particular case may include ion exchange, coordination, complexation, chelation, adsorption, and microprecipitation. Proper analysis and design of adsorption/biosorption separation processes requires relevant adsorption/biosorption equilibrium also as vital information for the adsorption process. In equilibrium, a certain relationship prevails between solute concentration in solution and adsorbed state (i.e., the amount of solute adsorbed per unit mass of adsorbent). The equilibrium concentrations of heavy metals are a function of temperature. Therefore, the adsorption equilibrium relationship at a given temperature is referred to as adsorption isotherm. Several adsorption isotherms originally used for gas-phase adsorption are readily adopted to correlate adsorption equilibriums in heavy metal biosorption. Some well-known ones are Freundlich, Langmuir, Redlich–Peterson, and Sips equations. The most widely used among them are Freundlich and Langmuir equations. The application of these isotherms on biosorbent-assisted heavy metal removal from water and wastewater is discussed in subsequent sections.

16.2.1   Freundlich Isotherm

Freundlich isotherm is an empirical equation widely used for the description of adsorption equilibrium. Freundlich isotherm is capable of describing the adsorption of organic and inorganic compounds on a wide variety of adsorbents including biosorbents and has the following form, Equation 16.1:

$$\log q_e = \log K_F + \frac{1}{n}\log C_e \qquad (16.1)$$

The plot of $\log q_e$ versus $\log C_e$ has a slope with the value of $1/n$ and an intercept magnitude of $\log K_F$. $\log K_F$ is equivalent to $\log q_e$ when C_e equals unity. However, in other cases when $n \neq 1$, the K_F value depends on the units in which q_e and C_e are expressed.

On average, a favorable adsorption tends to have Freundlich constant n between 1 and 10. A larger value of n (smaller value of $1/n$) implies a stronger interaction between biosorbent and heavy metal, while $1/n$ equal to 1 indicates a linear increase in the adsorption rate leading to identical adsorption energies for all sites [17].

Freundlich isotherm has the ability to fit nearly all experimental adsorption–desorption data and is especially excellent for fitting data from highly heterogeneous sorbent systems. Accordingly, this isotherm can adequately represent the biosorption isotherm for most of the systems. There are some cases where the Freundlich isotherm could not fit the experimental data well (by giving low correlation values) or was not even suitable to express biosorption equilibriums. Apart from the ability to represent well in most cases (by giving high correlation values), the physical meaning of $1/n$ was not clear in several studied systems [18].

As a precautionary note, the Freundlich equation is unable to predict adsorption equilibrium data at extreme concentrations [19]. Furthermore, this equation is not reduced to linear adsorption expression at very low concentrations.

16.2.2 LANGMUIR ISOTHERM

Another widely used model for describing heavy metals sorption to biosorbents is the Langmuir model. The Langmuir equation relates the coverage of molecules on a solid surface to the concentration of a medium at a fixed temperature. This isotherm is based on three assumptions, namely, adsorption is limited to monolayer coverage, all surface sites are alike and only can accommodate one adsorbed atom, and the ability of a molecule to be adsorbed on a given site is independent of its neighboring sites occupancy [20]. By applying these assumptions and a kinetic principle (rates of adsorption and desorption from the surface are equal), the Langmuir equation can be written in the following form, Equation 16.2:

$$\frac{C_e}{q_e} = \frac{1}{q_{max}} C_e + \frac{1}{K_L q_{max}} \tag{16.2}$$

Only narrow understanding of the data fitting process is required, and the calculation can be done as spreadsheets in Microsoft Excel. Within the Langmuir model, the saturation capacity q_{max} is supposed to coincide with the saturation of a fixed number of identical surface sites, and as such, it should logically be independent of temperature. This is contrary to real conditions where a small to modest increase [21] in saturation capacity is observed with the increase of temperature.

16.2.3 TEMKIN ISOTHERM

The Temkin equation was proposed to describe the adsorption of hydrogen on platinum electrodes from acidic solutions. The derivation of the Temkin isotherm was based on the assumption that the decline in the heat of adsorption as a function of temperature is linear rather than logarithmic, as implied in the Freundlich equation [22].

The Temkin isotherm has the following form, Equation 16.3:

$$q_e = \frac{RT}{b} \ln\left(aC_e\right) \tag{16.3}$$

The Temkin isotherm is incapable of predicting well the adsorption equilibriums. It is apparent that the Temkin equation is superior in the prediction of gas-phase equilibriums. On the other hand, in liquid-phase adsorption especially in heavy metal ions adsorption using biosorbent, this model falls short in representing the equilibrium data. Adsorption in the liquid phase is a more complex phenomenon than in the gas phase, since the adsorbed molecules do not necessarily organize in a tightly packed structure with identical orientation. In addition, the presence of solvent molecules and the formation of micelles from adsorbed molecules add to the complexity of liquid-phase adsorption. Numerous factors including pH, solubility of adsorbent in the solvent, temperature, and surface chemistry of the adsorbent influence the adsorption in liquid phase. Since the derivation of the Temkin equation is based on simple assumptions, the complex phenomena involved in liquid-phase adsorption are totally ignored [23]. As a result, this equation is often not suitable for the representation of experimental data in complex systems.

16.2.4 DUBININ–RADUSHKEVICH MODEL

This model was conceived for subcritical vapors in microporous solids where the adsorption process follows a pore-filling mechanism on an energetically nonuniform surface. The Dubinin–Radushkevich (DR) equation is excellent for interpreting organic compounds sorption equilibriums (in gas-phase condition) in porous solids. The DR equation is rarely applied to liquid-phase adsorption due to the complexities introduced by changes in factors such as pH and ionic equilibriums inherent in these systems. The solute–solvent interactions often render the bulk solution nonideal [24].

The mathematical expression for the DR equation in the liquid-phase system is given in Equation 16.4:

$$q_e = (q_s)\exp\left(-k_{ad}E_o^2\right) \tag{16.4}$$

By taking into account the energetically nonuniform surface, this equation is capable of describing the biosorption data well [25].

Within the DR equation, the characteristic energy can also be obtained as a parameter. One of the best features of the DR equation is that it is temperature dependent. If the adsorption data at different temperatures are plotted as the logarithm of the amount adsorbed ($\log q_e$) versus the square of potential energy (E_o^2), all suitable data shall in general lie on the same curve. This curve can later be utilized to determine the applicability of the DR equation in expressing the adsorption equilibrium data [26].

If the fitting by DR gives a high correlation but the characteristic curves obtained from the analyzed data do not lie in the same curve, the validity of the ascertained parameters is questionable. To this end, however, the characteristic curve of biosorption systems could be examined since all of the experiments are conducted at a single temperature [27].

16.2.5 FLORY–HUGGINS ISOTHERM

The Flory–Huggins (FH) isotherm is chosen in order to account for the degree of the surface coverage characteristic of the adsorbed species on the biosorbent [28]. The FH isotherm has the following form, Equation 16.5:

$$\log\frac{\theta}{C_o} = \log K_{FH} + n_{FH}\log(1-\theta) \tag{16.5}$$

16.3   KINETIC STUDIES OF HEAVY METAL BIOSORPTION USING VARIOUS BIOSORBENTS

Adsorption equilibrium studies are important to determine the efficiency of adsorption. However, it is also necessary to identify the adsorption mechanism of a given system. For the purpose of investigating the mechanism of biosorption and its potential rate-controlling steps that includes mass transport and chemical reaction processes, kinetic models have been used to test the experimental data. Information on the kinetics of metal uptake is necessary to select the optimum conditions for full-scale batch metal removal processes. Adsorption kinetics is an expression of the solute removal rate and controls the residence time of the adsorbent at the solid–solution interface. In practice, kinetic studies were carried out in batch reactions using various initial adsorbent concentrations, adsorbent doses, biosorbent particle sizes, agitation speeds, pH values, and temperatures along with different sorbent and adsorbent types. Then, linear regression was used to determine the best-fitting kinetic rate model. As an additional step, linear least-squares method can also be used to the linearly transformed kinetic rate equations for confirming the experimental data and kinetic rate equations using the coefficients of determination [20].

Several adsorption kinetic models have been developed to understand adsorption kinetics and rate-limiting steps. These include pseudo-first- and pseudo-second-order rate models, Weber and Morris sorption kinetic model, Adam–Bohart–Thomas relation [29], first-order reversible reaction model [30], external mass transfer model [31], first-order equation of Bhattacharya and Venkobachar [32], Elovich model, and Ritchie equation. The pseudo-first- and pseudo-second-order kinetic models are the most widely used models to study the biosorption kinetics of heavy metals and to quantify the extent of uptake in biosorption kinetics. A comprehensive review about the application of second-order models for adsorption systems is available [33]. The following sections focus on the kinetic modeling of heavy metal biosorption systems.

16.3.1   PSEUDO-FIRST-ORDER KINETIC

The Lagergren first-order rate expression based on solid capacity is generally expressed as follows, Equation 16.6:

$$\frac{dq}{dt} = k_1 \left(q_e - q \right) \tag{16.6}$$

Integration of Equation 16.6 with the boundary conditions $t = 0$, $q = 0$, and at $t = t$, $q = q$, gives (Equation 16.7)

$$\ln \left(q_e - q \right) = \ln q_e - k_1 t \tag{16.7}$$

The q_e value acquired by this method is then compared with the experimental value. If large discrepancies are observed, the reaction cannot be classified as first-order although this plot has high correlation coefficient from the fitting process.

Disagreement occurs for most systems, in which as-calculated q_e values are not equal to the experimental q_e, further indicating the inability of the pseudo-first-order model to fit the heavy metal biosorption data. The trend shows that the predicted q_e values seem to be lower than the experimental values. A time lag, probably caused by the presence of a boundary layer or external resistance controlling the beginning of the sorption process, was argued to be the responsible factor behind the discrepancy [34].

16.3.2 PSEUDO-SECOND-ORDER KINETIC

Predicting the rate of adsorption for a given system is among the most important factors in adsorption system design, as the system's kinetics determine adsorbent residence time and the reactor dimensions [20]. Although various factors govern the adsorption capacity (e.g., initial heavy metal concentration, temperature, pH of solution, biosorbent particle size, heavy metals nature), a kinetic model is concerned only with the effect of observable parameters on the overall rate [20]. The pseudo-second-order kinetic model derived on the basis of the sorption capacity of the solid phase is expressed as follows, Equation 16.8:

$$\frac{dq}{dt} = k_2 \left(q_e - q \right)^2 \tag{16.8}$$

Integration of Equation 16.8 with the boundary conditions of time and equilibrium gives the following linear form, Equation 16.9:

$$\frac{1}{q_e - q} = \frac{1}{q_e} + k_2 t \tag{16.9}$$

The pseudo-second-order rate constant, k_2, was determined experimentally by plotting $1/(q_e - q)$ against t.

Ho [20] conducted an evaluation of the linear and nonlinear formats of Equation 16.9 to determine the pseudo-second-order kinetic parameters. He chose cadmium as the heavy metal and tree fern as the biosorbent. The resulting kinetic parameters from four kinetic models using the linear equations showed discrepancies between the models. Among the linear methods, the pseudo-second-order kinetic model had the highest coefficient of determination. In contrast to the linear model, the resulting kinetic parameters from the nonlinear model were almost identical. As a result, the nonlinear method is considered to be a better method to ascertain the desired parameters.

16.3.3 FIRST-ORDER REVERSIBLE REACTION MODEL

This model is not widely used for the study of biosorption kinetics, even though it can describe the adsorption and desorption phenomena simultaneously using rate constant parameters.

To derive this model, the sorption of a metal on biosorbent is assumed to be a first-order reversible reaction, as expressed by the following equation (Equation 16.10) [35]:

$$A \Leftrightarrow B \qquad (16.10)$$

The rate equation for the reaction is expressed as follows (Equation 16.11):

$$\frac{dC_B}{dt} = -\frac{dC_A}{dt} = k_1^o C_A - k_2^o C_B = k_1^o \left(C_{Ao} - C_{Ao} X_A \right) - k_2^o \left(C_{Bo} - C_{Ao} \right) \qquad (16.11)$$

At equilibrium condition (Equation 16.12),

$$K_c = \frac{C_{Be}}{C_{Ae}} = \frac{k_2^o}{k_1^o} \qquad (16.12)$$

Integrating Equation 16.11 and applying the equilibrium conditions gives Equation 16.13:

$$\ln\left(\frac{-\left(C_{AO} + C_A \right)}{C_{AO} - C_{Ae}} \right) = -\left(k_1^o + k_2^o \right) t \qquad (16.13)$$

Baral and coworkers [30,35] evaluated several models to analyze metal ions biosorption experimental data, including one that was first-order reversible reaction model. The latter fit well their experimental data. The reduced rate constants and increasing equilibrium constants with increasing temperature signify that the biosorption of metal ions onto treated sawdust is exothermic. Since the adsorption process is exothermic as a rule, the rate constant value of k_1^o should decrease with increasing temperature. Based on Le Chatelier's principle, if the adsorption is exothermic, desorption would be endothermic. Therefore, the rate constant value of k_2^o should increase with temperature rise. As mentioned before, the sorption of heavy metals on any biosorbents takes place by physical bonding, ion exchange, complexation, coordination/chelation, or a combination of these.

16.4 AGRICULTURAL WASTE MATERIALS AS BIOSORBENTS

The discharge of wastewater from industrial activity normally releases effluent containing heavy metals such as Cu(II), Pb(II), Zn(II), Ni(II), and Cr(VI). Generally, industrial wastewater from different industries is divided into two categories: from the electroplating process and the other from the rinsing process. In developed countries, the removal of heavy metals in wastewater is normally achieved by advanced technologies such as ion exchange resins, vacuum evaporation, crystallization, solvent extraction, and membrane technology [36–38]. However, in developing countries, these treatments cannot be applied because of technical levels and insufficient funds [39]. Therefore, it is desired that simple and economical removal methods be established and utilized in developing countries. These by-products influence the flow and storage of water and the quality of available freshwater. It is evident that wastewater released from industrial activities is one of the major causes of environmental pollution, caused by the presence of these heavy metals [4].

16.4.1 RICE HUSK AS BIOSORBENT

Rice husks are the hard protective coverings on the grains of rice. The husk is made of hard materials, including opaline silica and lignin, in order to protect the seed during the growing season. The husk is mostly indigestible to humans. During the milling process, the husks are removed from the grains to create brown rice. The brown rice is then milled further to remove the bran layer to make white rice. Rice husks are a class "A" insulating material because they are difficult to burn and less likely to allow moisture to grow mold or fungi. When burned, rice husk produces significant amounts of silica. The very high content of amorphous silica in the husks (SiO_2 ~ 20 wt.%) after combustion are very valuable properties for excellent thermal insulation. Besides, rice husk contains abundant floristic fiber, protein, and some functional groups such as carboxyl and amide groups [40]. Rice husk as a low-value agricultural by-product can be used as a biosorbent material for different heavy metals and dyes removal. Currently, the study of rice husk as a low-cost biosorbent for removing heavy metals has regained attention.

Tarley and Arruda [41] evaluated the morphological characteristics and chemical composition of rice husks with different techniques such as spectroscopy and thermogravimetry. The husk, which is a by-product obtained from rice milling, was then investigated as a potential decontaminant of toxic heavy metals present in laboratory effluents. The study was performed using graduated glass columns, at room temperature, employing 100 mL of synthetic solutions containing Cd(II) and Pb(II) at 100 mg/L. The study evaluated the effects of pH, flow rate, and particle size of rice husk on Cd(II) and Pb(II) adsorption. At optimized conditions, the efficiencies of rice husks in removing Cd(II) and Pb(II) ions from 100 mL of laboratory solutions were 78% and 88%, respectively. The adsorption ability of other metals species such as Al(III), Cu(II), and Zn(II) was also studied by the authors.

The removal of Zn(II) from aqueous solution by different adsorbents was investigated by Bhattacharya et al. [42]. Clarified sludge (a steel industry waste), rice husk ash, neem bark, and activated alumina were used for adsorption studies. The influences of pH, contact time, initial metal concentration, adsorbent nature, and concentration on the selectivity and sensitivity of the removal process were investigated. The study showed that the adsorption of Zn(II) increased with the increasing concentration of adsorbents and reached maximum uptake at 10 g/L and pH between 5 and 7. The equilibrium time (t_{eq}) was achieved after 1 h for clarified sludge, 3 h for rice husk ash, and 4 h for activated alumina and neem bark. The kinetics of the adsorption process was found to follow first-order rate kinetics, and rate constant was evaluated at 30°C. Langmuir and Freundlich adsorption isotherms were used to analyze the experimental data. The adsorption capacity (q_{max}) calculated from Langmuir adsorption isotherm and the values of Gibbs free energy showed that clarified sludge has the largest capacity and affinity for the removal of Zn(II) compared to the other adsorbents.

Asadi et al. [43] examined rice hull and sawdust, as sorbents to remove Pb(II), Cd(II), Zn(II), Cu(II), and Ni(II) from synthetic solutions or wastewater samples. The different biosorbents were treated with HCl, NaOH, and heat to modify their efficiency. The sorption of the heavy metals from the synthetic solutions increased with increasing pH and initial concentration. At pH 5, Pb(II) and Cd(II) showed the highest sorption. Sorption capacity of rice hull was higher than that of sawdust. The treatments of the biosorbents changed the sorption capacity of the natural sorbents in the following order of treatment as base > heat > untreated > acid. The sorption isotherms of sorbents were best described by both the Freundlich and Langmuir models. The base-treated rice hull and sawdust, followed by the heat-treated rice hull, adsorbed the maximum amount of heavy metals from the industrial wastewater samples.

The adsorption of Cu(II) from aqueous solution with carbon prepared from rice husk through pyrolysis and steam activation was studied by Zhang et al. [44]. The rice husk carbon was characterized by Fourier transform infrared spectroscopy (FTIR) and scanning electron microscopy and pore size. After comparing the characteristics of the carbons prepared under different conditions and their adsorption abilities for Cu(II), the optimum temperatures for pyrolysis and steam activation were found to be 700°C and 750°C, respectively, using 3% (v/v) steam/nitrogen as the best activation gas mixture. It was found that the Cu(II) adsorption on the rice husk–derived carbons was dependent on pH and time, with an optimum pH value of 5 for 24 h. The adsorption kinetics and isotherms of Cu(II) on rice husk–derived carbon were fitted by the pseudo-second-order kinetic models and the Langmuir isotherm model under different temperatures. The mean free energy E (kJ/mol) obtained in the Dubinin–Radushkevich (D-R) adsorption isotherm equation indicated a chemical ion exchange mechanism.

Ding et al. [45] reported biosorption of aquatic cadmium (II) by rice straw. Rice straw has the potential to remove Cd from large-scale effluents contaminated by heavy metals since it exhibited a short adsorption equilibrium time of 5 min, high biosorption capacity (13.9 mg/g), and high removal efficiency in the pH range of 2–6. The main Cd biosorption mechanism was Cd^{2+} ion exchange with K^+, Na^+, Mg^{2+}, and Ca^{2+}, along with chelation to functional groups such as C=C, C–O, O–H, and carboxylic acids. When 0.5% (w/v) rice straw was exposed to 50 mg/mL $CdSO_4$ solution with stirring at 150 rpm for 3 h, about 80% of the aquatic Cd^{2+} was adsorbed and the Cd^{2+} content in rice straw reached 8–10 mg/g. This suggests that the metal-enriched rice straw could become high-quality bio-ore by virtue of the industrial mining grade of its metal content and easy metal recovery.

The removal of Cr(VI) ions from their aqueous solution with rice straw was investigated [46]. The optimal pH was found to be 2, and Cr(VI) removal rate increased with decreased Cr(VI) concentration and with increased temperature. The decrease in rice straw particle size led to an increase in Cr(VI) removal. Equilibrium was achieved in about 48 h. Adsorption isotherms showed that equilibrium sorption data were better represented by the Langmuir model, and the adsorption capacity of rice straw was found to be 3.15 mg/g.

Gonc et al. [47] reported about adsorption experiments using rice straw as a bio-sorbent for Cu(II), Zn(II), Cd(II), and Hg(II) ions from aqueous solutions at room temperature. The adsorption isotherms did fit the Freundlich model. Based on the experimental data and Freundlich model, the adsorption order on the rice straw was Cd(II) > Cu(II) > Zn(II) > Hg(II), and equilibrium was reached after 1.5 h, with maximum adsorption at pH 5.

Chockalingam and Subramanian [48] utilized rice husk as an adsorbent for metal ions such as Fe(III), Zn(II), and Cu(II) from acid mine wastewater. The adsorption isotherms exhibited Langmuirian behavior and were endothermic in nature. The free energy values for adsorption of Cu(II) and Zn(II) ions onto rice husk were highly negative, attesting to favorable interaction. Over 99% Fe^{3+}, 98% Fe^{2+} and Zn^{2+}, and 95% Cu^{2+} uptakes were achieved from acid mine water, with a concomitant increase in the pH value by two units using rice husk.

16.4.2 WHEAT BRAN AND RICE BRAN

Most grains, like wheat and oats, have a hard outer layer. When they are processed, this layer becomes a by-product and is called bran. In the case of processing wheat to make wheat flour, one gets miller's or wheat bran. It is packed with nutrition and offers many dietary benefits.

Batch adsorption experiments with wheat bran and rice bran were performed [49] for the removal of Congo red dye from aqueous solution at pH 8 and temperature of 25°C. The effects of contact time, adsorbent concentration, adsorbent treatment, and ion strength were investigated. The raw biomass and biomass loaded by Congo red dye were characterized using FTIR. The results showed that adsorption of Congo red dye on the different biosorbents follows the pseudo-second-order kinetic model. The Langmuir and Freundlich equations were applied to the adsorption isotherms data, and the observed maximum adsorption capacities were 22.73 and 14.63 mg/g for wheat bran and rice bran, respectively. Larger effects of adsorbent concentration and ionic strength on the Congo red adsorption were marked. The adsorption performance was significantly improved with rice bran modified with $Cu(NH_3)_4^{2+}$.

Jenkins and coauthors [50] reported the removal of troublesome elements in biomass to reduce slagging and fouling in furnaces and other thermal conversion systems by washing (leaching) the fuel with water. Rice straw and wheat straw were washed using various techniques and analyzed for their composition and ash fusibility. Potassium, sodium, and chlorine were easily removed by washing with both tap and distilled water. Total ash was reduced by about 10% in rice straw and up to 68% in wheat straw due to washing with water. Washing was more effective in increasing ash fusion temperatures for rice straw than for wheat straw due to the higher initial silica concentration in rice straw. Unwashed straw ash that fused below 1000°C was observed to become more refractory at higher temperatures when washed. Scanning electron microscopy of unwashed and washed rice straw ashed at 1000°C revealed that all unwashed ash particles were fused and glassy. While washed particles remained unfused, were heavily depleted in most elements other than Si, and displayed structures characteristic of original cellular morphology. The fusion temperature of the straw ash was consistent with predicted temperatures from alkali oxide–silica phase systems based on the concentrations of elements in the ash. A simple attempt at simulating a possible full-scale washing process was made by spraying the surface of a bed of straw with water for an arbitrary time of 1 min.

16.4.3 BURNED RICE HUSK ASH

Rice husk ash is generated by burning rice husk. On burning, cellulose and lignin are removed leaving behind silica ash. A controlled temperature and environment of burning yields better quality rice husk ash as its particle size and specific surface area are dependent on the burning condition. The ash produced by controlled burning of the rice husk between 550°C and 700°C incinerating temperature for 1 h transforms the silica content of the ash into amorphous phase. The biosorption of amorphous silica is directly proportional to the specific surface area of ash.

Batch adsorption studies on Pb(II) [51] were performed to evaluate the influences of pH, initial concentration, adsorbent dosage, contact time, and temperature. Optimum conditions for Pb(II) removal were pH 5 and adsorbent dosage of 5 g/L of solution, and equilibrium was attained after 1 h. Adsorption of Pb(II) followed pseudo-second-order kinetics, and the diffusion coefficient was 10^{-10} m²/s. The equilibrium adsorption isotherm was described by the Freundlich adsorption isotherm model. The adsorption capacity (q_{max}) of rice husk ash for Pb(II) ions in terms of monolayer adsorption was 91.74 mg/g. The changes in entropy (ΔS) and enthalpy (ΔH) were estimated at 0.132 and 28.923 kJ/mol, respectively. The negative value of Gibbs free energy (ΔG) indicates feasible and spontaneous adsorption of Pb(II) on rice husk ash. An application study was carried out to find the suitability of the process in wastewater treatment operation.

The removal of chlorophenol (CP) from water using carbon obtained from burning rice straw was evaluated by Wang et al. [52]. Rice straw was burned at 300°C in air to obtain rice straw carbon (RC). Scanning electron micrographs showed a highly porous structure of RC. NMR and FTIR spectroscopy showed an enhanced aromaticity in RC chemical structure and the presence of oxygen-containing functional groups. The adsorption of CP by RC was characterized by L-shaped nonlinear isotherm, indicating the presence of surface adsorption mechanism for CP removal rather than partitioning. Strong adsorption occurred when CP was in a neutral state. The adsorption decreased with increasing pH due to increased deprotonation of surface functional groups of RC, which led to dissociation of CP. The adsorption capacity values obtained by fitting data to the Langmuir model were 11.4, 12.9, 14.2, and 4.9 mg/g at pH 4, 6, 7, and 10, respectively. These results suggest that RC can be effectively used as a low-cost biosorbent to produce activated carbon for the removal of CPs from wastewater.

Hsu and coauthors [53] evaluated the removal of Cr(VI) ions from wastewater by carbon obtained by burning rice straw. Rice straw was burned in air to obtain RC, and then the removal of Cr(VI) ions by RC was investigated under various pH values and ionic strengths. After the experiments, the oxidation state of Cr ions bound to RC was analyzed using X-ray photoelectron spectroscopy, which revealed that Cr ions bound to RC were predominately in the form of Cr(III) ions. The results showed that upon reacting with RC, Cr(VI) was reduced to Cr(III), which was adsorbed on RC and released back into solution. The extent and rate of Cr(VI) removal increased with decreasing pH of the solution. This was due to Cr(VI) adsorption, and subsequent reduction of adsorbed Cr(VI) to Cr(III) occurs preferentially at low pH. The effect of ionic strength on the rates of Cr(VI) removal and Cr(III) adsorption was minor, which indicates the presence of specific interactions between Cr(VI)/Cr(III) and the surface binding sites on RC. The results demonstrate that rice straw–based carbon can be effectively used at low pH as a biosorbent for activated carbon source in the treatment of Cr(VI)-contaminated wastewater.

16.4.4 MODIFIED RICE HUSK

Wong et al. [54] reported the use of tartaric acid–modified rice husk (TARH) as a sorbent for the removal of Cu(II) and Pb(II) from semiconductor electroplating wastewater. The application of Langmuir isotherm indicated that there was no difference in the adsorption capacity of TARH for Cu(II) versus Pb(II) in synthetic solution and wastewater. In column studies, an increase in column depth yields longer treatment time, while increase in effluent concentration and flow rate results in faster biosorbent saturation. Column study studies the adsorption efficiency of the biosorbents when placed in a vertical column has a standard diameter and the contaminated water flows through the column by constant speed at constant temperature. The adsorption capacity of the TARH column for Cu(II) and Pb(II) agrees with the values obtained from batch equilibrium studies. Theoretical breakthrough curves at different column heights and flow rates generated using a two-parameter model agree well with experimental values in the treatment of semiconductor wastewater. Breakthrough curve plots the relation between C_t/C_0 and time of adsorption. The regeneration study showed that Cu(II) and Pb(II) can be recovered almost quantitatively by eluting the column with 0.1 M HCl. The repeatability test showed that the column filled with the biosorbent can be used repeatedly for five cycles without losing efficiency.

Dada et al. [55] studied the equilibrium adsorption with phosphoric acid–modified rice husk. The physicochemical properties of the modified rice husk were determined. The equilibrium sorption data were fitted to Langmuir, Freundlich, Temkin, and Dubinin–Radushkevich (DRK) isotherms. The study showed that R^2 value for the Langmuir adsorption isotherm was the highest. The maximum monolayer coverage (q_m) based on the Langmuir isotherm was determined to be 101 mg/g. Also the Freundlich adsorption constant (n), which indicates sorption tendency, and the correlation value (R^2) were 1.6 and 0.89, respectively. The heat of adsorption estimated from Temkin adsorption isotherm was 25.34 J/mol. The mean free energy estimated from the DRK isotherm model was 0.7 kJ/mol, which showed that the adsorption experiment followed a physisorption mechanism.

Kumar and Bandyopadhyay [56] used a fixed bed of sodium carbonate–treated rice husk for the removal of Cd(II) from wastewater. The material was found to be efficient for the removal of Cd(II) ions in continuous mode using fixed bed column. The column having a diameter of 2 cm, with bed depths of 10, 20, and 30 cm, can treat 2.96, 5.70, and 8.55 L of contaminated wastewater with 10 mg/L Cd(II) concentration and flow rate at 9.5 mL/min. Column design parameters, such as depth of exchange zone, adsorption rate, and adsorption capacity, were calculated. The effects of flow rate and initial concentration were studied. Theoretical breakthrough curve was drawn from the batch isotherm data and was compared with experimental breakthrough curve. Column regeneration and reusability studies were conducted for two cycles of adsorption–desorption processes.

Gong et al. [57] modified the rice straw with citric acid (CA) as etherifying agent. The two free carboxyl groups in the esterified rice straw were loaded with sodium ion to yield potentially biodegradable anionic biosorbent (biosorbent has negative charges). In order to investigate the effect of chemical modification on the cationic dye adsorption of rice straw, the removal capacities of native and modified rice straw (MRS) for adsorbing a cationic dye (malachite green [MG]) from aqueous solution were compared. The effects of initial pH, sorbent dose, dye concentration, and contact time were investigated. For MRS, the MG dye removal percentage reached the maximum value near pH 4. For the 250 mg/L concentration of MG dye solution, 1.5 g/L or higher concentrations of MRS can completely remove the dye from its aqueous solution. Under the condition of 2.0 g/L sorbent used, the percentage of MG dye adsorbed on MRS was above 93% for a range from 100 to 500 mg/L of MG dye concentration. The adsorption isotherms fitted the Langmuir or Freundlich model. The adsorption equilibriums were reached at about 10 h. The adsorption processes followed the pseudo-first-order rate kinetics. The results indicated that MRS was an excellent biosorbent for the removal of MG dye from aqueous solutions.

Biodegradable anionic biosorbent (biosorbent has negative charges) with high adsorption capacity toward basic dyes was prepared [58] by esterifying oxalic acid and rice straw, which was then loaded with sodium ions for enhancing its adsorption capacity for cationic dyes. The adsorption of two basic dyes, basic blue 9 (BB9) and basic green 4 (BG4), from aqueous solutions onto modified product was investigated. The effects of initial pH, biosorbent dosage, dye concentration, ion strength, and contact time were studied, and optimal experimental conditions were developed. The BB9 and BG4 removal ratios were maximum near pH 6. At 2.0 g/L or above, the adsorbent can completely remove BB9 and BG4 from a solution contaminated with 250 mg/L of the different dyes. The adsorbed ratios of BB9 and BG4 dyes were above 97% for wastewater with 50–250 mg/L of dyes, using 2 g/L of adsorbent. Increasing the ion strength of solution enhanced the adsorption of both BB9 and BG4 dyes. The isothermal data fit well to both the Langmuir and Freundlich models. The sorption process can be described by the pseudo-second-order kinetic model.

16.4.5 Nut Shells as Biosorbent

There is a great potential in using nut shells in several applications. Adsorption studies are one of these. Nut shells were utilized as a low-cost natural adsorbent for the removal of metal ions from industrial wastewater.

Abdulrasaq and Basiru [59] studied the removal of Cu(II), Fe(III), and Pb(II) ions from monocomponent simulated waste effluent by adsorption on coconut husk. Batch experiments were conducted to determine the effects of varying adsorbent loadings, pH, contact time, initial metal ion concentration, and temperature of adsorption. The adsorption of Pb(II) ions was found to be maximum (94%) at pH 5, temperature of 100°C, initial metal ion concentration of 30 ppm, and contact time of 30 min. The adsorption of Cu(II) and Fe(III) ions was highest (92% and 94%, respectively) in the pH range of 5–7, initial metal ion concentration of 50 ppm, temperature of 50°C, but at different contact times of 30 and 90 min, respectively. 1 g of the adsorbent material was sufficient for optimal removal of all the tested metal ions.

The Freundlich isotherm suitably fitted the adsorption of Cu(II) and Fe(III) ions, whereas the Langmuir isotherm gave the best fit for the adsorption of Pb(II) ions.

Biosorption of metal ions (Cd, Cr, and As) by coconut shell powder was investigated [60]. The study investigated the effect of coconut shell particle size (0.044–0.297 mm), initial metal concentration (20–1000 mg/L), and solution pH (2–9) in batch experiments. The experimental data for each metal ion were evaluated and fitted using different adsorption models. Biosorbents were analyzed using scanning electron microscopy and X-ray photoelectron spectroscopy before and after the adsorption experiments to confirm the presence of metal ions species.

Activated carbons with high adsorption capacities were produced from coconut shells activated with reagents such as $CaCO_3$, KOH, H_3PO_4, and $ZnCl_2$. Olowoyo and Garuba [61] reported the investigation of the adsorption of Cd(II) ions on such activated carbon prepared from coconut shells. The effects of contact time and pH in the presence of different activating reagents mentioned earlier on the adsorption capacity of the adsorbents were investigated. The study revealed that the adsorption of Cd(II) ions with respect to time in all activating agents had similar patterns. The order of decreasing the removal percentage of Cd(II) according to activating agents used was $H_3PO_4 > ZnCl_2 > KOH > CaCO_3$. The pH results indicated that adsorption of Cd(II) ions increased at a steady rate as the pH increased especially in the case of $CaCO_3$ and KOH activating agent solutions, which almost tends to 100% out of the four activating agents investigated.

Adsorption of Cu(II) and Cr(III) ions from aqueous solutions using peanut shell biomass was investigated by Krowiak et al. [62]. The optimum adsorption conditions such as initial pH, initial biomass concentration, and temperature were determined for each metal. The kinetics and equilibrium of biosorption were examined in detail. Four kinetic models, namely, pseudo-first-order, pseudo-second-order, power-function equation, and Elovich models, were used to correlate the experimental data and to determine the kinetic parameters. Four well-known adsorption isotherms were chosen to describe the biosorption equilibrium. The experimental data were analyzed using two two-parameter models (Langmuir and Freundlich) and two three-parameter models (Redlich, Peterson, and Sips). The equilibrium adsorption isotherms showed that peanut shells possess a high affinity and adsorption capacity for Cu(II) and Cr(III) ions, with monolayer adsorption capacities of 25.39 mg of Cu(II) and 27.86 mg of Cr(III) per 1 g of biomass. The results showed that peanut shell biomass is an attractive, low-cost alternative biosorbent for the removal of heavy metal ions from aqueous media.

Walnut shell after treatment with citric acid (WNS) (*Juglans regia*) was used as an adsorbent for the removal of Cr(VI) ions from aqueous solutions [63]. The walnut shell modification reaction variables, such as citric acid (5–10 g/L), reaction time (0–24 h), and temperature (110°C–130°C), were investigated in batch experiments (30 g of walnut shell in 1 L reaction mixture). The rate of adsorption was studied under various conditions including initial Cr(VI) ion concentration (0.1–1.0 mM), amount of adsorbent (0.02–0.20 g), pH (2–9), temperature, and contact time (10–240 min). In all cases, adsorption of Cr(VI) was pH dependent. Maximum adsorption at equilibrium was observed at pH of 2–3 for citric acid–modified walnut shell (CA-WNS). The applicability of the Langmuir, Freundlich, and Dubinin–Radushkevich adsorption isotherms was tested for the equilibrium and showed that the maximum adsorption capacities of CA-WNS and untreated WNS under experimental conditions were 0.59 and 0.15 mmol/g for Cr(VI) ions, respectively.

Batch adsorption of Cu(II) ions onto hazelnut shells was studied by Demirbaş et al. [64]. The adsorption capacity for the removal of copper ions from aqueous solution was investigated as a function of solution contact time (1–360 min), hazelnut shell particle size (0–75, 75–150, and 150–200 μm), temperature (25–60), pH (3–7), and zeta potential of particles at different initial pH (2–10). The equilibrium data were processed using Langmuir and Freundlich models and showed high adsorption capacity values toward Cu(II) ions. The adsorption kinetics was investigated, and the best fit was obtained by a second-order kinetic adsorption model.

16.4.6 COFFEE WASTE AS BIOSORBENT

Roasted coffee contains three essential groups of acids: aliphatic acid, chlorogenic acid (Figure 16.1), tannic acid (Figure 16.2), and phenolic acid (Figure 16.3) [65]. The specific types and concentrations of these acids in coffee vary depending on the type of bean and brewing condition.

Coffee contains at least two dozens of aliphatic acids including acetic, citric, lactic, malic, and pyruvic acids. In its green form, coffee beans contain only a negligible amount of these acids. But when roasted, the concentration of these acids increases exponentially. Next to the chlorogenic acids, aliphatic acids exist in the highest proportions in roasted coffee.

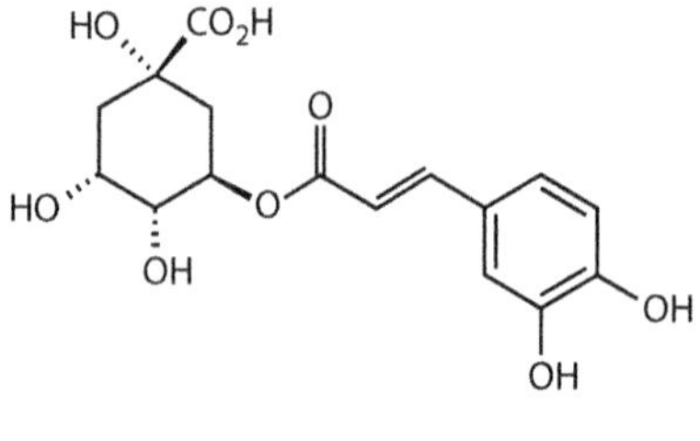

FIGURE 16.1 Chemical structure of chlorogenic acid.

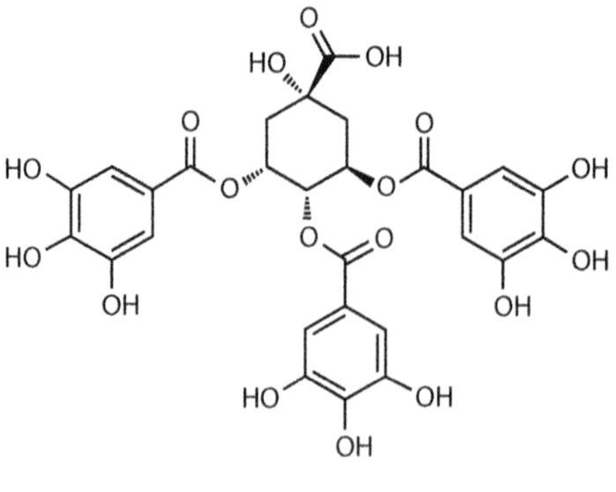

FIGURE 16.2 Chemical structure of tannic acid.

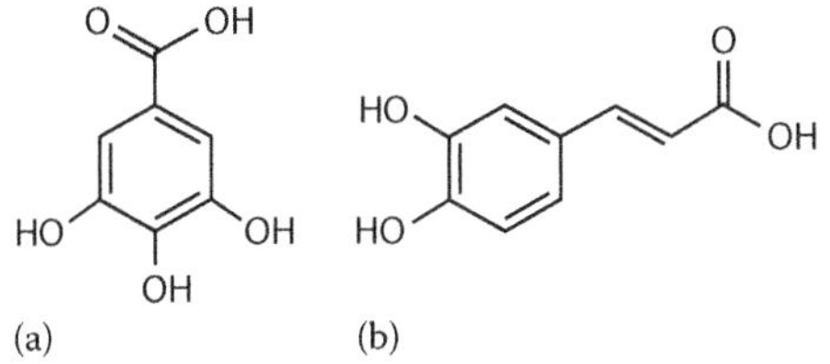

FIGURE 16.3   Chemical structure of (a) gallic acid and (b) caffeic acid.

Approximately 7% of the coffee's dry weight can be attributed to chlorogenic acids, which is responsible for coffee's acidic taste, or perceived acidity. Chlorogenic acids have an astringent taste and may contribute to a heightened "body" in the coffee. Over 17 different chlorogenic acids have been isolated in coffee beans, most of which degrade during the roasting process. Chlorogenic acids in coffee help the body to absorb and to utilize dietary glucose, thereby encouraging weight loss.

Tannic acid, also known as gallotannic acid, is an astringent plant product present in coffee. Its biological purpose is still unknown, according to *A Dictionary of Biology* [66]. Biologists and ecologists theorize that tannic acid may impart an unpalatable taste to discourage animals from grazing on the plants containing it and it may also shield the plants from attack by pathogens. The generic term "tannins" is used to describe any number of compounds containing hydroxy acids, phenolic acids, and glucosides.

Coffee is rich in phenolic acids such as caffeic, ferulic, and quinic acids. Phenolic acids are polyphenols and possess antioxidant properties. A 2009 study by Simões et al. [67] suggests that the caffeic acid in coffee may possess anticarcinogenic properties as well.

Untreated coffee residues such as those from coffee may also act as adsorbent for the removal of heavy metals. The removal of Cu(II) and Cr(VI) ions from aqueous solutions with commercial coffee wastes was studied by Kyzas [68]. The equilibrium data were successfully analyzed using the Langmuir, Freundlich, and Langmuir–Freundlich models. The maximum adsorption capacity of the coffee residues can reach 70 mg/g for the removal of Cu(II) and 45 mg/g for Cr(VI). The kinetic data were analyzed using pseudo-first-, pseudo-second-, and pseudo-third-order equations. Adsorption equilibrium was achieved after 120 min. The effect of pH on adsorption and desorption was investigated, as well as the influence of agitation rate. Ten adsorption–desorption cycles were carried out revealing the efficient reusability potential of these low-cost coffee residue adsorbents and confirmed their economic approach.

Exhausted ground coffee waste has been investigated as biosorbent for Cr(VI) ions from aqueous solution [69]. Maximum metal adsorption occurred at pH 3. Kinetic studies revealed that the initial metal ion uptake was quite rapid and took only 5 days to reach equilibrium. The value of the Langmuir maximum uptake was found to be 10.2 mg of Cr(VI) ion per 1 g of coffee waste. The sorbent was able to reduce hexavalent chromium to trivalent form. A solution of 1 M NaOH was the most effective desorption agent and after 24 h contact 42% of total Cr ions still adsorbed in hexavalent and trivalent oxidation states.

16.4.7 Live Plants and Dried Straw of Water Hyacinth as Biosorbents

Both live plants and dried straw of water hyacinth were employed to a sequential treatment of swine wastewater for nitrogen and phosphorus reduction [70]. In the fermentative tank, the straw behaved as a kind of adsorbent toward phosphorus. Its phosphorus removal rate varied considerably with contact time between the straw and the effluent. In the laboratory, the straw displayed a rapid total phosphorus reduction in a KH_2PO_4 solution. The adsorption efficiency was about 36% upon saturation. At the same time, the water hyacinth straw in the fermentative tank enhanced NH_3-N removal efficiency. The results demonstrated an economically feasible method to employ water hyacinth straw for the swine wastewater treatment.

Laboratory investigations of the potential of the biomass of nonliving and dried roots of water hyacinth (*Eichhornia crassipes*) to remove methylene blue and Victoria blue dyes from aqueous solutions were conducted [71]. Parameters investigated include pH, sorbent dosage, contact time, and initial dye concentration. The Langmuir isotherm well represented the measured adsorption data. Maximum adsorption capacities of water hyacinth roots for methylene blue and Victoria blue were 128.9 and 145.4 mg/g, respectively. The study revealed that the water hyacinth roots are a cheap source of biosorbent for adsorption of basic dyes, since they are readily available in great abundance.

16.5 BIOLOGICAL WASTE AS BIOSORBENTS

16.5.1 Chitosan

Chitosan (Figure 16.4), a biopolymer, is a good adsorbent to remove various kinds of anionic and cationic dyes as well as heavy metal ions. Chemical modifications that lead to the formation of chitosan derivatives such as: grafting chitosan and chitosan composites have gained much attention have been extensively studied and are widely reported in the literature [72].

Kyzas et al. [73] prepared chitosan sorbents, cross-linked and grafted with amido or carboxyl groups, and studied their adsorption properties for Cu(II) and Cr(VI) ions. Equilibrium sorption experiments were carried out at different pH values and initial ion concentrations. The equilibrium data were successfully analyzed with the Langmuir–Freundlich isotherm model. The calculated maximum adsorption capacity of the carboxyl-grafted sorbent for Cu(II) was found to be 318 mg/g at pH 6, while the respective capacity for Cr(VI) uptake onto the amido-grafted adsorbent was found to be 935 mg/g at pH 4. Thermodynamic parameters of the sorption

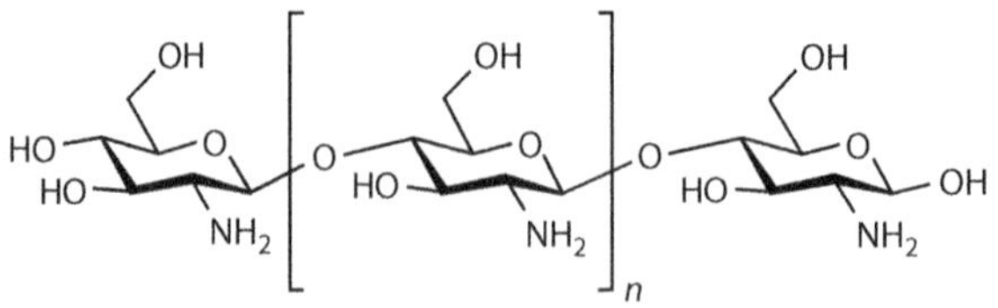

FIGURE 16.4 Chemical structure of chitosan biopolymer.

process such as ΔG, ΔH, and ΔS were calculated. The experimental kinetic data were successfully analyzed using a novel phenomenological diffusion–reaction model (DIFRE), which combines (1) mass transfer of the metal ions from the bulk solution on the sorbent surface, (2) diffusion of the ions through the swollen polymer particle, and (3) instantaneous local chelation (for cations) or electrostatic attraction (for anions) on the amino groups of the biopolymer. The regeneration of adsorbents was affirmed in four sequential adsorption–desorption cycle experiments, without significant loss in sorption capacity.

To increase the uptake capacity of mercury ions, several chemical modifications of chitosan beads (CB) cross-linked with glutaraldehyde were prepared [74]. Aminated chitosan bead was prepared through chemical reaction with ethylene diamine. This product has a high uptake capacity of about 2.3 mmol/g of dry mass at pH 7. The increased number of amine groups was confirmed by IR analysis and by measuring the saturation capacities for adsorption of HCl. The surface condition and existence of mercury ions on the beads were confirmed by environmental scanning electron microscopy and energy-dispersive X-ray spectroscopy analyses.

Cross-linked metal chitosan microparticles were prepared [75] from chitosan using four metals (Cu(II), Zn(II), Ni(II), and Pb(II) ions) as templates and epichloro-hydrin as a cross-linker. The microparticles were characterized by FTIR, solid-state ^{13}C-nuclear magnetic resonance spectroscopy, and energy-dispersive X-ray spectros-copy. Templates were used for comparative biosorption of Cu(II), Zn(II), Ni(II), and Pb(II) ions in an aqueous solution. The results showed that the adsorption capacities of Cu(II), Zn(II), Ni(II), and Pb(II) on the template microparticles increased from 25% to 74%, 13% to 46%, 41% to 57%, and 12% to 43%, respectively, as compared to the microparticles without metal ion templates. The dynamic study showed that the adsorption process fitted well the second-order kinetic equation. Three adsorp-tion models, Langmuir, Freundlich, and Dubinin–Radushkevich, were applied to the equilibrium isotherm data. Langmuir isotherm equation fitted well the adsorption processes. Furthermore, the microparticles can be regenerated and reused for metal removal.

Ngah and Fatinathan [76] synthesized chitosan–tripolyphosphate (CTPP) beads and used them for the adsorption of Pb(II) and Cu(II) ions from aqueous solution. The effects of initial pH, agitation period, adsorbent dosage, initial concentrations of heavy metal ions, and temperature were studied. Experimental data were correlated using the Langmuir, Freundlich, and Dubinin–Radushkevich isotherm models. The maximum adsorption capacities of Pb(II) and Cu(II) ions based on the Langmuir isotherm models were 57.33 and 26.06 mg/g, respectively. In addition, the beads showed higher selectivity toward Cu(II) over Pb(II) ions in a binary system. Various thermodynamic parameters such as enthalpy (ΔH), Gibbs free energy (ΔG), and entropy (ΔS) changes were computed. The results showed that the adsorption of Cu(II) and Pb(II) metal ions onto CTPP beads was spontaneous and endothermic in nature. The kinetic data were analyzed using the pseudo-first- and pseudo-second-order kinetics and intraparticle diffusion model. Infrared spectra were recorded to elucidate the mechanism of Pb(II) and Cu(II) ions adsorption onto CTPP beads.

The copper sorption capacity of raw CB was improved by chemical modifications such as protonation of chitosan beads (PCB), carboxylation of chitosan beads (CCB), and grafting to chitosan beads (GCB) [77]. These products showed an adsorption capacity of 52, 86, and 126 mg/g, respectively, while raw CB displayed only 40 mg/g. Among the sorbents, GCB showed the highest sorption capacity than CB, PCB, and CCB. Adsorption experiments were performed by varying contact time, pH, and presence of anions, initial copper concentrations, and temperature. The nature and morphology of the sorbents were determined using FTIR and SEM analyses. The copper adsorption onto PCB, CCB, and GCB obeys the Freundlich isotherm. Thermodynamic studies revealed that the nature of copper adsorption was spontaneous and endothermic.

Adsorption of Cr(VI) ions from aqueous solution was studied by Hu et al. [78] using ethylene diamine–modified cross-linked magnetic chitosan resin using a batch adsorption system. The results showed that Cr(VI) removal was pH dependent and the optimum adsorption was observed at pH 2. The adsorption rate was extremely fast, and equilibrium was established within 6–10 min. The adsorption data could be well interpreted using both the Langmuir and Temkin models. The maximum adsorption capacities obtained from the Langmuir model were 51.8 mg/g, 48.8 mg/g, and 45.9 mg/g at 293, 303, and 313 K, respectively. The adsorption process was analyzed using pseudo-second-order kinetic model. Thermodynamic parameters revealed the feasibility, spontaneity, and exothermic nature of the adsorption. The sorbents were successfully regenerated using a solution of 0.1 N NaOH.

Cestari and coworkers [79] utilized vanillin-modified thin chitosan membrane as adsorbents for the removal of Cu(II) from aqueous solutions. The results showed that an increase of temperature accelerates a mass transfer of Cu(II) to the membranes surface. The kinetic data were fitted to the Lagergren adsorption kinetic equations.

The kinetics of Cd(II) ion uptake was reported by Evans et al. [80] for chitosan-based crab shells. Crushed crab shells were chemically treated to convert the chitin present into chitosan. Three particle sizes of crushed shells with average diameters of 0.65, 1.43, and 3.38 mm, average pore diameters ranging from 30 to 54 mm, and a specific surface area of 30 m²/g were obtained. Batch experiments were performed to study the uptake equilibrium and kinetics of Cd(II) ions by chitosan. Adsorption equilibrium was analyzed using the Freundlich adsorption isotherm and was found to be independent of particle size indicating that adsorption takes place largely in the pore space. A high initial rate of Cd(II) ion uptake was followed by a slower uptake rate suggesting intraparticle diffusion as the rate-limiting step. The kinetic uptake data were modeled using a pore diffusion model incorporating nonlinear adsorption. The effect of boundary layer resistance was modeled through inclusion of a mass transfer expression at the outside boundary. Two fitting parameters, the tortuosity factor (t) and the mass transfer coefficient at the outside boundary, were used. These parameters were unique for all solute and sorbent concentrations. The tortuosity factor varied from 1.5 for large particles to 5.1 for small particles. The mass transfer coefficient varied from 2×10^{-7} m/s at 50 rpm to 2×10^{-3} m/s at 200 rpm. At agitation rates below 100 rpm, boundary layer resistance reduced the uptake rate significantly. The high adsorption capacity and relatively low production cost make chitosan an attractive adsorbent for the removal of heavy metals from waste streams.

16.5.2 Microbial Biomass Application in Heavy Metal Removal

Hannachi et al. [81] reported the characteristics of Cd(II) biosorption from aqueous solution using the brown alga (*Dictyota dichotoma*) as a function of pH, biomass dosage, contact time, and temperature. The Langmuir, Freundlich, and Dubinin–Radushkevich (D–R) models were applied to describe the biosorption isotherms of Cd(II) by *D. dichotoma* biomass. The monolayer biosorption capacity of *D. dichotoma* biomass for Cd(II) ions was found to be 75 mg/g. The mean free energy calculated from the D–R isotherm indicated that the biosorption of Cd(II) on *D. dichotoma* took place by chemisorption. Kinetic evaluation of the experimental data showed that the biosorption process followed pseudo-second-order kinetic model. The calculated thermodynamic parameters showed that the biosorption of Cd(II) on *D. dichotoma* biomass was feasible, spontaneous, and exothermic under the examined conditions. X-ray photoelectron spectroscopy and FTIR analyses of *D. dichotoma* revealed the chelating character of the ion coordination to carboxyl groups. It was confirmed that carboxyl, ether, alcoholic, and amino groups were responsible for binding of the metal ions.

The utility of a waste, dead fungal biomass for removal of various metal ions such as calcium, iron, nickel, and chromium was demonstrated by Sekhar et al. [82]. The tests were carried out on synthetic solutions of varying pH and metal ion concentrations. The maximum metal uptake was found to be dependent on solution pH (4–5 for Fe, 4–7 for Ca(II), 6–7 for Ni(II), and 6 for Cr(III)) and increased with biomass loading up to 10 g/L. The adsorption capacities of the biomass varied for the different metal ions and decreased in the order Ca(II) > Cr(III) > Ni(II) > Fe(III) > Cr(VI).

Nizam and Baznjaneh [83] tested *Saccharomyces cerevisiae* strain, which was grown in a suitable culture, for its ability to uptake heavy metals. A comparison was made on the ability of *S. cerevisiae* to uptake cadmium and cobalt under different conditions, where three concentrations were used for each metal, at three different pH values, and three temperatures. The grown *S. cerevisiae* was able to uptake both cobalt and cadmium individually and in mixture. The highest uptake for Co(II) was 1.99 ppm, and for Cd(II) it was 0.48 ppm. It was noticed that the uptake amount was best when the two metals were mixed, and at temperatures of 5°C and 28°C, and pH of 4 and 9. This biotechnology can be applied using *S. cerevisiae* to uptake heavy metals from industrial wastewater.

The batch adsorption of Ni(II) onto *sphagnum moss* peat was studied by Ho et al. [84]. The adsorption was pH dependent and the optimum range being 4–7. Langmuir and Freundlich isotherms were used to fit the data for various initial Ni(II) concentrations, and for a range of pH values. The analysis gave a single relationship between initial metal concentration, metal removal, and initial pH. The latter was found to control the efficiency of nickel removal. Kinetic data suggested involvement of a rate-limiting step, and a predictive relationship was derived relating nickel removal to peat dose. In comparison with other metals, nickel removal was poor, and the possible reasons were discussed.

A biofilm of *Arthrobacter viscosus* supported on granular activated carbon was used to remove chromium and organic compounds (CP, phenol, and cresol) from aqueous solutions [85]. The compounds were studied as single solutes and in different combinations in the presence of Cr(VI) ions. Optimum Cr(VI) adsorption was observed at a phenol concentration of 100 mg/L and at an initial metal ion concentration of 60 mg/L. The maximum values of biosorption of organic compounds were 9.94 mg/g for phenol, 9.70 mg/g for CP, and 13.99 mg/g for cresol. The removal order after 15 h experiment was as follows: phenol > chlorophenol > cresol > Cr(VI).

The feasibility of fish scales (*Labeo rohita*) as low-cost biosorbent for the removal of MG dye from aqueous solution was investigated by Chowdhury et al. [86]. Employing a batch experimental setup, the effect of operational parameters such as biosorbent dose, initial solution pH, contact time, and temperature on the dye removal process was studied. The equilibrium biosorption data followed both Langmuir and Freundlich isotherm models. The experimental kinetic data fitted well to the pseudo-second-order kinetic model. Thermodynamic study indicated a spontaneous and endothermic nature of the biosorption process. The results suggest that fish scales are an effective biosorbent for removal of MG dye from contaminated solutions.

16.6   SUMMARY

Heavy metal ions are considered one of the greatest pollutants in industrial wastewater. Heavy metal ions are present in aquatic systems due to mineral activities, chemical industries, battery manufacturing, and electroplating. Their presence in the aquatic system has a potent effect on plants, animals, and humans. Removal of heavy metal ions from the environment is very important. The presence of heavy metal ions in tribological systems including cutting fluids, grease industries, or lubricating agents decreases their efficiency in the lubrication process. Several methods have been used to remove these metal ions from the wastewater including chelation, precipitation, and ion exchange, but the costs of these methods are very high. The use of agricultural and plant wastes can help in the removal of these ions with almost no cost. Several agricultural wastes including rice husks, wheat husks, rice bran, wheat bran, and activated carbon obtained from different plant wastes were described here.

SYMBOLS

Symbol	Definition	Unit
C_0	The liquid-phase concentration at initial time	(mg/L)
C_t	The liquid-phase concentration at any time	(mg/L)
C_e	The liquid-phase concentration at equilibrium	(mg/L)
q_e	The adsorbed metal ion at equilibrium	(mg/g)
q_t	The adsorbed metal ion at any time	(mg/g)
t	Time	(h)
t_{eq}	Equilibrium time: time at which no more decrease in the adsorbed amount of the metal on the biosorbent	(h)
K_F	Freundlich constant related to sorption capacity and sorption intensity of the adsorbent and can be defined as the adsorption or distribution coefficient	(mg/g (l/mg)$^{1/n}$)
n	Freundlich constant related to sorption tendency of the adsorbent	(l/mg)

(Continued)

q_{max}	The maximum amount of the sorbent per unit weight of the adsorbent to form a complete monolayer on the surface	(mg/g)
K_L	Langmuir constant related to the affinity of the binding sites	(l/mg)
b	The Temkin constant related to heat of sorption	(J/mol)
a	The equilibrium binding constant corresponding to the maximum binding energy	(l/g)
R	The universal gas constant	(J/mol K)
T	The absolute solution temperature	(K)
q_s	The theoretical isotherm saturation capacity	(mg/g)
K_{ad}	Dubinin–Radushkevich isotherm constant	(mol^2/kJ^2)
E_o	Potential energy	
n_{FH}	Flory–Huggins isotherm model exponent	
K_{FH}	Flory–Huggins isotherm equilibrium constant	(L/g)
θ	Degree of surface coverage	
q	The adsorbed metal ion on adsorbent	(mg/g)
k_1	The adsorption rate constant of pseudo-first-order adsorption	(h^{-1})
k_2	The adsorption rate constant of pseudo-second-order adsorption	(g/mg h)

REFERENCES

1. S.T. Ong, P.S. Keng, M.S. Voon, and S.L. Lee, Application of durian peel (*Durio zibethinus murray*) for the removal of methylene blue from aqueous solution, *Asian J. Chem.*, *23*, 2898–2902 (2011).

2. S.T. Ong, P.S. Keng, and S.L. Lee, Basic and reactive dyes sorption enhancement of rice hull through chemical modification, *Am. J. Appl. Sci.*, *7*, 447–452 (2010).

3. F.D. Ardejani, K. Badii, N.Y. Limaee, N.M. Mahmoodi, M. Arami, S.Z. Shafaei, and A.R. Mirhabibi, Numerical modeling and laboratory studies on the removal of Direct Red 23 and Direct Red 80 dyes from textile effluents using orange peel, a low cost adsorbent, *Dyes Pigments*, *73*, 178–185 (2007).

4. B.H. Hameed, R.R. Krishni, and S.A. Sata, A novel agricultural waste adsorbent for the removal of cationic dyes from aqueous solutions, *J. Hazard. Mater.*, *162*, 305–311 (2009).

5. R. Gong, S. Zhu, D. Zhang, J. Chen, S. Ni, and R. Guan, Adsorption behavior of cationic dyes on citric acid esterifying wheat straw: Kinetic and thermodynamic profile, *Desalination*, *230*, 220–228 (2008).

6. G.L. Rorrer, Heavy metal ions removal from wastewater. In: *Encyclopedia of Environmental Analysis and Remediation*, R.A. Meyers (Ed.), Wiley, New York, Vol. 4, pp. 2102–2125 (1998).

7. D.M. Khalid, M.S. Amran, and W.A. Wan, Langmuir model application on solid-liquid adsorption using agricultural wastes: Environmental application review, *J. Purity Utility React. Environ.*, *1*, 120–126 (2012).

8. A. Kapoor and T. Viraraghavan, Fungal biosorption: An alternative treatment option for heavy metal bearing wastewaters: A review, *Bioresource Technol.*, *53*, 195–206 (1995).

9. S. Bailey, T.R. Olin, and M.A. Dean, A review of potentially low cost sorbents for heavy metals, *Water Res.*, *33*, 2469–2479 (1999).

10. K.S. Low, C.K. Lee, and S.C. Liew, Sorption of cadmium and lead from aqueous solutions by spent grain, *Process Biochem.*, *36*, 59–64 (2000).

11. S. Babel and T.A. Kurniawan, Low cost adsorbents for heavy metal uptake from contaminated water: A review, *J. Hazard. Mater.*, *97*, 219–243 (2003).

12. J.T. Matheickal and Q. Yu, Biosorption of lead(II) from aqueous solutions by *Phellinus badius*, *Miner. Eng.*, *10*, 947–957 (1997).

13. S. Ilhan, M.N. Noubakhsh, S. Kilicarslan, and H. Ozdag, Removal of chromium, lead and copper ions from industrial wastewater by *Staphylococcus saprophyticus*, *Turkish Electron. J. Biotechnol.*, *2*, 50–57 (2004).

14. A. Cabuk, S. Ilhan, C. Filik, and F. Caliskan, Pb^{2+} biosorption by pretreated fungal biomass, *Turkish J. Biol.*, *29*, 23–28 (2005).

15. P.S. Sheng, Y.P. Ting, J.P. Chen, and L. Hong, Sorption of lead, copper, cadmium, zinc and nickel by marine algae biomass: Characterization of biosorptive capacity and investigation of mechanisms, *J. Colloid Interface Sci.*, *275*, 131–141 (2004).

16. C.S. Freire-nordi, A.A. Vieira, and H. Nascimento, The metal binding capacity of *Anabaena spiroides* extracellular polysaccharide: An EPR study, *Process Biochem.*, *40*, 2215–2224 (2005).

17. S.A. Delle, Factors affecting sorption of organic compounds in natural sorbent/water systems and sorption coefficients for selected pollutants. A review, *J. Phys. Chem. Ref. Data*, *30*, 187–439 (2001).

18. R. Liu, W. Ma, C. Jia, L. Wang, and H. Li, Effect of pH on biosorption of boron onto cotton cellulose, *Desalination*, *207*, 257–263 (2007).

19. X. Wang, Y. Qin, and Z. Li, Kinetics and equilibrium studies, *Separ. Sci. Technol.*, *41*, 747–752 (2006).

20. Y.S. Ho, Isotherms for the sorption of lead onto peat: Comparison of linear and non-linear methods, *Polish J. Environ. Stud.*, *15* 81–86 (2006).

21. A.Y.A. Dursun, Comparative study on determination of the equilibrium, kinetic and thermodynamic parameters of biosorption of Cu(II) and Pb(II) ions onto pretreated *Aspergillus niger*, *Biochem. Eng. J.*, *28*, 187–195 (2006).

22. S. Basha, Z.V. Murthy, and B. Jha, Sorption of Hg(II) from aqueous solutions onto *Carica papaya*: Application of isotherms, *Ind. Eng. Chem. Res.*, *47*, 980–986 (2008).

23. M. Dundar, C. Nuhoglu, and Y. Nuhoglu, Biosorption of Cu(II) ions onto the litter of natural trembling poplar forest, *J. Hazard. Mater.*, *151*, 86–95 (2008).

24. B. Kiran and A. Kaushik, Chromium binding capacity of *Lyngbya putealis exopolysaccharides*, *Biochem. Eng. J.*, *38*, 47–54 (2008).

25. J.C. Igwe and A.A. Abia, Equilibrium sorption isotherm studies of Cd(II), Pb(II) and Zn(II) ions detoxification from wastewater using unmodified and EDTA-modified maize husk, *Electron. J. Biotechnol.*, *10*, 536–548 (2007).

26. A. Cabuk, T. Akar, S. Tunal, and S. Gedikli, Biosorption of Pb(II) by industrial strain of *Saccharomyces cerevisiae* immobilized on the biomatrix of cone biomass of *Pinus nigra*: Equilibrium and mechanism analysis, *Chem. Eng. J.*, *131*, 293–300 (2007).

27. R. Apiratikul, P. Pasavant, V. Sungkhum, P. Suthiparinyanont, S. Wattanachira, and T.F. Marhaba, Biosorption of Cu^{2+}, Cd^{2+}, Pb^{2+}, and Zn^{2+} using dried marine green macroalga *Caulerpa lentillifera*, *Bioresour. Technol.*, *97*, 2321–2329 (2006).

28. K. Vijayaraghavan, T.V. Padmesh, K. Palanivelu, and M. Velan, Biosorption of nickel(II) ions onto *Sargassum wightii*: Application of two-parameter and three-parameter isotherm models, *J. Hazard. Mater.*, *133*, 304–308 (2006).

29. R. Djeribi and O. Hamdaoui, Sorption of copper(II) from aqueous solutions by cedar sawdust and crushed brick, *Desalination*, *225*, 95–112 (2008).

30. S.S. Barsal and S.N. Das, Hexavalent chromium removal from aqueous solution by adsorption on treated sawdust, *Biochem. Eng. J.*, *31*, 216–222 (2006).

31. R. Apiratikul and P. Pavasant, Batch and column studies of biosorption of heavy metals by *Caulerpa lentillifera*, *Bioresour. Technol.*, *99*, 2766–2777 (2008).

32. Y. Sag and Y. Aktay, Kinetic studies on sorption of Cr(VI) and Cu(II) ions by chitin, chitosan and *Rhizopus arrhizus*, *Biochem. Eng. J.*, *12*, 143–153 (2002).

33. Y.S. Ho and G. McKay, Application of kinetic models to the sorption of Cu(II) onto peat, *Adsorpt. Sci. Technol.*, *20*, 797–815 (2002).

34. Y. Vijaya, S.R. Popuri, V.M. Boddu, and A. Krishnaiah, Modified chitosan and calcium alginate biopolymer sorbents for removal of Ni(II) through adsorption, *Carbohydr. Polym.* *72*, 261–271 (2008).

35. S.S. Baral, S.N. Das, P. Rath, G. Roy Chaudhury, and Y.V. Swamy, Removal of Cr(VI) from aqueous solution using waste weed: *Salvinia cucullata*, *Chem. Ecol.*, *23*, 105–117 (2007).

36. M. Regel-Rosocka, A review on methods of regeneration of spent pickling solutions from steel processing, *J. Hazard. Mater.*, *177*, 57–69 (2010).

37. A. Agrawal and K.K. Sahu, An overview of the recovery of acid from spent acidic solutions from steel and electroplating industries, *J. Hazard. Mater.*, *171*, 61–75 (2009).

38. R.K. Nagarale, G.S. Gohil, and V.K. Shahi, Recent developments on ion-exchange membranes and electro-membrane processes, *Adv. Colloid Int. Sci.*, *119*, 97–130 (2010).

39. R. Ferreira, Biomass adsorbent for removal of toxic metal ions from electroplating industry wastewater, *Electroplating*, *166*, 155–163 (2012).

40. W. Nakbanpote, P. Thiravetyan, and C. Kalambaheti, Preconcentration of gold by rice husk ash, *Miner. Eng.*, *13*, 391–400 (2000).

41. C.R.T. Tarley and M.A. Arruda, Biosorption of heavy metals using rice milling by-product: Characterization and application for removal of metals from aqueous effluents, *Chemosphere, 54*, 987–995 (2004).

42. A.K. Bhattacharya, S.N. Mandal, and S.K. Das, Adsorption of Zn(II) from aqueous solution by using different adsorbents, *Chem. Eng. J.*, *123*, 43–51 (2006).

43. F. Asadi, H. Shariatmadar, and N. Mirghaffari, Modification of rice hull and sawdust sorptive characteristics to remove heavy metals from synthetic solutions and wastewater, *J. Hazard. Mater.*, *154*, 451–458 (2008).

44. J. Zhang, H. Fu, X. Lv, J. Tang, and X. Xu, Removal of Cu(II) from aqueous solution using the rice husk carbons prepared by the physical activation process, *Biomass Bioenergy, 35*, 464–472 (2011).

45. Y. Ding, D. Jing, H. Gong, and X.Z. Yang, Biosorption of aquatic Cd(II) by unmodified rice straw, *Bioresour. Technol.*, *114*, 20–25 (2012).

46. H. Gao, Y. Liu, G. Zeng, W. Xu, T. Li, and W. Xia, Characterization of Cr(VI) removal from aqueous solutions by a surplus agricultural waste-Rice straw, *J. Hazard. Mater.*, *150*, 446–452 (2008).

47. C. Gonc, A. Rocha, D.A. Zaia, and R.V. Alfaya, Use of rice straw as biosorbent for removal of Cu(II), Zn(II), Cd(II) and Hg(II) ions in industrial effluents, *J. Hazard. Mater.*, *166*, 383–388 (2009).

48. E. Chockalingam and S. Subramanian, Studies on removal of metal ions and sulphate reduction using rice husk and *Desulfotomaculum nigrificans* with reference to remediation of acid mine drainage, *Chemosphere, 62*, 699–708 (2006).

49. X.S. Wang and J.P. Chen, Biosorption of Congo red from aqueous solution using wheat bran and rice bran: Batch studies, *Separ. Sci. Technol.*, *44*, 1452–1466 (2009).

50. B.M. Jenkins, R.R. Bakker, and J.B. Wei, On the properties of washed straw, *Biomass Bioenergy, 4*, 177–200 (1996).

51. T.K. Naiya, A.K. Bhattacharya, S. Mandal, and S.K. Das, The sorption of Pb(II) ions on rice husk ash, *J. Hazard. Mater.*, *163*, 1254–1264 (2009).

52. S. Wang, Y. Tzou, Y. Lu, and G. Sheng, Removal of 3-chlorophenol from water using rice-straw-based carbon, *J. Hazard. Mater.*, *147*, 313–318 (2007).

53. N. Hsu, S. Wang, Y. Liao, S. Huang, Y. Tzou, and Y. Huang, Removal of hexavalent chromium from acidic aqueous solutions using rice straw-derived carbon, *J. Hazard. Mater.*, *171*, 1066–1070 (2009).

54. K.K. Wong, C.K. Lee, K.S. Low, and M.J. Haron, Removal of Cu and Pb from electroplating wastewater using tartaric acid modified rice husk, *Process Biochem.*, *39*, 437–445 (2003).

55. A.F. El-Kafrawy, S.M. El-Saeed, R.K. Farag, H.A. El-Saied, and M.E. Abdel-Raouf, Adsorbents based on natural polymers for removal of some heavy metals from aqueous solution, *Egypt. J. Petrol.*, *26*, 23–32 (2017).

56. U. Kumar and M. Bandyopadhyay, Fixed bed column study for Cd(II) removal from wastewater using treated rice husk, *J. Hazard. Mater.*, *129*, 253–259 (2006).

57. R. Gong,Y. Jin, F. Chen, J. Chen, and Z. Liu, Enhanced malachite green removal from aqueous solution by citric acid modified rice straw, *J. Hazard. Mater.*, *137*, 865–870 (2006).

58. R. Gong, Y. Jin, J. Sun, and K. Zhong, Preparation and utilization of rice straw bearing carboxyl groups for removal of basic dyes from aqueous solution, *Dyes Pigments*, *76*, 519–524 (2008).

59. G. Abdulrasaq and G. Basiru, Removal of Cu(II), Fe(III) and Pb(II) ions from mono-component simulated waste effluent by adsorption on coconut husk, *Afr. J. Environ. Sci. Technol.*, *4*, 382–387 (2010).

60. G.H. Pino, L.M.S. Mesquita, M.L. Torem, and G.A.S. Pinto, Biosorption of heavy metals by powder of green coconut shell, *Separ. Sci. Technol.*, *41*, 3141–3153 (2006).

61. D. Olowoyo and A. Garuba, Adsorption of Cd(II) activated carbon prepared from coconut shell, *J. Food Sci. Technol.*, *1*, 81–84 (2012).

62. A.W. Krowiak, R.G. Szafran, and S. Modelski, Biosorption of heavy metals from aqueous solutions onto peanut shell as a low-cost biosorbent, *Desalination*, *265*, 126–134 (2011).

63. T. Altun and E. Pehlivan, Removal of Cr(VI) from aqueous solutions by modified walnut shells, *Food Chem.*, *132*, 693–700 (2012).

64. O. Demirbaş, A. Karadağ, M. Alkan, and M. Doğan, Removal of copper ions from aqueous solutions by hazelnut shell, *J. Hazard. Mater.*, *153*, 677–684 (2008).

65. P. Pappa, F.M. Pellera, and E. Gidarakos, Characterization of biochar produced from spent coffee waste. In: *Proceedings of Third International Conference on Industrial and Hazardous Waste Management*, Chania, Greece, pp. 1–8 (2012).

66. C.B. Kumar and K.N. Mohana, Phytochemical screening and corrosion inhibitive behavior of *Pterolobium hexapetalum* and *Celosia argenteaplant* extracts on mild steel in industrial water medium, *Egypt. J. Petrol.*, *23*, 201–211 (2014).

67. J. Simões, P. Madureira, F.M. Nunes, M.R. Domínguez, M. Vilanova, and M.A. Coimbra, Immunostimulatory properties of coffee mannans. *Mol. Nutr. Food Res.*, *53*, 1036–1043 (2009).

68. G.Z. Kyzas, Commercial coffee wastes as materials for adsorption of heavy metals from aqueous solutions, *Materials*, *5*, 1826–1840 (2012).

69. N. Fiol, C. Escudero, and I. Villaescusa, Re-use of exhausted ground coffee waste for Cr(VI) sorption, *Separ. Sci. Technol.*, *43*, 582–596 (2008).

70. X. Chen, X. Wan, B. Weng, and Q. Huang, Water hyacinth (*Eichhornia crassipes*) waste as an adsorbent for phosphorus removal from swine wastewater, *Bioresour. Technol.*, *101*, 9025–9030 (2010).

71. K.S. Low, C.K. Lee, and K.K. Tan, Biosorption of basic dyes by water hyacinth roots, *Bioresour. Technol.*, *52*, 79–83 (1995).

72. W.S. Ngah, L.C. Teong, and M.A. Hanafiah, Adsorption of dyes and heavy metal ions by chitosan composites: A review. *Carbohydr. Polym.*, *83*, 1446–1456 (2011).

73. G.Z. Kyzas, M. Kostoglou, and N.K. Lazaridis, Copper and chromium (VI) removal by chitosan derivatives—Equilibrium and kinetic studies, *Chem. Eng. J.*, *152*, 440–448 (2009).

74. C. Jeon and W.H. Holl, Chemical modification of chitosan and equilibrium study for mercury ion removal, *Water Res.*, *37*, 4770–4780 (2003).

75. C. Chen, C. Yang, and A. Chen, Biosorption of Cu(II), Zn(II), Ni(II) and Pb(II) ions by cross-linked metal-imprinted chitosan with epichlorohydrin, *J. Environ. Manage.*, *92*, 796–802 (2011).

76. W.S. Ngah and S. Fatinathan, Adsorption characterization of Pb(II) and Cu(II) ions onto chitosan-tripolyphosphate beads: Kinetic, equilibrium and thermodynamic studies, *J. Environ. Manage.*, *91*, 958–969 (2010).

77. M.R. Gandhi, G.N. Kousalya, N. Viswanathan, and S. Meenakshi, Sorption behavior of copper on chemically modified chitosan beads from aqueous solution, *Carbohydr. Polym.*, *83*, 1082–1087 (2011).

78. X. Hu, J. Wang, Y. Liu, X. Li, G. Zeng, W. Chen, and F. Long, Adsorption of chromium(VI) by ethylenediamine-modified crosslinked magnetic chitosan resin: Isotherms, kinetics and thermodynamics, *J. Hazard. Mater.*, *185*, 306–310 (2011).

79. A.R. Cestari, E.F. Vieira, J.D. Matos, and D.S. Anjos, Determination of kinetic parameters of Cu(II) interaction with chemically modified thin chitosan membranes, *J. Colloid Interface Sci.*, *285*, 288–295 (2005).

80. J.R. Evans, W.G. Davids, J.D. MacRae, and A. Amirbahman, Kinetics of cadmium uptake by chitosan-based crab shells, *Water Res.*, *36*, 3219–3226 (2002).

81. Y. Hannachi, A. Rezgui, A.B. Dekhil, and T. Boubaker, Removal of cadmium (II) from aqueous solutions by biosorption onto the brown macroalga (*Dictyota dichotoma*), *Desalin. Water Treat.*, *54*, 1663–1673 (2015).

82. K.C. Sekhar, S. Subramanian, J.M. Modak, and K.A. Natarajan, Removal of metal ions using an industrial biomass with reference to environmental control, *Int. J. Miner. Process.*, *53*, 107–120 (1998).

83. A. Nizam and R. Baznjaneh, Use of *Saccharomyces cerevisiae* in biological treatment for heavy metals uptake from industrial wastewater, *Najah Univ. J. Res. (Nat. Sci.)*, *26*, 101–118 (2012).

84. Y.S. Ho, D. John, and C.F. Forster, Batch nickel removal from aqueous solution by *sphagnum moss* peat, *Water. Res.*, *29*, 1327–1332 (1995).

85. C. Quintelas, E. Sousa, F. Silva, S. Neto, and T. Tavares, Competitive biosorption of ortho-cresol, phenol, chlorophenol and chromium(VI) from aqueous solution by a bacterial biofilm supported on granular activated carbon, *Process Biochem.*, *41*, 2087–2091 (2006).

86. S. Chowdhury, P.D. Saha, and U. Ghosh, Fish (*Labeo rohita*) scales as potential low-cost biosorbent for removal of malachite green from aqueous solutions, *Bioremed. J.*, *16*, 235–242 (2012).

Index

Printed and bound by CPI Group (UK) Ltd, Croydon, CR0 4YY

24/10/2024

01778306-0010